Charles C. Rhode



.95

# Linear and Integer Programming

## PRENTICE-HALL INTERNATIONAL SERIES IN MANAGEMENT

| | |
|---|---|
| Athos and Coffey | *Behavior in Organizations: A Multidimensional View* |
| Baumol | *Economic Theory and Operations Analysis, 2nd ed.* |
| Boot | *Mathematical Reasoning in Economics and Management Science* |
| Brown | *Smoothing, Forecasting, and Prediction of Discrete Time Series* |
| Chambers | *Accounting, Evaluation and Economic Behavior* |
| Churchman | *Prediction and Optimal Decision: Philosophical Issues of a Science of Values* |
| Clarkson | *The Theory of Consumer Demand: A Critical Appraisal* |
| Cohen and Cyert | *Theory of the Firm: Resource Allocation in a Market Economy* |
| Cyert and March | *A Behavioral Theory of the Firm* |
| Fabrycky and Torgersen | *Operations Economy: Industrial Applications of Operations Research* |
| Frank, Massy, and Wind | *Market Segmentation* |
| Green and Tull | *Research for Marketing Decisions, 2nd ed.* |
| Greenlaw, Herron, and Rawdon | *Business Simulation in Industrial and University Education* |
| Hadley and Whitin | *Analysis of Inventory Systems* |
| Holt, Modigliani, Muth, and Simon | *Planning Production, Inventories, and Work Force* |
| Hymans | *Probability Theory with Applications to Econometrics and Decision-Making* |
| Ijiri | *The Foundations of Accounting Measurement: A Mathematical, Economic, and Behavioral Inquiry* |
| Kaufmann | *Methods and Models of Operations Research* |
| Lesourne | *Economic Analysis and Industrial Management* |
| Mantel | *Cases in Managerial Decisions* |
| Massé | *Optimal Investment Decisions: Rules for Action and Criteria for Choice* |
| McGuire | *Theories of Business Behavior* |
| Miller and Starr | *Executive Decisions and Operations Research, 2nd ed.* |
| Montgomery and Urban | *Management Science in Marketing* |
| Montgomery and Urban | *Applications of Management Science in Marketing* |
| Morris | *Management Science: A Bayesian Introduction* |
| Muth and Thompson | *Industrial Scheduling* |
| Nelson (ed.) | *Marginal Cost Pricing in Practice* |
| Nicosia | *Consumer Decision Processes: Marketing and Advertising Decisions* |
| Peters and Summers | *Statistical Analysis for Business Decisions* |
| Pfiffner and Sherwood | *Administrative Organization* |
| Simonnard | *Linear Programming* |
| Singer | *Antitrust Economics: Selected Legal Cases and Economic Models* |
| Vernon | *Manager in the International Economy* |
| Wagner | *Principles of Operations Research with Applications to Managerial Decisions* |
| Zangwill | *Nonlinear Programming: A Unified Approach* |
| Zenoff and Zwick | *International Financial Management* |
| Zionts | *Linear and Integer Programming* |

Prentice-Hall, Inc.
Prentice-Hall International, Inc., *United Kingdom and Eire*
Prentice-Hall of Canada, Ltd., *Canada*
J. H. DeBussy, Ltd., *Holland and Flemish-Speaking Belgium*
Dunod Press, *France*
Maruzen Company, Ltd., *Far East*
Herrero Hermanos, Sucs, *Spain and Latin America*
R. Oldenbourg, Verlag, *Germany*
Ulrico Hoepli Editore, *Italy*

# LINEAR AND INTEGER PROGRAMMING

STANLEY ZIONTS

School of Management
State University of New York at Buffalo

**PRENTICE-HALL, INC.,** Englewood Cliffs, New Jersey

*Library of Congress Cataloging in Publication Data*

ZIONTS, STANLEY
Linear and integer programming.

(Prentice-Hall International series in management)
Bibliography: p.
1. Linear programming. 2. Integer programming.
I. Title.
T57.74.Z56 519.7 73-4214
ISBN 0-13-536763-8

*To my parents—*

**Samuel E. Zionts and Henrietta S. Zionts**



Printed in the United States of America

10 9 8 7 6 5 4 3 2

Prentice-Hall International, Inc., *London*
Prentice-Hall of Australia, Pty. Ltd., *Sydney*
Prentice-Hall of Canada, Ltd., *Toronto*
Prentice-Hall of India Private Limited, *New Delhi*
Prentice-Hall of Japan, Inc., *Tokyo*

# Contents

# Preface

This book was written to be a clear up-to-date text and reference on linear and integer programming. As a text it is suitable for the advanced undergraduate levels and graduate levels in economics, engineering, management, and mathematics. It is intended for either a two-semester course on linear and integer programming in which all of the chapters would be followed in sequence (with minor deletions), or it may be used for a one-semester course in linear programming as well as a one-semester course in integer programming. The linear programming course would include Chapters 1 through 11, possibly omitting some or all of the following: Chapters 2, 6 and 10, Sections 8.3 and 8.4, and Sections 9.3 through 9.10. A one-semester course completely covering Chapters 1 through 11 is possible but would require rapid coverage. A one-semester course in integer programming for students without previous training in linear programming would consist of Chapters 1, 2 (optional), 3 and 5, Sections 8.1 and 8.2, and Chapters 12 through 17.

The book is also intended for operations research practitioners as a reference book. Although the book is not intended for persons without any mathematical background, I feel that Chapter 1 (possibly excluding Section 1.4) Section 5.1, 7.1, Chapter 11, Section 12.1 and portions of Chapter 17 are understandable for those having minimal mathematical background who would like to know more about linear and integer programming.

Some time ago I observed that when I learned something well, I was then able to explain it in a clear and concise manner. Presumably, if I could explain a topic well verbally, I could express it well in a written form. This thought was my prime motivation for writing this book. A few months ago, a good friend and colleague at the State University of New York at Buffalo, Professor Phillip Ross, suggested what had never occurred to me (perhaps subconsciously intentionally), the contrapositive argument: If I can't state the material in an easily understandable way; then perhaps I don't understand the material that well. Let the reader decide!

Every author owes a debt of gratitude to those who have assisted him. I wish to thank Mr. J. Lee O'Nan of the Aluminum Company of America for contributing Section 11.10, Professor Der-San Chen, University of Alabama, for his collaboration in the writing of Chapter 14, Mr. Kenneth E. Kendall, State University of New York at Buffalo, for his collaboration in writing Section 16.5 and Professor R. E. D. Woolsey, Colorado School of Mines, for his contribution of Chapter 17. The following journals have permitted me to draw from my publications in them: *Management Science*, *Naval Research Logistics Quarterly*, *Operations Research* and *Opsearch*. I also wish to thank Professor G. B. Dantzig (Stanford), and R. M. Van Slyke (University of California at Berkeley) and the Academic Press, publishers of the *Journal of Computer and Systems Sciences*, for permission to draw extensively from their article in writing Section 8.3.

I appreciate the valuable comments, suggestions and questions provided by the many students who used numerous versions of the manuscript at the State University of New York at Buffalo (SUNY), and the comments on the manuscript by Professor E. Balas (Carnegie-Mellon University), Professor E. J. Bell (University of Washington), Professor J. C. G. Boot (SUNY), Professor W. W. Cooper (Carnegie-Mellon University), Professor A. Geoffrion (University of California, Los Angeles), Dr. W. C. Mylander (General Research Corporation, McLean, Va.), Professor H. Salkin (Case-Western Reserve University), Professor Linus Schrage (University of Chicago), and Professor W. Szwarc (University of Wisconsin, Milwaukee).

I also wish to acknowledge the support of the Ford Foundation and the School of Management, State University of New York at Buffalo.

Mr. Rashmi Thakkar (SUNY) provided some useful comments in his careful reading of an early draft of the manuscript and generated some of the problems and their solutions. Mr. Butch Lesniak (SUNY) did an outstanding job of reading numerous drafts of the manuscript and identifying passages that required improved exposition. He also carefully proofread the galleys and pages and prepared many of the illustrations and most of the problem solutions. Mrs. Irene Forster did the bulk of the typing with a substantial amount of help from Mrs. Eileen Asaro. They not only carefully translated my hieroglyphics into neat manuscript form, but they did it cheerfully and patiently.

Finally, I acknowledge the contribution of my wife Terri and our children David, Michael, Rebecca Jo, and Andrew: many nights and weekends without me. To them this book must have seemed a monster as evidenced by the children's continuing question, "Are you still working on that book, Daddy?"

# 1

# Introduction

Linear programming is one of the most important and most-used optimization techniques of operations research. Integer programming is currently of lesser importance, but is sufficiently general as a potential problem-solving technique and is gaining in importance as its methodology evolves. Before we define these two problem areas and introduce applications and solution methods, we briefly motivate the importance of constrained optimization and linear and integer programming.

## 1.1 CONSTRAINED OPTIMIZATION: LINEAR AND INTEGER PROGRAMMING

Most problems can be expressed as optimization problems subject to constraints. "Surely I can think of many complex problems where this is just not true," a skeptical reader may think. "What about complex *real* problems such as buying a house?" Let us consider that problem and argue that it is indeed a problem of constrained optimization. We ignore the negotiating aspects of the problem and assume the price and terms fixed. Such aspects could be included in the framework by further complicating the problem. The constraints involved include the following types:

1. Limits on price.
2. Limits on time to spend examining alternatives.
3. Limits on location:

a) Within a certain reasonable distance from one's place of employment.
b) Quality of neighborhood services, e.g., schools.

4. Limits on distance from or proximity of services.
5. Limits on number of rooms, size of lot, and so on.

The measure to be optimized in choosing the "best" house subject to the above constraints is a function of a set of variables, both subjective and objective, that are evaluated for each house that appears to satisfy the constraints. Examples of such variables are those included in the above constraints, as well as such items as total floor space, size of heating bills, outward appearance, and so on. The determination of a function of the indicated variables for evaluating a house is by no means a trivial matter, and in many instances the determination of such a function is an evolutionary process. There are many subjective variables that are hard to measure, and behavioral variables such as one's feelings and emotions may come into play. A further, not unusual, outcome which may occur is that no house may exist which satisfies the constraints (for example, when the constraints specify at least a two acre, seven bedroom, four bath house with a three car garage in Westchester County, New York, for under $25,000). The constraints must, in such a case, be relaxed by the formulator in order to find a feasible solution.

We can think of many other real problems of constrained optimization, for example:

1. A business which wants to maximize profits subject to certain constraints, such as factory capacities, market potential, raw material availability, budgetary limitations, and legal requirements.
2. A platoon leader on the field of battle who wishes to penetrate the enemy lines as deeply as possible subject to the constraints of his manpower ability, fire power, ammunition availability, and so on.
3. A building contractor who, having been awarded a construction contract, wants to minimize construction costs plus penalty costs (if any) subject to availability of equipment, men, materials, and money.

We shall restrict the constrained optimization problems to those in which all the variables can be objectively defined, and both the objectives and constraints can be formulated mathematically. Such problems are often referred to as mathematical programming problems.

In this book we treat two important classes of mathematical programming problems: linear programming problems and integer linear programming problems. Linear programming problems are characterized by objective functions and constraints where each incremental unit of a variable of the problem contributes the same amount (per unit) to the value of the objective, and consumes the same amounts (per unit) of the resources used in producing that variable. Integer linear programming problems are essentially the same kind of problem, with one important difference: some or all of the variables are restricted to integral values. To illustrate the integrality requirement, the number of airplanes of a given type produced by an airplane manufacturer must be zero or one or some other positive integer.

More precisely, in linear programming the objective function or mathematical statement of the objective and the constraints are linear.

A linear objective function is an expression of the form $c_1x_1 + c_2x_2 + \cdots + c_nx_n$ where $c_1, \ldots, c_n$ are known real constants and $x_1, \ldots, x_n$ are real variables whose optimal values (when they exist) satisfy the desired objective of maximizing or minimizing the objective function. Linear constraints are constraints on the problem variables and are of three forms

$$a_1x_1 + a_2x_2 + \cdots + a_nx_n \, (\leq, =, \geq) \, a_0$$

where $a_j, j = 0, 1, \ldots, n$ are known constants.

The symbol $\leq$ means "less than or equal to" (or not greater than) and the symbol $\geq$ means "greater than or equal to" (or not less than).

In the first eleven chapters we shall introduce linear programming, explore the significance of solving linear programming problems, and examine numerous ramifications of the theory and applications. In the remaining chapters we do the same for integer linear programming, an area which has been the subject of considerable research, and is becoming increasingly practical as a problem-solving technique.

## 1.2 THE GENERAL FRAMEWORK OF LINEAR PROGRAMMING PROBLEMS

As discussed above, linear programming can be used when a problem under consideration can be described by a linear objective function to be maximized or minimized subject to linear constraints which may be expressed as equalities or inequalities or a combination of the two. The problem, therefore, may be expressed in the following form* (where some or all of the inequalities may be reversed or replaced by equalities).

$$\text{Maximize} \quad z = c_1x_1 + \cdots + c_nx_n \qquad \text{Objective Function}$$

$$\text{(1.1)} \quad \text{subject to:} \quad \left.\begin{array}{l} a_{11}x_1 + \cdots + a_{1n}x_n \leq b_1 \\ a_{21}x_1 + \cdots + a_{2n}x_n \leq b_2 \\ \cdots \quad \cdots \quad \cdots \quad \cdots \\ a_{m1}x_1 + \cdots + a_{mn}x_n \leq b_m \end{array}\right\} \text{Explicit Constraints}$$

$$x_1, x_2, \ldots, x_n \geq 0 \qquad \text{(Implicit) Nonnegativity Constraints}$$

* We shall also consider objective functions which are to be minimized. Such functions can also be represented as above.

The numbers $a_{ij}$, $b_i$, and $c_j$ are known constants that describe the problem; the variables $x_j$ are to be chosen in such a way that the constraints are satisfied and the objective maximized (or minimized). The objective function value is $z$.

The first line of the statement of the problem indicates the objective function, which might be, for example, the profits of a firm where $c_1$ is the profit to be realized in producing a unit of product 1. The number of units of product 1 to be produced is a variable $x_1$, whose value is determined in the course of solution. The $m$ inequalities are the (explicit) constraints. They state that a maximum amount of each resource, such as resource $i$, is available, and that amount is $b_i$. Each product—say product $j$—requires $a_{ij}$ units of resource $i$ to produce a unit of product $j$. (Also required are $a_{1j}$ units of resource 1, $a_{2j}$ units of resource 2, and so on.) Thus the entire constraint states that the amount of resource used in production cannot exceed that which is available.

The last set of constraints are called the (implicit) nonnegativity constraints; they specify that the variables must not be negative. Explicit representation of the nonnegativity constraints is not required in the solution process. For some problems it may be desirable to have variables which take on negative values. That topic will be considered in a later chapter.

Any linear programming problem can be written in the form (1.1), although it may not generally seem sensible to do that. We shall show how such a transformation may be accomplished. As mentioned earlier, for an arbitrary problem, we can have three kinds of linear constraints: $\leq$, $\geq$, or $=$. If the constraint is of the "less than or equal to" type ($\leq$), there is nothing to be done. If the constraint is of the "greater than or equal to" type ($\geq$), then it should be multiplied through by $-1$ because this reverses the direction of the inequality. For example, a constraint of the form $a_{r1}x_1 + a_{r2}x_2 + \cdots + a_{rn}x_n \geq b_r$, when multiplied by $-1$, becomes

$$-a_{r1}x_1 - a_{r2}x_2 - \cdots - a_{rn}x_n \leq -b_r$$

Finally, an equality constraint can be represented as two inequality constraints. For example, the constraint

$$a_{p1}x_1 + a_{p2}x_2 + \cdots + a_{pn}x_n = b_p$$

becomes

$$a_{p1}x_1 + a_{p2}x_2 + \cdots + a_{pn}x_n \leq b_p$$

and

$$a_{p1}x_1 + a_{p2}x_2 + \cdots + a_{pn}x_n \geq b_p$$

Both constraints can be satisfied only when the corresponding equality is satisfied. To complete the transformation multiply the greater-than-or-equal-to inequality by $-1$ to get

$$\begin{aligned} a_{p1}x_1 + a_{p2}x_2 + \cdots + a_{pn}x_n &\leq b_p \\ -a_{p1}x_1 - a_{p2}x_2 - \cdots - a_{pn}x_n &\leq -b_p \end{aligned}$$

Later we shall make extensive use of a formulation having only equalities for our conceptual and theoretical framework. It can be shown that such a formulation is just as general as formulation (1.1). For solution purposes, any combination of equalities and inequalities that describes the problem may be used.

## 1.3 AN EXAMPLE

For illustrative purposes, we shall develop an example that will be used repeatedly to develop various aspects of linear programming. It is simplified, and we shall alter the problem as we proceed to illustrate various principles.

Suppose that a dog food manufacturer, Canine Products, Inc., produces two blends of dog food, Frisky Pup and Husky Hound. Two raw materials, cereal and meat, are available. Assuming the data given in Table 1.1 are applicable to the problem, the manufacturer wants to find a production mix that maximizes his profits. Frisky Pup dog food is a blend of 1 pound cereal and 1.5 pounds meat, and sells for 70 cents per 2.5 pound package. Husky Hound dog food is a blend of 2 pounds cereal and 1 pound meat and sells for 60 cents

**Table 1.1**

Data for Example Problem

| | FRISKY PUP | HUSKY HOUND |
|---|---|---|
| Contents of finished package | 2.5 lb. | 3.0 lb. |
| Sale price per package | $ .70 | $ .60 |
| Raw materials usage per package | | |
| Cereal | 1.0 lb. | 2.0 lb. |
| Meat | 1.5 lb. | 1.0 lb. |
| Purchase price of raw materials | | |
| Cereal | $0.10/lb. | $0.10/lb. |
| Meat | $0.20/lb. | $0.20/lb. |
| Blending, packaging, and other variable costs per package | $ .14 | $ .18 |

Resources available for production per month:

Raw materials

| | |
|---|---|
| Cereal | 240,000 lbs. |
| Meat | 180,000 lbs. |

Processing capacities

| | |
|---|---|
| Husky Hound blending, Husky Hound packaging, Frisky Pup blending | sufficient for production of any product mix within above raw material availability. |
| Frisky Pup packaging | A maximum of 110,000 packages per month. |

1. The marketing manager estimates that any feasible mix of the two blends, given the resource restrictions, can be sold at prices indicated above.
2. Canine Products has entered into long-term contracts with the suppliers of raw materials whereby every month Canine Products is required to purchase the quantities shown above.
3. Any unused quantity of raw materials left over at the end of a month is a total loss—it can neither be used in the following month nor can it be sold.

per 3 pound package. Cereal costs 10 cents per pound; meat costs 20 cents per pound. Husky Hound dog food costs 18 cents per package for packaging and Frisky Pup dog food costs 14 cents per package for packaging. Blending of the meat and cereal is accomplished automatically during the packaging. Frisky Pup requires the use of a special packaging machine.

In a one month period the company has available 240,000 pounds of cereal and 180,000 pounds of meat for which it has contracted at the above prices. If the meat or cereal is not completely used there is no alternate use for it. The special packaging machine for Frisky Pup dog food can package as many as 110,000 units per month. The Husky Hound packaging facility is of sufficient capacity to handle any mixture of products, given the raw material available. The marketing manager of Canine Products estimates that all of Canine production—whatever the mixture of products, given the raw materials available—will be sold at the indicated prices. Assuming that all of the variable costs of production are indicated above, what is the production plan that maximizes contribution to overhead and profits?

This problem is distinctly different from choosing a specific blend to accomplish a specific purpose, such as finding the lowest cost blend that satisfies certain nutritional requirements. That problem can also be formulated and solved as a linear programming problem, providing the constraints are linear.

To formulate the given problem, let $x_1$ and $x_2$ respectively represent the number of packages of Frisky Pup and Husky Hound dog foods produced in a month. The number of pounds of cereal required for production of Frisky Pup dog food is

$$1\,\frac{\text{pound}}{\text{package}} \cdot (x_1 \text{ packages}) = x_1 \text{ pounds}$$

and the number of pounds of cereal required for Husky Hound dog food is

$$2\,\frac{\text{pounds}}{\text{package}} \cdot (x_2 \text{ packages}) = 2x_2 \text{ pounds}$$

The total cereal required to produce an arbitrary program of production of $x_1$ packages of Frisky Pup and $x_2$ packages of Husky Hound is then

$$x_1 + 2x_2$$

Incorporating the cereal availability, we have the inequality constraint

$$x_1 + 2x_2 \leq 240{,}000 \tag{1.2}$$

In a similar manner, we determine that $1.5x_1$ and $1x_2$ pounds of meat are used to produce $x_1$ packages of Frisky Pup and $x_2$ packages of Husky Hound dog foods, respectively. Incorporating the meat availability for the month, we have the inequality constraint

$$1.5x_1 + x_2 \leq 180{,}000 \tag{1.3}$$

The packaging constraint for Frisky Pup is simply that, at most, 110,000 packages can be prepared in a given month, or

$$(1.4) \qquad x_1 \leq 110{,}000$$

Since the number of packages produced must be zero or positive, the following constraints must be satisfied:

$$(1.5) \qquad x_1 \geq 0,\ x_2 \geq 0$$

We now formulate the objective function. Our objective is to find the values of $x_1$ and $x_2$ which yield the maximum profit without violating the above constraints. Since a fixed cost is sunk, profits are maximized when we maximize the total contribution to overhead and profits (excess of selling price over marginal cost). We must therefore first determine the contribution of a unit of $x_1$ and of a unit of $x_2$. The computations are given in Table 1.2.

**Table 1.2**

Computation of Per Unit Contribution to Overhead and Profits For Example

| | FRISKY PUP | HUSKY HOUND |
|---|---|---|
| | $x_1$ | $x_2$ |
| | $ | $ |
| I. Selling price per package | 0.70 | 0.60 |
| Marginal cost: | | |
| *Material cost* | | |
| Cereal* | — | — |
| Meat* | — | — |
| II. Total variable costs of blending and packaging | 0.14 | 0.18 |
| Contribution (I − II) | 0.56 | 0.42 |

* In view of notes 2 and 3 to Table 1.1, cost of these materials is treated as fixed or sunk costs which are in total independent of the actual production.

We can now formulate our objective function as

$$(1.6) \qquad \text{Maximize Total Contribution } z = .56x_1 + .42x_2$$

Writing expressions (1.2) to (1.6), we have the complete linear programming problem, which is to

$$\begin{array}{lll}
(1.6) & \text{Maximize } z = .56x_1 + .42x_2 & \\
(1.2) & \text{subject to:} \quad x_1 + 2x_2 \leq 240{,}000 & \\
(1.3) & \qquad 1.5x_1 + x_2 \leq 180{,}000 & \\
(1.4) & \qquad x_1 \leq 110{,}000 & \\
(1.5) & \qquad x_1, x_2 \geq 0 &
\end{array}$$

By explanation, any solution to the problem that we may find must be *feasible*. A solution is said to be feasible if it *satisfies* all the constraints. For example, the following, among many others, are feasible solutions:

$$x_1 = 100{,}000,\ x_2 = 0;\ x_1 = 1{,}954,\ x_2 = 76{,}950;$$
$$x_1 = 50{,}000,\ x_2 = 50{,}000;\ x_1 = 0,\ x_2 = 0$$

On the other hand, $x_1 = 115{,}000$, $x_2 = 10{,}000$; and $x_1 = 50{,}000$, $x_2 = 100{,}000$ are *not* feasible solutions because they violate one or more constraints.

Of all possible feasible solutions to the problem posed above, only a subset can fulfill the objective, namely, to maximize the objective function, in this case profits. That solution (or solutions) is called an *optimal feasible solution*. For some problems there may be no solutions that satisfy the constraints. No feasible solution is said to exist for such problems. The occurrence of no feasible solution generally indicates one of two conditions:

1. An error has been made in formulating a problem for which a feasible solution exists.
2. An inconsistent problem has been formulated in the sense that it is impossible to satisfy all constraints simultaneously.

Another anomalous kind of solution that may occur is an optimal solution that is infinite. This is called an *unbounded solution* and usually indicates that the problem has not been correctly formulated. Such a condition would occur, for our example, if unlimited supplies of meat and cereal were available at the indicated prices. Altering the formulation accordingly, the optimal solution would be to produce 110,000 packages of Frisky Pup dog food and an infinite amount of Husky Hound. Such a solution, aside from being nonsensical, is incorrect for the following reasons:

1. Only a finite amount of meat and cereal has been contracted for at the indicated price. In general, as greater and greater quantities are purchased, the price of the materials might be expected to rise.
2. We have not considered limitations on blending capacity and Husky Hound packaging capacity, which are presumably finite.
3. The marketing manager's assumption about selling all that can be produced cannot be extended beyond certain limits.

For the example problem, it can be shown that the unique optimal feasible solution would be to produce 60,000 packages of Frisky Pup and 90,000 packages of Husky Hound per month. The reader should attempt to verify the optimality of this solution by trial and error. Total contribution to overhead and profits is $71,400 per month for the optimal solution. Later, methods for computing the optimal solution in general will be introduced.

## 1.4 AN INTUITIVE ALGEBRAIC APPROACH FOR SOLVING LINEAR PROGRAMMING PROBLEMS*

We shall now use the concepts that have been introduced to develop a method for solving linear programming problems. Consider the example problem of Canine Products, Inc.:

$$
\begin{aligned}
\text{Maximize } z &= .56x_1 + .42x_2 \\
\text{subject to:} \quad & x_1 + 2x_2 \leq 240{,}000 \\
& 1.5x_1 + x_2 \leq 180{,}000 \\
& x_1 \leq 110{,}000 \\
& x_1, x_2 \geq 0
\end{aligned}
$$

We first define some additional variables. Let $x_3$ be the amount of cereal not used to produce dog food. By definition then, $x_3 = 240{,}000 - x_1 - 2x_2$ or, equivalently, $x_1 + 2x_2 + x_3 = 240{,}000$. The variable $x_3$ is called a slack variable since it takes up the slack of an inequality. (The slack variable for a reverse inequality is called a surplus variable by some authors; our convention will be to call such variables slack variables.) We shall refer to nonslack variables as structural variables.

Similarly, we define $x_4$ as the amount of meat not used to produce dog food. This yields

$$1.5x_1 + x_2 + x_4 = 180{,}000$$

Finally, we denote $x_5$ as the amount of Frisky Pup packaging capacity that is not used for packaging Frisky Pup dog food. The third constraint then becomes the following:

$$x_1 + x_5 = 110{,}000$$

Putting all the constraints together with the objective function yields the following problem, in which $z$ denotes the value of the objective function; we have

$$
\begin{aligned}
\text{Maximize } z &= .56x_1 + .42x_2 \\
\text{subject to:} \quad & x_1 + 2x_2 + x_3 = 240{,}000 \\
& 1.5x_1 + x_2 + x_4 = 180{,}000 \\
& x_1 + x_5 = 110{,}000 \\
& x_1, x_2, x_3, x_4, x_5 \geq 0
\end{aligned}
$$

*This section introduces an algebraic approach essentially equivalent to the simplex method presented in Chapter 3. It may be omitted by those who do not require a mathematical motivation for the concept of the simplex method.

The above problem is called the equality form of a linear programming problem. It has the following form in general:

$$
\begin{aligned}
\text{Maximize } z &= c_1x_1 + c_2x_2 + \cdots + c_nx_n \\
\text{subject to:} \quad & a_{11}x_1 + a_{12}x_2 + \cdots + a_{1n}x_n = b_1 \\
& a_{21}x_1 + a_{22}x_2 + \cdots + a_{2n}x_n = b_2 \\
& \cdots \quad \cdots \quad \cdots \quad \cdots \quad \cdots \\
& a_{m1}x_1 + a_{m2}x_2 + \cdots + a_{mn}x_n = b_m \\
& x_1, \ldots, x_n \geq 0
\end{aligned}
$$

It is of interest to compare the above equality form with the inequality form given by expression (1.1). The values of $m$ and $n$ given above differ in general from those of the corresponding inequality form (1.1).

The above example problem now consists of three equalities and five variables. Such a system is in general underdefined and possesses many solutions. (When there are as many equality constraints as variables, there is usually a unique solution.)

In the case of an underdefined system like the one above, we may choose all but $m$ variables (where $m$ is the number of constraints) to take on specific values, and then solve for the remaining $m$ variables in $m$ equations. For which $m$ variables should we solve? Any arbitrary set may not be satisfactory because the remaining equalities may not be solvable. Therefore, it is useful to have the problem in what is called a canonical form, where for every equation there is one variable that appears with a coefficient of $+1$ only in that equation, and with a coefficient of zero in every other equation as well as the objective function. In the example we have $x_3$, $x_4$, and $x_5$, the slack variables, appearing in that manner, so we can choose $x_1$ and $x_2$ to take on any values we like, and then compute the values of $x_3$, $x_4$, and $x_5$ uniquely (given the chosen values of $x_1$ and $x_2$).

For simplicity, we first choose $x_1$ and $x_2$ to be zero. Note that we can conveniently determine the associated values of $x_3$, $x_4$, $x_5$, and $z$ from the statement of the problem

$$
\begin{aligned}
&\text{Maximize } z = .56\overset{0}{x_1} + .42\overset{0}{x_2} \\
&\text{subject to:} \quad x_1 + 2x_2 + x_3 = 240{,}000 \\
&\qquad\qquad 1.5x_1 + x_2 + x_4 = 180{,}000 \\
&\qquad\qquad x_1 + x_5 = 110{,}000
\end{aligned}
\tag{1.7}
$$

where the zeroes above $x_1$ and $x_2$ indicate that they have been set to zero. The associated solution is $x_3 = 240{,}000$, $x_4 = 180{,}000$, $x_5 = 110{,}000$, and $z = 0$. If this solution were optimal, the problem would be solved. How can we test its optimality? Let us set $x_1$ and $x_2$ to any other set of values to see whether a greater value of $z$ can be achieved. We can choose only nonnegative values of $x_1$ and $x_2$. In addition, $x_3$, $x_4$, and $x_5$ must turn out to be zero or greater. What about the change in the objective function as we increase $x_1$ or $x_2$ from zero? The value of $z$ is increased by 0.56 for each additional unit of $x_1$ and 0.42 for each additional unit of $x_2$. Since either one can be increased at least a small amount, our initial solution cannot be optimal. So far, we have found a feasible solution which is not optimal. At this time we really cannot say much about how to achieve an optimal solution. It is convenient, however, to have the problem in a canonical form so that optimality can be checked easily. Denoting variables $x_3$, $x_4$, and $x_5$ as basic variables and variables $x_1$ and $x_2$ as nonbasic variables, we can check the feasibility of the solution. We obtain the objective function value by specifying the value of nonbasic variables as zero and checking the optimality of the solution. Apparently, the optimality of such a solution is evident if all the coefficients in the objective function at that point are zero or negative.

There does not appear to be a systematic way of setting all the nonbasic variables simultaneously to optimal values. It is desirable to increase one or more of them from the value of zero, and it is desirable to maintain the canonical form. Therefore, let us choose the variable which increases the objective function *most* per unit (this choice is arbitrary) and increase it. Perhaps we can make it a nonzero (basic) variable and make one of the presently basic variables zero (nonbasic). The choice of the variable which increases the objective function most per unit depends, of course, on the units used. In the example problem, $x_1$ would be chosen because its coefficient of .56 is maximum. (On the other hand, if Husky Hound dog food production were measured in double packages, Husky Hound dog food would have a coefficient in the objective function—contribution to overhead and profits—of 0.84, and $x_2$ would be chosen to be increased.)

We now increase $x_1$, holding $x_2$ at zero. A partial table of values of the variables is given in Table 1.3. Apparently, the largest value of $x_1$ which can be attained preserving the nonnegativity of the basic variables is 110,000. Any

**Table 1.3**

Initial Basic Variables and Objective Function as a Function of $x_1$, Holding $x_2$ at Zero

| $x_1$ | $z$ | $x_3$ | $x_4$ | $x_5$ |
|---|---|---|---|---|
| 0 | 0 | 240,000 | 180,000 | 110,000 |
| 1 | .56 | 239,999 | 179,998.5 | 109,999 |
| 10,000 | 5,600 | 230,000 | 165,000 | 100,000 |
| 50,000 | 28,000 | 190,000 | 105,000 | 60,000 |
| 100,000 | 56,000 | 140,000 | 30,000 | 10,000 |
| 109,999 | 61,599.44 | 130,001 | 15,001.5 | 1 |
| 110,000 | 61,600 | 130,000 | 15,000 | 0 |
| 110,001 | 61,600.56 | 129,999 | 14,998.5 | −1 |
| 115,000 | 64,400 | 125,000 | 7,500 | −5,000 |

larger value of $x_1$ causes $x_5$ to be negative, so it seems appropriate to make $x_1$ a basic variable in place of $x_5$. We call $x_1$ the incoming variable and $x_5$ the outgoing variable. Therefore, we solve for $x_1$ in the equality in which $x_5$ appears, and substitute for it in every other constraint. Solving for $x_1$ in the third constraint, we have:

$$x_1 = 110{,}000 - x_5$$

Substituting for $x_1$ in every other constraint (except the third constraint) and the objective function in (1.7), we have

$$\begin{array}{lrcll}
\text{Maximize } z = .56(110{,}000 - x_5) + .42x_2 & & & & \\
\text{subject to:} \quad (110{,}000 - x_5) + & 2x_2 + x_3 & & & = 240{,}000 \\
1.5(110{,}000 - x_5) + & x_2 & + x_4 & & = 180{,}000 \\
x_1 & & & + x_5 & = 110{,}000
\end{array}$$

Simplifying, we have

$$(1.8) \quad \begin{array}{lrrrrl}
\text{Maximize } z = & .42x_2 & & - .56x_5 & + & 61{,}600 \\
\text{subject to:} & 2x_2 + x_3 & & - x_5 & = & 130{,}000 \\
& x_2 & + x_4 & - 1.5x_5 & = & 15{,}000 \\
x_1 & & & + x_5 & = & 110{,}000
\end{array}$$

which is a new canonical form with basic variables $x_1$, $x_3$, and $x_4$. We can visualize this as a completely new linear programming problem which we shall treat as we did the original. The nonbasic variables are $x_2$ and $x_5$. We can set the nonbasic variables to any arbitrary values not less than zero, and quickly determine the corresponding values of the basic variables and the value of the objective function. If we choose to set the nonbasic variables to zero (which is always a simple solution to explore), we can determine whether or not that solution is optimal. Setting the nonbasic variables to zero for the newly derived problem, we have $x_3 = 130{,}000$, $x_4 = 15{,}000$, $x_1 = 110{,}000$, and associated contribution to overhead and profits of \$61,600. It is not optimal because the coefficient of $x_2$ in the objective function is positive and it is possible to set $x_2$, a nonbasic variable, to values of 1, 2, or larger values, thereby increasing the objective function. Since there is only one positive coefficient in the objective function, the associated variable is the only nonbasic variable that can be increased from zero and increase the objective function from \$61,600. Table 1.4 presents a partial table of values of all basic variables and the objective function as a function of $x_2$, holding the other nonbasic variables—in this case only $x_5$—at zero.

The largest value of $x_2$ that can be achieved, given the basic and nonbasic variables, is 15,000. For any larger value of $x_2$, $x_4$ is negative. Therefore, $x_2$

**Table 1.4**

Basic Variables $x_1$, $x_3$, and $x_4$ and the Objective Function as a Function of the Nonbasic Variable $x_2$

| $x_2$ | $z$ | $x_1$ | $x_3$ | $x_4$ |
|---|---|---|---|---|
| 0 | 61,600 | 110,000 | 130,000 | 15,000 |
| 1 | 61,600.42 | 110,000 | 129,998 | 14,999 |
| 10,000 | 65,800 | 110,000 | 110,000 | 5,000 |
| 14,999 | 67,899.58 | 110,000 | 100,002 | 1 |
| 15,000 | 67,900 | 110,000 | 100,000 | 0 |
| 15,001 | 67,900.42 | 110,000 | 99,998 | −1 |
| 20,000 | 70,000 | 110,000 | 90,000 | −5,000 |

should be made basic, and $x_4$ should be made nonbasic. Solving for $x_2$ in the equation in which $x_4$ is basic, we have

$$x_2 = 15{,}000 - x_4 + 1.5x_5$$

and substituting the above for $x_2$ in every other constraint in (1.8) we have

$$\begin{aligned}
\text{Maximize } z = \quad & .42(15{,}000 - x_4 + 1.5x_5) - .56x_5 + 61{,}600 \\
\text{subject to:} \quad & 2(15{,}000 - x_4 + 1.5x_5) + x_3 - x_5 = 130{,}000 \\
& x_2 + x_4 - 1.5x_5 = 15{,}000 \\
& x_1 + x_5 = 110{,}000
\end{aligned}$$

or

$$\begin{aligned}
\text{Maximize } z = \quad & -.42x_4 + .07x_5 + 67{,}900 \\
\text{subject to:} \quad & x_3 - 2x_4 + 2x_5 = 100{,}000 \\
& x_2 + x_4 - 1.5x_5 = 15{,}000 \\
& x_1 + x_5 = 110{,}000
\end{aligned} \tag{1.9}$$

which is a new canonical form with basic variables $x_1$, $x_2$, and $x_3$, and nonbasic variables $x_4$ and $x_5$. We can treat this as still another linear programming problem and repeat the analysis. The solution is still not optimal, since the coefficient of $x_5$ is positive in the objective function. We find in the same manner as before that when $x_5$ is increased sufficiently, $x_3$ in the first constraint of formulation (1.9) becomes negative. Therefore, we solve for $x_5$ in the equation in which $x_3$ is basic to obtain

$$x_5 = 50{,}000 - (1/2)x_3 + x_4$$

and then substitute for $x_5$ in every other constraint. We always divide the equation by the appropriate *positive* number to assure that the coefficient of the basic variable in its constraint is + 1. However, in the earlier stages, the coefficient was +1 initially, and dividing an equation by + 1 did not alter it. Here the equation must be divided by + 2. Carrying out the above operations yields the following:

$$
\begin{aligned}
&\text{Maximize } z = && - .035x_3 - .35x_4 \qquad + 71{,}400 \\
(1.10)\quad &\text{subject to:} && .5x_3 - x_4 + x_5 = 50{,}000 \\
& && x_2 + .75x_3 - 0.5x_4 \qquad = 90{,}000 \\
& && x_1 \quad - .5x_3 + x_4 \qquad = 60{,}000
\end{aligned}
$$

which can be treated as yet another linear programming problem. Here $x_1$, $x_2$, and $x_5$ are the basic variables, and $x_3$ and $x_4$ are the nonbasic variables. Because none of the coefficients of the nonbasic variables in the objective function are positive, the above solution, with the nonbasic variables set equal to zero, would appear to be the optimal solution.

Thus the optimal solution is that $x_1$, the number of packages of Frisky Pup produced, is 60,000, $x_2$, the number of packages of Husky Hound produced, is 90,000, and $x_5$, the amount of unused Frisky Pup packaging capacity, is 50,000. The associated objective function value is $71,400.

### The Procedure

By manipulating the equalities of the problem in a logical algebraic manner, we have been able to find the optimal solution to the problem. We have not as yet really specified the method, nor have we explored the generality of the method that has been proposed. Let us define the method more precisely, in a sequence of steps, at the same time indicating possible problems which might arise in certain situations.

*Step 1:* Add slack variables where required. Then obtain a canonical form statement of the problem—a statement in which each constraint has appearing in it some variable with a coefficient of + 1, but which variable has a coefficient of 0 in every other constraint. Such variables are called basic variables. When the remaining variables (the nonbasic variables) are set to zero, the values of the basic variables, which are instantly seen, must be greater than or equal to zero.

*Possible Difficulty:* For the example problem a canonical form was available initially. In general, such a form may not be available, and some procedure is required to obtain a canonical form.

*Step 2:* Since the nonbasic variables may be freely specified to take on nonnegative values, check to see whether the objective function value can be made greater than that resulting from having the nonbasic variables take on zero values. More simply, if all coefficients of nonbasic variables in the objective function are nonpositive, the optimal solution is to set the existing nonbasic variables to zero. Otherwise the solution is not optimal, and one or more nonbasic variables must be basic in an optimal solution.

*Step 3:* Choose a nonbasic variable with a positive coefficient in the objective function. (We have been arbitrarily using the variable with the most positive coefficient in the objective function.) Determine the maximum value to which that variable can be set, holding the other nonbasic variables at zero. Corresponding to that maximum value, a basic variable will be reduced to zero.

*Possible Difficulty:* One of the basic variables will usually become zero, but we have not yet considered what to do if no variables become zero, or if two or more variables become zero simultaneously. Further, it may not be possible to increase the selected nonbasic variable *at all* without violating some constraint.

*Step 4:* Solve for the nonbasic variable selected in Step 3 in terms of the basic variable that was reduced to zero in Step 3. Designate the former variable as the incoming variable and the latter variable as the outgoing variable. Since the outgoing variable appears in only one equation at any step, the equation in which the incoming variable is solved for is always uniquely determined.

*Step 5:* Substitute for the incoming variable in every equation (including the objective function) except the equation in which the outgoing variable appeared originally. Divide the latter equation through by the coefficient of the incoming variable. The incoming variable is now a basic variable and the outgoing variable is now a nonbasic variable. Then go to Step 2.

We do not know as yet whether the above procedure is finite or that it will converge to an optimal solution. All that is known is that it appears intuitively reasonable, and that it works on our example problem. We shall consider this procedure further in Chapter 3.

## 1.5 SUMMARY

In this chapter we have introduced the concept of mathematical programming and defined what is meant by linear programming (maximizing or minimizing a linear objective function subject to a set of linear constraints) and integer linear programming (a linear programming problem in which some or all variables are constrained to be integers), and then formulated a linear programming problem in algebraic terms. Using this problem as an example, we developed some terminology for linear programming. Then we proceeded to develop, by means of an example, an intuitive algebraic method for solving at least some linear programming problems. Some problems to be resolved were identified.

## 1.6 PROBLEMS

For problems 1 through 5:

a. Graph the constraints and objective function.

b. Solve graphically.

c. Solve using the intuitive approach of section 1.4.

**1.** Maximize $z = 4x_1 + x_2$

subject to:
$$\begin{aligned} -6x_1 + x_2 &\leq 0 \\ -3x_1 + x_2 &\leq 2 \\ x_1 + x_2 &\leq 8 \\ x_1, x_2 &\geq 0 \end{aligned}$$

**2.** Maximize $z = 8x_1 + 15x_2$

subject to:
$$\begin{aligned} x_1 \qquad &\leq 6 \\ x_2 &\leq 10 \\ 10x_1 + 5x_2 &\leq 80 \\ x_1, x_2 &\geq 0 \end{aligned}$$

**3.** Maximize $z = -x_1 + 5x_2$

subject to:
$$\begin{aligned} -3x_1 + x_2 &\leq 4 \\ x_2 &\leq 6 \\ x_1 - 4x_2 &\leq 1 \\ \tfrac{9}{7}x_1 + x_2 &\leq 10 \\ x_1, x_2 &\geq 0 \end{aligned}$$

**4.** Minimize $z = -3x_1 + x_2$

subject to:
$$\begin{aligned} 4x_1 - x_2 &\leq 20 \\ x_1 - x_2 &\leq 3 \\ \tfrac{1}{3}x_1 - x_2 &\leq \tfrac{1}{3} \\ \tfrac{2}{3}x_1 + x_2 &\leq 8 \\ -x_1 + x_2 &\leq 4 \\ x_1, x_2 &\geq 0 \end{aligned}$$

**5.** Maximize $z = 3x_1 + 2x_2$

subject to:
$$\begin{aligned} x_1 + .5x_2 &\leq 15 \\ 2x_1 + 4x_2 &\leq 24 \\ x_1, x_2 &\geq 0 \end{aligned}$$

For problems 6 and 7:

Given the graphs of the shaded feasible region and the objective function as the bold line to be maximized in the indicated direction,

a. Solve the associated linear programming problem graphically.

b. Write the objective function and constraints and solve using the intuitive approach of section 1.4.

**6.**

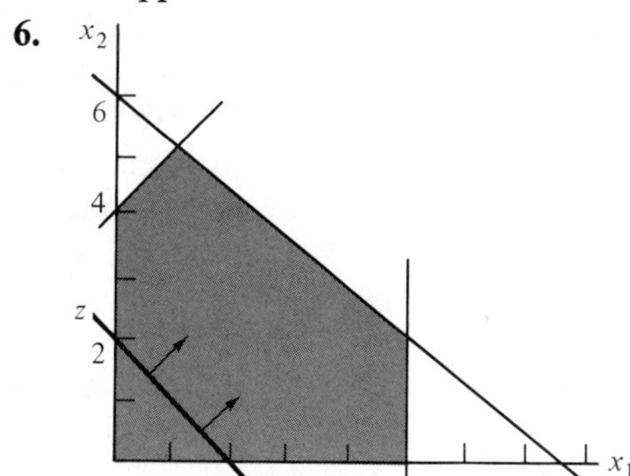


**7.**

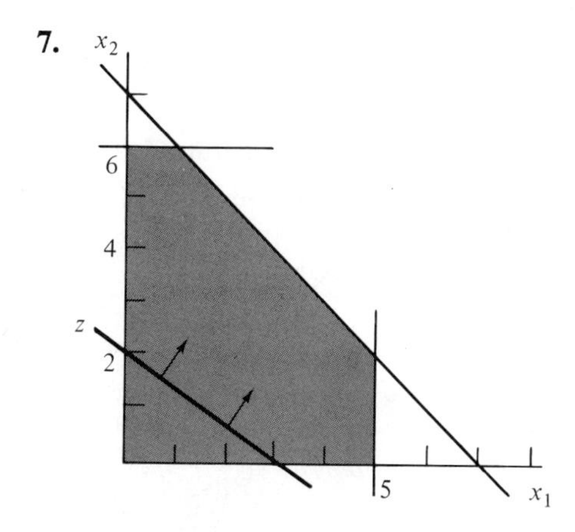


**8.** Formulate the problem whose solution is unbounded, as indicated in section 1.3 (p. 8).

**9.** a. In the text example, suppose that Canine Products, Inc., had not entered into a long-term contract for the raw materials, but the conditions were otherwise the same. How should the problem formulation be changed?

b. Suppose that the company had entered into a long-term contract but could sell unused cereal at \$0.10/lb. and meat at \$0.20/lb. Alter the formulation accordingly.

c. As in part b, suppose the company could resell the unused cereal at \$0.10/lb. and meat at \$0.15/lb. Alter the formulation accordingly.

d. In part c, what difference would it make if the maximum quantity of meat that could be resold were limited to 50 lbs.?

e. Compare the above formulations with that of the example in the text.

**10.** The Pasty Pizza Company (PPC) is considering franchising another outlet.

They have decided that Hudson, Ohio, is a good place for a pizza palace. Taking into consideration the small size of the town, PPC has rented a vacant store that measures 30′ × 40′. The PPC research department has found that each square foot of customer service area (service and table areas) increases weekly sales by 30¢, and that each square foot of production area (ovens, cutting boards) increases weekly sales by 80¢. Experience has shown that such a shop can be run by two people, a waitress for the customer service area and someone in the production area to do the cooking. Although the parent PPC can install various amounts of production facilities, efficiency studies have shown that one man can handle no more than 720 square feet of production area. Similarly, considering the waitress' efficiency, optimal table spacings, and so on, it was found that the customer service area had to be not smaller than 50 square feet less than 25 per cent of the production area.

How much of the store area should be devoted to customer service and how much should be devoted to production in order to maximize weekly earnings?

a. Formulate the Pasty Pizza Company problem.

b. Put it in canonical form.

c. Solve using the intuitive algebraic method given in section 1.4.

**11.** The proprietor of Plump Poultry Farm has retained Bluff Bluster and Company as consultants to solve the following poultry feed problem:

Given the following data, find the minimum cost feed.

| | MINIMUM DAILY REQUIREMENT | NO. OF MILLIGRAMS CONTAINED PER LB. OF FEED AVAILABLE IN THE MARKET | | |
|---|---|---|---|---|
| *Vitamin* | *Milligrams* | *Brand 1* | *Brand 2* | *Brand 3* |
| A | 10 | 5 | 3 | 25 |
| B | 60 | 80 | 20 | 15 |
| C | 30 | 10 | 40 | 5 |
| D | 5 | 1 | 10 | 10 |
| Cost per lb. | | \$.22 | \$.18 | \$.14 |

Formulate the above problem as a linear programming problem.

**12.** Colorful Textiles, Inc., is faced with the problem of optimally allocating the following quantities of cotton fiber of different counts among three varieties of cloth.

| FIBER | MAXIMUM QUANTITY AVAILABLE PER DAY LBS. | YARDS PER LB. | COST PER LB. |
|---|---|---|---|
| 1 | 5000 | 3 | \$3.00 |
| 2 | 3000 | 4 | 6.00 |
| 3 | 2500 | 5 | 10.00 |

| Style | SKIN TYPE 1 | 2 | 3 | 4 | 5 |
|---|---|---|---|---|---|
| A | 5 | — | — | 2 | — |
| B | — | 3 | — | 1 | 3 |
| C | — | — | 2 | — | 1 |
| Waste | 8 | 10 | 9 | 5 | 4 |

e cost of each type of skin is $30, $25, $20, $15, and $17, respectively. rmulate the following problems:

a. Minimize costs, specifying that at least the number of boots ordered be produced.
b. Any wasted portion of a skin is of no use to the company. Minimize the waste assuming that at least the number of boots ordered be produced.
c. Minimize the waste as in part b, subject to a minimum acceptable level of profits, $P$.
d. Minimize the costs as in part a, subject to a maximum acceptable waste level, $W$.

An order for 150 batches of apple flavored oatmeal mix has been received he Oats Cereal Company. The mix is of the instant variety and is made up ne following components (all dehydrated):

| INGREDIENT | SPECIFICATION |
|---|---|
| Apple chips | at least 3% |
| Vitamin supplements | between 2 and 3% |
| Sugar | at least 8% |
| Salt | less than 1% |
| Oats | the remainder |

oatmeal mix can be made from any of the already available mixtures below emainder is oats):

| xture | COMPOSITION (PER CENT) OF MIXTURES Apple Chips | Vitamins | Sugar | Salt | Cost/Batch ($) |
|---|---|---|---|---|---|
| 1 | 50 | 4 | 20 | 1.5 | 18 |
| 2 | 20 | 2.5 | — | .6 | 15 |
| 3 | — | 1 | 7 | .2 | 10 |
| 4 | 100 | — | — | — | 20 |

ompany's objective is to develop a blend that minimizes its cost. ormulate the problem as a linear programming problem.

The standards department of the company has determined th
dard blend for each variety of cloth, and the company sells
fixed price shown below.

| VARIETY NO. | STANDARD BLEND | SE |
|---|---|---|
| 1 | Not more than 30% of 2<br>Not more than 20% of 3 | |
| 2 | Not more than 60% of 1<br>Not less than 15% of 2<br>Not more than 10% of 3 | |
| 3 | Not more than 15% of 1<br>Not more than 30% of 2 | |

Assuming all other costs fixed, formulate the problem as a lin

**13.** Two designs of speedboats are made at a Marine Boa
outboard model takes four days to build; the inboard mod
to build. There are 30 days per month available for buildi
estimated that, at most, *four* inboards and *five* outboards can
Plant policy specifies that the number of outboards prod
maximum, five more than 50 per cent of the number of
The slowest department in the plant is the finishing dep
complete only seven boats a month. If inboards contribute
outboards $1500, how many of each type should be manufa
Boat plant in order to maximize contributions to profits? (I
are acceptable.)

a. Formulate as a linear programming problem.
b. Solve graphically.
c. Solve using intuitive approach.

**14.** The Leathercraft Company has received the followi
hunting boots:

| BOOT STYLE | NUMBER ORDE |
|---|---|
| A | 20 |
| B | 30 |
| C | 40 |

Their boot craft kits enclose a portion of animal skin to
the style's pattern. All patterns are cut from a full skin.
of skins are available and each type allows for a different
be cut, involving the following number of patterns and w

# 2

# Mathematical Foundations and Interpretations of Linear Programming

## 2.1 INTRODUCTION

Having introduced the linear programming problem, some terminology, and an intuitive algebraic method for solving linear programming problems, we shall now develop further the mathematical foundations necessary for analytic treatment of linear programming. Our approach will be to introduce vector space concepts, using geometry as a tool where appropriate, develop the necessary mathematical theory for linear programming using matrix algebra, and prove certain fundamental results. Finally, we shall explore certain geometrical interpretations of linear programming, including methods for solving certain linear programming problems geometrically. We shall provide intuitive interpretations throughout.

## 2.2 MATHEMATICAL PRELIMINARIES—MATRIX ALGEBRA

We shall develop some concepts of matrix algebra and then formulate a linear programming problem in matrix algebraic framework.

**Definition 2.1** (Matrix)
*A* matrix *is a rectangular array of numbers, which may be written as follows:*

$$A = \begin{pmatrix} a_{11} & a_{12} & \cdots & a_{1n} \\ a_{21} & a_{22} & \cdots & a_{2n} \\ \cdot & \cdot & \cdot & \cdot \\ \cdot & \cdot & \cdot & \cdot \\ \cdot & \cdot & \cdot & \cdot \\ a_{m1} & a_{m2} & \cdots & a_{mn} \end{pmatrix}$$

The matrix $A$ is said to be an $m \times n$ (or $m$ by $n$) matrix. Matrices will normally be represented as boldface capital letters (e.g., $A$, $B$) and elements will normally be represented by lower-case double subscripted letters (e.g., $a_{ij}$, $b_{ij}$). The letter used for the element will normally be the same as that used for the matrix. The first subscript indicates the row of the element, and the second subscript indicates the column. For example, $a_{2,3}$ is the element in the second row and the third column of matrix $A$. The notation $a_{ij}$ will indicate the element in row $i$ and column $j$ of matrix $A$ in general. Matrices are simply convenient means of representing tables of numbers. Matrix algebra is a useful algebra (using matrices as ordinary algebra uses scalars) for representing and solving systems of mathematical equations. Like ordinary algebra, there are certain rules and definitions. We present some of the more important ones, for our purposes, below.

**Definition 2.2** (Matrix equality)
*Two matrices $A$ and $B$ are said to be equal if $a_{ij} = b_{ij}$ for every $i$ and $j$.*

Obviously, for two matrices to be equal, they must have the same number of rows and the same number of columns. Further, if a matrix has the same number of rows as columns, it is called a *square matrix*.

**Matrix Addition**

Two matrices can be added if and only if they have the same number of rows and the same number of columns. The rules for addition are to add corresponding elements. Thus, symbolically, if

$$C = A + B$$

then $c_{ij} = a_{ij} + b_{ij}$ for all $i$ and $j$. Some examples of matrices, and matrix addition and equality, follow. Given

$$A = \begin{pmatrix} 3 & 4 & 1 & 2 \\ 0 & -2 & 3 & 1 \\ 1 & 0 & 4 & 2 \end{pmatrix} \quad B = \begin{pmatrix} -1 & 1 & 2 & 3 \\ 3 & 0 & 1 & 4 \\ 1 & 2 & -1 & 3 \end{pmatrix}$$

$$C = \begin{pmatrix} 3 & 4 & 1 & 2 \\ 0 & -2 & 3 & 1 \\ 1 & 0 & 4 & 2 \end{pmatrix} \quad D = \begin{pmatrix} 3 & 4 & 1 \\ 0 & -2 & 3 \\ 1 & 0 & 4 \end{pmatrix}$$

$$A + B = \begin{pmatrix} 2 & 5 & 3 & 5 \\ 3 & -2 & 4 & 5 \\ 2 & 2 & 3 & 5 \end{pmatrix}$$

Matrix $A$ is equal to matrix $C$. The computations $A + D$ and $B + D$ cannot be executed because $A$ and $B$ are 3 by 4 and $D$ is 3 by 3.

A matrix can be multiplied by a scalar by multiplying every entry of the matrix by the scalar. For example,

$$-2\begin{pmatrix} 2 & 1 \\ -1 & 3 \\ 4 & 0 \end{pmatrix} = \begin{pmatrix} -4 & -2 \\ 2 & -6 \\ -8 & 0 \end{pmatrix}$$

Subtraction of matrices may be accomplished by multiplying the subtrahend matrix by the scalar $-1$ and adding the result to the minuend matrix. Thus

$$E - F = E + (-F)$$

For example,

$$\begin{pmatrix} 1 & 2 & 4 \\ 3 & 1 & 7 \end{pmatrix} - \begin{pmatrix} 2 & 5 & 1 \\ -1 & 1 & 5 \end{pmatrix} =$$

$$\begin{pmatrix} 1 & 2 & 4 \\ 3 & 1 & 7 \end{pmatrix} + \begin{pmatrix} -2 & -5 & -1 \\ 1 & -1 & -5 \end{pmatrix} = \begin{pmatrix} -1 & -3 & 3 \\ 4 & 0 & 2 \end{pmatrix}$$

#### Matrix Multiplication

Multiplication of two matrices can only be accomplished if the number of columns of the left matrix is equal to the number of rows of the right matrix. Matrix multiplication is, in general, not commutative; that is, given two matrices $A$ and $B$ and their product $AB$, the product $BA$ may not exist and, if it does exist, *may not* equal $AB$. The product of two matrices has the number of rows of the left matrix and the number of columns of the right matrix.

EXAMPLE:

$$AB = C$$

where $A$ is $m$ by $n$, $B$ is $n$ by $p$, and $C$ is $m$ by $p$.

The elements of the product matrix are found by taking sums of products of entries in the two individual matrices. Specifically, $c_{ij}$ (an arbitrary element in matrix $C$ above) equals $a_{i1}b_{1j} + a_{i2}b_{2j} + \cdots + a_{in}b_{nj}$ where $n$ is the number of columns of $A$ and rows of $B$.

EXAMPLES:

1. $$A = \begin{pmatrix} 1 & 2 & -1 \\ 2 & 1 & 3 \end{pmatrix} \qquad B = \begin{pmatrix} 1 & 1 & 0 \\ 2 & -1 & 1 \\ 1 & 2 & 3 \end{pmatrix}$$

$$AB = \begin{pmatrix} 4 & -3 & -1 \\ 7 & 7 & 10 \end{pmatrix} \qquad (BA \text{ does not exist})$$

2. $$C = \begin{pmatrix} 1 & 2 & -1 \\ 2 & 1 & 3 \end{pmatrix} \qquad D = \begin{pmatrix} 1 & -2 \\ -1 & 1 \\ 1 & 3 \end{pmatrix}$$

$$CD = \begin{pmatrix} -2 & -3 \\ 4 & 6 \end{pmatrix} \qquad DC = \begin{pmatrix} -3 & 0 & -7 \\ 1 & -1 & 4 \\ 7 & 5 & 8 \end{pmatrix}$$

### Special Matrices

We shall define two special matrices that exist in matrix algebra.

**Definition 2.3** (Null matrix)
*A null matrix is a matrix that has all entries zero.*

**Definition 2.4** (Identity matrix)
*An identity matrix, denoted as **I**, is a square matrix that has 1's as entries on the principal diagonal (the diagonal that goes from upper left to lower right) and zeroes off the principal diagonal.*

EXAMPLES:

1. a 2 by 3 null matrix $$\begin{pmatrix} 0 & 0 & 0 \\ 0 & 0 & 0 \end{pmatrix}$$

2. a 4 by 2 null matrix $$\begin{pmatrix} 0 & 0 \\ 0 & 0 \\ 0 & 0 \\ 0 & 0 \end{pmatrix}$$

3. a 3 by 3 identity matrix $$\begin{pmatrix} 1 & 0 & 0 \\ 0 & 1 & 0 \\ 0 & 0 & 1 \end{pmatrix}$$

4. a 5 by 5 identity matrix $\begin{pmatrix} 1 & 0 & 0 & 0 & 0 \\ 0 & 1 & 0 & 0 & 0 \\ 0 & 0 & 1 & 0 & 0 \\ 0 & 0 & 0 & 1 & 0 \\ 0 & 0 & 0 & 0 & 1 \end{pmatrix}$

The null matrix is analogous to zero with respect to addition in scalar algebra, and the identity matrix is analogous to the number one with respect to multiplication in scalar algebra. In other words, any matrix plus a null matrix of the same order is the same matrix, and any matrix multiplied by an identity matrix of appropriate order is the same matrix. A matrix is said to be *premultiplied* by a matrix on the left side, and *postmultiplied* by a matrix on the right side. Thus in the product $\boldsymbol{AB}$, $\boldsymbol{B}$ has been premultiplied by $\boldsymbol{A}$ and $\boldsymbol{A}$ has been postmultiplied by $\boldsymbol{B}$.

**Definition 2.5** (Matrix transpose)
*The transpose of a matrix is the matrix obtained by interchanging the rows and columns of the original matrix such that column* j *of the original matrix becomes row* j *of the transpose, and row* i *of the original matrix becomes column* i *of the transpose.*

The transpose of a matrix $\boldsymbol{A}$ is indicated by the notation $\boldsymbol{A}^T$ or $\boldsymbol{A}'$:

EXAMPLE:

$$\boldsymbol{A} = \begin{pmatrix} 2 & 1 & 3 \\ 1 & 2 & 0 \end{pmatrix} \quad \boldsymbol{A}' = \begin{pmatrix} 2 & 1 \\ 1 & 2 \\ 3 & 0 \end{pmatrix}$$

**Definition 2.6** (Symmetric matrix)
*A square matrix is said to be symmetric if it is equal to its transpose.*

An example of a symmetric matrix is the following:

$$\boldsymbol{A} = \boldsymbol{A}' = \begin{pmatrix} 1 & 0 & 5 \\ 0 & 1 & 2 \\ 5 & 2 & -3 \end{pmatrix}$$

**Definition 2.7** (Skew symmetric matrix)
*A skew symmetric matrix is a square matrix which equals its negative transpose, i.e.,* $\boldsymbol{A} = -\boldsymbol{A}'$.
The principal diagonal (the diagonal elements of the matrix from upper left to lower right) of such a matrix is clearly zero.

At times it will be convenient for various reasons to partition matrices into submatrices. A partition is made by using horizontal and vertical lines to divide a matrix into submatrices. Operations are performed on partitioned matrices by individual algebraic operations. Of course, the matrices must be appropriately partitioned for computation. Partitioning is generally a convenient way to work with matrices when certain submatrices have a special form that can lead to simplification of computations.

EXAMPLE:

$$\boldsymbol{A} = \begin{pmatrix} 1 & 0 & 5 \\ 0 & 1 & 2 \\ 5 & 2 & -3 \end{pmatrix} \quad \text{and} \quad \boldsymbol{D} = \begin{pmatrix} 3 & 5 & 0 \\ 1 & 2 & 0 \\ 1 & 3 & 4 \end{pmatrix}$$

Letting $\boldsymbol{B} = \begin{pmatrix} 5 \\ 2 \end{pmatrix}$, $\boldsymbol{C} = (-3)$, we have

$$\boldsymbol{A} = \left(\begin{array}{c:c} \boldsymbol{I} & \boldsymbol{B} \\ \hdashline \boldsymbol{B}' & \boldsymbol{C} \end{array}\right)$$

Letting

$$\boldsymbol{E} = \begin{pmatrix} 3 & 5 \\ 1 & 2 \end{pmatrix}$$

$$\boldsymbol{F} = \begin{pmatrix} 0 \\ 0 \end{pmatrix}$$

$$\boldsymbol{G} = (1 \quad 3)$$

$$\boldsymbol{H} = (4)$$

we have

$$\boldsymbol{D} = \left(\begin{array}{c:c} \boldsymbol{E} & \boldsymbol{F} \\ \hdashline \boldsymbol{G} & \boldsymbol{H} \end{array}\right)$$

Now

$$\boldsymbol{AD} = \left(\begin{array}{c:c} \boldsymbol{I} & \boldsymbol{B} \\ \hdashline \boldsymbol{B}' & \boldsymbol{C} \end{array}\right)\left(\begin{array}{c:c} \boldsymbol{E} & \boldsymbol{F} \\ \hdashline \boldsymbol{G} & \boldsymbol{H} \end{array}\right) = \left(\begin{array}{c:c} \boldsymbol{E} + \boldsymbol{BG} & \boldsymbol{F} + \boldsymbol{BH} \\ \hdashline \boldsymbol{B}'\boldsymbol{E} + \boldsymbol{CG} & \boldsymbol{B}'\boldsymbol{F} + \boldsymbol{CH} \end{array}\right)$$

$$= \left(\begin{array}{c:c} \begin{pmatrix} 3 & 5 \\ 1 & 2 \end{pmatrix} + \begin{pmatrix} 5 & 15 \\ 2 & 6 \end{pmatrix} & \begin{pmatrix} 0 \\ 0 \end{pmatrix} + \begin{pmatrix} 20 \\ 8 \end{pmatrix} \\ \hdashline (17 \quad 29) + (-3 \quad -9) & (0) + (-12) \end{array}\right) = \left(\begin{array}{cc:c} 8 & 20 & 20 \\ 3 & 8 & 8 \\ \hdashline 14 & 20 & -12 \end{array}\right)$$

The above calculations can be verified by direct multiplication of $\boldsymbol{A}$ and $\boldsymbol{D}$. In this example no advantage is gained by partition, although we shall encounter situations in which partitioning is very useful.

### The Inverse of a Matrix

The division operation is not defined in matrix algebra. For certain square matrices, however, there exists another (unique) square matrix of the same order such that the product of the two matrices—in either order—is the identity matrix. Such a matrix is called the inverse of the first matrix. The inverse of a matrix is designated by the matrix with an exponent of $-1$.

EXAMPLES:

$$A = \begin{pmatrix} 2 & 3 \\ 1 & 2 \end{pmatrix}$$

$$A^{-1} = \begin{pmatrix} 2 & -3 \\ -1 & 2 \end{pmatrix}$$

$$AA^{-1} = \begin{pmatrix} 2 & 3 \\ 1 & 2 \end{pmatrix}\begin{pmatrix} 2 & -3 \\ -1 & 2 \end{pmatrix} = \begin{pmatrix} 1 & 0 \\ 0 & 1 \end{pmatrix} = I$$

$$A^{-1}A = \begin{pmatrix} 2 & -3 \\ -1 & 2 \end{pmatrix}\begin{pmatrix} 2 & 3 \\ 1 & 2 \end{pmatrix} = \begin{pmatrix} 1 & 0 \\ 0 & 1 \end{pmatrix} = I$$

$$B = \begin{pmatrix} 1 & 2 & 1 \\ 0 & 1 & 1 \\ 1 & 0 & 3 \end{pmatrix}, \quad B^{-1} = \begin{pmatrix} \frac{3}{4} & -\frac{3}{2} & \frac{1}{4} \\ \frac{1}{4} & \frac{1}{2} & -\frac{1}{4} \\ -\frac{1}{4} & \frac{1}{2} & \frac{1}{4} \end{pmatrix}, \quad B^{-1}B = BB^{-1} = I$$

### Computation of the Inverse

There are many methods of computing the inverse of a matrix; we shall treat one of them here. For more information on computation of inverses, see any good reference on linear algebra, e.g., [143]. We shall use a method based on row operations, which are equivalent to linear transformations of a matrix. There are three aspects to row operations:

1. Multiplying or dividing any row of a matrix by a nonzero value.
2. Adding or subtracting a multiple of one row to or from another.
3. The interchange of any two rows.

To find the inverse of an arbitrary (square) matrix $A$, we augment the matrix by an identity matrix of the same order, which yields the augmented matrix $(A\,|\,I)$. Then we proceed by means of row operations to transform the original matrix to an identity matrix while performing the row operations on the entire augmented matrix. The rules are as follows:

1. If the uppermost left element of the augmented matrix is nonzero, divide through the first row by that element; otherwise add any other row whose

first element is nonzero to the first row and repeat step 1. If no row has a nonzero first element, the matrix does not have an inverse. In this step the uppermost left element has been reduced to one. Designate this element as the reduced element.

2. Add multiples of the row containing the reduced element to every other row so that all other entries in the column containing the reduced element become zero.
3. Delete *for the purpose of step 1 only* the row(s) and column(s) containing the reduced element(s). If all rows have been deleted, the computations are complete. Otherwise, go to step 1.

When computations are completed, the identity matrix has been transformed into the inverse matrix.

EXAMPLE:

$$\boldsymbol{B} = \begin{pmatrix} 1 & 2 & 1 \\ 0 & 1 & 1 \\ 1 & 0 & 3 \end{pmatrix}$$

The augmented matrix $(\boldsymbol{B}|\boldsymbol{I})$ is

$$\left(\begin{array}{ccc:ccc} 1 & 2 & 1 & 1 & 0 & 0 \\ 0 & 1 & 1 & 0 & 1 & 0 \\ 1 & 0 & 3 & 0 & 0 & 1 \end{array}\right)$$

Step 1. Divide through the first row by the entry in the first row, first column (=1).

$$\left(\begin{array}{ccc:ccc} 1 & 2 & 1 & 1 & 0 & 0 \\ 0 & 1 & 1 & 0 & 1 & 0 \\ 1 & 0 & 3 & 0 & 0 & 1 \end{array}\right)$$

Step 2. Add multiples of the first row to convert other elements of the first column to zero. Add zero times the first row to the second row. Add (−1) times the first row to the third row.

$$\left(\begin{array}{ccc:ccc} 1 & 2 & 1 & 1 & 0 & 0 \\ 0 & 1 & 1 & 0 & 1 & 0 \\ 0 & -2 & 2 & -1 & 0 & 1 \end{array}\right)$$

Step 3. Delete (by checkmark) row 1 and column 1 for step 1 only.

$$\begin{array}{c} \quad\checkmark \\ \checkmark \left(\begin{array}{ccc:ccc} 1 & 2 & 1 & 1 & 0 & 0 \\ 0 & 1 & 1 & 0 & 1 & 0 \\ 0 & -2 & 2 & -1 & 0 & 1 \end{array}\right) \end{array}$$

Step 1. Divide through the second row by one.

$$\begin{array}{c} \\ \checkmark \\ \\ \\ \end{array}\begin{array}{c} \begin{array}{ccc} \checkmark & \phantom{-2} & \phantom{2} \end{array} \\ \left(\begin{array}{ccc|ccc} 1 & 2 & 1 & 1 & 0 & 0 \\ 0 & 1 & 1 & 0 & 1 & 0 \\ 0 & -2 & 2 & -1 & 0 & 1 \end{array}\right) \end{array}$$

Step 2. Add multiples of the second row to convert other elements of the second column to zero. Add $-2$ times the second row to the first row. Add 2 times the second row to the third row.

$$\begin{array}{c} \\ \checkmark \\ \\ \\ \end{array}\begin{array}{c} \begin{array}{ccc} \checkmark & \phantom{0} & \phantom{-1} \end{array} \\ \left(\begin{array}{ccc|ccc} 1 & 0 & -1 & 1 & -2 & 0 \\ 0 & 1 & 1 & 0 & 1 & 0 \\ 0 & 0 & 4 & -1 & 2 & 1 \end{array}\right) \end{array}$$

Step 3. Delete by checkmark row 2 and column 2 for step 1 only.

$$\begin{array}{c} \\ \checkmark \\ \checkmark \\ \\ \end{array}\begin{array}{c} \begin{array}{ccc} \checkmark & \checkmark & \phantom{-1} \end{array} \\ \left(\begin{array}{ccc|ccc} 1 & 0 & -1 & 1 & -2 & 0 \\ 0 & 1 & 1 & 0 & 1 & 0 \\ 0 & 0 & 4 & -1 & 2 & 1 \end{array}\right) \end{array}$$

Step 1. Divide through the third row by 4.

$$\begin{array}{c} \\ \checkmark \\ \checkmark \\ \\ \end{array}\begin{array}{c} \begin{array}{ccc} \checkmark & \checkmark & \phantom{-1} \end{array} \\ \left(\begin{array}{ccc|ccc} 1 & 0 & -1 & 1 & -2 & 0 \\ 0 & 1 & 1 & 0 & 1 & 0 \\ 0 & 0 & 1 & -\frac{1}{4} & \frac{1}{2} & \frac{1}{4} \end{array}\right) \end{array}$$

Step 2. Add multiples of the third row to convert other elements of the third column to zero. Add 1 times the third row to the first row. Add $-1$ times the third row to the second row.

$$\begin{array}{c} \\ \checkmark \\ \checkmark \\ \\ \end{array}\begin{array}{c} \begin{array}{ccc} \checkmark & \checkmark & \phantom{0} \end{array} \\ \left(\begin{array}{ccc|ccc} 1 & 0 & 0 & \frac{3}{4} & -\frac{3}{2} & \frac{1}{4} \\ 0 & 1 & 0 & \frac{1}{4} & \frac{1}{2} & -\frac{1}{4} \\ 0 & 0 & 1 & -\frac{1}{4} & \frac{1}{2} & \frac{1}{4} \end{array}\right) \end{array}$$

Step 3. Delete by checkmark row 3 and column 3 for step 1 only. Since all rows have now been checkmarked, the computations are complete.

$$\begin{array}{c} \\ \checkmark \\ \checkmark \\ \checkmark \end{array}\begin{array}{c} \begin{array}{ccc} \checkmark & \checkmark & \checkmark \end{array} \\ \left(\begin{array}{ccc|ccc} 1 & 0 & 0 & \frac{3}{4} & -\frac{3}{2} & \frac{1}{4} \\ 0 & 1 & 0 & \frac{1}{4} & \frac{1}{2} & -\frac{1}{4} \\ 0 & 0 & 1 & -\frac{1}{4} & \frac{1}{2} & \frac{1}{4} \end{array}\right) \end{array} \quad \boldsymbol{B}^{-1} = \begin{pmatrix} \frac{3}{4} & -\frac{3}{2} & \frac{1}{4} \\ \frac{1}{4} & \frac{1}{2} & -\frac{1}{4} \\ -\frac{1}{4} & \frac{1}{2} & \frac{1}{4} \end{pmatrix}$$

## 2.3 VECTORS AND VECTOR SPACES

Vectors constitute a class of matrices for which it is useful to develop additional characterizations. Vectors will usually be represented by boldface lower-case letters, e.g., $\boldsymbol{a}$, $\boldsymbol{b}$, or $\boldsymbol{c}$. A subscripted vector usually indicates that the vector is the corresponding column of the matrix of the same letter, whereas a superscripted vector indicates that the vector is the corresponding row of the matrix of the same letter. Subscripted or superscripted vectors may also be used without reference to a matrix.

**Definition 2.8** (Vector)
*A vector is defined as a matrix having only one row or only one column.*

EXAMPLES:

$$(1 \quad 4 \quad 2 \quad 3), \quad \begin{pmatrix} 2 \\ 0 \\ 3 \end{pmatrix}, \quad (a_1 \quad a_2 \quad a_3 \quad a_4 \quad a_5), \quad \begin{pmatrix} b_1 \\ b_2 \\ b_3 \\ b_4 \end{pmatrix}$$

By further example, vector $\boldsymbol{a}_6$ is column six of matrix $\boldsymbol{A}$, and vector $\boldsymbol{a}^3$ is row three of matrix $\boldsymbol{A}$. Special vectors which will be useful as we proceed are the following:

**Definition 2.9** (Null vector)
*A null vector, usually written as* $\mathbf{0}$, *is a vector consisting of all zeroes.*

**Definition 2.10** (Unit vector)
*A unit vector, written as* $\boldsymbol{e}_j$, *is a vector with all entries zero except entry* j, *which is a one.*

**Definition 2.11** (Sum vector)
*A sum vector, written as* $\mathbf{1}$, *is a vector all of whose entries are ones.*

EXAMPLES:

$$\text{Null vectors:} \quad \begin{pmatrix} 0 \\ 0 \\ 0 \\ 0 \\ 0 \end{pmatrix}, \quad (0 \quad 0 \quad 0), \quad (0 \quad 0 \quad 0 \quad 0 \quad 0 \quad 0)$$

Sum vector: $(1 \quad 1 \quad 1 \quad 1)$

Unit Vectors (of order 4): $e_1 = \begin{pmatrix} 1 \\ 0 \\ 0 \\ 0 \end{pmatrix}, \quad e_2 = \begin{pmatrix} 0 \\ 1 \\ 0 \\ 0 \end{pmatrix}, \quad e_3 = \begin{pmatrix} 0 \\ 0 \\ 1 \\ 0 \end{pmatrix}, \quad e_4 = \begin{pmatrix} 0 \\ 0 \\ 0 \\ 1 \end{pmatrix}$

As indicated above, vectors are a special class of matrices; therefore, all algebraic operations on vectors are precisely the same as those on matrices. A vector can always be represented as a sum of scalar products of unit vectors, for example:

$$\begin{pmatrix} 4 \\ 6 \\ -1 \end{pmatrix} = 4\begin{pmatrix} 1 \\ 0 \\ 0 \end{pmatrix} + 6\begin{pmatrix} 0 \\ 1 \\ 0 \end{pmatrix} + (-1)\begin{pmatrix} 0 \\ 0 \\ 1 \end{pmatrix}$$

The concept of a vector space is important in linear programming.

**Definition 2.12** (Vector space)
*A vector space is a set of vectors having the property that for any two vectors of the set,* $\boldsymbol{a}$ *and* $\boldsymbol{b}$*, the vectors* $\boldsymbol{c} = \boldsymbol{a} + \boldsymbol{b}$ *and* $\lambda\boldsymbol{a}$ *(where* $\lambda$ *is a scalar) are also vectors of the set.*

A Euclidean vector space, which is of particular importance, is defined by Birkhoff and MacLane [34]:

**Definition 2.13** (Euclidean vector space)
*A Euclidean vector space is a vector space* $E^n$ *with real scalars, such that to any vectors* $\boldsymbol{a}$ *and* $\boldsymbol{b}$ *in* $E^n$ *corresponds a (real) "inner product"* $\boldsymbol{a}'\boldsymbol{b}$ *which is*
*symmetric* $(\boldsymbol{a}'\boldsymbol{b} = \boldsymbol{b}'\boldsymbol{a})$,
*bilinear* $((\boldsymbol{a}' + \boldsymbol{b}')(\boldsymbol{c} + \boldsymbol{d}) = \boldsymbol{a}'(\boldsymbol{c} + \boldsymbol{d}) + \boldsymbol{b}'(\boldsymbol{c} + \boldsymbol{d}) = (\boldsymbol{a} + \boldsymbol{b})'\boldsymbol{c} + (\boldsymbol{a} + \boldsymbol{b})'\boldsymbol{d})$
*and positive* $((\boldsymbol{a}'\boldsymbol{a}) > 0$ *unless* $\boldsymbol{a} = \boldsymbol{0})$.

In a Euclidean vector space, a vector may be thought of as a point (i.e., the tip of an arrow—corresponding to a vector—whose tail is at the origin).

In certain vector spaces, the "distance" between points, the "length" of a vector, or, more correctly, a metric, is defined. The Euclidean vector space is one such space, and its metric is the usual measure of distance, that is, distance is defined as follows:

$$||\boldsymbol{a}|| = (\boldsymbol{a}'\boldsymbol{a})^{1/2} = \left(\sum_{i=1}^{n} (a_i)^2\right)^{1/2}$$

where $a_i$ is the $i^{th}$ component of an $n$-dimensional vector $\boldsymbol{a}$. There are many other metrics as well.

**Linear Dependence**

The concept of linear dependence among vectors is important in vector spaces.

**Definition 2.14** (Linear dependence)
*A set of vectors* $\boldsymbol{x}_i$ $(i = 1, \ldots, k)$, *is said to be linearly dependent if there exists a set of multipliers* $\lambda_i$, *one for each vector, not all zero, such that the sum of the scalar products of the multipliers times the vectors is the null vector.*

That is, if there exists a set of multipliers $\lambda_i$ not all zero such that

$$\lambda_1 \boldsymbol{x}_1 + \lambda_2 \boldsymbol{x}_2 + \cdots + \lambda_k \boldsymbol{x}_k = \mathbf{0}$$

then the set of vectors is linearly dependent.

A set of vectors which is not linearly dependent is said to be *linearly independent*. If the null vector is included in any set of vectors, then such a set is linearly dependent, since any nonzero multiple of the null vector is always null. A set of multipliers yielding a null vector may always be found by choosing all multipliers zero except that of the null vector, whose multiplier can be chosen as any nonzero constant.

EXAMPLES:

(1)

$$\boldsymbol{x}_1 = \begin{pmatrix} 1 \\ 1 \\ 2 \end{pmatrix} \quad \text{and} \quad \boldsymbol{x}_2 = \begin{pmatrix} 2 \\ 2 \\ 4 \end{pmatrix}$$

are linearly *dependent* because

$$-2\begin{pmatrix} 1 \\ 1 \\ 2 \end{pmatrix} + 1\begin{pmatrix} 2 \\ 2 \\ 4 \end{pmatrix} = \begin{pmatrix} 0 \\ 0 \\ 0 \end{pmatrix}$$

(2)

$$\boldsymbol{x}_3 = \begin{pmatrix} 1 \\ 2 \end{pmatrix}, \quad \boldsymbol{x}_4 = \begin{pmatrix} 2 \\ 3 \end{pmatrix}, \quad \text{and} \quad \boldsymbol{x}_5 = \begin{pmatrix} 1 \\ 1 \end{pmatrix}$$

are linearly *dependent* because

$$1\begin{pmatrix} 1 \\ 2 \end{pmatrix} - 1\begin{pmatrix} 2 \\ 3 \end{pmatrix} + 1\begin{pmatrix} 1 \\ 1 \end{pmatrix} = \begin{pmatrix} 0 \\ 0 \end{pmatrix}$$

(3)

$$x_6 = \begin{pmatrix} 1 \\ 0 \\ 0 \end{pmatrix} \quad x_7 = \begin{pmatrix} 2 \\ 0 \\ 0 \end{pmatrix}, \quad \text{and} \quad x_8 = \begin{pmatrix} 0 \\ 2 \\ 1 \end{pmatrix}$$

are linearly *dependent* because

$$2\begin{pmatrix} 1 \\ 0 \\ 0 \end{pmatrix} - 1\begin{pmatrix} 2 \\ 0 \\ 0 \end{pmatrix} + 0\begin{pmatrix} 0 \\ 2 \\ 1 \end{pmatrix} = \begin{pmatrix} 0 \\ 0 \\ 0 \end{pmatrix}$$

(4)

$$x_9 = \begin{pmatrix} 1 \\ 2 \\ 0 \end{pmatrix} \quad x_{10} = \begin{pmatrix} 1 \\ 1 \\ 1 \end{pmatrix}, \quad \text{and} \quad x_{11} = \begin{pmatrix} 2 \\ 1 \\ 5 \end{pmatrix}$$

are linearly *independent* because

$$\lambda_1\begin{pmatrix} 1 \\ 2 \\ 0 \end{pmatrix} + \lambda_2\begin{pmatrix} 1 \\ 1 \\ 1 \end{pmatrix} + \lambda_3\begin{pmatrix} 2 \\ 1 \\ 5 \end{pmatrix} = \begin{pmatrix} 0 \\ 0 \\ 0 \end{pmatrix}$$

only for

$$\lambda_1 = \lambda_2 = \lambda_3 = 0$$

If a set of vectors is linearly dependent, then one vector of the set is equal to a linear combination of the other vectors. (Conversely, if at least one vector in a set is a linear combination of other vectors in the set, then the set is dependent.) This can be seen by writing a statement of linear dependence $\sum \lambda_i x_i = 0$. Choose one vector, $x_j$, with a nonzero multiplier, $\lambda_j$, and multiply the equation by the reciprocal of its multiplier (i.e., $1/\lambda_j$). By rearranging the equation, we can show that the vector $x_j$ is a linear combination of the others. In example (3) above, for instance, $x_6 = \frac{1}{2}x_7$. It is also possible to generate an "equivalent" linearly independent set of vectors in the sense that the same vectors may be generated as linear combinations of the set by deleting any one of the dependent vectors from the set and repeating the cycle until a linearly independent set remains.

This idea may be illustrated using the preceding examples. Example (1) is a set of linearly dependent vectors. Since $x_2$ is equal to $2x_1$, $x_2$ can be dropped from the set, and $x_1$ by itself is a linearly independent set of vectors. This example is trivial in the sense that any one nonnull vector is by definition *not* linearly dependent. Either vector can be dropped in this case. In example (2), any one of the vectors $x_3$, $x_4$, and $x_5$ can be dropped, and the remaining two vectors form a linearly independent set. By contrast, in the third example, either $\boldsymbol{x}_6$ or $\boldsymbol{x}_7$ can be dropped to have an equivalent linearly independent set, but $\boldsymbol{x}_8$ cannot be dropped since it is not a dependent vector. In example (4), no vector can be dropped, because example (4) is a set of linearly independent vectors. A few additional definitions are now given.

**Definition 2.15** (Rank)
*The rank of a matrix is defined as the maximum number of linearly independent rows or the maximum number of linearly independent columns of the matrix.*

It can be shown that any square matrix that possesses an inverse is of rank equal to the number of rows (or columns). If a matrix has its rank equal to the smaller of its orders (i.e., $m$ or $n$), it is said to be of *full* rank.

**Definition 2.16** (Orthogonality)
*Two vectors* $\mathbf{x}$ *and* $\mathbf{y}$ *(of the same order) are said to be orthogonal if their scalar product* $\mathbf{x}'\mathbf{y}$ *is zero.*

Intuitively, the orthogonal condition is analogous to vectors being perpendicular in two dimensions.

### Basis

In the last chapter we introduced terminology involving basic variables and nonbasic variables. These terms are related to the concept of a basis which is central to linear algebra.

**Definition 2.17** (Spanning set)
*Given an n-dimensional Euclidean space* $E^n$*, a set of vectors which can be used to express any arbitrary vector in* $E^n$ *(or the space* $E^n$*) is called a spanning set.*

A spanning set of vectors need not necessarily be linearly independent. Of the set of examples on pages 32–33, examples (2) and (4) form a spanning set of their respective spaces.

**Definition 2.18** (Basis)
*The smallest spanning set for a space is linearly independent and is called a basis.*

Of the examples on pages 32–33, it can be shown that the vectors of example (2) (having deleted any one vector) and example (4) each constitute a basis. It can be shown that a basis for an $n$-dimensional Euclidean space consists of precisely $n$ linearly independent vectors. The $n$ unit vectors $\boldsymbol{e}_j$, $j = 1, \ldots, n$ are linearly independent; consequently, there must be *at least n* linearly independent vectors in an $n$-space; secondly, any arbitrary point in an $n$-space may be represented as a linear combination of unit vectors, as follows:

$$\boldsymbol{a}_j = \begin{pmatrix} a_{1j} \\ a_{2j} \\ a_{3j} \\ \cdot \\ \cdot \\ \cdot \\ a_{nj} \end{pmatrix} = a_{1j} \begin{pmatrix} 1 \\ 0 \\ 0 \\ \cdot \\ \cdot \\ \cdot \\ 0 \end{pmatrix} + a_{2j} \begin{pmatrix} 0 \\ 1 \\ 0 \\ \cdot \\ \cdot \\ \cdot \\ 0 \end{pmatrix} + \cdots + a_{nj} \begin{pmatrix} 0 \\ 0 \\ 0 \\ \cdot \\ \cdot \\ \cdot \\ 1 \end{pmatrix}$$

Hence, there may be at most $n$ linearly independent vectors in an $n$-space. Therefore, there must be precisely $n$ vectors in an $n$-dimensional basis.

**THEOREM 2.1**

ANY VECTOR IN A VECTOR SPACE MAY BE REPRESENTED AS A LINEAR COMBINATION OF THE BASIS (OR BASIC VECTORS). FURTHERMORE, THAT COMBINATION IS UNIQUE.

PROOF:
The first part follows by showing that any $n$-vector $\boldsymbol{y}$ can be represented in terms of the basis $\boldsymbol{B}$ by virtue of the expressions $\boldsymbol{Bx} = \boldsymbol{y}$ and $\boldsymbol{x} = \boldsymbol{B}^{-1}\boldsymbol{y}$. To prove uniqueness, let $\boldsymbol{a} = \sum_{j=1}^{n} \lambda_j \boldsymbol{b}_j$ where $\boldsymbol{b}_j$, $j = 1, \ldots, n$ are the basic vectors. Assume that the representation is not unique. That is, $\boldsymbol{a} = \sum_{j=1}^{n} \alpha_j \boldsymbol{b}_j$, $\alpha_j \neq \lambda_j$ for at least some $j$. Then by subtracting one expression for $\boldsymbol{a}$ from the other we have

$$\boldsymbol{a} - \boldsymbol{a} = \boldsymbol{0} = \sum_{j=1}^{n} (\lambda_j - \alpha_j)\boldsymbol{b}_j$$

Now define $\gamma_j = \lambda_j - \alpha_j$. All $\gamma_j$'s are not zero; hence the set of vectors $\boldsymbol{b}_1, \ldots, \boldsymbol{b}_n$ is linearly dependent, a contradiction.

## 2.4 LINEAR EQUATIONS

A constraint set for a linear programming problem can be expressed in the form

$$\boldsymbol{Ax} = \boldsymbol{b} \tag{2.1}$$

where $\boldsymbol{A}$ is an $m$ by $n$ matrix, $\boldsymbol{x}$ and $\boldsymbol{b}$ are vectors of order $n$ by 1 and $m$ by 1, respectively. Let us first suppose that $m = n$. There are as many equations as unknowns and, providing that the matrix $\boldsymbol{A}$ has an inverse, a solution $\boldsymbol{x}$ is uniquely determined by left-multiplying equation (2.1) by $\boldsymbol{A}^{-1}$ to yield

$$\boldsymbol{x} = \boldsymbol{A}^{-1}\boldsymbol{b}$$

Recall that in section 2.2 we did not explore thoroughly the conditions under which a square matrix has an inverse. Consequently the conditions under which simultaneous linear equations are solvable may not be evident. We shall now characterize which matrices correspond to such equations.

A square matrix possesses an inverse if and only if its columns (or rows) are linearly independent or, equivalently, they form a basis in $m$-dimensions, or the matrix is of full rank. A square matrix which does not possess an inverse has linearly dependent columns (or rows), and therefore the matrix cannot form a basis. Such a matrix is said to be *singular*. Equivalently, a set of $m$ equations in $m$ unknowns has a unique solution if and only if the set of equations is consistent and independent. Consistency means that there is at least one solution to the equations, and independence indicates that the columns are linearly independent.

Suppose now that $m < n$. In general, there are an infinite number of solutions, providing there exists more than one set of $m$ linearly independent columns of $\boldsymbol{A}$ (any equations stated more than once having already been eliminated). Select an arbitrary linearly independent set of column vectors from $\boldsymbol{A}$. Denote this set, which forms a basis, as the submatrix $\boldsymbol{B}$, which consists of $\boldsymbol{A}$ less all columns that do not form the basis. $\boldsymbol{B}$ will usually be referred to as the basis. Further denote the vector of associated variables as $\boldsymbol{x}_B$. Then, the unique solution to the set of equations

$$\boldsymbol{B}\boldsymbol{x}_B = \boldsymbol{b}$$

which is

$$\boldsymbol{x}_B = \boldsymbol{B}^{-1}\boldsymbol{b}$$

is called a basic solution, and consists of setting all variables of $\boldsymbol{x}$ not associated with the columns of $\boldsymbol{B}$ to zero and solving for the remaining variables. Representing the linear programming constraint set as in expression (2.1), there are as many basic solutions to a linear programming problem as there are sets of $m$ linearly independent columns of the matrix $\boldsymbol{A}$.

EXAMPLE 2.1:

Consider the constraints of the example problem of Chapter 1:

$$\begin{aligned}
x_1 + 2x_2 + x_3 \qquad\qquad\quad &= 240{,}000 \\
1.5x_1 + \; x_2 \qquad + x_4 \qquad &= 180{,}000 \\
x_1 \qquad\qquad\qquad\qquad + x_5 &= 110{,}000
\end{aligned}$$

Designate the vectors as $\boldsymbol{a}_1\boldsymbol{a}_2\boldsymbol{a}_3\boldsymbol{a}_4\boldsymbol{a}_5$. Possible bases are

$$\boldsymbol{a}_1\boldsymbol{a}_2\boldsymbol{a}_3,\quad \boldsymbol{a}_1\boldsymbol{a}_2\boldsymbol{a}_4,\quad \boldsymbol{a}_1\boldsymbol{a}_2\boldsymbol{a}_5,\quad \boldsymbol{a}_1\boldsymbol{a}_3\boldsymbol{a}_4,\quad \boldsymbol{a}_1\boldsymbol{a}_3\boldsymbol{a}_5$$

$$\boldsymbol{a}_1\boldsymbol{a}_4\boldsymbol{a}_5,\quad \boldsymbol{a}_2\boldsymbol{a}_3\boldsymbol{a}_4,\quad \boldsymbol{a}_2\boldsymbol{a}_3\boldsymbol{a}_5,\quad \boldsymbol{a}_2\boldsymbol{a}_4\boldsymbol{a}_5,\quad \boldsymbol{a}_3\boldsymbol{a}_4\boldsymbol{a}_5$$

which total ten in number {$n$ variables taken $m$ at a time $\binom{n}{m} = \dfrac{n!}{m!\,(n-m)!}$ or, specifically $\binom{5}{3} = \dfrac{5\cdot4\cdot3\cdot2\cdot1}{3\cdot2\cdot1\cdot2\cdot1} = 10$}.

Of the ten, nine are bases. Can you tell by inspection which possible basis is not a basis? ($\boldsymbol{a}_2\boldsymbol{a}_3\boldsymbol{a}_4$ is not a basis since the vectors are not linearly independent.)

EXAMPLE 2.2:

We shall examine the basic solution corresponding to the basis $\boldsymbol{a}_2\boldsymbol{a}_4\boldsymbol{a}_5$. The corresponding equation $\boldsymbol{B}\boldsymbol{x}_B = \boldsymbol{b}$ is

$$\begin{pmatrix} 2 & 0 & 0 \\ 1 & 1 & 0 \\ 0 & 0 & 1 \end{pmatrix}\begin{pmatrix} x_2 \\ x_4 \\ x_5 \end{pmatrix} = \begin{pmatrix} 240{,}000 \\ 180{,}000 \\ 110{,}000 \end{pmatrix}$$

$$\boldsymbol{B}^{-1} = \begin{pmatrix} \frac{1}{2} & 0 & 0 \\ -\frac{1}{2} & 1 & 0 \\ 0 & 0 & 1 \end{pmatrix}$$

and $\boldsymbol{x}_B = \boldsymbol{B}^{-1}\boldsymbol{b}$, and

$$\begin{pmatrix} x_2 \\ x_4 \\ x_5 \end{pmatrix} = \begin{pmatrix} \frac{1}{2} & 0 & 0 \\ -\frac{1}{2} & 1 & 0 \\ 0 & 0 & 1 \end{pmatrix}\begin{pmatrix} 240{,}000 \\ 180{,}000 \\ 110{,}000 \end{pmatrix} = \begin{pmatrix} 120{,}000 \\ 60{,}000 \\ 110{,}000 \end{pmatrix}$$

Further, of the basic solutions, we are only interested in those which are feasible, i.e., have all variables nonnegative. We shall subsequently show that, generally, one of the basic feasible solutions must be optimal. Therefore a (cumbersome) means of solving linear programming problems would be to enumerate all possible $\binom{n}{m}$ basic solutions, evaluate each, and choose the optimum.

We shall instead use a procedure that examines a subset of possible basic solutions in a well-specified sequence, such that the last solution examined will be optimal, provided a finite optimum exists.

## 2.5 SOME GEOMETRIC CONSIDERATIONS

In order to develop the well-specified sequence of basic solutions, we need certain geometrical foundations. We have been using the concept of a set, which consists of a collection of objects. Point sets are collections of points or vectors in $E^n$. The sets contain either a finite or an infinite number of points, but for the present we shall assume that they contain an infinite number of points. (The number of points is finite, e.g., when there is a unique feasible solution to the constraint set, or as is often the case in integer linear programming.) Point sets are generally characterized by constraints. The constraints with which we are primarily concerned are linear. Furthermore, we shall generally be concerned with closed sets (sets which include the boundary points) as opposed to open sets (sets which do not include the boundary points). In the present context, closed sets consist of sets of constraints having weak inequalities ($\geq, \leq$) and equalities, whereas open sets consist of sets of constraints having strong inequalities ($>, <, \neq$).

Linear equations in two dimensions are graphs of straight lines; linear equations in three dimensions are graphs of planes. In more than three dimensions, graphs of linear equations are hyperplanes. We shall use the term hyperplane to denote the locus of points satisfying a linear equation, regardless of the number of variables. The points lying on the hyperplane form a closed set. Designate an arbitrary hyperplane as $\mathbf{A}_1\mathbf{x} = \mathbf{b}_1$. The hyperplane divides the space into two parts or open half-spaces, $\mathbf{A}_1\mathbf{x} > \mathbf{b}_1$ and $\mathbf{A}_1\mathbf{x} < \mathbf{b}_1$. The boundary of an inequality constraint is a hyperplane. The locus of points which satisfy an inequality constraint is a hyperplane $\mathbf{A}_1\mathbf{x} = \mathbf{b}_1$ and an open half-space $\mathbf{A}_1\mathbf{x} < \mathbf{b}_1$ *together* or, equivalently, a closed half-space $\mathbf{A}_1\mathbf{x} \leq \mathbf{b}_1$.

We shall now show that all points satisfying linear equalities and inequalities form a *convex set.*

**Definition 2.19** (Convexity)

*A point set is said to be convex if and only if a straight line segment connecting any two points of the set contains only points of the set.*

Intuitively, the concept of convexity can be indicated by the absence of "holes" within the sets and by no (hyper) plane tangent to the set intersecting the set except at the point(s) of tangency (see Fig. 2.1). Given two points $\mathbf{x}_1$ and $\mathbf{x}_2$ (written as vectors) of a convex set, any point on the line segment connecting them

$$\mathbf{x}_3 = \lambda\mathbf{x}_1 + (1 - \lambda)\mathbf{x}_2 \quad 0 < \lambda < 1$$

must also be contained in the convex set. We can thereby show that the points satisfying an inequality constitute a convex set. Let the inequality be $\mathbf{A}_1\mathbf{x} \leq \mathbf{b}_1$. Then by supposition $\mathbf{A}_1\mathbf{x}_1 \leq \mathbf{b}_1$ and $\mathbf{A}_1\mathbf{x}_2 \leq \mathbf{b}_1$. Multiplying the first inequality by $\lambda$ and the second by $1 - \lambda$ we have

$$\mathbf{A}_1\lambda\mathbf{x}_1 \leq \lambda\mathbf{b}_1$$

and

$$\mathbf{A}_1(1 - \lambda)\mathbf{x}_2 \leq (1 - \lambda)\mathbf{b}_1$$

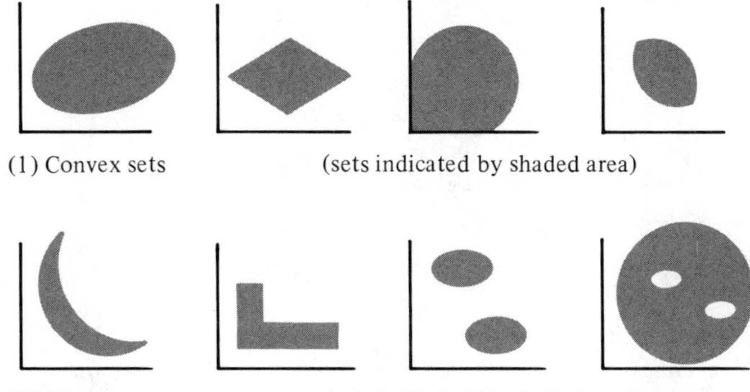


**Figure 2.1**
Some Convex Sets and Some Nonconvex Sets in Two Dimensions

Adding the equations, we have

$$A_1(\lambda x_1 + (1 - \lambda)x_2) \leq b_1$$

or

$$A_1 x_3 \leq b_1$$

which proves that a linear inequality constraint corresponds to a convex set.

Similarly, for an equality constraint $A_2 x = b_2$, let $x_1$ and $x_2$ be two points of the set, and let $x_3 = \lambda x_1 + (1 - \lambda)x_2$, $0 < \lambda < 1$, be an arbitrary point on the line segment between $x_1$ and $x_2$. Now $A_2 x_1 = b_2$ and $A_2 x_2 = b_2$, and

$$A_2 \lambda x_1 = \lambda b_2$$

and

$$A_2(1 - \lambda)x_2 = (1 - \lambda)b_2$$

Adding the two we get

$$A_2 x_3 = b_2$$

which proves that an equality constraint also describes a convex set.

The intersection of two or more convex sets is also a convex set. This may be shown by considering any two points which are contained in all of the sets whose intersection is the set of interest, and showing that any point lying on the line segment between the two points is also contained in each of the sets whose intersection is the set of interest. Since the constraint set of a linear programming problem is just the intersection of a set of constraints, the set of points satisfying the constraints is a convex set.

**Definition 2.20** (Convex combination)
*A convex combination of a number of points* $\boldsymbol{x}_1, \ldots, \boldsymbol{x}_n$ *is defined as*

$$\boldsymbol{x} = \lambda_1 \boldsymbol{x}_1 + \cdots + \lambda_n \boldsymbol{x}_n$$

*such that* $\lambda_1, \ldots, \lambda_n \geq 0$ *and*

$$\lambda_1 + \cdots + \lambda_n = 1$$

**Definition 2.21** (Convex hull)
*The convex hull of a (possibly infinite) number of points is the set of all convex combinations of these points.*

**Definition 2.22**
*A positive combination of a number of points* $\boldsymbol{x}_1, \ldots, \boldsymbol{x}_n$ *is defined as*

$$\boldsymbol{x} = \lambda_1 \boldsymbol{x}_1 + \cdots + \lambda_n \boldsymbol{x}_n$$

*such that*

$$\lambda_1, \ldots, \lambda_n \geq 0$$

The set of points $\boldsymbol{x}$ which satisfies the constraints of a linear programming problem is a convex set. Further, a convex hull of a finite number of points is called a convex polyhedron. A three-dimensional convex polyhedron is shown in Figure 2.2. Any linear programming problem which is bounded from above (e.g., every variable is not allowed to exceed some specified upper bound) has a convex set which is a convex polyhedron.

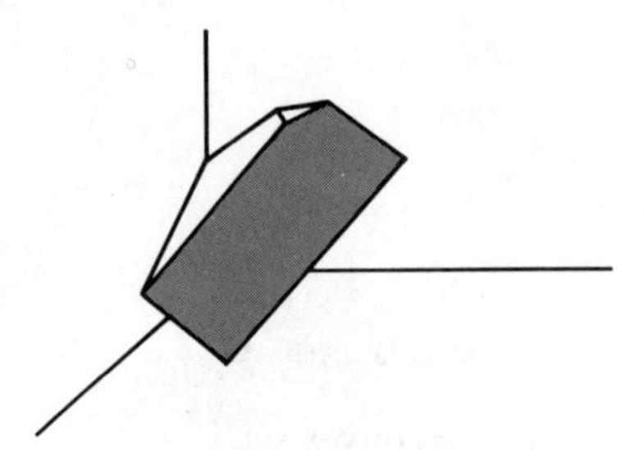

**Figure 2.2**
A Convex Polyhedron

**Definition 2.23** (Extreme point)
*A point is an extreme point of a convex set if and only if there are not two other distinct points in the set such that the extreme point can be expressed as a convex combination of the other two points.*

Extreme points are the corner points of the convex polyhedron of feasible solutions.

An additional definition—that of a *cone*—is required in order to proceed.

**Definition 2.24** (Cone)
*A cone is the set of points with the property that if* $\mathbf{x}$ *is a member of the set, so is* $p\mathbf{x}$ *for all scalars* $p \geq 0$*. The vector* $\mathbf{0}$ *is an element of every cone.*

A convex cone is a cone which is a convex set. It consists of all points generated by a set of vectors and their positive combinations. Some examples of cones are shown graphically in Figure 2.3.

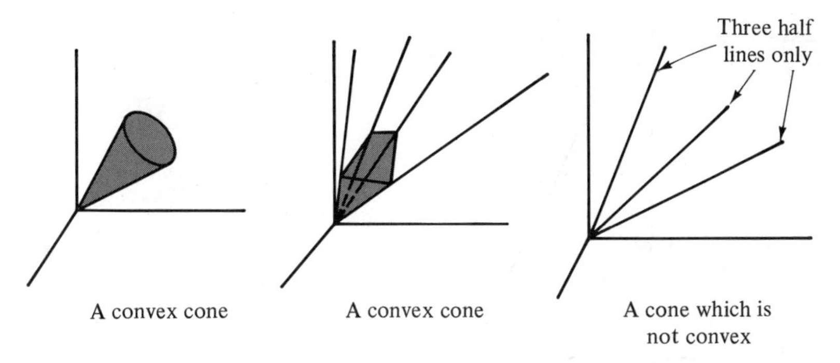


**Figure 2.3**
Some Examples of Cones

## 2.6 SOME GEOMETRIC INTERPRETATIONS OF LINEAR PROGRAMMING

Having introduced certain necessary geometric foundations, we are now in a position to consider some geometric interpretations of linear programming. We shall first introduce what is referred to as the activities space (or solution space) geometric representation. For convenience, we restate the example problem of Chapter 1 (here it is convenient to use the constraints in their inequality form):

Maximize $z = .56x_1 + .42x_2$

| | | | |
|---|---|---|---|
| (2.2) | subject to: | $x_1 + 2x_2 \leq 240{,}000$ | cereal |
| (2.3) | | $1.5x_1 + x_2 \leq 180{,}000$ | meat |
| (2.4) | | $x_1 \leq 110{,}000$ | Frisky Pup packaging capacity |
| (2.5) | | $x_1 \geq 0$ | nonnegativity constraint on Frisky Pup |
| (2.6) | | $x_2 \geq 0$ | nonnegativity constraint on Husky Hound |

By activities we mean the variables in the problem; hence the activity space approach uses the dimensions to represent the quantities of each activity. In Figure 2.4, we have plotted each of the five constraints. The bold lines are the hyperplanes which form the boundaries of the constraint set, and the set itself is indicated by the shaded area (which corresponds to the set of feasible solutions to the problem). The numbers on each constraint correspond to the numbers used above.

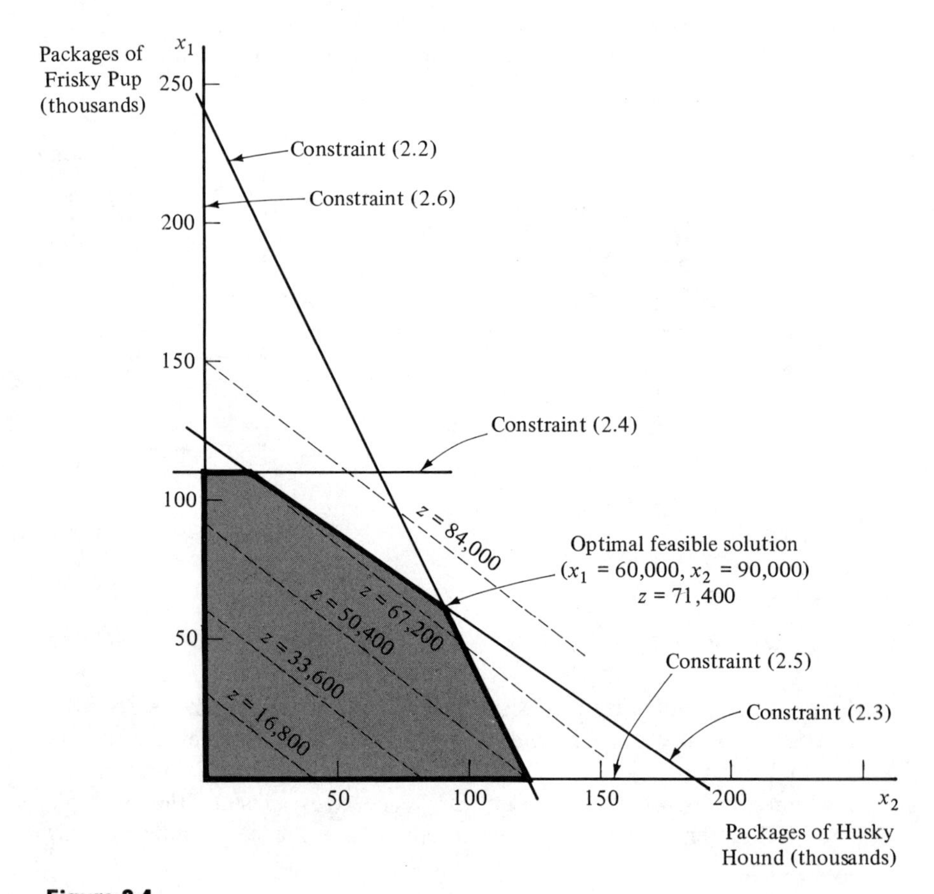


**Figure 2.4**

Activity Space Representation of Example Problem

The dotted lines (hyperplanes in general) in Figure 2.4 are lines of constant contribution to overhead and profit, as indicated. It is desired to find the line whose profit is greatest and which contains at least one point in the shaded area. The line corresponding to greatest profit and intersecting the feasible solution set is the line with profit equal to $71,400, and the solution is shown as 60,000 packages of Frisky Pup and 90,000 packages of Husky Hound dog food.

This method can only be used for two or three activity problems with any number of constraints. Though the method is not of any value of itself (except for solving small problems), it is a useful way of interpreting a problem of any dimension in an abstract sense.

The way in which the simplex method* (presented in an intuitive form in Chapter 1 and formally in Chapter 3) works is to begin with an extreme point—usually the origin. Then it is determined whether an adjacent extreme point (one which has precisely one basic variable different from the current extreme point) has a larger objective function value. If not, the solution is optimal. If an adjacent extreme point is better, then one such solution is selected and evaluated. The procedure is repeated until an optimum is found. For the example there are only two possible sequences:

a. $x_2 = 0,\ x_1 = 0,\ z = 0;$

$x_2 = 120{,}000,\ x_1 = 0,\ z = 50{,}400;$

$x_2 = 90{,}000,\ x_1 = 60{,}000,\ z = 71{,}400;$

b. $x_2 = 0,\ x_1 = 0,\ z = 0;$

$x_2 = 0,\ x_1 = 110{,}000,\ z = 61{,}600;$

$x_2 = 15{,}000,\ x_1 = 110{,}000,\ z = 67{,}900;$

$x_2 = 90{,}000,\ x_1 = 60{,}000,\ z = 71{,}400$

A second way of interpreting a linear programming problem geometrically is referred to as the requirements space method. The vector ***b*** is often referred to as the requirements vector (or stipulations vector) because it gives the requirements or limitations of the constraints. In this method it is convenient to use the constraint set in the equality form. Each axis represents the consumption of a resource (or satisfaction of a commitment). Vectors represent the consumption of resources and satisfaction of commitments associated with a unit of the corresponding activity. A feasible solution corresponds to any positive combination of activity vectors which *equals* the vector of available resources (the ***b*** vector). A basic feasible solution corresponds to a positive combination of $m$ activity vectors equaling the vector of available resources where $m$ is the number of constraints. (In degenerate cases, the number of activities may be less than $m$.) Another interpretation is that the convex cone formed by a set of basic vectors corresponding to a feasible solution must contain the ***b*** vector. It is desired to find the feasible basis whose corresponding objective function value is maximized.

For the example problem we drop the constraint on Frisky Pup packaging capacity (which in this problem does not limit the optimal solution) in order to give a two-dimensional representation. A two-dimensional plot for the example

* In this discussion we assume nondegeneracy, which is defined in Chapter 3. That chapter also treats and resolves the problems of degeneracy.

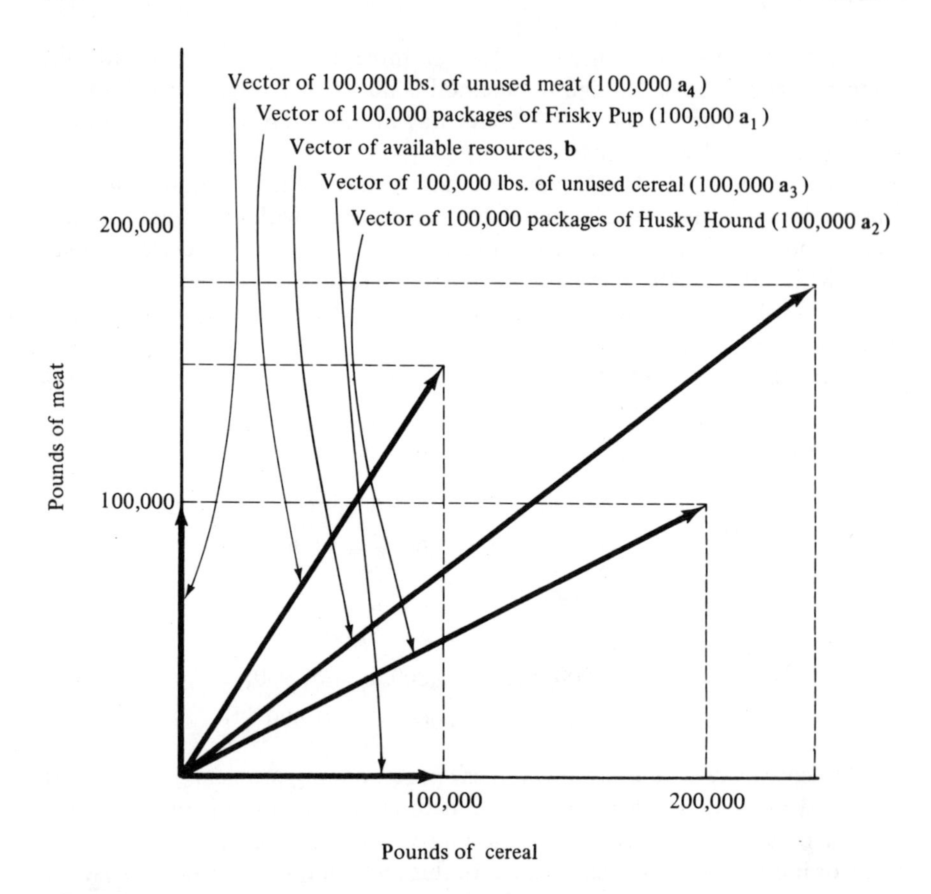


**Figure 2.5**

A Requirements Space Representation of the Dog Food Problem, Ignoring the Constraint on Frisky Pup Packaging Capacity

is shown in Figure 2.5, in which the requirements vector and each of the four activity vectors are plotted. The activity vector $\boldsymbol{a}_j$ corresponds to the variable $x_j$. The activity vectors, $\boldsymbol{a}_1$, $\boldsymbol{a}_2$, $\boldsymbol{a}_3$, and $\boldsymbol{a}_4$, generate a convex cone. Vectors $\boldsymbol{a}_3$ and $\boldsymbol{a}_4$ form a feasible basis (since $\boldsymbol{b}$ lies in the convex cone defined by $\boldsymbol{a}_3$ and $\boldsymbol{a}_4$), while $\boldsymbol{a}_1$ and $\boldsymbol{a}_4$ do not form a feasible basis (since $\boldsymbol{b}$ does not lie in the convex cone defined by $\boldsymbol{a}_1$ and $\boldsymbol{a}_4$). We have chosen to indicate 100,000 $\boldsymbol{a}_1$, 100,000 $\boldsymbol{a}_2$, and so on, because the lengths of $\boldsymbol{a}_1$, $\boldsymbol{a}_2$, and so on, would be insignificant in Figure 2.5.

It may be useful for the reader to examine some of the feasible bases. For example, 240,000 $\boldsymbol{a}_3$ plus 180,000 $\boldsymbol{a}_4$ equals $\boldsymbol{b}$. Similarly, 60,000 $\boldsymbol{a}_1$ plus 90,000 $\boldsymbol{a}_2$ equals $\boldsymbol{b}$. Hence two feasible bases are $\boldsymbol{a}_3$, $\boldsymbol{a}_4$ and $\boldsymbol{a}_1$, $\boldsymbol{a}_2$.

At this point it is useful to introduce an additional dimension—the objective function $z$. Each of the activity vectors will have a component (possibly zero)

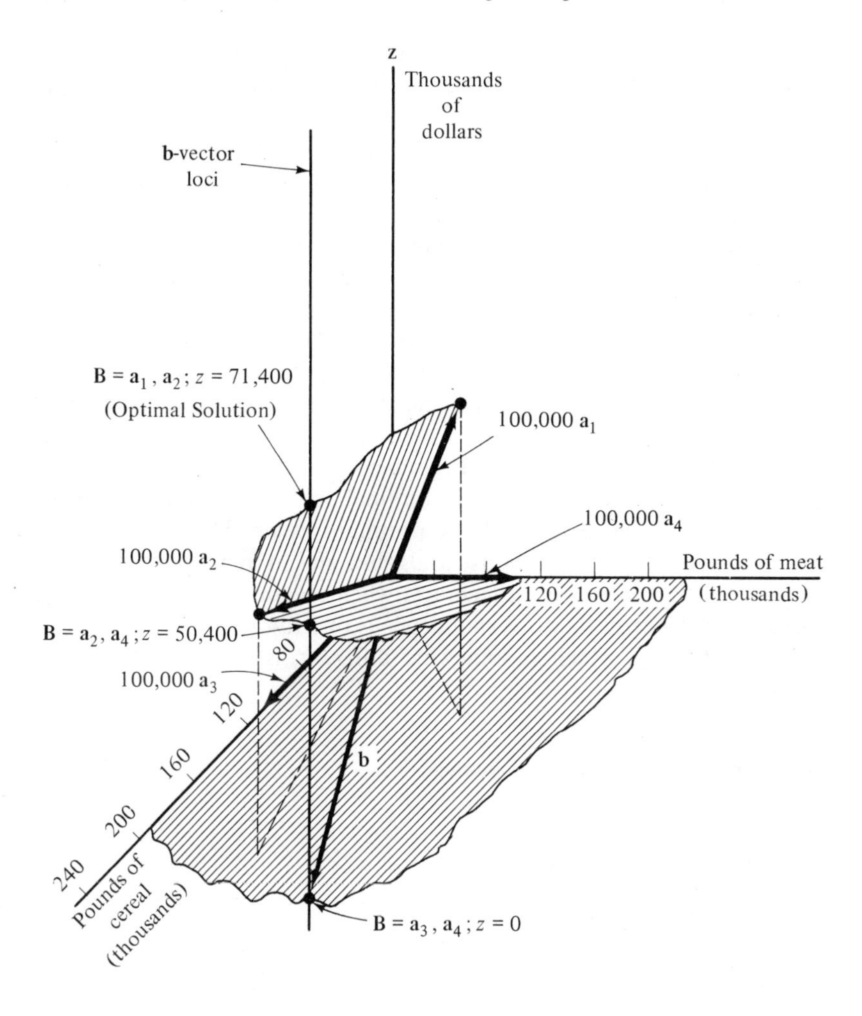


**Figure 2.6**

Requirement Space Representation of Example Problem Including Objective Function

in the $z$ direction, and the $\boldsymbol{b}$ vector is replaced by a set of vectors having all but one component equal to the components of $\boldsymbol{b}$ and the component in the $z$ direction equal to any finite value. Such a representation of the requirements space is given in Figure 2.6. Remember that this representation is the same as that in Figure 2.5 with the $z$-dimension added. The convex cone generated by a basis (in this case, a plane) which intersects the line corresponding to the $\boldsymbol{b}$ vector at the greatest $z$ value contains the optimal solution. Cones for three feasible bases are illustrated in Figure 2.6, together with their associated generating vectors. Recall, for purposes of comparison, that in the activities or solutions space, we move the objective function hyperplane parallel to itself until we obtain the greatest value of the objective function for which the corresponding hyperplane intersects the convex polyhedron of feasible solutions.

Interpreting the simplex method in the requirements space, having added the $z$-dimension, a starting basis (where possible, a slack basis with $z = 0$) is used. Then, of the nonbasic activity vectors which lie "above" the basis cone in the $z$ direction, one is selected to enter the basis. The vector selected to leave is chosen so that the $\boldsymbol{b}$ vector (with suitable $z$ value) is contained in the cone generated by the basis. The process continues until no nonbasic activity vector lies "above" the basis cone in the $z$ direction, at which time an optimal solution is at hand. For the example problem the sequence of solutions might be: $\boldsymbol{a}_3, \boldsymbol{a}_4; \boldsymbol{a}_2, \boldsymbol{a}_4; \boldsymbol{a}_2, \boldsymbol{a}_1$. Those bases, and their associated cones and solutions, are all shown in Figure 2.6.

Given these fundamentals, a number of authors (see, for example, Hadley [144]) proceed by showing that an optimal solution occurs at at least one extreme point, provided that a finite optimum exists. Then the correspondence between basic feasible solutions in algebra and extreme points in geometry is proven. (In the absence of degeneracy, when one or more of the basic variables turns out to be zero, the correspondence is one-to-one.)

Our strategy, which is somewhat different, is developed in the following chapter.

**Selected Supplemental References**

General
[33], [139], [282]

## 2.7 PROBLEMS

**1.** Show that the inverse of

$$\begin{pmatrix} 0 & 1 \\ 2 & 4 \end{pmatrix} \text{ is } \begin{pmatrix} -2 & .5 \\ 1 & 0 \end{pmatrix}$$

**2.** Show that the matrix

$$\begin{pmatrix} 1 & 1 & 1 \\ 2 & 3 & 3 \\ 0 & 1 & 1 \end{pmatrix}$$

does not possess an inverse.

**3.** Find the inverse of the following matrices:

$$\begin{pmatrix} 1 & 0 & 1 \\ 0 & 1 & 1 \\ 0 & 0 & 1 \end{pmatrix} \quad \begin{pmatrix} 1 & 1 & 1 \\ 0 & 0 & 1 \\ 0 & -4 & 1 \end{pmatrix} \quad \begin{pmatrix} 0 & 1 & 1 \\ 0 & 0 & 1 \\ 1 & -4 & 1 \end{pmatrix}$$

**4.** Prove that the method given in section 2.2 for computing the inverse does indeed find the inverse. Hint: The row operations are equivalent to a multiplication by a matrix.

**5.** $A$ is defined to be an orthogonal matrix if $AA' = I$ (i.e., $A' = A^{-1}$). Prove that if $A$ is orthogonal, then the $i^{\text{th}}$ row of $A$ and the $j^{\text{th}}$ column of $A'$ are orthogonal if $i \neq j$, and not orthogonal if $i = j$.

**6.** Show that the inverse of a symmetric matrix is symmetric.

**7.** Prove that if a square matrix is of full rank, then it possesses an inverse.

**8.** Prove that if a square matrix possesses an inverse, then it is of full rank.

**9.** Prove the statement on page 43 that the convex cone formed by the basic vectors corresponding to a feasible solution must contain the $\boldsymbol{b}$ vector.

**10.** Illustrating Definition 2.17, show that the vector $\begin{pmatrix} 3 \\ 4 \\ 5 \end{pmatrix}$ can be expressed in terms of the vectors of example (4) but not of example (3) on page 33.

**11.** Solve the Pasty Pizza Company problem of Chapter 1 (problem 10) graphically.

**12.**

$$\begin{aligned} \text{Maximize } z &= .5x_1 + 1.5x_2 \\ \text{subject to:} \quad x_1 + x_2 &\leq 100{,}000 \\ 1.25x_1 + 2x_2 &\leq 150{,}000 \\ x_2 &\leq 70{,}000 \\ x_1, x_2 &\geq 0 \end{aligned}$$

Given the above objective function and constraints:

a. Graph the constraints by the activities space method.

b. What is the area of feasible solutions?

c. What solutions are basic feasible solutions?

d. Draw in the required objective function and conceptualize its moving outward to an optimal solution.

e. How would the graph in subproblem (a) change if the first constraint were changed to $x_1 + x_2 \geq 50{,}000$?

**13.** Solve the following linear programming problem graphically:

$$\begin{aligned} \text{Maximize } z &= .75x_1 + .5x_2 \\ \text{subject to:} \quad x_1 + x_2 &\leq 4 \\ 1.5x_1 + .75x_2 &\leq 3 \\ x_1, x_2 &\geq 0 \end{aligned}$$

**14.** Solve the following linear programming problem graphically:

$$\begin{aligned} \text{Maximize } z = &\ .5x_1 + .75x_2 \\ \text{subject to:} \quad & 3x_1 + x_2 \leq 5 \\ & 1.5x_1 + 2x_2 \leq 15 \\ & x_1 + x_2 \leq 3 \\ & x_1, x_2 \geq 0 \end{aligned}$$

**15.** Solve the following linear programming problem graphically:
a. in the activity space.
b. in the requirement space.

$$\begin{aligned} \text{Maximize } z = &\ 1.1x_1 + x_2 \\ \text{subject to:} \quad & 3x_1 + 2x_2 \leq 12 \\ & x_1 \geq 1 \\ & x_1, x_2 \geq 0 \end{aligned}$$

**16.** Solve the following linear programming problem graphically:
a. in the activity space.
b. in the requirement space.

$$\begin{aligned} \text{Maximize } z = &\ .5x_1 + .75x_2 \\ \text{subject to:} \quad & 2x_1 - x_2 \leq -3 \\ & -x_1 + x_2 \leq -1 \\ & x_1, x_2 \geq 0 \end{aligned}$$

**17.** Solve the following linear programming problem graphically:
a. in the activity space.
b. in the requirement space.

$$\begin{aligned} \text{Maximize } z = &\ 1.5x_1 + .5x_2 \\ \text{subject to:} \quad & 3x_1 + x_2 \leq 5 \\ & x_1 + x_2 \leq 3 \\ & x_1, x_2 \geq 0 \end{aligned}$$

**18.** Solve the following linear programming problem graphically:
a. in the activity space.
b. in the requirement space.

$$\begin{aligned} \text{Minimize } z &= x_1 + x_2 \\ \text{subject to:} \quad 3x_1 + 2x_2 &\geq 6 \\ x_1 + .5x_2 &\geq 3 \\ x_1, x_2 &\geq 0 \end{aligned}$$

**19.** Solve the following problem graphically:

a. in the activity space.

b. in the requirement space.

$$\begin{aligned} \text{Maximize } z &= x_1 + x_2 \\ \text{subject to:} \quad 2x_1 + x_2 &\leq 7 \\ .5x_1 + x_2 &\leq 4 \\ x_1, x_2 &\geq 0 \end{aligned}$$

**20.** Explain the solution depicted in Figure 2.6 in terms of the bases $\boldsymbol{a}_3\boldsymbol{a}_4$, $\boldsymbol{a}_2\boldsymbol{a}_4$, $\boldsymbol{a}_2\boldsymbol{a}_1$. Show the tests for optimality.

# 3

# The Simplex Method

## 3.1 INTRODUCTION

In Chapter 1, an intuitive algebraic method for solving linear programming problems was proposed. The method seemed sensible, and worked for the example solved, but it was not examined for generality, convergence, and so on. Briefly, what the method required was having the problem in a canonical form (in each equation there is one variable with a $+1$ coefficient that has a zero coefficient in every other equation, including the objective function), with the $\boldsymbol{b}$ vector entries not less than zero. Then, by setting the nonbasic variables equal to zero, a feasible solution is available. However, so long as any coefficients in the objective function to be maximized are positive, that solution is known not to be optimal, since the objective function value can be increased by setting such a (nonbasic) variable to some positive value. The largest value that variable can take on (holding all other nonbasic variables at zero) is determined when some basic variable becomes zero. (If the nonbasic variable were to be increased to some larger value, then the basic variable would become negative, and the corresponding solution would violate one or more nonnegativity constraints.) The incoming nonbasic variable is solved for in terms of the basic variable that became zero (the outgoing basic variable), and the relationship is used to eliminate the incoming variable from all other equations. The equation in which the incoming variable was solved for is then divided by the coefficient of the incoming variable in that equation. Then the incoming variable is designated as a basic variable, the outgoing variable is designated as a nonbasic variable, and a new canonical form is obtained. Then the process is repeated. In this chapter we shall formalize the procedure; the formalization to be obtained is known as the *Simplex Method.*

## 3.2 A MATRIX REPRESENTATION

Any linear programming problem may be written in matrix form as follows:

$$\text{(3.1)} \qquad \begin{aligned} \text{Maximize} \quad & z = \boldsymbol{c}'\boldsymbol{x} \\ \text{subject to:} \quad & \boldsymbol{A}\boldsymbol{x} = \boldsymbol{b} \\ & \boldsymbol{x} \geq 0 \end{aligned}$$

where $\boldsymbol{c}$ and $\boldsymbol{x}$ are $n$ by 1, $\boldsymbol{b}$ is $m$ by 1, and $\boldsymbol{A}$ is $m$ by $n$. Every basic solution can be defined in terms of its basic (and nonbasic) variables, so that the basic solution is the unique result of setting to zero all the nonbasic variables. Associate with each variable (of the vector $\boldsymbol{x}$) its column in the original matrix (that is, the column of $\boldsymbol{A}$). That column or vector is referred to as a basic or nonbasic vector, according to whether the corresponding variable is basic or nonbasic. Thus, for any basic solution, we can rearrange the columns to separate the basic vectors (and variables) from the nonbasic vectors (and variables). Denoting the basis (the matrix of basic vectors) as $\boldsymbol{B}$ and the matrix of nonbasic vectors as $\boldsymbol{N}$, and $\boldsymbol{x}_B$ and $\boldsymbol{x}_N$ as vectors of the corresponding variables, we have

$$\text{(3.2)} \qquad \boldsymbol{B}\boldsymbol{x}_B + \boldsymbol{N}\boldsymbol{x}_N = \boldsymbol{b}$$

EXAMPLE 3.1:

From the constraints of the original problem, indicate some of the bases $\boldsymbol{B}$ and the corresponding partitions of the matrix.

$$\text{(3.3)} \qquad \begin{aligned} x_1 + 2x_2 + x_3 \qquad\qquad\quad &= 240{,}000 \\ 1.5x_1 + x_2 \qquad + x_4 \qquad &= 180{,}000 \\ x_1 \qquad\qquad\qquad\qquad + x_5 &= 110{,}000 \end{aligned}$$

Note that $\boldsymbol{A} = \begin{pmatrix} 1 & 2 & 1 & 0 & 0 \\ 1.5 & 1 & 0 & 1 & 0 \\ 1 & 0 & 0 & 0 & 1 \end{pmatrix}$

Basis 1:

$$\boldsymbol{B} = \begin{pmatrix} 1 & 0 & 0 \\ 0 & 1 & 0 \\ 0 & 0 & 1 \end{pmatrix} \quad \boldsymbol{N} = \begin{pmatrix} 1 & 2 \\ 1.5 & 1 \\ 1 & 0 \end{pmatrix}$$

$$x_B = \begin{pmatrix} x_3 \\ x_4 \\ x_5 \end{pmatrix} \qquad x_N = \begin{pmatrix} x_1 \\ x_2 \end{pmatrix}$$

Basis 2:

$$B = \begin{pmatrix} 1 & 2 & 0 \\ 1.5 & 1 & 0 \\ 1 & 0 & 1 \end{pmatrix} \quad N = \begin{pmatrix} 1 & 0 \\ 0 & 1 \\ 0 & 0 \end{pmatrix}$$

$$x_B = \begin{pmatrix} x_1 \\ x_2 \\ x_5 \end{pmatrix} \qquad x_N = \begin{pmatrix} x_3 \\ x_4 \end{pmatrix}$$

Basis 3:

$$B = \begin{pmatrix} 1 & 1 & 0 \\ 1.5 & 0 & 1 \\ 1 & 0 & 0 \end{pmatrix} \quad N = \begin{pmatrix} 2 & 0 \\ 1 & 0 \\ 0 & 1 \end{pmatrix}$$

$$x_B = \begin{pmatrix} x_1 \\ x_3 \\ x_4 \end{pmatrix} \qquad x_N = \begin{pmatrix} x_2 \\ x_5 \end{pmatrix}$$

We shall always use the convention that $\boldsymbol{B}$ is the *current* basis. (At certain times we introduce notation for an additional basis.) The order of the columns in a basis or a set of nonbasic vectors is not important. However, the order of the columns in the basis (or set of nonbasic vectors) must correspond to the order of the variables in the associated vector $\boldsymbol{x}_B$ (or $\boldsymbol{x}_N$).

Left-multiplying the original matrix equation (3.2) by $\boldsymbol{B}^{-1}$, we obtain the canonical form corresponding to the basis $\boldsymbol{B}$, namely

$$\boldsymbol{B}^{-1}\boldsymbol{B}\boldsymbol{x}_B + \boldsymbol{B}^{-1}\boldsymbol{N}\boldsymbol{x}_N = \boldsymbol{B}^{-1}\boldsymbol{b}, \quad \text{or}$$

$$\boldsymbol{x}_B + \boldsymbol{B}^{-1}\boldsymbol{N}\boldsymbol{x}_N = \boldsymbol{B}^{-1}\boldsymbol{b} \tag{3.4}$$

Obviously, with $\boldsymbol{x}_N = \boldsymbol{0}$, $\boldsymbol{x}_B = \boldsymbol{B}^{-1}\boldsymbol{b}$.

EXAMPLE 3.2:

Using the results of (3.4), express the bases of Example 3.1 in a canonical form. For convenience we repeat

$$\boldsymbol{b} = \begin{pmatrix} 240{,}000 \\ 180{,}000 \\ 110{,}000 \end{pmatrix}$$

Basis 1:

$$B = \begin{pmatrix} 1 & 0 & 0 \\ 0 & 1 & 0 \\ 0 & 0 & 1 \end{pmatrix} \quad N = \begin{pmatrix} 1 & 2 \\ 1.5 & 1 \\ 1 & 0 \end{pmatrix}$$

$$B^{-1} = \begin{pmatrix} 1 & 0 & 0 \\ 0 & 1 & 0 \\ 0 & 0 & 1 \end{pmatrix}$$

Substituting these into equation (3.4) we obtain:

$$\begin{pmatrix} x_3 \\ x_4 \\ x_5 \end{pmatrix} + \begin{pmatrix} 1 & 2 \\ 1.5 & 1 \\ 1 & 0 \end{pmatrix} \begin{pmatrix} x_1 \\ x_2 \end{pmatrix} = \begin{pmatrix} 240{,}000 \\ 180{,}000 \\ 110{,}000 \end{pmatrix}$$

Basis 2:

$$B = \begin{pmatrix} 1 & 2 & 0 \\ 1.5 & 1 & 0 \\ 1 & 0 & 1 \end{pmatrix} \quad N = \begin{pmatrix} 1 & 0 \\ 0 & 1 \\ 0 & 0 \end{pmatrix}$$

$$B^{-1} = \begin{pmatrix} -.5 & 1 & 0 \\ .75 & -.5 & 0 \\ .5 & -1 & 1 \end{pmatrix}$$

Substituting these into (3.4) we obtain

$$\begin{pmatrix} x_1 \\ x_2 \\ x_5 \end{pmatrix} + \begin{pmatrix} -.5 & 1 & 0 \\ .75 & -.5 & 0 \\ .5 & -1 & 1 \end{pmatrix} \begin{pmatrix} 1 & 0 \\ 0 & 1 \\ 0 & 0 \end{pmatrix} \begin{pmatrix} x_3 \\ x_4 \end{pmatrix} = \begin{pmatrix} -.5 & 1 & 0 \\ .75 & -.5 & 0 \\ .5 & -1 & 1 \end{pmatrix} \begin{pmatrix} 240{,}000 \\ 180{,}000 \\ 110{,}000 \end{pmatrix}$$

or

$$\begin{pmatrix} x_1 \\ x_2 \\ x_5 \end{pmatrix} + \begin{pmatrix} -.5 & 1 \\ .75 & -.5 \\ .5 & -1 \end{pmatrix} \begin{pmatrix} x_3 \\ x_4 \end{pmatrix} = \begin{pmatrix} 60{,}000 \\ 90{,}000 \\ 50{,}000 \end{pmatrix}$$

Basis 3:

$$B = \begin{pmatrix} 1 & 1 & 0 \\ 1.5 & 0 & 1 \\ 1 & 0 & 0 \end{pmatrix} \quad N = \begin{pmatrix} 2 & 0 \\ 1 & 0 \\ 0 & 1 \end{pmatrix}$$

$$B^{-1} = \begin{pmatrix} 0 & 0 & 1 \\ 1 & 0 & -1 \\ 0 & 1 & -1.5 \end{pmatrix}$$

Substituting these into (3.4) we obtain

$$\begin{pmatrix} x_1 \\ x_3 \\ x_4 \end{pmatrix} + \begin{pmatrix} 0 & 0 & 1 \\ 1 & 0 & -1 \\ 0 & 1 & -1.5 \end{pmatrix}\begin{pmatrix} 2 & 0 \\ 1 & 0 \\ 0 & 1 \end{pmatrix}\begin{pmatrix} x_2 \\ x_5 \end{pmatrix} = \begin{pmatrix} 0 & 0 & 1 \\ 1 & 0 & -1 \\ 0 & 1 & -1.5 \end{pmatrix}\begin{pmatrix} 240{,}000 \\ 180{,}000 \\ 110{,}000 \end{pmatrix}$$

or

$$\begin{pmatrix} x_1 \\ x_3 \\ x_4 \end{pmatrix} + \begin{pmatrix} 0 & 1 \\ 2 & -1 \\ 1 & -1.5 \end{pmatrix}\begin{pmatrix} x_2 \\ x_5 \end{pmatrix} = \begin{pmatrix} 110{,}000 \\ 130{,}000 \\ 15{,}000 \end{pmatrix}$$

We have thus far partitioned the basis from the nonbasic set of vectors and solved for the basic variables in terms of the nonbasic variables. A similar analysis can be carried out for the objective function. Denote $\boldsymbol{c}_B$ and $\boldsymbol{c}_N$ as the vector of objective function coefficients for activities $\boldsymbol{x}_B$ and $\boldsymbol{x}_N$ respectively. We may then write the objective function as

$$\text{(3.5)} \qquad \begin{aligned} &\text{Maximize } z = \boldsymbol{c}'_B\boldsymbol{x}_B + \boldsymbol{c}'_N\boldsymbol{x}_N \\ &\text{subject to:} \quad \boldsymbol{B}\boldsymbol{x}_B + \boldsymbol{N}\boldsymbol{x}_N = \boldsymbol{b} \end{aligned}$$

($\boldsymbol{x}_B \geq \boldsymbol{0}$, and $\boldsymbol{x}_N \geq \boldsymbol{0}$ will be assumed and not written.) Equation (3.4) can be solved for $\boldsymbol{x}_B$ as follows:

$$\text{(3.6)} \qquad \boldsymbol{x}_B = \boldsymbol{B}^{-1}\boldsymbol{b} - \boldsymbol{B}^{-1}\boldsymbol{N}\boldsymbol{x}_N$$

Using equation (3.6), $\boldsymbol{x}_B$ can be eliminated from the objective function altogether, yielding

$$\text{(3.7)} \qquad \begin{aligned} &\text{Maximize } z = \boldsymbol{c}'_B(\boldsymbol{B}^{-1}\boldsymbol{b} - \boldsymbol{B}^{-1}\boldsymbol{N}\boldsymbol{x}_N) + \boldsymbol{c}'_N\boldsymbol{x}_N \\ &\text{subject to:} \ \boldsymbol{x}_B + \boldsymbol{B}^{-1}\boldsymbol{N}\boldsymbol{x}_N = \boldsymbol{B}^{-1}\boldsymbol{b} \end{aligned}$$

or

$$\text{(3.8)} \qquad \begin{aligned} &\text{Maximize } z = (\boldsymbol{c}'_N - \boldsymbol{c}'_B\boldsymbol{B}^{-1}\boldsymbol{N})\boldsymbol{x}_N + \boldsymbol{c}'_B\boldsymbol{B}^{-1}\boldsymbol{b} \\ &\text{subject to:} \ \boldsymbol{x}_B + \boldsymbol{B}^{-1}\boldsymbol{N}\boldsymbol{x}_N = \boldsymbol{B}^{-1}\boldsymbol{b} \end{aligned}$$

which is *precisely* the matrix equivalent to the canonical form for an arbitrary basis which we obtained in Chapter 1. For $\boldsymbol{x}_N = \boldsymbol{0}$, the solution can be seen without further calculation to be $\boldsymbol{x}_B = \boldsymbol{B}^{-1}\boldsymbol{b}$, and $z = \boldsymbol{c}'_B\boldsymbol{B}^{-1}\boldsymbol{b}$. Recall that $\boldsymbol{B}$ may be any basis of the problem.

Before proceeding, it is useful to present the following definition and theorem.

**Definition 3.1**

*A basic solution is degenerate if and only if one or more components of* $\boldsymbol{B}^{-1}\boldsymbol{b}$ *are zero.*

**THEOREM 3.1**

IF A VECTOR $\boldsymbol{b}$ CAN BE REPRESENTED ONLY AS A LINEAR COMBINATION OF $m$ LINEARLY INDEPENDENT VECTORS OF THE $\boldsymbol{A}$ MATRIX, AND NOT AS A LINEAR COMBINATION OF ANY SUBSET OF $m - 1$ LINEARLY INDEPENDENT VECTORS, THEN NO BASIC SOLUTION OF THE PROBLEM WILL BE DEGENERATE.

PROOF:

Assume to the contrary that some degenerate basic solution exists. Denote the corresponding basis as $\boldsymbol{B}$, and by assumption at least one component of the corresponding vector of $\boldsymbol{x}_B$ is zero. We then have $\boldsymbol{b}$ expressed as multipliers of $m - 1$ vectors contrary to the premise of the theorem. This completes the proof.

We shall henceforth assume that all linear programming problems are nondegenerate, and will later present methods that ensure that all problems—whether degenerate or not—are effectively nondegenerate.

We shall next introduce the simplex tableau, which is simply the detached coefficients of what we have been writing in matrix format (and earlier in algebraic terms), with one important difference. We do not rearrange the basic and nonbasic variables at each iteration to preserve, in a visible sense, the partition between $\boldsymbol{B}$ and $\boldsymbol{N}$.

In addition, for reasons which will become clear, we reverse the sign of the objective function. From expression (3.8) we have

$$\text{Maximize } z = (\boldsymbol{c}'_N - \boldsymbol{c}'_B\boldsymbol{B}^{-1}\boldsymbol{N})\boldsymbol{x}_N + \boldsymbol{c}'_B\boldsymbol{B}^{-1}\boldsymbol{b}$$

but we shall instead write

$$-z + \boldsymbol{c}'_B\boldsymbol{B}^{-1}\boldsymbol{b} = (\boldsymbol{c}'_B\boldsymbol{B}^{-1}\boldsymbol{N} - \boldsymbol{c}'_N)\boldsymbol{x}_N \tag{3.9}$$

For any basic solution with $\boldsymbol{x}_N = \boldsymbol{0}$, $z = \boldsymbol{c}'_B\boldsymbol{B}^{-1}\boldsymbol{b}$. The detached coefficients of expression (3.9), as well as the detached coefficients of the constraints of (3.8), which we write below for convenience

$$\boldsymbol{B}^{-1}\boldsymbol{b} = \boldsymbol{x}_B + \boldsymbol{B}^{-1}\boldsymbol{N}\boldsymbol{x}_N \tag{3.10}$$

form what is known as a *simplex tableau*, or the complete schematic representation of a basis. A general tableau as well as a specific tableau are given in Figure 3.1. The first row of the tableau is sometimes referred to as the evaluator row.

| | OBJECTIVE FUNCTION VALUE AND VALUES OF BASIC VARIABLES | BASIC VARIABLES | NONBASIC VARIABLES |
|---|---|---|---|
| $z$ | $c'_B B^{-1} b$ | $0'$ | $(c'_B B^{-1} N - c'_N)$ |
| $x_B$ | $B^{-1} b$ | $I$ | $B^{-1} N$ |

A simplex tableau in matrix form given a partition of the basic and nonbasic variables. This partition is not usually preserved.

| | | $x_3$ | $x_4$ | $x_5$ | $x_1$ | $x_2$ |
|---|---|---|---|---|---|---|
| $z$ | 0 | 0 | 0 | 0 | −.56 | −.42 |
| $x_3$ | 240,000 | 1 | 0 | 0 | 1 | 2 |
| $x_4$ | 180,000 | 0 | 1 | 0 | 1.5 | 1 |
| $x_5$ | 110,000 | 0 | 0 | 1 | 1 | 0 |

The first tableau of the example problem.

**Figure 3.1**

A General Simplex Tableau and an Example

The leftmost entry in the first row is the objective function value for the current basic feasible solution. The entries for the basic variables are always zero. The entries for the nonbasic variables, the vector $(c'_B B^{-1} N - c'_N)$, are the evaluator entries which are scanned to determine whether or not a particular basic feasible solution is optimal. The evaluator entries are often referred to as the $z_j - c_j$, or reduced costs. The entries of the main portion of the tableau are, from left to right:

1. the updated **b** vector $B^{-1} b$, which is the vector of values of the variables of the basic solution;
2. the coefficients of the basic variables (an identity matrix); and
3. the coefficients of the nonbasic variables $(B^{-1} N)$, that is, the amount by which each basic variable will be reduced for a unit increment is a nonbasic variable.

To the left of the tableau are labels indicating the variables basic in each equation. Above the tableau are the labels of all problem variables. Variations of the tableau have the first row last, the first column last, or the row of evaluators replaced by their negatives. Still another variation of the tableau omits the identity matrix. However, in this book we will use the format described above. In some cases we omit the identity matrix.

## 3.3 THE SIMPLEX METHOD

In this section we present a method for solving general linear programming problems. It requires a starting basic feasible solution and generates a sequence of basic feasible solutions. If the sequence is completed, the last solution is optimal; otherwise, evidence is generated to show that the problem does not have a finite optimum.

We have asserted above that a basic feasible solution is optimal if a finite optimum exists. The following theorem proves this result.

**THEOREM 3.2**

PROVIDING THAT A FINITE OPTIMUM EXISTS, A BASIC FEASIBLE SOLUTION IS OPTIMAL.

PROOF:

Assume to the contrary that no basic feasible solution is optimal. Because of the assumption of a finite optimum, a feasible nonbasic solution must then be optimal. As is shown elsewhere (see, for example, Hadley [144]), any feasible nonbasic solution may be expressed as a convex combination of basic feasible solutions. Designate the basic feasible solutions to the constraint set (3.1) as $\boldsymbol{x}_1, \ldots, \boldsymbol{x}_p$. Let the assumed optimal solution be $\boldsymbol{x}_0$. Now,

$$\boldsymbol{x}_0 = \sum_{k=1}^{p} \lambda_k \boldsymbol{x}_k$$

and

$$\sum_{k=1}^{p} \lambda_k = 1 \quad \lambda_k \geq 0 \quad \text{for } k = 1, \ldots, p$$

By assumption

$$\boldsymbol{c}'\boldsymbol{x}_0 > \boldsymbol{c}'\boldsymbol{x}_j \text{ for } j = 1, \ldots, p$$

or

$$\sum_{k=1}^{p} \lambda_k \boldsymbol{c}'\boldsymbol{x}_k > \boldsymbol{c}'\boldsymbol{x}_j \quad \text{for } j = 1, \ldots, p$$

which says that a weighted average of a set of scalars is strictly greater than any one of them and is a contradiction, thereby proving the theorem.

The simplex solution procedure is as follows:

1. Evaluate the current solution. If the solution is optimal, stop. (The solution is optimal if increasing any nonbasic variable an infinitesimal amount does not increase the objective function.) Otherwise, go to step 2.

2. Select a variable to enter the basis. Choose one (usually the one variable that increases the objective function *most* per unit) which, if increased from zero, would increase the objective function.
3. Determine which variable will leave the basis. Express the problem in terms of the new basis and go to step 1.

In step 1, we check to see if all the elements $z_j - c_j$ are nonnegative. If they are, the solution is optimal. (In Chapter 1 we checked to see whether all components of $c_j - z_j$ were nonpositive, which is equivalent to the above.)

In step 2, we choose the most negative $z_j - c_j$ and designate the corresponding variable to enter the basis. (As with step 1, this was also done in Chapter 1.)

In step 3, we find the variable that is to leave the basis. Recall from Chapter 1 that we decided which variable was to leave the basis by trial and error—we increased the nonbasic variable until a basic variable became zero. The basic variable that became zero was then selected to leave the basis. An equivalent but simpler method is to consider only equations in which the coefficients of the incoming variable are positive, whereby increasing the nonbasic variable *decreases* the basic variables. Determine for each such equation the maximum value that the nonbasic variable could take on before the basic variable (in each equation) became zero. For each equation this value is simply the value of the updated $\boldsymbol{b}$ vector entry (an entry of $\boldsymbol{B}^{-1}\boldsymbol{b}$) divided by the appropriate (positive) entry of the updated column of the incoming nonbasic variable (i.e., an entry of $\boldsymbol{B}^{-1}\boldsymbol{N}$ which is the rate of substitution of a nonbasic variable for a basic variable). The minimum quotient is the value of the incoming basic variable in the new basic solution. The equation in which the minimum quotient is found is the equation whose basic variable now becomes nonbasic.

If it ever occurs that all the entries in a column whose variable is to be introduced into the basis are nonpositive, then there exists no finite optimal solution. As indicated in Chapter 1, such a condition usually indicates an error in formulation, such as omission of a necessary constraint.

| | | $x_1$ | $x_2$ | $\cdots$ | $x_n$ |
|---|---|---|---|---|---|
| $z$ | $a_{00}$ | $a_{01}$ | $a_{02}$ | $\cdots$ | $a_{0n}$ |
| $x_{B1}$ | $a_{10}$ | $a_{11}$ | $a_{12}$ | $\cdots$ | $a_{1n}$ |
| $x_{B2}$ | $a_{20}$ | $a_{21}$ | $a_{22}$ | $\cdots$ | $a_{2n}$ |
| $\vdots$ | $\vdots$ | $\vdots$ | $\vdots$ | $\vdots$ | $\vdots$ |
| $x_{Bm}$ | $a_{m0}$ | $a_{m1}$ | $a_{m2}$ | $\cdots$ | $a_{mn}$ |

**Figure 3.2**

Another Representation of the Simplex Tableau

We shall now indicate a simplex tableau in a slightly different manner, thereby ceasing to distinguish between basic and nonbasic variables. We could use superscripts on the coefficients to distinguish between iterations, but we shall not. The tableau is shown in Figure 3.2; $\boldsymbol{x}_{Bi}$ represents the variable basic in row $i$, $a_{i0}$ represents an entry of $\boldsymbol{B}^{-1}\boldsymbol{b}$, $a_{0j}$ represents either 0 or a component of $\boldsymbol{c}'_B\boldsymbol{B}^{-1}\boldsymbol{N} - \boldsymbol{c}'_N$, according to whether $x_j$ is basic or not, respectively, and $a_{00}$ is the value of the objective function for the basic solution.

A flow chart for the simplex method is contained in Figure 3.3. To omit the

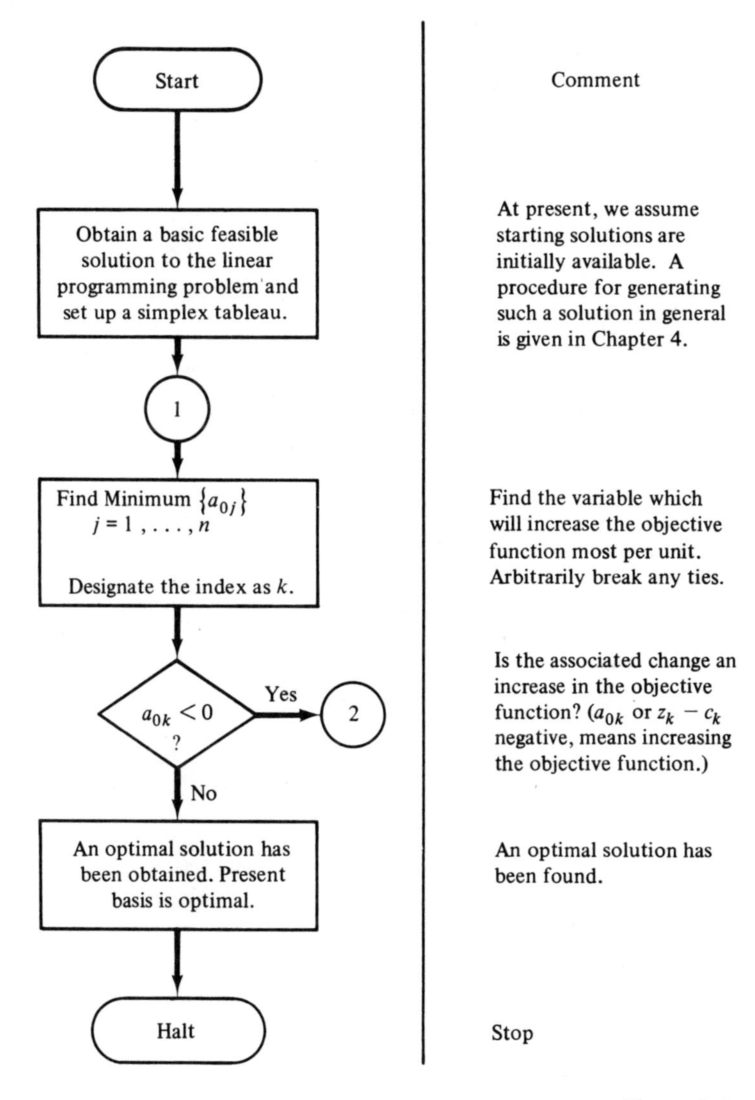


**Figure 3.3**

A Flow Chart of the Simplex Method

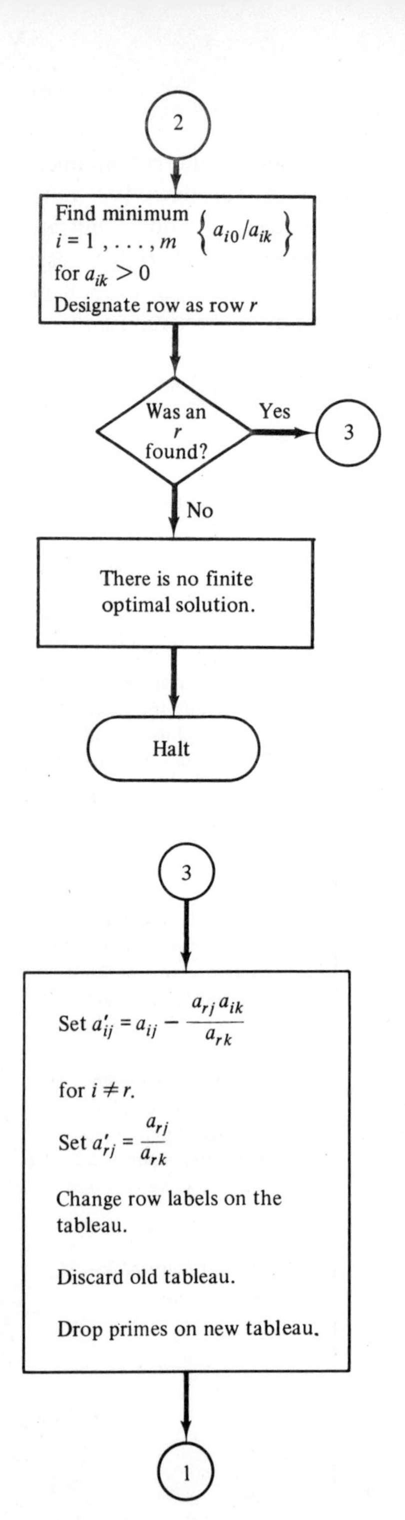


Figure 3.3 (concluded)

Determine which variable, $x_{Br}$, will leave the basis. We shall see later that assuming nondegeneracy prevents ties here.

Check to see if there is at least one $a_{ik} > 0$. Otherwise no variable will leave the basis as $x_k$ is increased.

No finite optimal solution exists. $x_k$ can be increased infinitely without bound and no basic variable decreases. The solution is said to be unbounded.

Compute new tableau using row operations. $a_{rk}$ is reduced to unity, while $a_{ik}$, $i \neq r$, is reduced to zero. All entries in the tableau undergo the same row operations. This process produces the simplex tableau for the next basis.

identity matrix in the simplex method, it is only necessary to alter the tableau slightly by changing the last block in the flow chart in Figure 3.3. We shall leave that as an exercise to be posed in section 3.6.

The following example is the simplex method applied to the dog food blending problem.

EXAMPLE 3.3:

The tableaus for the example problem and description follow.

**Tableau 1**

| | | $x_1$ ↓ | $x_2$ | $x_3$ | $x_4$ | $x_5$ | |
|---|---|---|---|---|---|---|---|
| $z$ | 0 | $-.56$ | $-.42$ | 0 | 0 | 0 | |
| $x_3$ | 240,000 | 1 | 2 | 1 | 0 | 0 | |
| $x_4$ | 180,000 | 1.5 | 1 | 0 | 1 | 0 | |
| $x_5$ | 110,000 | [1] | 0 | 0 | 0 | 1 | → |

The arrows indicate the incoming and outgoing variables. The entry $a_{rk}$ is enclosed in a block (□) and is called the pivot entry. $x_1$ is selected since the minimum value of $a_{0j}$ is $-.56$ for $j = 1$; therefore, $k = 1$. The variable $x_5$ is chosen to leave the basis because $a_{30}/a_{31}$ is minimum for any positive $a_{i1}$. Some specific entries of the new tableau are calculated, as follows:

$$a'_{00} = a_{00} - \frac{a_{30}a_{01}}{a_{31}} = 0 - \frac{(110{,}000)(-.56)}{1} = +61{,}600$$

$$a'_{02} = a_{02} - \frac{a_{32}a_{01}}{a_{31}} = -.42 - \frac{0(-.56)}{1} = -.42$$

$$a'_{10} = a_{10} - \frac{a_{30}a_{11}}{a_{31}} = 240{,}000 - \frac{(110{,}000)(1)}{1} = 130{,}000$$

$$a'_{01} = a_{01} - \frac{a_{31}a_{01}}{a_{31}} = 0$$

$$a'_{30} = \frac{a_{30}}{a_{31}} = \frac{110{,}000}{1} = 110{,}000$$

$$a'_{25} = a_{25} - \frac{a_{35}a_{21}}{a_{31}} = 0 - \frac{1(1.5)}{1} = -1.5$$

Tableaus 2, 3, and 4 follow in the same manner.

**Tableau 2**

| | | $x_1$ | $x_2$ ↓ | $x_3$ | $x_4$ | $x_5$ | |
|---|---|---|---|---|---|---|---|
| $z$ | 61,600 | 0 | −.42 | 0 | 0 | .56 | |
| $x_3$ | 130,000 | 0 | 2 | 1 | 0 | −1 | |
| $x_4$ | 15,000 | 0 | [1] | 0 | 1 | −1.5 | → |
| $x_1$ | 110,000 | 1 | 0 | 0 | 0 | 1 | |

**Tableau 3**

| | | $x_1$ | $x_2$ | $x_3$ | $x_4$ | $x_5$ ↓ | |
|---|---|---|---|---|---|---|---|
| $z$ | 67,900 | 0 | 0 | 0 | .42 | −.07 | |
| $x_3$ | 100,000 | 0 | 0 | 1 | −2 | [2] | → |
| $x_2$ | 15,000 | 0 | 1 | 0 | 1 | −1.5 | |
| $x_1$ | 110,000 | 1 | 0 | 0 | 0 | 1 | |

**Tableau 4**

| | | $x_1$ | $x_2$ | $x_3$ | $x_4$ | $x_5$ |
|---|---|---|---|---|---|---|
| $z$ | 71,400 | 0 | 0 | .035 | .35 | 0 |
| $x_5$ | 50,000 | 0 | 0 | .5 | −1 | 1 |
| $x_2$ | 90,000 | 0 | 1 | .75 | −.5 | 0 |
| $x_1$ | 60,000 | 1 | 0 | −.5 | 1 | 0 |

Tableau 4 is optimal, since all $a_{0j} \geq 0$. The iterations should be compared with those obtained using the intuitive method in Chapter 1. More generally, the procedures might be compared to see that they are the same as well.

It is well to consider the solution obtained. The solution says it is optimal to produce $x_1 = 60{,}000$ packages of Frisky Pup dog food and $x_2 = 90{,}000$ packages of Husky Hound dog food with associated monthly profits of $71,400. However, there is additional information. First, we are obviously producing $x_5 = 50{,}000$ units of (idle) packaging capacity for Frisky Pup dog food. Less obvious is the interpretation of the $a_{0j}$ entries. The values of the $a_{0j}$ for the basic variables, $x_1$, $x_2$, and $x_5$, are, of course, zero. Considering the values for the nonbasic variables, we can see that if either variable $x_3$ or $x_4$ is greater than zero, profits will be reduced by the amount indicated per unit increase of the variable. Thus, if one pound of cereal (constraint 1) were lost and not utilized to produce dog food, $x_3$ would be 1 instead of zero and profits would be reduced by \$.035. This would not be true for all unit increments, however. Intuitively, we would expect that the per unit cost would increase, and we shall discuss this further when we get to sensitivity analysis and parametric programming in Chapter 7. In a similar manner, the figure for $x_4$, \$.35, can be interpreted as the reduction in contribution to overhead and profits per pound of meat not utilized. The entries in the simplex tableau, for the nonbasic variables, give the amount by which the appropriate basic variable must be *reduced* for each unit *increment* in the nonbasic variable for the given basic solution. For example, from Tableau 4 the coefficient of $x_3$ in the second equation (in which $x_2$ is basic) is .75. This coefficient indicates that for each pound of cereal not utilized for dog food production, making the necessary adjustments, the production of Husky Hound dog food ($x_2$) would be decreased by $\frac{3}{4}$ of one pound, for the indicated basic solution.

The following is another example solved by the simplex method.

EXAMPLE 3.4:

$$\begin{aligned} \text{Maximize} \quad z &= 3x_1 + 2x_2 \\ \text{subject to:} \quad x_1 + .5x_2 &\leq 15 \\ 2x_1 + 4x_2 &\leq 24 \end{aligned}$$

Adding slack variables, and putting into tableau form, we have Tableau 1.

**Tableau 1**

| | | $x_1$ ↓ | $x_2$ | $x_3$ | $x_4$ | |
|---|---|---|---|---|---|---|
| $z$ | 0 | −3 | −2 | 0 | 0 | |
| $x_3$ | 15 | 1 | .5 | 1 | 0 | |
| $x_4$ | 24 | [2] | 4 | 0 | 1 | ⟶ |

$x_1$ enters the basis, replacing $x_4$, and thereby yielding Tableau 2, an optimal solution, for which the objective function is 36, with $x_1 = 12$ and $x_3 = 3$.

**Tableau 2**

| | | $x_1$ | $x_2$ | $x_3$ | $x_4$ |
|---|---|---|---|---|---|
| $z$ | 36 | 0 | 4 | 0 | 1.5 |
| $x_3$ | 3 | 0 | −1.5 | 1 | −.5 |
| $x_1$ | 12 | 1 | 2 | 0 | .5 |

## 3.4 CONVERGENCE OF THE SIMPLEX METHOD

We have refined our intuitive procedure into the simplex method, but we have not really considered whether it will find a finite optimum, if one exists. In this section we shall consider such problems. First we shall prove that a tie will not occur in choosing a variable to leave the basis if a problem is nondegenerate.

**THEOREM 3.3**

USING THE SIMPLEX METHOD IN A NONDEGENERATE PROBLEM, THERE CANNOT BE A TIE FOR DETERMINING THE VARIABLE TO LEAVE THE BASIS.

PROOF:

Assume to the contrary that there is a tie. Let the tied rows be $i$ and $r$. Assume that the variable in row $r$ is chosen to leave the basis. Then the new value of the variable basic in row $i$ is

$$a'_{i0} = a_{i0} - \frac{a_{r0}a_{ik}}{a_{rk}}$$

which is zero, because $a_{r0}/a_{rk} = a_{i0}/a_{ik}$. Thus we have reached a contradiction, thereby proving the theorem.

**THEOREM 3.4**

THE SIMPLEX METHOD, AS PRESENTED ABOVE, CONVERGES TO A FINITE OPTIMAL SOLUTION (FOR A NONDEGENERATE PROBLEM) IF ONE EXISTS, OR ELSE IT IS SHOWN THAT NO FINITE OPTIMAL SOLUTION EXISTS.

PROOF:

First, as was shown earlier, there are a finite number of basic solutions (at most, $n$ things taken $m$ at a time). Further, some of these solutions will not be feasible, so the number of basic feasible solutions is a subset of the set of basic solutions. Given a basic feasible solution, we can always test to determine whether or not it is optimal by checking for possible negative evaluators ($a_{0j}$). If there are no negative evaluators, an optimum has been attained. If there are any negative evaluators, an optimum has not yet been attained, and another iteration introducing into the basis a variable whose evaluator is negative must be undertaken. Two outcomes are possible: either the incoming variable can be increased without limit, in which case there is no finite optimum, or a new basic solution is achieved. It can be shown that the new solution is better, in terms of the objective function, than the old solution, for consider the calculation of $a'_{00}$.

$$a'_{00} = a_{00} - \frac{a_{r0}a_{0k}}{a_{rk}}$$

$a_{r0}$ and $a_{rk}$ are both positive and $a_{0k}$ is negative, therefore, $a'_{00} > a_{00}$, which means that the objective function is larger than before. Therefore, a basis cannot be repeated. The process can only be repeated a finite number of times until at most all basic feasible solutions are examined. Convergence of the simplex method, assuming nondegeneracy, has been proved.

**THEOREM 3.5** (Alternate Optimal Solutions)

IF, AT ANY OPTIMAL SOLUTION, ANY OF THE NONBASIC VARIABLES—SUCH AS $x_k$—HAS A ZERO VALUE OF $a_{0k}$, THEN AN ITERATION CAN BE TAKEN IN WHICH $x_k$ ENTERS THE BASIS AND REPLACES SOME BASIC VARIABLE. THE RESULT OF SUCH AN ITERATION IS AN OPTIMAL SOLUTION.

PROOF:

Let $a_{0k}$ be zero for a nonbasic variable $x_k$. Since the solution is optimal, all $a_{0k} \geq 0$. Suppose $a_{rk}$ is the pivot element for the iteration introducing $x_k$. Then, for any $j$,

$$a'_{0j} = a_{0j} - \frac{a_{rj}a_{0k}}{a_{rk}}, \quad \text{and } a'_{00} = a_{00} - \frac{a_{r0}a_{0k}}{a_{rk}}$$

since $a_{0k} = 0$, $a'_{0j} = a_{0j}$, and $a'_{00} = a_{00}$.

COROLLARY:

Any convex combination of optimal solutions is optimal.

## 3.5 DEGENERACY AND ITS RESOLUTION

In proving convergence of the simplex method, we have made the assumption that all linear programming problems were nondegenerate. This assumption is in general *not* fulfilled, and there is no easy way, to the author's knowledge, of testing a problem to determine whether or not it is degenerate. To illustrate the potential difficulty of degeneracy, consider a basic variable that has a value of zero. When an iteration is taken, the objective function may not change if the new basic variable replaces the one previously in the basis at a zero level. It is conceivable that still another variable may enter the basis at a zero level, and so on, until the earlier basis is encountered once again. The process of repeating a basis in such a manner is called cycling (also circling). If cycling were to occur in a problem, the optimal solution might never be attained. Although the resolution of cycling possibilities is important as a matter of basic theory, it does not appear to have been a problem of great importance in actual practice, even though most practical linear programming problems are degenerate. This is because most real problems do not cycle. (Massively degenerate problems have been known to stall or near-cycle, which is effectively the same as cycling.)

A number of problems have been constructed that cycle when prescribed "rules-of-thumb" replacement procedures are used to resolve ties. A.J. Hoffman [155] first constructed an example of a degenerate problem that cycles in such circumstances. An example by E.M.L. Beale [20] is given below.

$$\begin{aligned}
&\text{Maximize } z = \tfrac{3}{4}x_1 - 150x_2 + \tfrac{1}{50}x_3 - 6x_4 \\
&\text{subject to: } \tfrac{1}{4}x_1 - 60x_2 - \tfrac{1}{25}x_3 + 9x_4 + x_5 = 0 \\
&\qquad \tfrac{1}{2}x_1 - 90x_2 - \tfrac{1}{50}x_3 + 3x_4 + x_6 = 0 \\
&\qquad x_3 + x_7 = 1
\end{aligned} \tag{3.11}$$

The starting basic solution is encountered after six iterations, when the outgoing variable is designated by the lowest numbered row in case of a tie, using the simplex method. The iterations thereby generated are shown in Figure 3.4.

### The Perturbation Methods

Methods have been developed to prevent cycling. One involves the use of a polynomial perturbation [46], and the other utilizes lexicographic ordering conventions [67]. The two are formally very similar,* and for convenience we refer to them both as the "perturbation methods."

We shall present a variation of these methods. Add to the ***b*** vector a vector of polynomials for which an entry in row $i$ is of the form $a_{i1}\epsilon + a_{i2}\epsilon^2 + a_{i3}\epsilon^3 + \cdots + a_{i,n}\epsilon^n$, where the $a_{ij}$ elements are the elements of the original matrix and $\epsilon > 0$ is a sufficiently small real number. (Note that for $\epsilon$ sufficiently small $\epsilon \gg \epsilon^2 \gg \epsilon^3 \gg \epsilon^4 \gg \cdots \gg \epsilon^n$, where $\gg$ means much greater than.) If, instead of assigning a real value to $\epsilon$, we simply calculate the coefficients of $\epsilon^k$, $k = 1, \ldots, n$ each iteration, we have for every iteration a tableau row entry of the form

| $a_{i0} + a_{i1}\epsilon + a_{i2}\epsilon^2 + \cdots + a_{i,n}\epsilon^n$ | $a_{i1}$ | $a_{i2}$ | $\cdots$ | $a_{i,n}$ |
|---|---|---|---|---|

where the $a_{ij}$ is the updated coefficient in row $i$, column $j$. The coefficients of powers of $\epsilon$ in the updated ***b*** vector are exactly the coefficients of the matrix

* See Charnes and Cooper [47], pp. 427–438, for a discussion of this equivalence. A variant of the method utilizes only the slack or artificial vectors in a suitable arrangement. See G.B. Dantzig [62], pp. 231–234.

**Tableau 1**

| | | ↓ $x_1$ | $x_2$ | $x_3$ | $x_4$ | $x_5$ | $x_6$ | $x_7$ | |
|---|---|---|---|---|---|---|---|---|---|
| $z$ | 0 | −.75 | 150 | −.02 | 6 | 0 | 0 | 0 | |
| $x_5$ | 0 | [.25] | −60 | −.04 | 9 | 1 | 0 | 0 | → |
| $x_6$ | 0 | .5 | −90 | −.02 | 3 | 0 | 1 | 0 | |
| $x_7$ | 1 | 0 | 0 | 1 | 0 | 0 | 0 | 1 | |

**Tableau 2**

| | | $x_1$ | ↓ $x_2$ | $x_3$ | $x_4$ | $x_5$ | $x_6$ | $x_7$ | |
|---|---|---|---|---|---|---|---|---|---|
| $z$ | 0 | 0 | −30 | −.14 | 33 | 3 | 0 | 0 | |
| $x_1$ | 0 | 1 | −240 | −.16 | 36 | 4 | 0 | 0 | |
| $x_6$ | 0 | 0 | [30] | .06 | −15 | −2 | 1 | 0 | → |
| $x_7$ | 1 | 0 | 0 | 1 | 0 | 0 | 0 | 1 | |

**Tableau 3**

| | | $x_1$ | $x_2$ | ↓ $x_3$ | $x_4$ | $x_5$ | $x_6$ | $x_7$ | |
|---|---|---|---|---|---|---|---|---|---|
| $z$ | 0 | 0 | 0 | −.08 | 18 | 1 | 1 | 0 | |
| $x_1$ | 0 | 1 | 0 | [.32] | −84 | −12 | 8 | 0 | → |
| $x_2$ | 0 | 0 | 1 | .002 | −.5 | −.067 | .003 | 0 | |
| $x_7$ | 1 | 0 | 0 | 1 | 0 | 0 | 0 | 1 | |

**Figure 3.4**

An Example of a Problem That Cycles

itself. So long as there is no tie in determining the variable to leave the basis, the simplex method works as if the polynomials were not there at all. If ties would otherwise occur, the polynomials work so as to break them. The result is embodied in the following theorem.

**THEOREM 3.6**

AN ARBITRARY LINEAR PROGRAMMING PROBLEM FOR WHICH A BASIC FEASIBLE SOLUTION EXISTS WILL NEVER BE DEGENERATE USING THE POLYNOMIALS INVOLVING $\epsilon$.

**Tableau 4**

| | | $x_1$ | $x_2$ | $x_3$ | $x_4$ ↓ | $x_5$ | $x_6$ | $x_7$ | |
|---|---|---|---|---|---|---|---|---|---|
| $z$ | 0 | .25 | 0 | 0 | −3 | −2 | −3 | 0 | |
| $x_3$ | 0 | 3.12 | 0 | 1 | −263 | −37.5 | 25 | 0 | |
| $x_2$ | 0 | −.006 | 1 | 0 | **.025** | $\frac{1}{120}$ | −.016 | 0 | → |
| $x_7$ | 1 | 3.12 | 0 | 0 | 263 | 37.5 | −25 | 1 | |

**Tableau 5**

| | | $x_1$ | $x_2$ | $x_3$ | $x_4$ | $x_5$ ↓ | $x_6$ | $x_7$ | |
|---|---|---|---|---|---|---|---|---|---|
| $z$ | 0 | −.5 | 120 | 0 | 0 | −1 | −1 | 0 | |
| $x_3$ | 0 | 62.5 | 10,500 | 1 | 0 | **50** | 50 | 0 | → |
| $x_4$ | 0 | .25 | 40 | 0 | 1 | .33 | −.67 | 0 | |
| $x_7$ | 1 | 62.5 | −10,500 | 0 | 0 | −50 | 150 | 1 | |

**Tableau 6***

| | | $x_1$ | $x_2$ | $x_3$ | $x_4$ | $x_5$ | $x_6$ ↓ | $x_7$ | |
|---|---|---|---|---|---|---|---|---|---|
| $z$ | 0 | −1.75 | 330 | .02 | 0 | 0 | −2 | 0 | |
| $x_5$ | 0 | −1.25 | 210 | .02 | 0 | 1 | −3 | 0 | |
| $x_4$ | 0 | .167 | −30 | .0067 | 1 | 0 | **.33** | 0 | → |
| $x_7$ | 1 | 0 | 0 | 1 | 0 | 0 | 0 | 1 | |

* The tableau following Tableau 6 is Tableau 1.

**Figure 3.4 (concluded)**

PROOF:

Assume, to the contrary, that the problem is degenerate. Then, for some basis, a basic variable must be identically zero. For that basis, for the appropriate row $r$ we have

$$a_{r0} + a_{r1}\epsilon + a_{r2}\epsilon^2 + \cdots + a_{r,n}\epsilon^n = 0$$

which can be true if and only if $a_{rj} = 0$, for all $j = 0, 1, \ldots, n$. In other words, one row is identically zero, and therefore the problem matrix is not of full rank, contrary to the hypothesis that a basic feasible solution exists, and thereby proving the theorem.

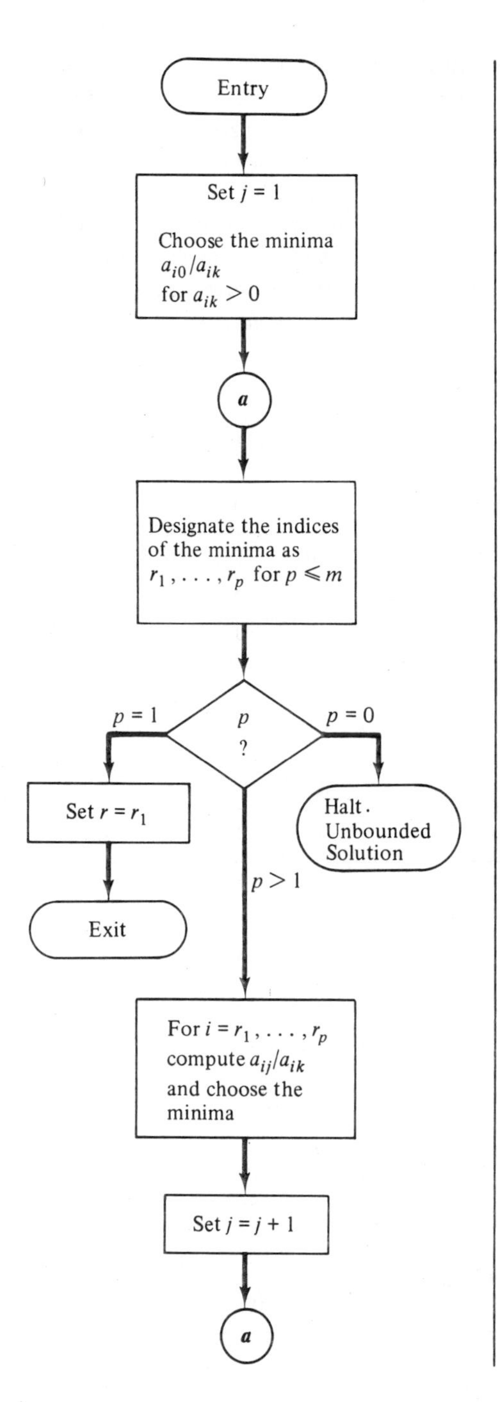


Entry corresponds to ② on page 60.

$k$ is the index of the incoming variable. Choose all minimum ratios of $a_{i0}$ to $a_{ik}$ for positive $a_{ik}$

How many minima are there?

If zero, halt; there is no finite optimum.

If it is unique, the designated index corresponds to the row whose basic variable is to leave the basis.

Exit corresponds to ③ on page 60.

(For the previously tied rows, choose the minimum ratios for the $j^{th}$ term of the polynomial to $a_{ik}$.)

**Figure 3.5**

Selection of the Outgoing Basic Variable Using the Perturbation Method

It might appear that the perturbation method is cumbersome because of the use of the polynomials in $\epsilon$. However, the polynomials need not be used directly; the coefficients in the matrix can be used instead. The flow chart in Figure 3.5 (which is to be superimposed upon the flow chart of Fig. 3.3) demonstrates how selection of the outgoing basic variable is achieved. Note that Theorem 3.6 guarantees that an exit from the routine of Figure 3.5 will be achieved in every case. The problem that was shown to cycle in Figure 3.4 has been solved using the perturbation method, and is shown in Figure 3.6. The optimal solution, reached after five iterations, is given in Tableau 6.

**Tableau 1**

| | | ↓ $x_1$ | $x_2$ | $x_3$ | $x_4$ | $x_5$ | $x_6$ | $x_7$ | |
|---|---|---|---|---|---|---|---|---|---|
| $z$ | 0 | $-\frac{3}{4}$ | 150 | $-\frac{1}{50}$ | 6 | 0 | 0 | 0 | |
| $x_5$ | 0 | $\boxed{\frac{1}{4}}$ | $-60$ | $-\frac{1}{25}$ | 9 | 1 | 0 | 0 | → |
| $x_6$ | 0 | $\frac{1}{2}$ | $-90$ | $-\frac{1}{50}$ | 3 | 0 | 1 | 0 | |
| $x_7$ | 1 | 0 | 0 | 1 | 0 | 0 | 0 | 1 | |

**Tableau 2**

| | | $x_1$ | ↓ $x_2$ | $x_3$ | $x_4$ | $x_5$ | $x_6$ | $x_7$ | |
|---|---|---|---|---|---|---|---|---|---|
| $z$ | 0 | 0 | $-30$ | $-\frac{7}{50}$ | 33 | 3 | 0 | 0 | |
| $x_1$ | 0 | 1 | $-240$ | $-\frac{4}{25}$ | 36 | 4 | 0 | 0 | |
| $x_6$ | 0 | 0 | $\boxed{30}$ | $\frac{3}{50}$ | $-15$ | $-2$ | 1 | 0 | → |
| $x_7$ | 1 | 0 | 0 | 1 | 0 | 0 | 0 | 1 | |

**Tableau 3**

| | | $x_1$ | $x_2$ | ↓ $x_3$ | $x_4$ | $x_5$ | $x_6$ | $x_7$ | |
|---|---|---|---|---|---|---|---|---|---|
| $z$ | 0 | 0 | 0 | $-\frac{2}{25}$ | 18 | 1 | 1 | 0 | |
| $x_1$ | 0 | 1 | 0 | $\frac{8}{25}$ | $-84$ | $-12$ | 8 | 0 | |
| $x_2$ | 0 | 0 | 1 | $\boxed{\frac{1}{500}}$ | $-\frac{1}{2}$ | $-\frac{1}{15}$ | $\frac{1}{30}$ | 0 | → |
| $x_7$ | 1 | 0 | 0 | 1 | 0 | 0 | 0 | 1 | |

**Figure 3.6**

Solution of the Problems of Figure 3.4 by Means of the Perturbation Method

**Tableau 4**

| | | $x_1$ | $x_2$ | $x_3$ | $x_4$ ↓ | $x_5$ | $x_6$ | $x_7$ | |
|---|---|---|---|---|---|---|---|---|---|
| $z$ | 0 | 0 | 40 | 0 | $-2$ | $-\frac{5}{3}$ | $\frac{7}{3}$ | 0 | |
| $x_1$ | 0 | 1 | $-160$ | 0 | $-4$ | $-\frac{4}{3}$ | $-4$ | 0 | |
| $x_3$ | 0 | 0 | 500 | 1 | $-250$ | $-\frac{100}{3}$ | $\frac{50}{3}$ | 0 | |
| $x_7$ | 1 | 0 | $-500$ | 0 | $\boxed{250}$ | $\frac{100}{3}$ | $-\frac{50}{3}$ | 1 | → |

**Tableau 5**

| | | $x_1$ | $x_2$ | $x_3$ | $x_4$ | $x_5$ ↓ | $x_6$ | $x_7$ | |
|---|---|---|---|---|---|---|---|---|---|
| $z$ | $\frac{1}{125}$ | 0 | 36 | 0 | 0 | $-\frac{7}{5}$ | $\frac{11}{5}$ | $\frac{1}{125}$ | |
| $x_1$ | $\frac{2}{125}$ | 1 | $-168$ | 0 | 0 | $-\frac{4}{5}$ | $\frac{12}{5}$ | $\frac{2}{125}$ | |
| $x_3$ | 1 | 0 | 0 | 1 | 0 | 0 | 0 | 1 | |
| $x_4$ | $\frac{1}{250}$ | 0 | $-2$ | 0 | 1 | $\boxed{\frac{2}{5}}$ | $-\frac{1}{15}$ | $\frac{1}{250}$ | → |

**Tableau 6**

| | | $x_1$ | $x_2$ | $x_3$ | $x_4$ | $x_5$ | $x_6$ | $x_7$ |
|---|---|---|---|---|---|---|---|---|
| $z$ | $\frac{1}{20}$ | 0 | 15 | 0 | $\frac{21}{2}$ | 0 | $\frac{3}{2}$ | $\frac{1}{20}$ |
| $x_1$ | $\frac{1}{25}$ | 1 | $-180$ | 0 | 6 | 0 | 2 | $\frac{1}{25}$ |
| $x_3$ | 1 | 0 | 0 | 1 | 0 | 0 | 0 | 1 |
| $x_5$ | $\frac{3}{100}$ | 0 | $-15$ | 0 | $\frac{15}{2}$ | 1 | $-\frac{1}{2}$ | $\frac{3}{100}$ |

**Figure 3.6 (concluded)**

By way of explanation, consider the first iteration, in which $x_1$ is to enter the basis. There is a tie between $x_5$ and $x_6$ in determining which variable is to leave the basis.

$$x_5: \quad 0/.25 = 0$$
$$x_6: \quad 0/.5 = 0$$

Hence the first column is tested for each tied row, and again there is a tie:

$$x_5: \quad .25/.25 = 1$$
$$x_6: \quad .5/.5 = 1$$

(The incoming column will always exhibit a tie and would normally not be considered.) Next the second column is tested:

$$x_5: \quad -60/.25 = -240$$
$$x_6: \quad -90/.5 = -180$$

Since the ratio is smaller for $x_5$, $x_5$ is selected to leave the basis. Other ties are resolved in the same way.

### A Geometric Interpretation of Degeneracy, Cycling, and the Resolution of Degeneracy

By reference to expression (3.1), a basic feasible solution corresponds to $n - m$ nonbasic variables being set equal to zero. For a nondegenerate solution, the other $m$ basic variables are strictly positive. In the case of a degenerate solution, some of the $m$ basic variables are zero. Geometrically, more than $n - m$ hyperplanes, corresponding to inequality constraints, intersect at the same point. A simple example of this can be seen in Figure 3.7, in which a two explicit constraint problem (given by expression [3.12]) is graphed; the shaded area is feasible.

$$\begin{aligned} 2x_1 + 3x_2 &\leq 20 \\ 3x_1 + 2x_2 &\leq 20 \\ x_1, x_2 &\geq 0 \end{aligned} \tag{3.12}$$

Consider point $p$, $x_1 = 4$, $x_2 = 4$. In Figure 3.7(b) the same problem has an additional constraint which does not affect the feasible solution space except to make point $p$ degenerate, namely

$$x_1 + x_2 \leq 8 \tag{3.13}$$

Designating the slacks on the constraints as follows:

$$\begin{aligned} 2x_1 + 3x_2 + x_3 \qquad\qquad &= 20 \\ 3x_1 + 2x_2 \qquad + x_4 \qquad &= 20 \\ x_1 + x_2 \qquad\qquad + x_5 &= 8 \end{aligned} \tag{3.14}$$

it can be seen that three distinct bases of equations (3.14) correspond to point $p$:

1. $x_1$, $x_2$, and $x_3$ as basic variables;
2. $x_1$, $x_2$, and $x_4$ as basic variables; and
3. $x_1$, $x_2$, and $x_5$ as basic variables.

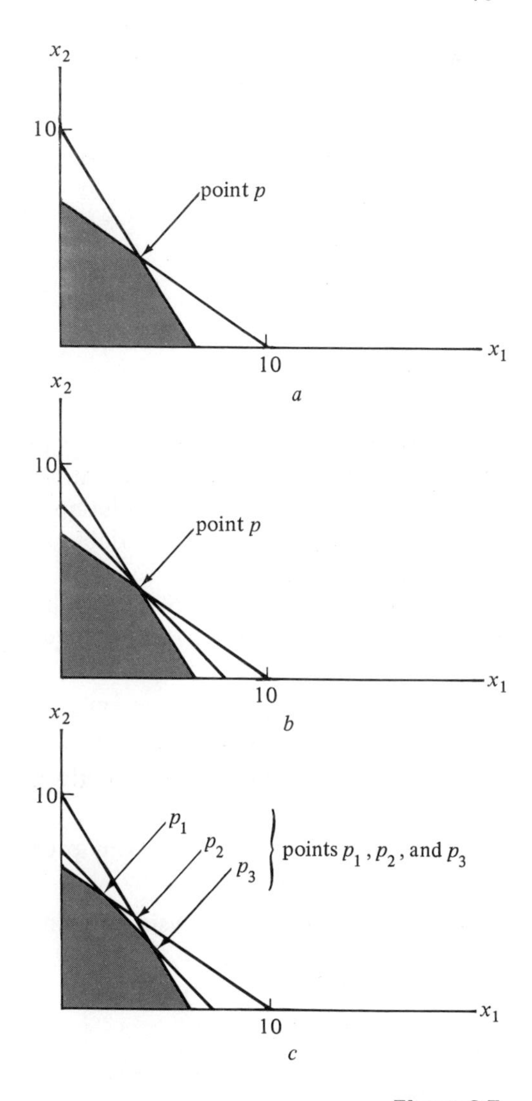


**Figure 3.7**

Graphs of a Degenerate Condition Resolved by Using the Perturbation Method

Since all three bases correspond to the same solution point, the possibility of cycling exists. The perturbation method has the effect of moving the constraints an infinitesimal amount. An exaggerated example of this is shown in Figure 3.7(c) where, instead of point $p$, there are three distinct points: $p_1$, $p_2$, and $p_3$. Only $p_1$ and $p_3$ are feasible. Since they are all distinct, the problem is no longer degenerate.

**Selected Supplemental References**

Section 3.3
[67]
Section 3.5
[20], [46], [155], [206], [274]
General
[286]

## 3.6 PROBLEMS

**1.** The simplex method criterion for selecting the variable to enter the basis is to select the variable with most negative $a_{0j}$.

Is it possible to obtain the solution by:

a. selecting the entering variable at random from among those having $a_{0j} < 0$?

b. selecting as the entering variable one having the largest of the strictly negative $a_{0j}$? (i.e., the negative entry which is smallest in absolute value.)

**2.** Prove the corollary to Theorem 3.5.

**3.** Develop the changes to the simplex method to omit the identity matrix.

**4.** Why does $a_{ik}$ have to be strictly greater than 0 in determining which variable should leave the basis?

**5.** Solve the Pasty Pizza Company problem of Chapter 1 (problem 10) using the simplex method.

Solve problems 6 through 9 using the simplex method.

**6.**

$$\begin{aligned}
\text{Maximize } z &= 5x_1 + 3x_2 + x_3 - x_4 \\
\text{subject to: } & 3x_1 + x_2 + 2x_3 - x_4 \leq 9 \\
& 4x_1 + 2x_2 + x_3 + 2x_4 \leq 6 \\
& x_1 + x_2 - x_3 + x_4 \geq -8 \\
& x_j \geq 0 \; (j = 1, 2, 3, 4)
\end{aligned}$$

**7.**

$$\begin{aligned}
\text{Maximize } z &= 200x_1 + 240x_2 \\
\text{subject to: } & 1.4x_1 + x_2 \leq 3360 \\
& 0.5x_1 + x_2 \leq 1500 \\
& x_1 \leq 2200 \\
& x_2 \leq 1450 \\
& x_1, x_2 \geq 0
\end{aligned}$$

**8.** Maximize $z = 3x_1 + x_2 + 5x_3 + 2x_4$

subject to:

$$\begin{aligned} x_1 + 2x_2 + 3x_3 - x_4 &\leq 10 \\ 2x_1 + 3x_2 - x_3 + x_4 &\leq 7 \\ x_1 + x_2 + 3x_3 + 2x_4 &\leq 8 \\ x_j \geq 0 \ (j = 1, 2, 3, 4) \end{aligned}$$

**9.** Maximize $z = 2x_1 + 0.5x_2 + x_3$

subject to:

$$\begin{aligned} x_1 + x_4 - 2x_5 + 3x_6 &= 6 \\ x_2 - x_4 + 4x_5 + 5x_6 &= 5 \\ x_3 + 2x_4 - x_5 + x_6 &= 3 \\ x_j &\geq 0 \\ (j = 1, 2, 3, 4, 5, 6) \end{aligned}$$

**10.** G.L. Thompson showed that a problem could be made much more difficult to solve by altering the constraint set in a trivial way. If the constraint set is $\boldsymbol{Ax} \leq \boldsymbol{b}$, then replace the constraint set by the following:

$$\begin{aligned} \boldsymbol{Ax} - \boldsymbol{b}y &\leq \boldsymbol{0} \\ y \leq 1, \ y &\geq 0 \end{aligned}$$

where $y$ is a scalar. Show that Beale's problem which cycles (section 3.5) is of this form. When put into the simpler form, the solution of the problem has no associated difficulties.

a. Write and solve the simpler form of Beale's problem.

b. Can you devise in the above manner another problem which cycles?

**11.** Solve problems 1, 2, 3, 4, 5, 9, and 13 of Chapter 1 using the simplex method.

**12.** Solve problems 12, 13, 14, 17, and 19 of Chapter 2 using the simplex method.

**13.** Explain the simplex method in the requirements space using the dog food example and Figure 2.6.

# 4

# Implementing the Simplex Method

There are a number of aspects of the simplex method that we have not yet fully resolved. In addition, there are some important devices that are fundamental to the simplex method, and there are also extensions and some important variations of the simplex method itself. In this chapter we shall consider certain topics in the above categories, and thereby proceed to extend the development of the simplex method.

## 4.1 OBTAINING A STARTING BASIC FEASIBLE SOLUTION TO A LINEAR PROGRAMMING PROBLEM—THE USE OF ARTIFICIAL VARIABLES

In every problem we have considered thus far there has always been a basic feasible solution available at the outset. Although we have started each problem's solution from a canonical form, it is sufficient to know only the basic variables corresponding to a basic feasible solution and then develop the canonical form using row operations. There do exist problems for which no basic feasible solution is known a priori. Consequently, a procedure is needed to obtain a basic feasible solution in all cases.

Three methods for finding basic feasible solutions are presented; they encompass most of the schemes used for this purpose. (The Lemke Start, which is described in Charnes and Cooper [47], p. 482, is one of a number of other methods that have been developed, but, to the author's knowledge, have not been used much in practice.) Most of the methods employ what are known as artificial variables. (See Chapter 6 for a method which does not use artificial variables but which is not generally applicable to the simplex method.)

**Definition 4.1**

*An artificial variable is a variable which has been added to one or more constraints of the problem in order to obtain a starting basic solution, and has no real meaning with respect to the original problem.*

Artificial variables are added to equality constraints and inequality constraints of the greater than or equal to ($\geq$) variety. The artificial variables serve the same purpose slack variables have served in our example problems. They give a convenient starting basis. Because they are added for convenience and without regard to the statement of the problem, artificial variables must be zero in an optimal solution in order that the solution be meaningful in the original problem context.

The artificial variables are added in such a manner that a basic feasible solution to the problem including the artificial variables (which we shall call the augmented problem) is immediately available. Then a procedure is followed that either finds a basic feasible solution to the original problem or generates evidence that such a solution does not exist. The use of artificial variables in this manner will be demonstrated when the particular methods are developed.

### The Big-M Method

The first method that we shall discuss is called the Big-$M$ method and was developed by Charnes and Cooper ([47] p. 176). The procedure for using this method begins with writing all the constraints so that the $b_i$ entries are nonnegative, multiplying any that are not by $-1$. *Remember that multiplying an inequality by a negative number reverses the direction of the inequality.* Then add slack variables to the inequalities. To every equality (to which no slack was added) and to every inequality to which a negative slack variable was added (that is, an inequality of the form $\Sigma a_{ij}x_j \geq b_i$ which became an equality) add an artificial variable with a coefficient of $+1$. The coefficient of each artificial variable in the objective function is $-M$, where $M$ is a sufficiently large positive number. The effect is to make the artificial variables sufficiently unattractive to be in any optimal solution, providing, of course, that a feasible solution to the original problem exists. Row operations must then be utilized in order to obtain a complete basis representation (i.e., to reduce the entries of the artificial variables in the objective function to zero). Then the simplex method is used until an optimal feasible solution is achieved. An artificial variable that leaves the basis at any stage may be dropped, since it has served its purpose and is no longer needed. If all artificial variables leave the basis, then a basic feasible solution to the original problem has been found.

However, it is possible that an optimal solution may be found before all artificial variables have been dropped. Then, provided that the optimal solution contains *any* artificial variables at *positive* levels, no feasible solution to the original problem exists. This can be proven by noting that any particular feasible solution will have a finite objective function. In contrast, an optimal solution to the augmented problem with a positive artificial variable will have an objective function value that can be made infinitely negative, since $M$ can be made arbitrarily large.

It is possible for artificial variables to be basic at zero levels in an optimal solution. This means that some equality constraint has been stated more than once (and of course could have been dropped). Provided that all artificial variables are zero, the optimal solution is optimal with respect to the original problem. The following is an example of the use of the Big-$M$ method.

EXAMPLE 4.1:

$$\begin{aligned}
\text{Maximize } z &= 2x_1 + 3x_2 \\
\text{subject to:} \quad & x_1 + x_2 \leq 4 \\
& -x_1 + x_2 - x_3 \leq -2 \\
& x_1 \geq 1 \\
& x_1 + x_2 + x_3 = 5 \\
& x_1, x_2, x_3 \geq 0
\end{aligned}$$

Multiply all negative $b_i$ constraints by $-1$ and add slack variables.

$$\begin{aligned}
\text{Maximize } z &= 2x_1 + 3x_2 \\
\text{subject to:} \quad & x_1 + x_2 + x_4 = 4 \\
& x_1 - x_2 + x_3 - x_5 = 2 \\
& x_1 - x_6 = 1 \\
& x_1 + x_2 + x_3 = 5
\end{aligned}$$

The last three equations require artificial variables. We shall call those variables $x_a$, $x_b$, and $x_c$, respectively.

$$\begin{aligned}
\text{Maximize } z = & \\
& 2x_1 + 3x_2 - Mx_a - Mx_b - Mx_c \\
\text{subject to:} \quad & x_1 + x_2 + x_4 = 4 \\
& x_1 - x_2 + x_3 - x_5 + x_a = 2 \\
& x_1 - x_6 + x_b = 1 \\
& x_1 + x_2 + x_3 + x_c = 5
\end{aligned}$$

Putting the problem into the simplex tableau format yields Tableau 0.

**Tableau 0**

| | | $x_1$ | $x_2$ | $x_3$ | $x_4$ | $x_5$ | $x_6$ | $x_a$ | $x_b$ | $x_c$ |
|---|---|---|---|---|---|---|---|---|---|---|
| $z$ | 0 | −2 | −3 | 0 | 0 | 0 | 0 | $M$ | $M$ | $M$ |
| $x_4$ | 4 | 1 | 1 | 0 | 1 | 0 | 0 | 0 | 0 | 0 |
| $x_a$ | 2 | 1 | −1 | 1 | 0 | −1 | 0 | [1] | 0 | 0 |
| $x_b$ | 1 | 1 | 0 | 0 | 0 | 0 | −1 | 0 | [1] | 0 |
| $x_c$ | 5 | 1 | 1 | 1 | 0 | 0 | 0 | 0 | 0 | [1] |

Performing row operations on the dashed pivotal elements to complete the canonical form yields the starting tableau, Tableau 1. Instead of setting $M$ to some value, we shall leave it as $M$, but shall consider it to be 1000 for computational purposes. If that is not large enough, we can make it larger subsequently.

**Tableau 1**

| | | $x_1$ ↓ | $x_2$ | $x_3$ | $x_4$ | $x_5$ | $x_6$ | $x_a$ | $x_b$ | $x_c$ | |
|---|---|---|---|---|---|---|---|---|---|---|---|
| $z$ | $-8M$ | $-3M-2$ | −3 | $-2M$ | 0 | $M$ | $M$ | 0 | 0 | 0 | |
| $x_4$ | 4 | 1 | 1 | 0 | 1 | 0 | 0 | 0 | 0 | 0 | |
| $x_a$ | 2 | 1 | −1 | 1 | 0 | −1 | 0 | 1 | 0 | 0 | |
| $x_b$ | 1 | [1] | 0 | 0 | 0 | 0 | −1 | 0 | 1 | 0 | → |
| $x_c$ | 5 | 1 | 1 | 1 | 0 | 0 | 0 | 0 | 0 | 1 | |

Introducing $x_1$ to replace $x_b$ yields Tableau 2.

**Tableau 2**

| | | $x_1$ | $x_2$ | $x_3$ | $x_4$ | $x_5$ | $x_6$ ↓ | $x_a$ | $x_b$ | $x_c$ | |
|---|---|---|---|---|---|---|---|---|---|---|---|
| $z$ | $-5M+2$ | 0 | −3 | $-2M$ | 0 | $M$ | $-2M-2$ | 0 | $3M+2$ | 0 | |
| $x_4$ | 3 | 0 | 1 | 0 | 1 | 0 | 1 | 0 | −1 | 0 | |
| $x_a$ | 1 | 0 | −1 | 1 | 0 | −1 | [1] | 1 | −1 | 0 | → |
| $x_1$ | 1 | 1 | 0 | 0 | 0 | 0 | −1 | 0 | 1 | 0 | |
| $x_c$ | 4 | 0 | 1 | 1 | 0 | 0 | 1 | 0 | −1 | 1 | |

Dropping $x_b$, since it is no longer needed, we introduce $x_6$ to replace $x_a$, which yields Tableau 3.

**Tableau 3**

| | | $x_1$ | $x_2$ ↓ | $x_3$ | $x_4$ | $x_5$ | $x_6$ | $x_a$ | $x_c$ | |
|---|---|---|---|---|---|---|---|---|---|---|
| $z$ | $-3M+4$ | 0 | $-2M-5$ | 2 | 0 | $-M-2$ | 0 | $2M+2$ | 0 | |
| $x_4$ | 2 | 0 | [2] | $-1$ | 1 | 1 | 0 | $-1$ | 0 | → |
| $x_6$ | 1 | 0 | $-1$ | 1 | 0 | $-1$ | 1 | 1 | 0 | |
| $x_1$ | 2 | 1 | $-1$ | 1 | 0 | $-1$ | 0 | 1 | 0 | |
| $x_c$ | 3 | 0 | 2 | 0 | 0 | 1 | 0 | $-1$ | 1 | |

Dropping $x_a$, since it is no longer needed, we introduce $x_2$ in place of $x_4$, which yields Tableau 4.

**Tableau 4**

| | | $x_1$ | $x_2$ | $x_3$ ↓ | $x_4$ | $x_5$ | $x_6$ | $x_c$ | |
|---|---|---|---|---|---|---|---|---|---|
| $z$ | $-M+9$ | 0 | 0 | $-M-\frac{1}{2}$ | $-M+\frac{5}{2}$ | $\frac{1}{2}$ | 0 | 0 | |
| $x_2$ | 1 | 0 | 1 | $-\frac{1}{2}$ | $\frac{1}{2}$ | $\frac{1}{2}$ | 0 | 0 | |
| $x_6$ | 2 | 0 | 0 | $\frac{1}{2}$ | $\frac{1}{2}$ | $-\frac{1}{2}$ | 1 | 0 | |
| $x_1$ | 3 | 1 | 0 | $\frac{1}{2}$ | $\frac{1}{2}$ | $-\frac{1}{2}$ | 0 | 0 | |
| $x_c$ | 1 | 0 | 0 | [1] | $-1$ | 0 | 0 | 1 | → |

Introducing $x_3$ to replace $x_c$ yields Tableau 5.

**Tableau 5**

| | | $x_1$ | $x_2$ | $x_3$ | $x_4$ | $x_5$ | $x_6$ | $x_c$ |
|---|---|---|---|---|---|---|---|---|
| $z$ | 9.5 | 0 | 0 | 0 | 2 | $\frac{1}{2}$ | 0 | $M+\frac{1}{2}$ |
| $x_2$ | $\frac{3}{2}$ | 0 | 1 | 0 | 0 | $\frac{1}{2}$ | 0 | $\frac{1}{2}$ |
| $x_6$ | $\frac{3}{2}$ | 0 | 0 | 0 | 1 | $-\frac{1}{2}$ | 1 | $-\frac{1}{2}$ |
| $x_1$ | $\frac{5}{2}$ | 1 | 0 | 0 | 1 | $-\frac{1}{2}$ | 0 | $-\frac{1}{2}$ |
| $x_3$ | 1 | 0 | 0 | 1 | $-1$ | 0 | 0 | 1 |

At this point $x_c$ could be dropped, since it has left the basis. A basic feasible solution has been found which, for this example, also happens to be optimal.

A second example is now presented.

EXAMPLE 4.2:

$$\begin{aligned} \text{Maximize } z &= -x_1 - 2x_2 \\ \text{subject to:} \quad & 2x_1 + 3x_2 \geq 3 \\ & x_1, x_2 \geq 0 \end{aligned}$$

Adding a slack variable, $x_3$, and an artificial variable, $x_a$, we have the sequence of iterations given by the following tableaus:

**Tableau 0**

| | | $x_1$ | $x_2$ | $x_3$ | $x_a$ |
|---|---|---|---|---|---|
| $z$ | 0 | 1 | 2 | 0 | $M$ |
| $x_a$ | 3 | 2 | 3 | $-1$ | [1] |

**Tableau 1**

| | | $x_1$ | $x_2$ ↓ | $x_3$ | $x_a$ | |
|---|---|---|---|---|---|---|
| $z$ | $-3M$ | $-2M+1$ | $-3M+2$ | $M$ | 0 | |
| $x_a$ | 3 | 2 | [3] | $-1$ | 1 | → |

**Tableau 2**

| | | $x_1$ ↓ | $x_2$ | $x_3$ | $x_a$ | |
|---|---|---|---|---|---|---|
| $z$ | $-2$ | $-\frac{1}{3}$ | 0 | $\frac{2}{3}$ | $M-\frac{2}{3}$ | ($x_a$ is being dropped.) |
| $x_2$ | 1 | [$\frac{2}{3}$] | 1 | $-\frac{1}{3}$ | $\frac{1}{3}$ | → |

**Tableau 3**

| | | $x_1$ | $x_2$ | $x_3$ | |
|---|---|---|---|---|---|
| $z$ | $-\frac{3}{2}$ | 0 | $\frac{1}{2}$ | $\frac{1}{2}$ | *Optimal Solution* |
| $x_1$ | $\frac{3}{2}$ | 1 | $\frac{3}{2}$ | $-\frac{1}{2}$ | |

The initial solution is $x_1 = x_2 = 0$, which is not feasible. After a sequence of iterations—for this example only one is necessary—the artificial variable has left the basis, and the feasible solution $x_2 = 1$ is obtained (Tableau 2). Then $x_a$ is dropped and after one more iteration the optimal solution, as indicated in Tableau 3, is obtained. It may be useful for the reader to plot the solution graphically.

### The Two-Phase Method

For the Big-$M$ method let $M$ be so large that it is essentially infinite. The method thereby obtained is the two-phase method. The effect of such a large $M$ is that attention is initially focused *only* on removing artificials from the basis. (Smaller values of $M$ blend the objectives of optimality and feasibility.)

Professor G.B. Dantzig [62] proposed that $M$ may be treated as essentially infinite, without being assigned some real value. To do this, the two-phase method has two objective functions:

1. For phase 1, to maximize the negative of the sum of the artificial variables (or, equivalently, to minimize the sum of the artificials). The phase 1 objective is maximized before proceeding to phase 2.
2. For phase 2, to maximize the objective function of the original problem.

To simplify matters, the two objective functions are maintained in the tableau during the first phase; the first row corresponds to the phase 1 objective function, and the second row corresponds to the phase 2 or original objective function. During phase 1, the phase 2 objective function is simply updated. It is not used for determining which variable should enter the basis, nor is it considered in determining the variable to leave the basis. When the phase 1 objective function is maximized, i.e., when its value equals zero, the first row of the tableau is dropped and calculations continue, using the phase 2 objective function. An example follows.

EXAMPLE 4.3:

$$\begin{aligned} \text{Maximize } z = {} & x_1 + x_2 \\ \text{subject to:} \quad & 3x_1 + 2x_2 \le 20 \\ & 2x_1 + 3x_2 \le 20 \\ & x_1 + 2x_2 \ge 2 \\ & x_1, x_2 \ge 0 \end{aligned}$$

Adding slack and artificial variables, we have

$$\begin{aligned} &\text{Phase 1: Maximize } z_1 = && -x_a \\ &\text{Phase 2: Maximize } z_2 = x_1 + x_2 \\ &\qquad \text{subject to:} \quad 3x_1 + 2x_2 + x_3 && = 20 \\ &\qquad\qquad 2x_1 + 3x_2 \quad + x_4 && = 20 \\ &\qquad\qquad x_1 + 2x_2 \quad - x_5 + x_a && = 2 \end{aligned}$$

Putting the problem into tableau form yields Tableau 0.

**Tableau 0**

| | | $x_1$ | $x_2$ | $x_3$ | $x_4$ | $x_5$ | $x_a$ |
|---|---|---|---|---|---|---|---|
| $z$(phase 1) | 0 | 0 | 0 | 0 | 0 | 0 | 1 |
| $z$(phase 2) | 0 | −1 | −1 | 0 | 0 | 0 | 0 |
| $x_3$ | 20 | 3 | 2 | 1 | 0 | 0 | 0 |
| $x_4$ | 20 | 2 | 3 | 0 | 1 | 0 | 0 |
| $x_a$ | 2 | 1 | 2 | 0 | 0 | −1 | [1] |

The basic representation must be made complete; hence, the partial iteration on the dashed pivotal element must be undertaken to yield Tableau 1.

**Tableau 1**

| | | $x_1$ | $x_2$ ↓ | $x_3$ | $x_4$ | $x_5$ | $x_a$ | |
|---|---|---|---|---|---|---|---|---|
| $z$(phase 1) | −2 | −1 | −2 | 0 | 0 | 1 | 0 | |
| $z$(phase 2) | 0 | −1 | −1 | 0 | 0 | 0 | 0 | |
| $x_3$ | 20 | 3 | 2 | 1 | 0 | 0 | 0 | |
| $x_4$ | 20 | 2 | 3 | 0 | 1 | 0 | 0 | |
| $x_a$ | 2 | 1 | [2] | 0 | 0 | −1 | 1 | → |

Using the phase 1 objective function, $x_2$ is selected to enter the basis, replacing $x_a$ and yielding Tableau 2.

**Tableau 2**

| | | $x_1$ | $x_2$ | $x_3$ | $x_4$ | $x_5$ ↓ | $x_a$ | |
|---|---|---|---|---|---|---|---|---|
| $z$(phase 1) | 0 | 0 | 0 | 0 | 0 | 0 | 1 | |
| $z$(phase 2) | 1 | $-\frac{1}{2}$ | 0 | 0 | 0 | $-\frac{1}{2}$ | $\frac{1}{2}$ | |
| $x_3$ | 18 | 2 | 0 | 1 | 0 | 1 | −1 | |
| $x_4$ | 17 | $\frac{1}{2}$ | 0 | 0 | 1 | [$\frac{3}{2}$] | $-\frac{3}{2}$ | → |
| $x_2$ | 1 | $\frac{1}{2}$ | 1 | 0 | 0 | $-\frac{1}{2}$ | $\frac{1}{2}$ | |

Variable $x_a$ can now be dropped. Since the phase 1 objective function is zero, we can also drop the phase 1 objective function and proceed to phase 2. Introducing $x_5$ (choosing $x_5$ arbitrarily instead of $x_1$) into the basis to replace $x_4$, and then introducing $x_1$ into the basis to replace $x_3$, leads to the following tableaus.

**Tableau 3**

| | | $x_1$ ↓ | $x_2$ | $x_3$ | $x_4$ | $x_5$ | |
|---|---|---|---|---|---|---|---|
| $z$(phase 2) | $\frac{20}{3}$ | $-\frac{1}{3}$ | 0 | 0 | $\frac{1}{3}$ | 0 | |
| $x_3$ | $\frac{20}{3}$ | $\boxed{\frac{5}{3}}$ | 0 | 1 | $-\frac{2}{3}$ | 0 | ⟶ |
| $x_5$ | $\frac{34}{3}$ | $\frac{1}{3}$ | 0 | 0 | $\frac{2}{3}$ | 1 | |
| $x_2$ | $\frac{20}{3}$ | $\frac{2}{3}$ | 1 | 0 | $\frac{1}{3}$ | 0 | |

**Tableau 4**

| | | $x_1$ | $x_2$ | $x_3$ | $x_4$ | $x_5$ |
|---|---|---|---|---|---|---|
| $z$(phase 2) | 8 | 0 | 0 | $\frac{1}{5}$ | $\frac{1}{5}$ | 0 |
| $x_1$ | 4 | 1 | 0 | $\frac{3}{5}$ | $-\frac{2}{5}$ | 0 |
| $x_5$ | 10 | 0 | 0 | $-\frac{1}{5}$ | $\frac{4}{5}$ | 1 |
| $x_2$ | 4 | 0 | 1 | $-\frac{2}{5}$ | $\frac{3}{5}$ | 0 |

Tableau 4 is optimal.

### An Improved Method for Assigning Artificial Variables to Inequalities

This method may be used in conjunction with either of the above methods. It reduces the number of artificial variables required when there is more than one constraint of the form

$$\sum a_{ij}x_j \geq b_i \quad (b_i > 0) \tag{4.1}$$

The reduction is accomplished as follows:

1. Add artificial variables, one per constraint, to all *equality* constraints.
2. Add slack variables to all inequalities.
3. Use one additional artificial variable, adding that variable to every constraint originally of the form (4.1). Of the constraints originally of the form (4.1), choose one with $b_i$ maximum. Designate the artificial variable as basic in that constraint.
4. The starting basis consists of artificial variables for equality constraints (one for each constraint) plus one additional artificial variable basic in one

of the constraints of the form (4.1) as selected in step 3. All other basic variables are slack variables.

The constraints of the form (4.1) in which slacks are basic must be multiplied by $-1$ to obtain the starting tableau. The procedure is illustrated in the following example using the Big-$M$ method.

EXAMPLE 4.4:

Solve the problem of Example 4.1, first recognizing that the second and third constraints are of the form (4.1). Adding artificials $x_c$ and $x_d$ in the manner described yields the following problem:

$$\begin{aligned}
\text{Maximize } z = 2x_1 + 3x_2 \qquad\qquad\qquad & - Mx_d - Mx_c \\
\text{subject to:} \quad x_1 + x_2 + x_4 \qquad\qquad\qquad & = 4 \\
x_1 - x_2 + x_3 - x_5 + x_d \qquad & = 2 \\
x_1 - x_6 + x_d \qquad & = 1 \\
x_1 + x_2 + x_3 + x_c & = 5
\end{aligned}$$

Setting up the tableau yields Tableau 000. Variable $x_d$ is chosen to be basic in the second constraint because the right-hand element is the larger (of the right-hand elements of constraints two and three). Pivoting $x_d$ into the indicated row yields Tableau 00.

**Tableau 000**

| | | $x_1$ | $x_2$ | $x_3$ | $x_4$ | $x_5$ | $x_6$ | $x_d$ | $x_c$ |
|---|---|---|---|---|---|---|---|---|---|
| $z$ | 0 | $-2$ | $-3$ | 0 | 0 | 0 | 0 | $M$ | $M$ |
| $x_4$ | 4 | 1 | 1 | 0 | 1 | 0 | 0 | 0 | 0 |
| $x_d$ | 2 | 1 | $-1$ | 1 | 0 | $-1$ | 0 | [1] | 0 |
| $x_6$ | 1 | 1 | 0 | 0 | 0 | 0 | $-1$ | 1 | 0 |
| $x_c$ | 5 | 1 | 1 | 1 | 0 | 0 | 0 | 0 | 1 |

**Tableau 00**

| | | $x_1$ | $x_2$ | $x_3$ | $x_4$ | $x_5$ | $x_6$ | $x_d$ | $x_c$ |
|---|---|---|---|---|---|---|---|---|---|
| $z$ | $-2M$ | $-M-2$ | $M-3$ | $-M$ | 0 | $M$ | 0 | 0 | $M$ |
| $x_4$ | 4 | 1 | 1 | 0 | 1 | 0 | 0 | 0 | 0 |
| $x_d$ | 2 | 1 | $-1$ | 1 | 0 | $-1$ | 0 | 1 | 0 |
| $x_6$ | $-1$ | 0 | 1 | $-1$ | 0 | 1 | $-1$ | 0 | 0 |
| $x_c$ | 5 | 1 | 1 | 1 | 0 | 0 | 0 | 0 | [1] |

Pivoting $x_c$ where indicated yields Tableau 0.

**Tableau 0**

| | | $x_1$ | $x_2$ | $x_3$ | $x_4$ | $x_5$ | $x_6$ | $x_d$ | $x_c$ |
|---|---|---|---|---|---|---|---|---|---|
| $z$ | $-7M$ | $-2M-2$ | $-3$ | $-2M$ | 0 | $M$ | 0 | 0 | 0 |
| $x_4$ | 4 | 1 | 1 | 0 | 1 | 0 | 0 | 0 | 0 |
| $x_d$ | 2 | 1 | $-1$ | 1 | 0 | $-1$ | 0 | 1 | 0 |
| $x_6$ | $-1$ | 0 | 1 | $-1$ | 0 | 1 | $-1$ | 0 | 0 |
| $x_c$ | 5 | 1 | 1 | 1 | 0 | 0 | 0 | 0 | 1 |

Multiplying the constraint containing $x_6$ by $-1$ gives the starting basis in Tableau 1. The above steps can be taken in any order. They are not complete iterations.

**Tableau 1**

| | | $x_1$ ↓ | $x_2$ | $x_3$ | $x_4$ | $x_5$ | $x_6$ | $x_d$ | $x_c$ | |
|---|---|---|---|---|---|---|---|---|---|---|
| $z$ | $-7M$ | $-2M-2$ | $-3$ | $-2M$ | 0 | $M$ | 0 | 0 | 0 | |
| $x_4$ | 4 | 1 | 1 | 0 | 1 | 0 | 0 | 0 | 0 | |
| $x_d$ | 2 | [1] | $-1$ | 1 | 0 | $-1$ | 0 | 1 | 0 | → |
| $x_6$ | 1 | 0 | $-1$ | 1 | 0 | $-1$ | 1 | 0 | 0 | |
| $x_c$ | 5 | 1 | 1 | 1 | 0 | 0 | 0 | 0 | 1 | |

Now $x_1$ replaces $x_d$ in the basis of Tableau 1, yielding Tableau 2.

**Tableau 2**

| | | $x_1$ | $x_2$ | $x_3$ | $x_4$ | $x_5$ | $x_6$ | $x_d$ | $x_c$ |
|---|---|---|---|---|---|---|---|---|---|
| $z$ | $-3M+4$ | 0 | $-2M-5$ | 2 | 0 | $-M-2$ | 0 | $2M+2$ | 0 |
| $x_4$ | 2 | 0 | 2 | $-1$ | 1 | 1 | 0 | $-1$ | 0 |
| $x_1$ | 2 | 1 | $-1$ | 1 | 0 | $-1$ | 0 | 1 | 0 |
| $x_6$ | 1 | 0 | $-1$ | 1 | 0 | $-1$ | 1 | 0 | 0 |
| $x_c$ | 3 | 0 | 2 | 0 | 0 | 1 | 0 | $-1$ | 1 |

After dropping $x_d$ because it is no longer needed, Tableau 2 is the same (except for the second and third rows being reversed) as Tableau 3 of Example 4.1. The iterations continue exactly as in Example 4.1.

## 4.2 THE REVISED SIMPLEX METHOD

As shown in Chapter 3, an iteration in the simplex method consists of three steps:

1. Choose the most "profitable" nonbasic variable, that is, the one with the most negative element of $\boldsymbol{c}'_B\boldsymbol{B}^{-1}\boldsymbol{N} - \boldsymbol{c}'_N$, to enter the basis. If it is zero or greater the problem is solved, i.e., the optimal solution has been achieved. Otherwise, let the variable selected have the index $k$. Denote the associated column vector (of $\boldsymbol{N}$) as $\boldsymbol{a}_k$.
2. Choose the variable to leave the basis. In this step we consider the column of $\boldsymbol{N}$ which corresponds to the element selected in step 1. The transformed column of $\boldsymbol{N}$ is $\boldsymbol{B}^{-1}\boldsymbol{a}_k$. Compute the ratios of the elements of $\boldsymbol{B}^{-1}\boldsymbol{b}$ to the corresponding (positive) elements of $\boldsymbol{B}^{-1}\boldsymbol{a}_k$. Designate the row with the minimum ratio as row $\boldsymbol{r}$. If all coefficients of $\boldsymbol{B}^{-1}\boldsymbol{a}_k$ are negative or zero, then no finite optimal solution exists.
3. Perform the iteration, and compute the new tableau.

Although the simplex method develops an entire tableau of coefficients, essentially only $\boldsymbol{B}^{-1}$ plus the initial coefficients are needed in the simplex method. Using the simplex tableau, of course, all of the calculations made in the above three steps are part of the simplex method. The idea of the revised simplex method is to update only $\boldsymbol{B}^{-1}$ and a few other entries of the simplex tableau. Any other entries are developed using the expressions given above. We now explore this procedure.

In Chapter 3, we partitioned matrix $\boldsymbol{A}$ as $(\boldsymbol{B} \mid \boldsymbol{N})$, rearranging columns so that the basis is always first and the updated form (left-multiplying by $\boldsymbol{B}^{-1}$) is $(\boldsymbol{I} \mid \boldsymbol{B}^{-1}\boldsymbol{N})$. Consider now the initial partition $(\boldsymbol{G} \mid \boldsymbol{I})$, where the last $m$ columns correspond to the initial basis, whether artificial or not. Without rearranging columns, the updated form is $(\boldsymbol{B}^{-1}\boldsymbol{G} \mid \boldsymbol{B}^{-1})$. Since the preceding statement holds true for every tableau, it is possible to compute $\boldsymbol{B}^{-1}$ for each tableau by beginning with the initial $\boldsymbol{B}^{-1}$, which is $\boldsymbol{I}$, and updating only $\boldsymbol{B}^{-1}$ each iteration.

There are two forms of the revised simplex method:

1. The explicit form of the inverse.
2. The product form of the inverse.

Conceptually they are identical, but computationally they are not. We shall consider both methods.

#### Using the Explicit Form of the Inverse

The explicit-form revised simplex method maintains the inverse explicitly. Since we have already discussed the nature of the algorithm, we proceed directly to a flow chart of the method, given in Figure 4.1.

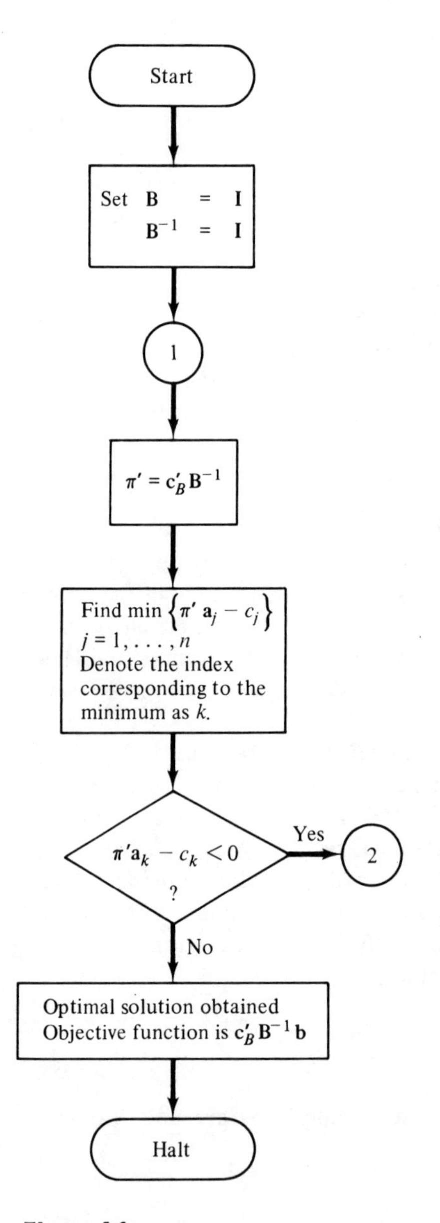


The starting basis is an identity matrix. We are assuming that a starting basic feasible solution is available. If not, the procedures of section 4.1 can be used in the same framework.

Compute updated "pricing vector" $\pi'$, which values a unit of each resource.

Find the minimum $\pi' \mathbf{a}_j - c_j$ (or $z_j - c_j$) where $\mathbf{a}_j$ is column $j$ in the original problem.

Is it negative?
(Solution not optimal)

Stop.
Optimal Solution.

**Figure 4.1**

Flow Chart of the Revised Simplex Method

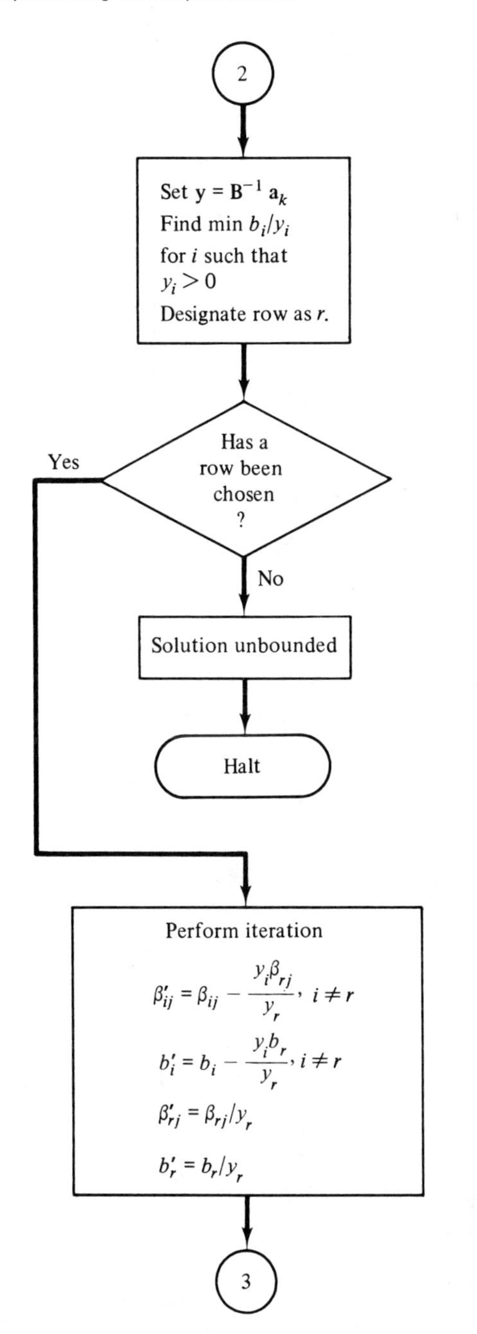


Compute the updated column of coefficients for the incoming variable. Determine which variable will leave the basis. (Initially, **b** is the right-hand side vector; subsequently, it is the updated right-hand side vector.)

Was at least one component of **y** positive?

No finite optimum exists

Halt

Update inverse and **b** vector. ($\beta_{ij}$ represents the elements of $\mathbf{B}^{-1}$.)

**Figure 4.1 (continued)**

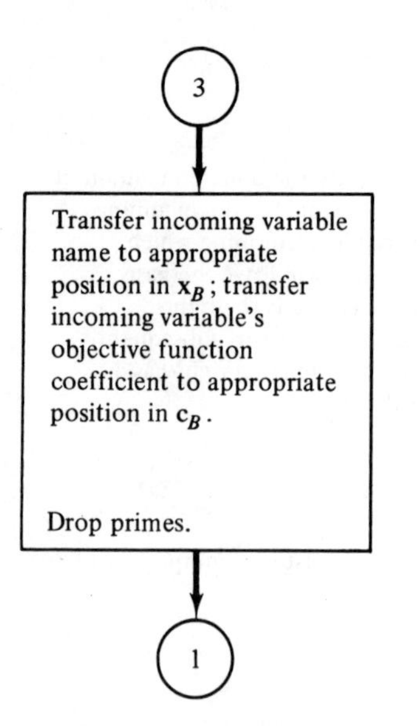


Prepare for evaluation phase of next iteration.

**Figure 4.1 (concluded)**

If no feasible solution is available at the outset, then one of the methods from section 4.1 should be employed in the framework of the revised simplex method, with relatively few alterations.

As an example of an application of the revised simplex method, we solve our dog food production problem.

EXAMPLE 4.5:

$$\begin{aligned} \text{Maximize } z &= .56x_1 + .42x_2 \\ \text{subject to: } \quad x_1 + 2x_2 + x_3 &= 240{,}000 \\ 1.5x_1 + x_2 + x_4 &= 180{,}000 \\ x_1 + x_5 &= 110{,}000 \\ x_j \geq 0, j &= 1, \ldots, 5 \end{aligned}$$

Then we may indicate the following vectors and matrices:

$$A = \begin{pmatrix} 1 & 2 & 1 & 0 & 0 \\ 1.5 & 1 & 0 & 1 & 0 \\ 1 & 0 & 0 & 0 & 1 \end{pmatrix} \quad b = \begin{pmatrix} 240{,}000 \\ 180{,}000 \\ 110{,}000 \end{pmatrix}$$

$$c' = (.56 \quad .42 \quad 0 \quad 0 \quad 0)$$

$$B = \begin{pmatrix} 1 & 0 & 0 \\ 0 & 1 & 0 \\ 0 & 0 & 1 \end{pmatrix} \quad x_B = \begin{pmatrix} x_3 \\ x_4 \\ x_5 \end{pmatrix} \quad c_B = \begin{pmatrix} 0 \\ 0 \\ 0 \end{pmatrix} \quad B^{-1} = \begin{pmatrix} 1 & 0 & 0 \\ 0 & 1 & 0 \\ 0 & 0 & 1 \end{pmatrix}$$

*Iteration 1:* Compute

$$\boldsymbol{\pi}' = \boldsymbol{c}_B'\boldsymbol{B}^{-1} = (0 \quad 0 \quad 0)\begin{pmatrix} 1 & 0 & 0 \\ 0 & 1 & 0 \\ 0 & 0 & 1 \end{pmatrix} = (0 \quad 0 \quad 0)$$

We need only evaluate nonbasic variables:

$$\text{for } x_1\text{: } z_1 - c_1 = \boldsymbol{\pi}'\boldsymbol{a}_1 - c_1 = (0 \quad 0 \quad 0)\begin{pmatrix} 1 \\ 1.5 \\ 1 \end{pmatrix} - .56 = -.56$$

$$\text{for } x_2\text{: } z_2 - c_2 = \boldsymbol{\pi}'\boldsymbol{a}_2 - c_2 = (0 \quad 0 \quad 0)\begin{pmatrix} 2 \\ 1 \\ 0 \end{pmatrix} - .42 = -.42$$

Choose $x_1$ to enter the basis, since $-.56$ is Min $(\boldsymbol{\pi}'\boldsymbol{a}_j - c_j)$. Now compute

$$\boldsymbol{y} = \boldsymbol{B}^{-1}\boldsymbol{a}_1 = \begin{pmatrix} 1 & 0 & 0 \\ 0 & 1 & 0 \\ 0 & 0 & 1 \end{pmatrix}\begin{pmatrix} 1 \\ 1.5 \\ 1 \end{pmatrix} = \begin{pmatrix} 1 \\ 1.5 \\ 1 \end{pmatrix}$$

Recalling that

$$\boldsymbol{b} = \begin{pmatrix} 240{,}000 \\ 180{,}000 \\ 110{,}000 \end{pmatrix}$$

for positive $y_j$, compute $\theta = b_i/y_i$; choose the minimum

$$\begin{aligned} i = 1, \quad \theta &= 240{,}000/1 = 240{,}000 \\ i = 2, \quad \theta &= 180{,}000/1.5 = 120{,}000 \\ i = 3, \quad \theta &= 110{,}000/1 = 110{,}000 \longrightarrow \textit{Choose Minimum} \end{aligned}$$

Therefore set $r = 3$.
Determine new inverse and new $\boldsymbol{b}$.

| $c_B$ | $x_B$ | $b$ | *Old Inverse* | | | $y$ |
|---|---|---|---|---|---|---|
| 0 | $x_3$ | 240,000 | 1 | 0 | 0 | 1 |
| 0 | $x_4$ | 180,000 | 0 | 1 | 0 | 1.5 |
| 0 | $x_5$ | 110,000 | 0 | 0 | 1 | $\boxed{1}$ |

The entries for the new $\boldsymbol{b}$ and the new inverse are found in precisely the same manner as in the simplex method—by using row operations.

| $c_B$ | $x_B$ | *New* $\boldsymbol{b}$ | *New Inverse* | $y$ | *y for next iteration** |
|---|---|---|---|---|---|
| 0 | $x_3$ | 130,000 | $\begin{pmatrix} 1 & 0 & -1 \\ 0 & 1 & -1.5 \\ 0 & 0 & 1 \end{pmatrix}$ | $\begin{pmatrix} 0 \\ 0 \\ 1 \end{pmatrix}$ | $\begin{pmatrix} 2 \\ \boxed{1} \\ 0 \end{pmatrix}$ |
| 0 | $x_4$ | 15,000 | | | |
| .56 | $x_1$ | 110,000 | | | |

*See below for computation.

*Iteration 2:* Compute

$$\boldsymbol{\pi}' = \boldsymbol{c}'_B \boldsymbol{B}^{-1} = (0 \quad 0 \quad .56)\begin{pmatrix} 1 & 0 & -1 \\ 0 & 1 & -1.5 \\ 0 & 0 & 1 \end{pmatrix} = (0 \quad 0 \quad .56)$$

Evaluate nonbasic variables:

$$\text{for } x_2\text{: } z_2 - c_2 = \boldsymbol{\pi}'\boldsymbol{a}_2 - c_2 = (0 \quad 0 \quad .56)\begin{pmatrix} 2 \\ 1 \\ 0 \end{pmatrix} - .42 = -.42$$

$$\text{for } x_5\text{: } z_5 - c_5 = \boldsymbol{\pi}'\boldsymbol{a}_5 - c_5 = (0 \quad 0 \quad .56)\begin{pmatrix} 0 \\ 0 \\ 1 \end{pmatrix} - 0 = .56$$

Choose $x_2$ to enter the basis, since $-.42$ is the minimum $(z_j - c_j)$. Since $-.42 < 0$, compute

$$\boldsymbol{y} = \boldsymbol{B}^{-1}\boldsymbol{a}_2 = \begin{pmatrix} 1 & 0 & -1 \\ 0 & 1 & -1.5 \\ 0 & 0 & 1 \end{pmatrix}\begin{pmatrix} 2 \\ 1 \\ 0 \end{pmatrix} = \begin{pmatrix} 2 \\ 1 \\ 0 \end{pmatrix}$$

Now $x_4$ is chosen to leave the basis. The iteration is performed to yield the following:

| $c_B$ | $x_B$ | *New* $\boldsymbol{b}$ | *New Inverse* | $y$ | *y for next iteration** |
|---|---|---|---|---|---|
| 0 | $x_3$ | 100,000 | $\begin{pmatrix} 1 & -2 & 2 \\ 0 & 1 & -1.5 \\ 0 & 0 & 1 \end{pmatrix}$ | $\begin{pmatrix} 0 \\ 1 \\ 0 \end{pmatrix}$ | $\begin{pmatrix} \boxed{2} \\ -1.5 \\ 1 \end{pmatrix}$ |
| .42 | $x_2$ | 15,000 | | | |
| .56 | $x_1$ | 110,000 | | | |

*See below for computation.

*Iteration 3:* Compute

$$\pi' = c_B'B^{-1} = (0 \quad .42 \quad .56)\begin{pmatrix} 1 & -2 & 2 \\ 0 & 1 & -1.5 \\ 0 & 0 & 1 \end{pmatrix} = (0 \quad .42 \quad -.07)$$

Evaluate nonbasic variables:

$$\text{for } x_4\text{: } z_4 - c_4 = \pi' a_4 - c_4 = (0 \quad .42 \quad -.07)\begin{pmatrix} 0 \\ 1 \\ 0 \end{pmatrix} - 0 = .42$$

$$\text{for } x_5\text{: } z_5 - c_5 = \pi' a_5 - c_5 = (0 \quad .42 \quad -.07)\begin{pmatrix} 0 \\ 0 \\ 1 \end{pmatrix} - 0 = -.07$$

Choose $x_5$ to enter the basis. Compute

$$y = B^{-1}a_5 = \begin{pmatrix} 1 & -2 & 2 \\ 0 & 1 & -1.5 \\ 0 & 0 & 1 \end{pmatrix}\begin{pmatrix} 0 \\ 0 \\ 1 \end{pmatrix} = \begin{pmatrix} 2 \\ -1.5 \\ 1 \end{pmatrix}$$

Next $x_3$ is chosen to leave the basis, because its ratio $b_i/y_i$ for $y_i > 0$ is minimum.
The iteration is now performed to yield the following:

| $c_B$ | $x_B$ | *New b* | *New Inverse* | $y$ |
|---|---|---|---|---|
| 0 | $x_5$ | 50,000 | .5 −1 1 | 1 |
| .42 | $x_2$ | 90,000 | .75 −.5 0 | 0 |
| .56 | $x_1$ | 60,000 | −.5 1 0 | 0 |

*Because the solution is optimal there is no* $y$ *for the next iteration.*

*Iteration 4:* Compute

$$\pi' = c_B'B^{-1} = (0 \quad .42 \quad .56)\begin{pmatrix} .5 & -1 & 1 \\ .75 & -.5 & 0 \\ -.5 & 1 & 0 \end{pmatrix} = (.035 \quad .35 \quad 0)$$

Evaluate nonbasic variables:

$$\text{for } x_3\text{: } z_3 - c_3 = \pi' a_3 - c_3 = (.035 \quad .35 \quad 0)\begin{pmatrix} 1 \\ 0 \\ 0 \end{pmatrix} - 0 = .035$$

$$\text{for } x_4\text{: } z_4 - c_4 = \pi' a_4 - c_4 = (.035 \quad .35 \quad 0)\begin{pmatrix} 0 \\ 1 \\ 0 \end{pmatrix} - 0 = .35$$

The solution is optimal; no further iterations are required. The optimal solution has the objective function value

$$c'_B B^{-1} b = (0 \quad .42 \quad .56) \begin{pmatrix} 50{,}000 \\ 90{,}000 \\ 60{,}000 \end{pmatrix} = 71{,}400$$

This solution should be compared for similarities with the solutions of the example problem in Chapters 1 and 3.

We stress that the revised simplex method is conceptually the same as the simplex method, the real difference being that only a partial tableau is maintained. (The displays of $\boldsymbol{c}_B$, $\boldsymbol{x}_B$, $\boldsymbol{b}$, $\boldsymbol{B}^{-1}$, and $\boldsymbol{y}$ are called tableaux by many authors.) As far as computational comparisons between the two methods are concerned, it may appear from the example that the efficiency of the revised simplex method, using the explicit form of the inverse, is questionable. However, where the number of nonbasic variables is larger than the number of basic variables, the revised simplex method using the explicit form of the inverse is generally more efficient than the simplex method.

For computer computation using the revised simplex method form just presented, $m^2$ entries are required to store an $m$ constraint problem inverse. Since the calculations are performed using the inverse, the solving of large problems is cumbersome because of the large size of the inverse. Accordingly, a scheme more economical in its requirements for computer storage of the inverse was sought. The method described in the next subsection achieves that objective.

### Using the Product Form of the Inverse

The revised simplex method using the product form of the inverse is more economical than the revised simplex method using the explicit form of the inverse in its demands for computer space, particularly core storage. It differs from the previously described method *only* in the form in which the inverse is recorded.

To provide some motivation for the development of the product form of an inverse, consider the problem of Example 4.5. After the first iteration, the basis inverse is the following matrix:

$$\boldsymbol{B}^{-1} = \begin{pmatrix} 1 & 0 & -1 \\ 0 & 1 & -1.5 \\ 0 & 0 & 1 \end{pmatrix}$$

In this case the inverse is an identity matrix except for one column.

**Definition 4.2**

*An elementary matrix is a nonsingular square matrix that is an identity matrix except for one column.*

**THEOREM 4.1**

ANY INVERSE OF AN ELEMENTARY MATRIX IS AN ELEMENTARY MATRIX.

The proof of this theorem is asked in the problems at the end of the chapter.

That any basis one iteration after the original is an elementary matrix appears to be of limited value. Starting with an initial basis which is an identity matrix, the inverse of the second basis is an elementary matrix, but the inverses of third and subsequent bases, except in fortuitous circumstances, are not. We can, however, redefine the problem after the first iteration so that the second basis is treated as an original basis for a new problem. Thus, the third basis (and its inverse) is an elementary matrix in terms of the second. In the same way, each basis is an elementary matrix in terms of the immediately preceding basis. We may now state the following theorem.

**THEOREM 4.2**

ANY BASIS OR ITS INVERSE (OR ANY NONSINGULAR SQUARE MATRIX, FOR THAT MATTER) MAY BE REPRESENTED AS A PRODUCT OF ELEMENTARY MATRICES.

We give the proof informally. In finding an inverse of matrix $\boldsymbol{B}$ using row operations shown in Chapter 2, reduction of the first column is accomplished by the equivalent of left-multiplying (or premultiplying) by an elementary matrix $\boldsymbol{E}_1$:

$$\boldsymbol{E}_1 = \begin{pmatrix} \dfrac{1}{b_{11}} & 0 & \cdots & 0 & 0 \\ \dfrac{-b_{21}}{b_{11}} & 1 & \cdots & 0 & 0 \\ \cdot & \cdot & \cdot & \cdot & \cdot \\ \cdot & \cdot & \cdot & \cdot & \cdot \\ \cdot & \cdot & \cdot & \cdot & \cdot \\ \dfrac{-b_{m1}}{b_{11}} & 0 & \cdots & 0 & 1 \end{pmatrix}$$

Similarly, the second column reduction is accomplished by left-multiplying the result by

$$\boldsymbol{E}_2 = \begin{pmatrix} 1 & \dfrac{-b'_{12}}{b'_{22}} & 0 & \cdots & 0 \\ 0 & \dfrac{1}{b'_{22}} & 0 & \cdots & 0 \\ 0 & \dfrac{-b'_{32}}{b'_{22}} & 1 & \cdots & 0 \\ \cdot & \cdot & \cdot & \cdot & \cdot \\ \cdot & \cdot & \cdot & \cdot & \cdot \\ \cdot & \cdot & \cdot & \cdot & \cdot \\ 0 & \dfrac{-b'_{m2}}{b'_{22}} & 0 & \cdots & 1 \end{pmatrix}$$

where the primes indicate the matrix after premultiplication by $E_1$. Other steps in the reduction follow in the same manner.

The inverse $B^{-1} = E_m E_{m-1} \cdots E_2 E_1$. In other words, successively left-multiplying an identity matrix by the matrices $E_1, E_2, \ldots, E_m$ we obtain $B^{-1}$. It also follows that

$$B = E_1^{-1} E_2^{-1} \cdots E_m^{-1}$$

We now change notation slightly. Let $E_1$ be the inverse of the second basis with respect to the first, and $E_2$ be the inverse of the third basis with respect to the second, and so on. Then, using an argument closely related to the above, the inverse of a basis after $p$ iterations is

$$B^{-1} = E_p E_{p-1} \cdots E_1$$

It would hardly be sensible to multiply the elementary matrices at each iteration to form $B^{-1}$; the explicit-form revised simplex method would clearly be superior. Instead, a multiplication by $B^{-1}$ is made by multiplying the sequence of elementary matrices one at a time. Multiplication by an elementary matrix is particularly simple, and multiplying by an inverse consists of a number of multiplications by elementary matrices. Hence, a multiplication of $B^{-1}a_r$ becomes $E_p E_{p-1} \cdots E_2 E_1 a_r$, which is much simpler than it appears.

To illustrate the simplicity of multiplying by an elementary matrix, a left-multiplication of a vector by an elementary matrix is equivalent to adding a multiple of the pivotal row element to every other element, and replacing the pivotal row element by a designated multiple of itself.

EXAMPLE 4.6:

Row 3 is the pivotal row, since column 3 of the elementary matrix is not a unit vector.

$$\begin{pmatrix} 1 & 0 & .25 & 0 \\ 0 & 1 & .5 & 0 \\ 0 & 0 & .4 & 0 \\ 0 & 0 & 1 & 1 \end{pmatrix} \begin{pmatrix} 3 \\ 2 \\ 1 \\ 4 \end{pmatrix} = \begin{pmatrix} 3 + .25(1) \\ 2 + .5(1) \\ .4(1) \\ 1(1) + 4 \end{pmatrix} = \begin{pmatrix} 3.25 \\ 2.5 \\ .4 \\ 5 \end{pmatrix}$$

In a similar manner, the right-multiplication of a vector by an elementary matrix is shown to leave every entry of the vector unchanged except the pivotal element, which is replaced by a linear combination of all the previous elements of the vector.

EXAMPLE 4.7:

Column 3 is the pivotal column.

$$(3 \quad 2 \quad 1 \quad 4) \begin{pmatrix} 1 & 0 & .25 & 0 \\ 0 & 1 & .5 & 0 \\ 0 & 0 & .4 & 0 \\ 0 & 0 & 1 & 1 \end{pmatrix} = \begin{pmatrix} 3 \\ 2 \\ 3(.25) + 2(.5) + (.4) + 4(1) \\ 4 \end{pmatrix}' = \begin{pmatrix} 3 \\ 2 \\ 6.15 \\ 4 \end{pmatrix}'$$

Perhaps the utility of the product form of the inverse is now evident. For computer computations it is necessary to store only a sequence of vectors and scalars, each pair corresponding to an elementary matrix, specifically the nonunit column and the index of that column. When it is desired to multiply some vector by the inverse, each elementary matrix is retrieved and used to multiply in turn the desired vector. Because of the buildup of elementary matrices that occurs as iterations are undertaken, periodic reinversions to current bases are worthwhile to reduce iteration time and to prevent the accumulation of round-off error from becoming excessive. The example problem is now solved by the revised simplex method using the product form of the inverse.

EXAMPLE 4.8:

$$\begin{aligned} \text{Maximize } z &= .56x_1 + .42x_2 \\ \text{subject to: } \quad x_1 + 2x_2 + x_3 &= 240{,}000 \\ 1.5x_1 + x_2 + x_4 &= 180{,}000 \\ x_1 + x_5 &= 110{,}000 \\ x_1, \dots, x_5 &\geq 0 \end{aligned}$$

Using the vectors and matrices of Example 4.5, we have the following vectors and matrices defined:

$$\boldsymbol{A} = \begin{pmatrix} 1 & 2 & 1 & 0 & 0 \\ 1.5 & 1 & 0 & 1 & 0 \\ 1 & 0 & 0 & 0 & 1 \end{pmatrix} \quad \boldsymbol{b} = \begin{pmatrix} 240{,}000 \\ 180{,}000 \\ 110{,}000 \end{pmatrix} \quad \boldsymbol{c} = \begin{pmatrix} .56 \\ .42 \\ 0 \\ 0 \\ 0 \end{pmatrix}$$

$$\boldsymbol{x}_B = \begin{pmatrix} x_3 \\ x_4 \\ x_5 \end{pmatrix} \quad \boldsymbol{c}_B = \begin{pmatrix} 0 \\ 0 \\ 0 \end{pmatrix} \quad \boldsymbol{B} = \begin{pmatrix} 1 & 0 & 0 \\ 0 & 1 & 0 \\ 0 & 0 & 1 \end{pmatrix} \quad \boldsymbol{B}^{-1} = \begin{pmatrix} 1 & 0 & 0 \\ 0 & 1 & 0 \\ 0 & 0 & 1 \end{pmatrix}$$

The flow chart of Figure 4.2 is applicable, except that we use the product form of the inverse instead of the explicit form. In addition, to get the desired elementary matrices, an identity matrix is updated instead of making use of the old inverse.

$$\boldsymbol{\pi}' = \boldsymbol{c}_B'\boldsymbol{B}^{-1} = (0 \quad 0 \quad 0)$$

Compute $\boldsymbol{\pi}'\boldsymbol{a}_j - c_j$ for $x_j$ nonbasic

$$x_1\colon -.56$$
$$x_2\colon -.42$$

Choose $x_1$ to enter the basis.

$$y = B^{-1}a_1 = a_1 = \begin{pmatrix} 1 \\ 1.5 \\ 1 \end{pmatrix}$$

$$\text{Min } (a_{i0}/y_i) \text{ for } y_i > 0 = 110{,}000/1$$

The minimum is for $i = 3$; therefore $x_5$ leaves the basis.

| $c_B$ | $x_B$ | $b$ | *Identity Matrix* | *Incoming Variable* $x_1$ |
|---|---|---|---|---|
| 0 | $x_3$ | 240,000 | 1 0 0 | 1 |
| 0 | $x_4$ | 180,000 | 0 1 0 | 1.5 |
| 0 | $x_5$ | 110,000 | 0 0 1 | $\boxed{1}$ |

Updating as in the simplex method, the identity matrix is transformed into the first elementary matrix.

| $c_B$ | $x_B$ | $b$ | *First Elementary Matrix* $E_1$ | *New Incoming Variable (see below)* $x_2$ | *New Identity Matrix* |
|---|---|---|---|---|---|
| 0 | $x_3$ | 130,000 | 1 0 −1 | 2 | 1 0 0 |
| 0 | $x_4$ | 15,000 | 0 1 −1.5 | $\boxed{1}$ | 0 1 0 |
| .56 | $x_1$ | 110,000 | 0 0 1 | 0 | 0 0 1 |

We next compute $\pi' = c_B'B^{-1}$, which is $c_B'E_1$ or $\pi' = (0 \quad 0 \quad .56)\, E_1 = (0 \quad 0 \quad .56)$. Calculating $\pi' a_j - c_j$ for $x_j$ nonbasic, we have

$$x_2\colon -.42$$

$$x_5\colon \quad .56$$

and hence $x_2$ is selected to enter the basis in place of $x_4$.

$$y = B^{-1}a_2 = E_1 a_2 = \begin{pmatrix} 2 \\ 1 \\ 0 \end{pmatrix}$$

The iteration indicated above is then performed, updating the identity matrix into the second elementary matrix, $E_2$, as indicated below.

| $c_B$ | $x_B$ | $b$ | *New Identity Matrix* | *Incoming Variable (see below)* $x_5$ | *Second Elementary Matrix* $E_2$ |
|---|---|---|---|---|---|
| 0 | $x_3$ | 100,000 | $\begin{pmatrix}1 & 0 & 0\\0 & 1 & 0\\0 & 0 & 1\end{pmatrix}$ | $\boxed{2}$ | $\begin{pmatrix}1 & -2 & 0\\0 & 1 & 0\\0 & 0 & 1\end{pmatrix}$ |
| .42 | $x_2$ | 15,000 | | $-1.5$ | |
| .56 | $x_1$ | 110,000 | | 1 | |

The new inverse is $E_2 E_1$ (multiply it out and compare with the inverse in the revised simplex method, Example 4.4). We now compute $\pi'$ for the next iteration

$$\pi' = c'_B B^{-1} = c'_B E_2 E_1 = \left[(0 \quad .42 \quad .56)\begin{pmatrix}1 & -2 & 0\\0 & 1 & 0\\0 & 0 & 1\end{pmatrix}\right]\begin{pmatrix}1 & 0 & -1\\0 & 1 & -1.5\\0 & 0 & 1\end{pmatrix}$$

$$= (0 \quad .42 \quad .56)\begin{pmatrix}1 & 0 & -1\\0 & 1 & -1.5\\0 & 0 & 1\end{pmatrix}$$

or $\pi' = (0 \quad .42 \quad -.07)$ and $\pi' a_j - c_j$ yields for

$$x_4:\quad .42$$
$$x_5:\ -.07$$

hence $x_5$ is selected to enter the basis to replace $x_3$.

$$y = B^{-1}a_5 = E_2 E_1 a_5 = \begin{pmatrix}1 & -2 & 0\\0 & 1 & 0\\0 & 0 & 1\end{pmatrix}\begin{pmatrix}1 & 0 & -1\\0 & 1 & -1.5\\0 & 0 & 1\end{pmatrix}\begin{pmatrix}0\\0\\1\end{pmatrix}$$

$$= \begin{pmatrix}1 & -2 & 0\\0 & 1 & 0\\0 & 0 & 1\end{pmatrix}\begin{pmatrix}-1\\-1.5\\1\end{pmatrix} = \begin{pmatrix}2\\-1.5\\1\end{pmatrix}$$

The iteration above is performed, yielding the following:

| $c_B$ | $x_B$ | $b$ | *Third Elementary Matrix* $E_3$ |
|---|---|---|---|
| 0 | $x_5$ | 50,000 | $\begin{pmatrix}.5 & 0 & 0\\.75 & 1 & 0\\-.5 & 0 & 1\end{pmatrix}$ |
| .42 | $x_2$ | 90,000 | |
| .56 | $x_1$ | 60,000 | |

*There is no identity matrix here since this solution is optimal.*

The new inverse is $E_3 E_2 E_1$. We now calculate $\pi'$ for the next iteration.

$$\boldsymbol{\pi}' = \boldsymbol{c}'_B \boldsymbol{B}^{-1} = \boldsymbol{c}'_B \boldsymbol{E}_3 \boldsymbol{E}_2 \boldsymbol{E}_1$$

$$= (0 \quad .42 \quad .56)\begin{pmatrix} .5 & 0 & 0 \\ .75 & 1 & 0 \\ -.5 & 0 & 1 \end{pmatrix}\begin{pmatrix} 1 & -2 & 0 \\ 0 & 1 & 0 \\ 0 & 0 & 1 \end{pmatrix}\begin{pmatrix} 1 & 0 & -1 \\ 0 & 1 & -1.5 \\ 0 & 0 & 1 \end{pmatrix}$$

or

$$\boldsymbol{\pi}' = (.035 \quad .42 \quad .56)\begin{pmatrix} 1 & -2 & 0 \\ 0 & 1 & 0 \\ 0 & 0 & 1 \end{pmatrix}\begin{pmatrix} 1 & 0 & -1 \\ 0 & 1 & -1.5 \\ 0 & 0 & 1 \end{pmatrix} = (.035 \quad .35 \quad 0)$$

$\boldsymbol{\pi}'\boldsymbol{a}_j - c_j$ yields for

$$x_3: \quad .035$$
$$x_4: \quad .35$$

Since they are both positive, the optimal solution has been found.

## 4.3 MISCELLANEOUS TOPICS

In this section we consider a number of miscellaneous aspects of linear programming which are of varying importance both in theory and in practice.

### Complete Regularization

Complete regularization, which we shall refer to as regularization, is a device used by Charnes and Cooper [47] to ensure that, for both theoretical and practical purposes, no linear programming problem will be unbounded in the stage of computation. The process of regularization is to add to a problem a constraint of the form $\sum x_j \leq M$, where $M$ is sufficiently large. The regularization constraint provides an upper limit of $M$ on the sum of all variables, so that the feasible solution space is bounded. Then, when the problem has been solved, the regularization constraint is tested to see whether it is binding or not. If it is not binding, a finite optimum exists and has been found; otherwise there is no finite optimal solution.

### Variables Unrestricted in Sign

Thus far all of the problems which have been considered have had variables assumed to be nonnegative (i.e., $x_j \geq 0$). Many real problems contain variables which may be either positive or negative. Such variables appear to be more complicated to represent than nonnegative variables, but their representation actually results in a simplification of the problem. After adding slacks to convert all inequalities to equalities, the procedure is to solve for an unrestricted variable in *any* constraint in which it appears. Use that solution to eliminate it from all other constraints and the objective function. (In other words, introduce it into the basis. It may be convenient to introduce the variable into the basis via the usual simplex method rules in order to preserve feasibility.) Then drop the constraint in which the unrestricted variable was solved for, and solve the problem. This is analogous to treating it as a secondary constraint (i.e., an ignored constraint) which can never be binding. Once the optimal solution has been obtained using the simplex method, use the dropped equation to determine the optimal value of the unrestricted variable. Note that every (linearly independent) unrestricted variable treated in this way thereby eliminates one constraint and one variable. That a variable unrestricted in sign may be treated in this manner may be proven in a number of ways, one of which is to show that the reason any variable cannot in general be so eliminated is to preserve its nonnegativity. (See, for example, Dantzig [62], p. 86.)

EXAMPLE 4.9 (taken from Hadley [144], p. 169):

$$\begin{aligned} \text{Maximize } z = \; & 3x_1 + 2x_2 + x_3 \\ \text{subject to:} \quad & 2x_1 + 5x_2 + x_3 = 12 \\ & 3x_1 + 4x_2 \qquad = 11 \\ & \qquad x_2, x_3 \geq 0 \\ & x_1 \text{ is unrestricted in sign} \end{aligned}$$

Solve for $x_1$ in the first constraint and substitute into the objective function and the other constraint (we could have solved for $x_1$ in the second constraint and proceeded):

$$x_1 = 6 - \tfrac{5}{2}x_2 - \tfrac{1}{2}x_3 \tag{4.2}$$

This yields the following problem:

$$\begin{aligned} \text{Maximize } z = \; & 18 - \tfrac{11}{2}x_2 - \tfrac{1}{2}x_3 \\ \text{subject to:} \quad & \tfrac{7}{2}x_2 + \tfrac{3}{2}x_3 = 7 \\ & x_2, x_3 \geq 0 \end{aligned}$$

which can be solved by inspection to yield $x_3 = 14/3$, $x_2 = 0$.

Variable $x_1$ is thus seen to be $11/3$ by solving equation (4.2). It should be stressed that this procedure will work *only* with *variables unrestricted in sign* and not with nonnegative variables.

### Minimization Problems

Our treatment of the simplex method has been to maximize an objective function subject to constraints. Via a symmetry argument, an objective function can also be minimized using the simplex method. By changing the rules for selecting variables to *enter* the basis, we choose the variable with the most positive evaluator. A problem is then optimal when all values of evaluators are nonpositive. An equivalent method is to maximize the negative of the objective function to be minimized, since it is easy to show that

$$\text{Minimum } (z) = -\text{Maximum } (-z)$$

EXAMPLE 4.10:
Consider the problem

$$\begin{aligned} \text{Minimize } z &= x_1 + 2x_2 \\ \text{subject to:} \quad 2x_1 + 3x_2 &\geq 3 \\ x_1, x_2 &\geq 0 \end{aligned}$$

Minimize $z = x_1 + 2x_2$ is equivalent to the objective function

$$\text{Maximize } z' = -x_1 - 2x_2$$

and the problem becomes the same as that of Example 4.2.

### Redundant Constraints and Extraneous Variables

There are certain constraints and variables which may be omitted from linear programming problems without affecting the problem solution. To understand the nature of these special constraints and variables, we provide some definitions. First assume that the linear programming problem is in the form

$$\begin{aligned} \text{Maximize } z &= \boldsymbol{c}'\boldsymbol{x} \\ \text{subject to:} \quad \boldsymbol{Ax} &\leq \boldsymbol{b} \\ \boldsymbol{x} &\geq \boldsymbol{0} \end{aligned} \tag{4.3}$$

where $\boldsymbol{c}$, $\boldsymbol{x}$, $\boldsymbol{b}$, and $A$ are vectors and matrices of appropriate order. The constraints $\boldsymbol{Ax} \leq \boldsymbol{b}$, $\boldsymbol{x} \geq \boldsymbol{0}$, define a convex set which may be denoted by $X = \{\boldsymbol{x} \mid \boldsymbol{Ax} \leq \boldsymbol{b}, \boldsymbol{x} \geq \boldsymbol{0}\}$ to mean that $X$ is the polyhedral convex set of all $\boldsymbol{x}$ that satisfy the conditions on the right side of the vertical stroke. To ensure that $X$ has the indicated properties, we require that its solution set be nonempty and bounded.

We designate any one constraint of the set that defines $X$ by $\boldsymbol{a}^i\boldsymbol{x} \leq b_i$, where $\boldsymbol{a}^i$ is the $i^{\text{th}}$ row of $\boldsymbol{A}$. Then we may denote the remainder of the constraints $\bar{\boldsymbol{A}}\boldsymbol{x} \leq \bar{\boldsymbol{b}}$, $\boldsymbol{x} \geq \boldsymbol{0}$. By reference to these symbols, we then introduce:

**Definition 4.3** (Redundancy)
*Let $X_1 = \{\boldsymbol{x} \mid \boldsymbol{A}\boldsymbol{x} \leq \boldsymbol{b},\ \boldsymbol{x} \geq \boldsymbol{0}\}$ and $X_2 = \{\boldsymbol{x} \mid \bar{\boldsymbol{A}}\boldsymbol{x} \leq \bar{\boldsymbol{b}},\ \boldsymbol{x} \geq \boldsymbol{0}\}$. If $X_1$ and $X_2$ are identical sets, then $\boldsymbol{a}^i\boldsymbol{x} \leq b_i$ is redundant.*

**Definition 4.4**
*Let $X_1$ and $X_2$ be as in Definition 4.3. Then $\boldsymbol{a}^i\boldsymbol{x} \leq b_i$ is essential if $X_1$ and $X_2$ are not identical sets.*

> REMARK:
> Consider an otherwise nonredundant constraint of (4.3). Restate the constraint set of (4.3) so that such a constraint is stated twice. Then, of the identical constraints, either one can be shown to be redundant, and the other essential.

**Definition 4.5**
*A constraint that is not redundant is said to be binding at an optimum if it is satisfied as an equality at that optimum.*

**Definition 4.6**
*A constraint that is binding at no optimal solution is said to be a nonbinding constraint.*

Obviously, every redundant constraint is nonbinding. The converse is not true, however.

**Definition 4.7**
*A variable that is positive in no optimal solution is said to be extraneous.*

There are a number of tests for determining whether constraints are redundant, nonbinding, or neither, and whether or not variables are extraneous. Such variables and constraints can be omitted from a problem. We shall not discuss any of these methods here, but some tests for such variables and constraints are included in the problems at the end of the chapter. (See Thompson, Tonge, and Zionts [279] for more information.)

The effect of not removing such constraints and variables is generally to increase computation time. Some evidence of this is given in Zionts [313]. In fact, as is shown there (pp. 137–139), the problem that was shown to cycle in Chapter 3 converges without using the perturbation methods by eliminating certain easy to detect nonbinding constraints and extraneous variables. However, the detection method used does not guarantee convergence in general without the perturbation methods.

### Secondary Constraints and Variables

G.B. Dantzig [65] has defined a secondary constraint as one which is by some evaluation thought not likely to be binding at an optimum. Such constraints are therefore omitted from the problem and the problem is solved. (A regularization constraint may be employed to prevent an unbounded solution from occurring.) Then the omitted constraints are tested, and if they are all satisfied, the optimal solution to the reduced problem is optimal for the entire problem. Otherwise, the violated constraints must be appended to the problem and the process repeated until all constraints are satisfied. (The dual simplex method, covered in Chapter 5, may be utilized for this purpose.) Obviously, constraints should not be treated as secondary unless it is believed likely that they will be satisfied without their being incorporated.

A similar analysis and procedure can be employed for variables thought to be extraneous. Such variables can be omitted from the problem and treated as secondary. Once the reduced problem is solved, the evaluators of the secondary variables are found by using the inverse and proceeding as in the revised simplex method. If it is found that they are nonnegative and the solution is therefore optimal, nothing further need be done. Otherwise, variables with negative evaluators should be appended to the problem and the simplex method (or the revised simplex method) used to find an optimal solution. The process is repeated as with secondary constraints until an optimal solution is achieved.

### Alternative Criteria for Pivot Selection

In the simplex method and its variations, the most negative evaluator $(z_j - c_j)$ was used to determine the variable that was to enter the basis. Other methods have been considered and tried. An extensive study was carried out by Wolfe and Cutler [298] in which the authors identified some of the superior alternatives as being the computation of the greatest improvement in objective function (which tends to be good for reducing the number of iterations, but expensive computationally) and the positive normalized procedures, proposed by Dickson and Frederick [75], which ". . . aim at eliminating the effects of bad scaling of the problem data by dividing the reduced costs ($z_j - c_j$ in our notation) used in choosing the pivot column, by the sum of the positive coefficients $a_{ij}$" (Wolfe and Cutler [298], p. 187). They also found that, for the problems they solved, the product-form revised simplex method appears to be superior to the explicit form, which in turn appears to be superior to the ordinary simplex method.

Various strategies are used for product-form algorithms; one which is in wide use is called multiple pricing. Some number of nonbasic variables (usually a set having the most negative reduced costs) are updated. Only that set of variables is considered as candidates for basis entry, for some specified number of iterations, and the corresponding iterations are very much like ordinary simplex iterations. Once such iterations have been undertaken, all nonbasic variables are again considered for basis entry and the process is repeated.

Research on computational efficiency of various algorithms is continuing, with apparently diminishing returns; therefore, it would appear that the question of finding the computationally most efficient algorithm is far from being resolved.

### Selected Supplemental References

Section 4.2
[48]
Section 4.3
[65], [75], [221], [256], [279], [298], [313]

## 4.4 PROBLEMS

**1.** What is the $\boldsymbol{\pi}$ vector?

**2.** How can we represent the basis in the revised simplex method as having $m + 1$ variables (the $m$ we have been considering and $z$)? Using the two-phase method we would have $m + 2$ variables in the basis. How does this change the revised simplex method?

**3.** Prove Theorem 4.1, that the inverse of an elementary matrix is an elementary matrix.

**4.** Show that, at any iteration, the updated form of the initial identity matrix is the inverse of the current basis.

**5.** Solve the following problem:

$$\begin{aligned} \text{Minimize } z &= x_1 + 3x_2 - x_3 - 2x_4 \\ \text{subject to: } & 2x_1 + x_2 + 3x_3 + 5x_4 = 15 \\ & 3x_1 - x_2 + 2x_3 + 6x_4 = 11 \\ & x_j \geq 0 \ (j = 1, 2, 3, 4) \end{aligned}$$

**6.** Solve the following problems of Chapter 2 using artificial variables: 15, 16, 18.

**7.** Solve the following problems of Chapter 1 using the revised simplex method, both explicit form and product form of the inverse: 1, 2, 3, 4, 5, 9, 10, 13.

**8.** Solve the following problems of Chapter 2 using the revised simplex method, both explicit form and product form of the inverse: 12, 13, 14, 15, 16, 17, 18, 19.

**9.** Solve the following problems of Chapter 3 using the revised simplex method, both explicit form and product form of the inverse: 6, 7, 8, 9.

**10.** Show that an alternative, but less efficient, way of handling a variable unrestricted in sign, as opposed to the method given in section 4.3, is to replace every variable $x_j$ by the difference of two variables $x_{ja} - x_{jb}$. Because the vectors corresponding to these variables are dependent, only one can be basic at a time.

**11.** Solve Example 4.9 using the method proposed in problem 10.

**12.** Show that the following constraint forms imply the indicated results.

a.
$$\sum_{j=1}^{m} a_{ij}x_j = b_i$$

where

$$a_{ik} > 0,\ b_i \geq 0$$
$$a_{ij} \leq 0,\ j \neq k$$

implies $x_k$ is basic in some optimal solution. Such a constraint is called a definitional constraint.

b.
$$\sum_{j=1}^{m} a_{ij}x_j = 0$$

where

$$a_{ij} \geq 0,\ j = 1, \ldots, m$$

implies for $a_{ik} > 0$, $x_k = 0$ in every feasible solution.

# 5

# Duality and Its Significance

In earlier chapters we considered various aspects of linear programming problem structure and solutions. In this chapter, we consider another linear programming problem—referred to as the dual of the original problem—and explore the relationship between the dual and the original problem. This remarkable relationship is of interest from both mathematical and economic points of view.

## 5.1 STATEMENT OF THE DUAL PROBLEM—A PRICING PROBLEM

The dual problem can be stated as finding a set of prices or rents for resources such that the total value of all the resources is minimized subject to constraints requiring that the rents or prices be set in a manner consistent with alternative uses for the resources. In short, it is a pricing problem which can best be introduced by an example. Therefore, we shall state the example of Canine Products, Inc., and proceed to formulate the dual problem. Recall that Frisky Pup dog food is a blend of one pound of cereal and one and a half pounds of meat, that it utilizes one unit of Frisky Pup packaging capacity, and that it generates a contribution to overhead and profits of $.56 per package. Similarly, Husky Hound is a blend of two pounds of cereal and one pound of meat and generates a contribution to overhead and profits of $.42 per package. The associated resources are 240,000 pounds of cereal, 180,000 pounds of meat, and packaging equipment that can package at most 110,000 packages of Frisky Pup dog food a month. The associated problem developed is

$$\begin{aligned} \text{Maximize } z = .56x_1 + .42x_2 & \\ \text{subject to:} \quad x_1 + 2x_2 &\leq 240{,}000 \\ 1.5x_1 + x_2 &\leq 180{,}000 \\ x_1 &\leq 110{,}000 \\ x_1, x_2 &\geq 0 \end{aligned}$$

Earlier, we found that the optimal solution is to produce 60,000 packages of Frisky Pup dog food ($x_1$) and 90,000 packages of Husky Hound dog food ($x_2$). The contribution to overhead and profits thereby generated is \$71,400 per month.

Given the same setting as the above problem, we now wish to focus on the dual problem, a pricing problem. We seek to determine prices at which we should value the resources so that we can determine the minimum total value at which we would be willing to lease or sell the resources, as appropriate. We would be willing to sell cereal or meat, and we would be willing to rent capacity on the Frisky Pup packaging equipment.

We designate as $y_1$ and $y_2$ the prices per pound to be charged for cereal and meat respectively, and as $y_3$ the rent per unit to be charged for Frisky Pup packaging capacity. Given the availability of the resources, the total monthly sales and rentals are

$$240{,}000y_1 + 180{,}000y_2 + 110{,}000y_3$$

The lowest value of this objective is desired so that we may intelligently view any bids to buy or lease all the resources as a total package; therefore, we wish to minimize the sum of rentals and raw materials sales.

Consider next the constraints. The prices (henceforth we shall use the term *prices* to include both prices and rents) should all be zero or greater. Obviously, no resource should have a negative price, since any resource sold (we shall henceforth use the verb *sell* to mean either sell or rent) at a negative price could be more profitably left idle. Accordingly, the following constraints must be satisfied:

$$y_1, y_2, y_3 \geq 0$$

The other conditions to be satisfied in the other constraints are that the prices should be competitive with available alternatives. For example, since one pound of cereal plus one and a half pounds of meat plus one unit of Frisky Pup packaging capacity can be employed to produce one package of Frisky Pup dog food, the value in terms of resource prices of a unit of Frisky Pup dog food is

$$y_1 + 1.5y_2 + y_3$$

That price should be at least as great as what is obtained when a unit of Frisky Pup dog food is produced—namely, a contribution to overhead and profit of \$.56. In other words,

$$y_1 + 1.5y_2 + y_3 \geq .56$$

Similarly, two pounds of cereal together with one pound of meat can be employed to produce a package of Husky Hound dog food and thereby generate a contribution to overhead and profits of \$.42. Hence, the following inequality should also be satisfied:

$$2y_1 + y_2 \geq .42$$

There are no further restrictions, so we now formulate the objective function, which is to find the total minimum value of all the resources. The problem then is

$$\begin{aligned} \text{Minimize } z &= 240{,}000y_1 + 180{,}000y_2 + 110{,}000y_3 \\ \text{subject to: } & y_1 + 1.5y_2 + y_3 \geq .56 \\ & 2y_1 + y_2 \geq .42 \\ & y_1, y_2, y_3 \geq 0 \end{aligned}$$

For the solution, we shall first formulate it as a maximizing problem by maximizing the negative of the objective function. In other words, the objective function becomes

$$\text{Maximize } z = -240{,}000y_1 - 180{,}000y_2 - 110{,}000y_3$$

Using the method presented in section 4.1, that uses a single artificial variable in all constraints having greater than or equal inequalities, we add an artificial variable ($y_a$). The resulting problem, including slack and artificial variables, is as follows:

$$\begin{aligned} \text{Maximize } z &= -240{,}000y_1 - 180{,}000y_2 - 110{,}000y_3 - My_a \\ \text{subject to: } & y_1 + 1.5y_2 + y_3 - y_4 + \boxed{y_a} = .56 \\ & 2y_1 + y_2 \quad \boxed{-y_5} + y_a = .42 \end{aligned}$$

Pivoting $y_a$ into the first equation and then multiplying the second equation by $-1$ yields the starting tableau. The solution is given in Figure 5.1, where the entries in the original objective function have been divided by 100,000 for convenience.

Tableau 3 is optimal, and the corresponding solution is that the optimal price of cereal, $y_1$, be \$.035 per pound, and that the price of meat, $y_2$, be \$.35 per pound. The optimal price of Frisky Pup packaging capacity, $y_3$, which is not in the basis, is \$.00. Finally note that the minimum monthly value of the

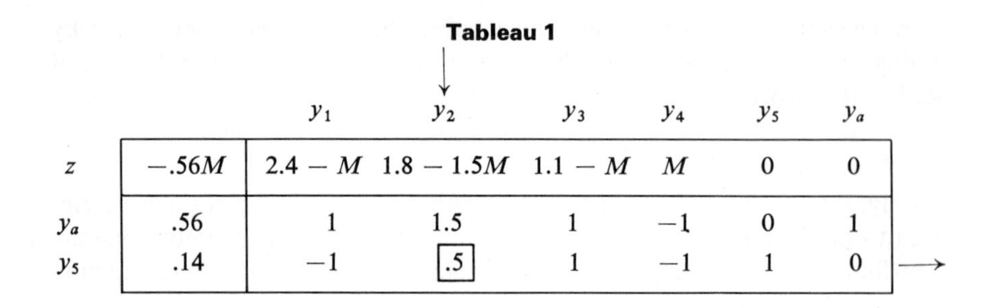

**Tableau 1**

| | | $y_1$ | ↓ $y_2$ | $y_3$ | $y_4$ | $y_5$ | $y_a$ | |
|---|---|---|---|---|---|---|---|---|
| $z$ | $-.56M$ | $2.4 - M$ | $1.8 - 1.5M$ | $1.1 - M$ | $M$ | 0 | 0 | |
| $y_a$ | .56 | 1 | 1.5 | 1 | $-1$ | 0 | 1 | |
| $y_5$ | .14 | $-1$ | [.5] | 1 | $-1$ | 1 | 0 | → |

**Tableau 2**

| | | ↓ $y_1$ | $y_2$ | $y_3$ | $y_4$ | $y_5$ | $y_a$ | |
|---|---|---|---|---|---|---|---|---|
| | $-.504 - .14M$ | $6 - 4M$ | 0 | $-2.5 + 2M$ | $-3.6 - 2M$ | $-3.6 + 3M$ | 0 | |
| $y_a$ | .14 | [4] | 0 | $-2$ | 2 | $-3$ | 1 | → |
| $y_2$ | .28 | $-2$ | 1 | 2 | $-2$ | 2 | 0 | |

**Tableau 3**

| | | $y_1$ | $y_2$ | $y_3$ | $y_4$ | $y_5$ | $y_a$ |
|---|---|---|---|---|---|---|---|
| | $-.714$ | 0 | 0 | .5 | .6 | .9 | $M - 1.5$ |
| $y_1$ | .035 | 1 | 0 | $-.5$ | .5 | $-.75$ | .25 |
| $y_2$ | .35 | 0 | 1 | 3 | $-1$ | .5 | .5 |

**Figure 5.1**

The Simplex Solution to the Pricing Problem

resources is $71,400, exactly the same as the maximum profit attainable by producing dog food. *This is no coincidence*! The relationship between the original linear program (henceforth called the primal) and the dual is so strong that they are in a sense two sides of the same problem. In fact, when either problem is solved using standard methods, e.g., the simplex method, the solution to the other problem is obtained as a by-product.

To illustrate this last point, refer to the optimal solution to the primal problem in Chapter 3 (p. 62). The reduced cost of $x_3$ is $.035 per pound. Now $x_3$ is the amount of unused cereal, and $.035 per pound is the marginal value or *shadow price* (or dual variable $y_1$) of cereal. Similarly, the optimal reduced cost of $x_4$ (unused meat) is $.35 per pound, corresponding to the dual variable $y_2$, and the optimal reduced cost of $x_5$ (unused Frisky Pup packaging capacity) is zero, corresponding to the dual variable $y_3$. In exactly the same manner the primal solution may be seen from the optimal solution for the dual problem. Thus, the solution to the primal problem may be seen as the reduced costs of the $y$'s in Tableau 3 in Figure 5.1. (Remember to multiply by 100,000.)

Returning to the example, it has been shown that the prices determined in the formulation above are the minimum prices that should be accepted for units of material and packaging capacity. Were those prices to prevail in the market, the managers of Canine Products, Inc., would be indifferent between the alternatives of producing dog food and selling resources. If the market prices were higher than the solution prices, then the managers should prefer to sell resources, and if the market prices were lower, then the managers should prefer to buy resources. In other words, the prices solved for give a measure for evaluating the marginal value of additional capacity or other resources. To illustrate, suppose that the current market price of cereal per pound is \$.08, and the current market price of meat per pound is \$.24. Canine Products should then be willing to sell some cereal (even though it is at a loss compared to the original purchase price), since \$.045 can be earned on each pound of cereal sold (\$.08 per pound from the sale minus \$.035 per pound in reduced profits). Similarly, they should be willing to buy some meat, since \$.11 can be earned on each pound of meat purchased (\$.35 per pound in increased profits less \$.24 per pound as the purchase price). Of course, a corresponding change must be made in the production schedule. Note that *neither* set of prices here has any relationship to the original purchase price of meat and cereal. That is not unusual because, *except* for market influences, there is no systematic relationship between purchase price and market price for goods.

The prices above indicate whether units of resources should be purchased or sold, but the question of amount has not yet been considered. The prices are valid as given if the entire set of assets are sold at those prices. If the assets are sold piecemeal, then the solution prices are valid only over some range (which may be small or large, depending on the structure of the problem). Holding everything else in the problem constant, the marginal value of a resource (such as pounds of cereal) would remain constant as the availability of the resource was increased, and then—once the upper limit of its range of validity was reached—the value would drop. The reason is that the availability of some other resource usually becomes limiting once enough of the first resource is made available. In a similar manner, as the availability decreases below the range of validity, the value of that resource increases in value. We are observing what economists call "decreasing returns to scale" or, more correctly, nonincreasing returns to scale. To illustrate this, note first that the optimal value of Frisky Pup packaging capacity is \$0.00, because it is not all utilized; i.e., an idle resource is of zero value. If, however, the capacity were decreased to less than 60,000 units, then, holding everything else constant, the value of additional capacity would jump to some positive value because production of Frisky Pup (which is 60,000 packages per month) would have to be reduced. Detailed studies evaluating the relative merits of altering capacity (or changing objective function coefficients and matrix coefficients) can be made using linear programming methods, and we shall discuss such techniques (known as parametric programming techniques) in Chapter 7.

## 5.2 THE DUAL PROBLEM—A FORMAL TREATMENT

We shall now treat the duality question more formally. Any linear programming problem can be written as

$$\begin{aligned} &\text{Maximize } z = \boldsymbol{c}'\boldsymbol{x} \\ &\text{subject to:} \quad \boldsymbol{A}\boldsymbol{x} \leq \boldsymbol{b} \\ &\qquad\qquad\qquad \boldsymbol{x} \geq \boldsymbol{0} \end{aligned} \tag{5.1}$$

(Note that this is the inequality form of the linear programming problem, and that $\boldsymbol{x}$ differs in general from the $\boldsymbol{x}$ used in the equality form.) We shall define this problem as the primal problem. In this chapter, all primal problems will be of this form, but we shall consider the effect of equality constraints upon the dual problem.

Next we state and prove the dual theorem of linear programming.

**THEOREM 5.1** (Duality)

GIVEN TWO LINEAR PROGRAMMING PROBLEMS

a) Maximize $z = \boldsymbol{c}'\boldsymbol{x}$ subject to: $\boldsymbol{A}\boldsymbol{x} \leq \boldsymbol{b}$, $\boldsymbol{x} \geq \boldsymbol{0}$

or alternatively

(a) Minimize $z' = -\boldsymbol{c}'\boldsymbol{x}$ subject to: $-\boldsymbol{A}\boldsymbol{x} \geq -\boldsymbol{b}$, $\boldsymbol{x} \geq \boldsymbol{0}$

AND

b) Minimize $z = \boldsymbol{b}'\boldsymbol{y}$ subject to: $\boldsymbol{A}'\boldsymbol{y} \geq \boldsymbol{c}$, $\boldsymbol{y} \geq \boldsymbol{0}$

or alternatively

(b) Maximize $z' = -\boldsymbol{b}'\boldsymbol{y}$ subject to: $-\boldsymbol{A}'\boldsymbol{y} \leq -\boldsymbol{c}$, $\boldsymbol{y} \geq \boldsymbol{0}$

THEN IF ONE PROBLEM HAS A FINITE OPTIMUM, SO DOES THE OTHER, AND THE CORRESPONDING OPTIMAL VALUES OF THE OBJECTIVE FUNCTION ARE EQUAL. IF ONE PROBLEM HAS AN UNBOUNDED SOLUTION, THEN THE OTHER HAS NO FEASIBLE SOLUTION.

PROOF:

Suppose that problem (a) has a feasible solution and a finite optimum $\boldsymbol{x}^*$. Then we may write the constraints of problem (a) as $\boldsymbol{A}\boldsymbol{x} + \boldsymbol{I}\boldsymbol{s} = \boldsymbol{b}$ (where

$s$ is a vector of slack variables). Further, we may partition the variables as $Bx_B + Nx_N = b$ where $x$ and $s$ together are partitioned as $x_B$ and $x_N$. The basic solution is, as usual, $x_B = B^{-1}b$, $x_N = 0$.
As we have done earlier, we write the objective function as

$$z = c'_B B^{-1} b + (c'_N - c'_B B^{-1} N)x^*_N + (c'_B - c'_B B^{-1} B)x^*_B$$

Since $x^*$ is an optimal solution, it is known that

$$c'_N - c'_B B^{-1} N \leq 0$$
$$c'_B - c'_B B^{-1} B \leq 0$$

(Since $B^{-1}B = I$, the second set of constraints states that $c'_B - c'_B \leq 0$.)
Rewriting these inequalities yields

$$c'_B B^{-1} N \geq c'_N$$
$$c'_B B^{-1} B \geq c'_B$$

or

(5.2) $$c'_B B^{-1}(N \,\vdots\, B) \geq (c'_N \,\vdots\, c'_B)$$

Matrices $N$ and $B$ are just matrices $A$ and $I$ in a different arrangement; in the same way, $c_N$ and $c_B$ are the same elements as $c$ and $0$. Hence we may write (5.2) as follows:

(5.3) $$c'_B B^{-1}(A \,\vdots\, I) \geq (c' \,\vdots\, 0')$$

The transpose of (5.3) is

(5.4) $$A'(c'_B B^{-1})' \geq c$$
$$I'(c'_B B^{-1})' \geq 0$$

Define $y^* = (c'_B B^{-1})'$. Therefore $y^*$ is a feasible solution to the dual problem, since (5.4) indicates that $y^*$ satisfies the dual constraints. The objective function of the dual problem for the solution $y^*$ is

$$b'y^* = (y^{*\prime}b)' = (c'_B B^{-1} b)' = c'_B B^{-1} b$$

Is it possible that this solution to the dual problem is optimal? Consider the primal constraints

$$Ax \leq b$$

Left-multiplying the primal constraints by $y'$ yields

(5.5) $$y'Ax \leq y'b$$

Similarly, consider the dual constraints

$$A'y \geq c$$

and their transpose

$$y'A \geq c'$$

Right-multiplying the dual constraints by $x$ yields

$$y'Ax \geq c'x \tag{5.6}$$

Combining (5.5) and (5.6) yields

$$y'b \geq y'Ax \geq c'x \tag{5.7}$$

which says that no feasible solution to the primal problem can have an objective function value greater than that of any feasible solution to the dual problem. Similarly, no feasible solution to the dual can have an objective function value less than that of any feasible solution to the primal problem. The optimal solution to the primal problem—namely $x^*$—has an objective function value $c_B'B^{-1}b$, and the constructed solution to the dual $y^*$ has the same objective function value. By (5.7), we conclude that $y^*$ is an optimal solution to the dual. This completes the first part of the proof. Considering the second part of the theorem, suppose that problem (a) has an infinite optimum. This means that $c'x$ can be made arbitrarily large for an associated feasible solution $x$. Next assume that there exists a feasible solution $y$ to the dual. By expression (5.7) we have

$$b'y \geq c'x$$

Because $c'x$ can be made arbitrarily large, the objective function to any feasible solution to the dual cannot be finite contrary to hypothesis. The second part of the theorem is thereby proven. We may prove the other two parts of the theorem—that is, if problem (b) has a finite optimum then so does problem (a), and if problem (b) has an infinite optimum then problem (a) has no feasible solution—in exactly the same manner by noting that each problem has an alternate form and the proofs just given apply to the alternate forms.

COROLLARY:
The dual of the dual is the primal.

PROOF:
The proof follows from the definition of the dual applied to the alternate form of (b).

It might inferred that if one problem has no feasible solution, then its dual has an infinite solution. That this is not necessarily true is demonstrated by the following problem, for which neither the primal nor the dual problem has a feasible solution.

EXAMPLE 5.1:

*Primal problem*

$$\begin{aligned} \text{Maximize } z &= x_1 + 2x_2 \\ \text{subject to: } \quad x_1 - x_2 &\leq -2 \\ -x_1 + x_2 &\leq 1 \\ x_1, x_2 &\geq 0 \end{aligned}$$

*Corresponding dual problem*

$$\begin{aligned} \text{Minimize } z &= -2y_1 + y_2 \\ \text{subject to: } \quad y_1 - y_2 &\geq 1 \\ -y_1 + y_2 &\geq 2 \\ y_1, y_2 &\geq 0 \end{aligned}$$

Neither problem has a feasible solution.

As a device to aid memory, many authors use a formulation of a linear programming problem (given in Fig. 5.2) to indicate the duality relationship. The primal is read horizontally, and the dual is read vertically.

Primal variables

$$\begin{array}{cc} & (x_1 \quad x_2 \quad \cdots \quad x_n) & & \text{Minimize} \\ \text{Dual variables} \begin{pmatrix} y_1 \\ y_2 \\ \cdot \\ \cdot \\ \cdot \\ y_m \end{pmatrix} & \begin{pmatrix} a_{11} & a_{12} & \cdots & a_{1n} \\ a_{21} & a_{22} & \cdots & a_{2n} \\ \cdot & \cdot & \cdot & \cdot \\ \cdot & \cdot & \cdot & \cdot \\ \cdot & \cdot & \cdot & \cdot \\ a_{m1} & a_{m2} & \cdots & a_{mn} \end{pmatrix} & \leq & \begin{pmatrix} b_1 \\ b_2 \\ \cdot \\ \cdot \\ \cdot \\ b_m \end{pmatrix} \\ & \geq & & \\ \text{Maximize} & (c_1 \quad c_2 \quad \cdots \quad c_n) & & \end{array}$$

**Figure 5.2**

A Matrix Representation of the Primal and Dual Problems Using Only One Matrix

EXAMPLE 5.2:

For a primal problem

$$\begin{aligned} \text{Maximize } z &= 3x_1 + 2x_2 + x_3 \\ \text{subject to: } \quad x_1 - x_2 + 2x_3 &\leq 10 \\ .5x_1 + x_2 + 3x_3 &\leq 15 \\ x_1, x_2, x_3 &\geq 0 \end{aligned}$$

the corresponding dual problem is

$$\begin{aligned} \text{Minimize } z &= 10y_1 + 15y_2 \\ \text{subject to: } \quad y_1 + .5y_2 &\geq 3 \\ -y_1 + \phantom{.5}y_2 &\geq 2 \\ 2y_1 + 3y_2 &\geq 1 \\ y_1, y_2 &\geq 0 \end{aligned}$$

**An Alternate Interpretation**

There is an additional way of interpreting duality in linear programming. In the formulation of problem (5.1) we have $m$ structural constraints $\boldsymbol{Ax} \leq \boldsymbol{b}$, and $n$ nonnegativity constraints $\boldsymbol{x} \geq \boldsymbol{0}$. A basic solution can be found by adding $m$ slack variables to the structural constraints and setting all but $m$ variables (whose vectors of coefficients are linearly independent) to zero. The equivalent alternate interpretation of a basic solution is to select $n$ constraints of the set $(\boldsymbol{Ax} \leq \boldsymbol{b}, \boldsymbol{x} \geq \boldsymbol{0})$ to be satisfied as equalities. The vectors of the coefficients of the resulting set of equalities must be linearly independent. We may look upon one formulation as a primal and the other as a dual, although both express only the primal variables. By reference to Example 5.2, consider the solution with $x_1$ and $x_2$ basic. That is the same solution as that with three primal inequalities (all but the nonnegativity constraints on $x_1$ and $x_2$) satisfied as equalities.

**Equality Constraints and Variables Unrestricted in Sign**

It is of interest to consider the dual of an equality constraint. By representing an equality $\boldsymbol{a}^i\boldsymbol{x} = \boldsymbol{b}_i$ as $\boldsymbol{a}^i\boldsymbol{x} \leq \boldsymbol{b}_i$ and $-\boldsymbol{a}^i\boldsymbol{x} \leq -\boldsymbol{b}_i$, we obtain in every dual constraint two dual variables, both nonnegative, having the same coefficients but opposite signs. Instead of treating these two variables separately, the difference can be treated as a variable unrestricted in sign. Hence the dual of an equality constraint is a variable *unrestricted in sign*. Conversely, the dual of an unrestricted variable is an equality constraint. Equivalently, the optimal value of a dual variable corresponding to an equality constraint may turn out to be of either sign, and thus the constraint may be inhibiting in either direction. As indicated in Chapter 4, a variable unrestricted in sign can be eliminated from the problem, since such a variable will always be basic in an optimal solution. Example 5.3 illustrates the treatment of equality constraints.

EXAMPLE 5.3:

A primal problem in the form

$$\text{Maximize } z = x_1 + 1.5x_2$$

$$\begin{aligned} \text{subject to:} \quad 2x_1 + 3x_2 &\leq 25 \\ x_1 + x_2 &\geq 1 \\ x_1 - 2x_2 &= 1 \\ x_1, x_2 &\geq 0 \end{aligned}$$

may be rewritten in the form

$$\begin{aligned} \text{Maximize } z &= x_1 + 1.5x_2 \\ \text{subject to:} \quad 2x_1 + 3x_2 &\leq 25 \\ -x_1 - x_2 &\leq -1 \\ x_1 - 2x_2 &\leq 1 \\ -x_1 + 2x_2 &\leq -1 \\ x_1, x_2 &\geq 0 \end{aligned}$$

by multiplying the second constraint by $-1$ and representing the third constraint as two inequalities. The dual of the rewritten problem is then the following:

$$\begin{aligned} \text{Minimize } z &= 25y_1 - y_2 + y_3 - y_4 \\ \text{subject to:} \quad 2y_1 - y_2 + y_3 - y_4 &\geq 1 \\ 3y_1 - y_2 - 2y_3 + 2y_4 &\geq 1.5 \\ y_1, y_2, y_3, y_4 &\geq 0 \end{aligned}$$

Since $y_3 - y_4$ appears in every constraint and objective function, it can be replaced with an unrestricted variable ($y_5$), as follows:

$$\begin{aligned} \text{Minimize } z &= 25y_1 - y_2 + y_5 \\ \text{subject to:} \quad 2y_1 - y_2 + y_5 &\geq 1 \\ 3y_1 - y_2 - 2y_5 &\geq 1.5 \\ y_1, y_2 &\geq 0, \; y_5 \text{ unrestricted in sign} \end{aligned}$$

As indicated in section 4.3, $y_5$ could be introduced into the basis, and the constraint in which it is basic could be dropped.

## 5.3 COMPLEMENTARY SLACKNESS

In the dual formulation of the linear programming problem, there is a variable corresponding to every structural constraint of the primal, and a structural constraint corresponding to every structural variable of the primal. (By structural variables we mean nonslack variables; by structural constraints, those excluding the nonnegativity constraints.) What we now wish to develop is referred to as principle of complementary slackness (See G.B. Dantzig [62]), and is closely related to the Theorem of the Alternative (see, e.g., Charnes and Cooper [47]).

**THEOREM 5.2** (Complementary slackness)

FOR A PRIMAL CONSTRAINT AND THE CORRESPONDING DUAL VARIABLE (OR, EQUIVALENTLY, A DUAL CONSTRAINT AND THE CORRESPONDING PRIMAL VARIABLE), THE FOLLOWING STATEMENTS CONCERNING AN OPTIMAL SOLUTION MUST BE TRUE:

1. IF A CONSTRAINT IS OVERLY SATISFIED (I.E., THE CORRESPONDING SLACK IS GREATER THAN ZERO), THEN THE CORRESPONDING DUAL VARIABLE IS ZERO.
2. IF A DUAL VARIABLE IS POSITIVE, THEN THE CORRESPONDING PRIMAL CONSTRAINT IS EXACTLY SATISFIED.

Note that we do not exclude the (degenerate) case in which both the dual variable is zero and the associated constraint is exactly satisfied.

PROOF:

In proving the duality theorem we established that at an optimum $\boldsymbol{c}'\boldsymbol{x}^* = \boldsymbol{y}^{*\prime}\boldsymbol{A}\boldsymbol{x}^* = \boldsymbol{y}^{*\prime}\boldsymbol{b}$, where $\boldsymbol{x}^*$ and $\boldsymbol{y}^*$ are the vectors of optimal solutions of the primal and dual problems, respectively. The equality $\boldsymbol{y}^{*\prime}\boldsymbol{A}\boldsymbol{x}^* = \boldsymbol{y}^{*\prime}\boldsymbol{b}$ is the vector of dual variables times the primal constraints fully satisfied. We can rewrite the equality as $\boldsymbol{y}^{*\prime}(\boldsymbol{b} - \boldsymbol{A}\boldsymbol{x}^*) = \boldsymbol{0}$. Recall that the vector $\boldsymbol{b} - \boldsymbol{A}\boldsymbol{x}$ in formulation (5.1) is the vector of slack variables. Since $\boldsymbol{b} - \boldsymbol{A}\boldsymbol{x}^* \geq \boldsymbol{0}$ (that is, $\boldsymbol{A}\boldsymbol{x}^* \leq \boldsymbol{b}$), and $\boldsymbol{y}^* \geq \boldsymbol{0}$, we see that in order to satisfy the above equality the following statement must be true: if $y_i^* > 0$, then $b_i - \boldsymbol{a}^i\boldsymbol{x}^* = 0$; or if $b_i - \boldsymbol{a}^i\boldsymbol{x} > 0$, then $y_i^* = 0$. The proof for the primal variables and the associated dual constraints follows, using the same analysis on $\boldsymbol{c}'\boldsymbol{x}^* = \boldsymbol{y}^{*\prime}\boldsymbol{A}\boldsymbol{x}^*$, $\boldsymbol{x} \geq \boldsymbol{0}$ and $\boldsymbol{A}'\boldsymbol{y} \geq \mathbf{c}$.

Let us state the above theorem in an intuitive manner in order to gain further insight. Our first statement says that if a resource is not fully utilized in an optimal solution, then the marginal value of that resource is zero. We should therefore be willing to sell units of that resource at positive prices, however small, but unwilling to pay to someone a positive price, however small, if he offered to us the use of such resource. Similarly, the second statement says that if we value a resource at a positive price in an optimal solution, we should be utilizing it completely in that solution. For the dual problem, we can express the following two statements:

1. If, at an optimum, the value of resources required to produce a unit of product exceeds the contribution to overhead and profits of producing that product (i.e., the imputed cost), then we should produce zero units of that product.

2. If, at an optimum, we produce some positive number of units of a product, then the value of resources required to produce a unit of the product (i.e., the imputed cost) should precisely equal the contribution to overhead and profits earned from producing that product.

The complementary slackness properties can be cast in yet another way. Given a pair of associated primal and dual variables (i.e., a dual variable and its associated primal slack variable, or a primal variable and its associated dual slack variable), at most one can be greater than zero.

EXAMPLE 5.4:

It is of interest to check the solution to the dog food example for complementary slackness. The solutions to the primal and dual problems are stated in the table for ease of reference.

PRIMAL SOLUTION

| | | |
|---|---|---|
| $x_1$ | Packages of Frisky Pup produced | 60,000 |
| $x_2$ | Packages of Husky Hound produced | 90,000 |
| $x_3$ | Pounds of cereal not utilized | 0 |
| $x_4$ | Pounds of meat not utilized | 0 |
| $x_5$ | Frisky Pup packaging capacity (in packages) not utilized | 50,000 |

DUAL SOLUTION

| | | |
|---|---|---|
| $y_1$ | Imputed price or value per pound of cereal | $.035 |
| $y_2$ | Imputed price or value per pound of meat | $.35 |
| $y_3$ | Imputed price or value per unit of Frisky Pup packaging capacity | $ 0 |
| $y_4$ | *Imputed cost* per package* of Frisky Pup dog food *in excess* of contributions to overhead and profits | $ 0 |
| $y_5$ | *Imputed cost* per package* of Husky Hound dog food *in excess* of contributions to overhead and profits | $ 0 |

* Note the interpretation of dual slack variables.

The correspondence or duality of the variables is indicated below:

$$x_1 \longleftrightarrow y_4$$

$$x_2 \longleftrightarrow y_5$$

$$x_3 \longleftrightarrow y_1$$

$$x_4 \longleftrightarrow y_2$$

$$x_5 \longleftrightarrow y_3$$

Let us now explore the various forms of complementary slackness.

1. The imputed price of any resource not fully utilized (i.e., its slack variable is positive), should be zero. Since $x_5$, the excess Frisky Pup packaging capacity, is positive, the corresponding imputed price, $y_3$, is zero.
2. If a positive price is imputed to a resource, then that resource should be fully utilized. For example, a positive imputed price per pound of cereal ($y_1$) implies that all cereal will be utilized and that $x_3 = 0$.
3. If the imputed costs of resources required to produce a unit of product exceed the contribution to overhead and profits of that product, then that product should not be produced. No example of this occurs in our problem. We can easily add another dog food, Prancing Poodle, to the product line to demonstrate the point. Prancing Poodle dog food requires 1.5 pounds of meat and 1.5 pounds of cereal and generates a contribution to overhead and profit of $.50. The imputed cost of this dog food is 1.5 (.35) for the meat plus 1.5 (.035) for the cereal, or .5775, which is in excess of the contribution to overhead and profits of .50. Hence, it is expected that there will be no Prancing Poodle produced in an optimal solution. By incorporating new variables in the problems—$x_6$ in the primal and $y_6$ in the dual—we would find zero Prancing Poodle dog food produced in an optimal solution and $y_6 = .0775$. However, if the imputed value were less than the contribution to overhead and profits, then the associated solution would not be optimal and more iterations would have to be utilized to achieve an optimum.
4. If the amount of product produced (at an optimum) is positive, then the imputed value of the resources used to produce a product is precisely equal to the contribution to overhead and profit. Frisky Pup dog food is produced ($x_1 > 0$); therefore, we should find that $y_4$, the imputed price per package of Frisky Pup in excess of contributions to overhead and profits, should be zero, which it is.

The use of the word *resource* may be a trifle confusing, in the sense that any constraint should be viewed as limiting a resource. If, for example, a constraint of the form of a sales commitment stipulates that $x_1 \geq 25$, then that would have to be interpreted as a resource. An interpretation can easily be made for that form of constraint. The dual variable gives the marginal value of relaxing (or loosening) the constraint.

## 5.4 THE DUAL SIMPLEX METHOD

It was shown in Chapter 4 that when a starting feasible solution to a linear programming problem is not available, we can proceed by adding artificial variables and using the simplex method.

Suppose now that for a particular problem no starting feasible solution to the primal is available, but the solution to the dual problem is feasible. Then, instead of using artificial variables, write the given problem in its dual form and use the simplex method to solve for an optimum in the usual fashion. We could instead leave the problem in the primal form, perform precisely the same sequence of iterations on the tableau, and obtain the optimal solution. This is what the dual simplex method (developed by Lemke [193]) does, and we shall spell out the rules of the dual simplex algorithm. Given a linear programming problem in the following form (with $\boldsymbol{b} \geq 0$):

$$\begin{aligned} \text{Minimize } & z = \boldsymbol{b}'\boldsymbol{y} \\ \text{subject to: } & \boldsymbol{A}'\boldsymbol{y} \geq \boldsymbol{c} \\ & \boldsymbol{y} \geq \boldsymbol{0} \end{aligned}$$

we may add slack variables $\boldsymbol{s}$ to obtain the following basic (but usually *not* feasible) solution (with $\boldsymbol{y} = \boldsymbol{0}$):

$$\begin{aligned} \text{Minimize } & z = \boldsymbol{b}'\boldsymbol{y} \\ \text{subject to: } & -\boldsymbol{A}'\boldsymbol{y} + \boldsymbol{I}\boldsymbol{s} = -\boldsymbol{c} \end{aligned} \tag{5.8}$$

More generally, we can treat $\boldsymbol{s}$ as the vector of basic variables and $\boldsymbol{y}$ as the vector of nonbasic variables, relabeling as necessary after each iteration to preserved the above format. We are using a seemingly cumbersome notation to preserve the one-to-one correspondence between the dual simplex method and the simplex method used on the dual problem. Alternatively, we could use the following notation for the constraints:

$$\boldsymbol{P}\boldsymbol{y} + \boldsymbol{I}\boldsymbol{s} = \boldsymbol{g}$$

In the description of the method we shall indicate in parenthesis where appropriate the parameter corresponding to the second notational form for constraints, in addition to using the first notational form. The algorithm follows:

1. Choose the most negative element of $-\boldsymbol{c}$ (of $\boldsymbol{g}$). This corresponds to the most violated constraint. If there is no negative element, then the present solution is feasible (and optimal), so halt. Otherwise designate the corresponding basic variable to leave the basis and go to step 2.
2. Designating the outgoing row as $r$, compute $b_j/a_{rj}$ (compute $g_j/-p_{rj}$) for $j$ such that $-a_{rj} < 0$ ($p_{rj} < 0$), and choose the minimum. Designate the value of $j$ corresponding to the minimum as $k$ and go to step 3. If all $-a_{rj} \geq 0$ (all $p_{rj} \geq 0$), there is no feasible solution to the problem, so halt.
3. Perform an iteration using row operations transforming $-a_{rk}$ (alternatively $p_{rk}$) to unity and $-a_{ik}$ (alternatively $p_{ik}$) ($i \neq r$) to zero. Rearrange the variables so that $\boldsymbol{y}$ are the nonbasic variables and $\boldsymbol{s}$ are the basic variables, $\boldsymbol{b}$ is the vector of reduced costs, $-\boldsymbol{c}$ (alternatively $\boldsymbol{g}$) is the vector of the values of the new basic variables, and $-\boldsymbol{A}$ (alternatively $\boldsymbol{P}$) is the updated matrix of coefficients. Then go to step 1.

As stated above, this method is equivalent to the simplex method used to solve the dual problem; a proof of the dual simplex method may be constructed along those lines. We shall now illustrate the dual simplex method applied to a problem.

EXAMPLE 5.5:

Solve the following problem using the dual simplex method:

$$\begin{aligned} \text{Maximize } z &= -2x_1 - 2x_2 \\ \text{subject to:} \quad 2x_1 + x_2 &\geq 6 \\ x_1 + 2x_2 &\geq 6 \\ x_1, x_2 &\geq 0 \end{aligned}$$

Denoting the slack variables as $x_3$ and $x_4$, we have the starting tableau, Tableau 1.

**Tableau 1**

|  |  | $x_1$ ↓ | $x_2$ | $x_3$ | $x_4$ |  |
|---|---|---|---|---|---|---|
| $z$ | 0 | 2 | 2 | 0 | 0 |  |
| $x_3$ | −6 | [−2] | −1 | 1 | 0 | → |
| $x_4$ | −6 | −1 | −2 | 0 | 1 |  |

The problem is dual feasible, i.e., all $a_{0j} \geq 0$, but not primal feasible, i.e., some $a_{i0} < 0$. [Remember, Tableau 1 is in the form of expression (5.8).] Hence we can use the dual simplex method. We choose $x_3$ to leave the basis. The choice of variables to enter the basis is $x_1$ because

$$\frac{b_1}{|a_{r1}|} = \frac{2}{|-2|} < \frac{b_2}{|a_{r2}|} = \frac{2}{|-1|}$$

Performing the iteration leads to Tableau 2.

**Tableau 2**

|  |  | $x_1$ | $x_2$ ↓ | $x_3$ | $x_4$ |  |
|---|---|---|---|---|---|---|
| $z$ | −6 | 0 | 1 | 1 | 0 |  |
| $x_1$ | 3 | 1 | .5 | −.5 | 0 |  |
| $x_4$ | −3 | 0 | [−1.5] | −.5 | 1 | → |

In Tableau 2, $x_4$ is chosen to leave the basis and $x_2$ is the variable which must enter because

$$\frac{b_2}{|a_{r2}|} = \frac{1}{|-1.5|} < \frac{b_3}{|a_{r3}|} = \frac{1}{|-.5|}$$

The indicated iteration is performed, leading to Tableau 3, for which the solution is feasible and therefore optimal.

**Tableau 3**

| | | $x_1$ | $x_2$ | $x_3$ | $x_4$ |
|---|---|---|---|---|---|
| $z$ | $-8$ | 0 | 0 | $\frac{2}{3}$ | $\frac{2}{3}$ |
| $x_1$ | 2 | 1 | 0 | $-\frac{2}{3}$ | $\frac{1}{3}$ |
| $x_2$ | 2 | 0 | 1 | $\frac{1}{3}$ | $-\frac{2}{3}$ |

A useful exercise for the reader is to write the dual of this problem, solve it by the (primal) simplex method, and show that the iterations are identical to those given above. That exercise is included in the problems at the end of the chapter.

The importance of the dual simplex method is, of course, that it permits solving a problem for which a dual feasible solution is available without first having to formulate the dual problem explicitly. As we shall see, this aspect of the dual simplex method is made use of in parametric programming and in integer programming.

**Selected Supplemental References**

Section 5.4
[193]
General
[140], [287], [288]

## 5.5 PROBLEMS

**1.** Show that any linear programming problem can be written in either form given in Theorem 5.1.

**2.** Solve Example 5.5 using artificial variables and compare the solution with that of the dual simplex method of section 5.4.

**3.** The dual problem presented in Theorem 5.1 is called the symmetric dual.

The dual of any other form of primal problem may be found by converting the problem to the form (5.1), and passing to the dual. Using the above, show that the dual of

$$\begin{aligned} \text{Maximize } z &= \boldsymbol{c}'\boldsymbol{x} \\ \text{subject to:} \quad \boldsymbol{A}\boldsymbol{x} &= \boldsymbol{b} \\ \boldsymbol{x} &\geq \boldsymbol{0} \end{aligned}$$

is

$$\begin{aligned} &\text{Minimize } z = \boldsymbol{b}'\boldsymbol{y} \\ &\text{subject to:} \quad \boldsymbol{A}'\boldsymbol{y} \geq \boldsymbol{c};\ \boldsymbol{y} \text{ unrestricted in sign} \end{aligned}$$

**4.** The Minkowski-Farkas-Weyl lemma states: For every matrix $\boldsymbol{A}$ and vector $\boldsymbol{b}$, one and only one of the following two statements is true:

a. The system of linear equations $\boldsymbol{A}\boldsymbol{x} = \boldsymbol{b}$ has a nonnegative solution $\boldsymbol{x} \geq \boldsymbol{0}$;

b. The system of linear equalities

$$\begin{aligned} \boldsymbol{A}'\boldsymbol{y} &\geq \boldsymbol{0} \\ \boldsymbol{b}'\boldsymbol{y} &< \boldsymbol{0} \end{aligned}$$

has a solution $\mathbf{y}$.

Use the results of Theorem 5.1 to prove the above result. (Hint: Set up the dual linear programs

$$\begin{aligned} \text{Maximize } z &= \boldsymbol{0}'\boldsymbol{x} \\ \text{subject to:} \quad \boldsymbol{A}\boldsymbol{x} &= \boldsymbol{b} \\ \boldsymbol{x} &\geq \boldsymbol{0} \end{aligned}$$

and

$$\begin{aligned} &\text{Minimize } z = \boldsymbol{b}'\boldsymbol{y} \\ &\text{subject to: } \boldsymbol{A}'\boldsymbol{y} \geq \boldsymbol{0};\ \boldsymbol{y} \text{ unrestricted in sign} \end{aligned}$$

using the results of problem 3.)

**5.** Solve the dual to the linear programming problem of Example 5.5 by the simplex method, and observe the correspondence between iterations of the solution and the dual simplex method.

**6.** Given the following problem:

$$\begin{aligned} \text{Maximize } z = -5x_1 - 3x_2 - 2x_3 & \\ \text{subject to:} \qquad x_2 + .5x_3 &\leq 15 \\ 2x_1 - x_2 + x_3 &\geq 4 \\ x_1, x_2, x_3 &\geq 0 \end{aligned}$$

a. Solve the primal using the simplex method.
b. Write the corresponding dual problem.
c. Solve the dual problem.
d. Solve the primal problem using the dual simplex method.

**7.** Discuss the effect on the solution to a primal problem and to the corresponding dual problem of:
a. Adding a constraint.
b. Adding a variable.

**8.** For the following problem:

$$\begin{aligned} \text{Minimize } z &= 8x_1 + 4x_2 + 6x_3 \\ \text{subject to:} \quad & 5x_1 - 3.5x_2 + 1.5x_3 = 7 \\ & 4x_1 + 3x_2 + 2x_3 = 5 \\ & -x_1 + 2x_2 + 3x_3 = 8 \\ & x_1, x_3 \geq 0; \; x_2 \text{ unrestricted in sign} \end{aligned}$$

a. Determine the dual problem.
b. Determine the dual problem so that the dual variables are nonnegative.

**9.** For the following problem:

$$\begin{aligned} \text{Maximize } z &= 5x_1 + x_2 + 2x_3 + 3x_4 \\ \text{subject to:} \quad & 3.5x_1 + 5x_2 + 1.5x_3 + 4x_4 \leq 10 \\ & 2x_1 + 3x_2 + 4x_3 + x_4 \leq 8 \\ & x_1, x_3 \geq 0; \; x_2, x_4 \text{ unrestricted in sign} \end{aligned}$$

a. Determine the dual problem.
b. Determine the dual problem so that no equalities appear in the dual constraints.

**10.** For the following problem:

$$\begin{aligned} \text{Minimize } z &= 3x_1 + 2.5x_2 + 2x_3 \\ \text{subject to:} \quad & 3x_1 + x_2 + 2x_3 \leq 3 \\ & x_1 - 2x_2 + 5x_3 \geq 5 \\ & 4x_1 + 1.5x_2 + 7x_3 \leq 8 \\ & x_1, x_2, x_3 \geq 0 \end{aligned}$$

Determine the dual problem.

**11.** Solve the following linear programming problem by the dual simplex method.

$$\begin{aligned}
\text{Minimize } z = \quad & x_1 + 4x_2 + 2x_3 \\
\text{subject to:} \quad & -2x_1 + x_2 + 5x_3 \geq 20 \\
& x_1 + 3x_2 - 2x_3 \geq 12 \\
& 4x_1 - 2x_2 + 3x_3 \geq 15 \\
& x_1, x_2, x_3 \geq 0
\end{aligned}$$

**12.** Develop the dual of the perturbation methods (from Chapter 3) for use in the dual simplex method.

**13.** Solve the following problem using the dual simplex method (no feasible solution).

$$\begin{aligned}
\text{Minimize } z = \quad & 5x_1 + 2x_2 \\
\text{subject to:} \quad & 3x_1 + 2x_2 \leq 12 \\
& x_1 - x_2 \geq 6 \\
& x_1, x_2 \geq 0
\end{aligned}$$

**14.** Solve each of these problems using the dual simplex method, and write the dual of the problem.

a. Problem 16 of Chapter 2.

b. Problem 18 of Chapter 2.

**15.** For each of the following problems of Chapter 1, write the dual problem, solve the dual problem using the dual simplex method, and state the complementary slackness conditions: 1, 2, 3, 4, 5, 9, 10, 13.

**16.** For each of the following problems of Chapter 2, write the dual problem, solve the dual problem using the dual simplex method, and state the complementary slackness conditions: 12, 13, 14, 17, 19.

**17.** For each of the following problems of Chapter 3, write the dual problem, solve the dual problem using the dual simplex method, and state the complementary slackness conditions: 6,7,8,9.

**18.** For each of the following problems of Chapter 1, write the dual problem, interpret the dual variables, and state the complementary slackness conditions, both in mathematical terms and in words: 10, 11, 12, 13, 14, 15.

**19.** Show that the following nonbasic variable forms for any basic feasible solutions imply the indicated results. (The proof of this is quite simple, using duality and the results of Chapter 4 problem 12.)

a.
$$z_j - c_j \geq 0$$

all $a_{ij}$ in updated form $\geq 0$ implies $x_j = 0$ in some optimal solution.

b.
$$z_j - c_j = 0$$

all $a_{ij}$ in updated form $\leq 0$ implies that variables basic in constraint $i$ for which $a_{ij} < 0$ are basic in an optimal solution.

# 6

# Primal-Dual Algorithms and the Criss-Cross Method

The possibility of solving both the primal and dual linear programming problems at the same time—in the sense of moving both problems directly toward optimality (or feasibility) as opposed to first solving one of the problems (either the primal or the dual) and obtaining the other as a by-product—has been proposed by a number of researchers. Their methods are usually referred to as primal-dual methods. In contrast, the simplex method begins with a primal feasible problem (even an artificial variable solution is primal feasible with respect to the expanded set of variables) and moves systematically toward dual feasibility, or primal optimality. In this chapter we shall consider methods which alternately attack the primal and dual problems, which Lemke and Charnes [194] labeled the "pincers" approach, and methods which create and preserve certain primal and/or dual feasibility conditions.

## 6.1 THE DANTZIG-FORD-FULKERSON PRIMAL-DUAL ALGORITHM

The first algorithm to be considered is the primal-dual algorithm of Dantzig, Ford, and Fulkerson [66]. The method was developed as an extension of a primal-dual method for solving the transportation problem which is presented in Chapter 9. To initiate the solution process, a starting feasible solution to the dual problem must be available, although there is a procedure for generating such a solution. Artificial variables are added as necessary to obtain an artificially feasible primal solution. By means of a sequence of "extended" and "restricted" primal problems, which correspond to considering and not considering certain variables as candidates for entry into the basis, iterations are taken that preserve the dual feasibility of the problem while attaining true primal feasibility by systematically eliminating artificial variables. When a primal feasible solution is attained, the problem is solved.

The primal-dual algorithm can be made formally equivalent to an altered Big-$M$ method. (The Big-$M$ method was described in Chapter 4.) To make this analogy, $M$ is treated as a parameter that is adjusted after certain iterations. During what is called the extended primal, the largest value of $M$ that preserves dual feasibility (that is, so that all $z_j - c_j \geq 0$) is determined. At such a point at least one nonbasic variable will have $z_j - c_j = 0$; otherwise $M$ could have been set to some larger value. The values of $z_j - c_j$ may be written in the form $\alpha M + \beta$. For variables with $z_j - c_j = 0$ it can be seen that the coefficient of $M$ (that is, $\alpha$) in $z_j - c_j$ must be nonpositive, and that $\beta$ must be nonnegative. Then, appending to the basis only those variables for which $z_j - c_j = 0$ for the given value of $M$, we have a problem called the restricted primal (basis plus nonbasic variables with $z_j - c_j = 0$ for the current value of $M$). Iterations are then taken in the restricted primal, utilizing only the coefficients of $M$ in the evaluator row for choosing variables to enter the basis. Given the value of $M$ selected in the previous extended primal, the value of the objective function does not change with restricted primal iterations, because the iterations correspond simply to moving among alternate optima for that value of $M$. Accordingly, the values of other reduced costs for that value of $M$ do not change either. Once the restricted primal has been solved, the coefficients of $M$ in $z_j - c_j$ in the restricted primal problem will all be nonnegative. The restricted primal phase will usually be very short, often consisting of only one iteration, since frequently only one nonbasic variable will have $z_j - c_j = 0$. We then proceed to another extended primal phase in which $M$ can be increased to some *larger* value. None of the nonbasic variables of the restricted primal at the end of the restricted primal phase can inhibit increasing the value of $M$. (Why?) Since all other nonbasic variable evaluators remain unchanged (for the previous value of $M$) at positive values, $M$ is strictly monotonically increasing between extended primal problems. When an extended primal problem is reached in which $M$ can be made arbitrarily large without affecting the optimality conditions, the problem has been solved. However, the solution may not be optimal with respect to the initial problem. It is optimal only if no artificial variables are in the optimal basis at positive levels; otherwise, as before, there is no feasible solution to the original problem. An unbounded solution is recognized in the usual way. A simplified flow chart of the procedure is given in Figure 6.1.

In order to illustrate the method, consider an example.

EXAMPLE 6.1:

$$\begin{aligned} \text{Maximize } z &= -x_1 - x_2 \\ \text{subject to:} \quad 3x_1 + 2x_2 &\leq 20 \\ x_1 + 2x_2 &\geq 2 \\ x_1, x_2 &\geq 0 \end{aligned}$$

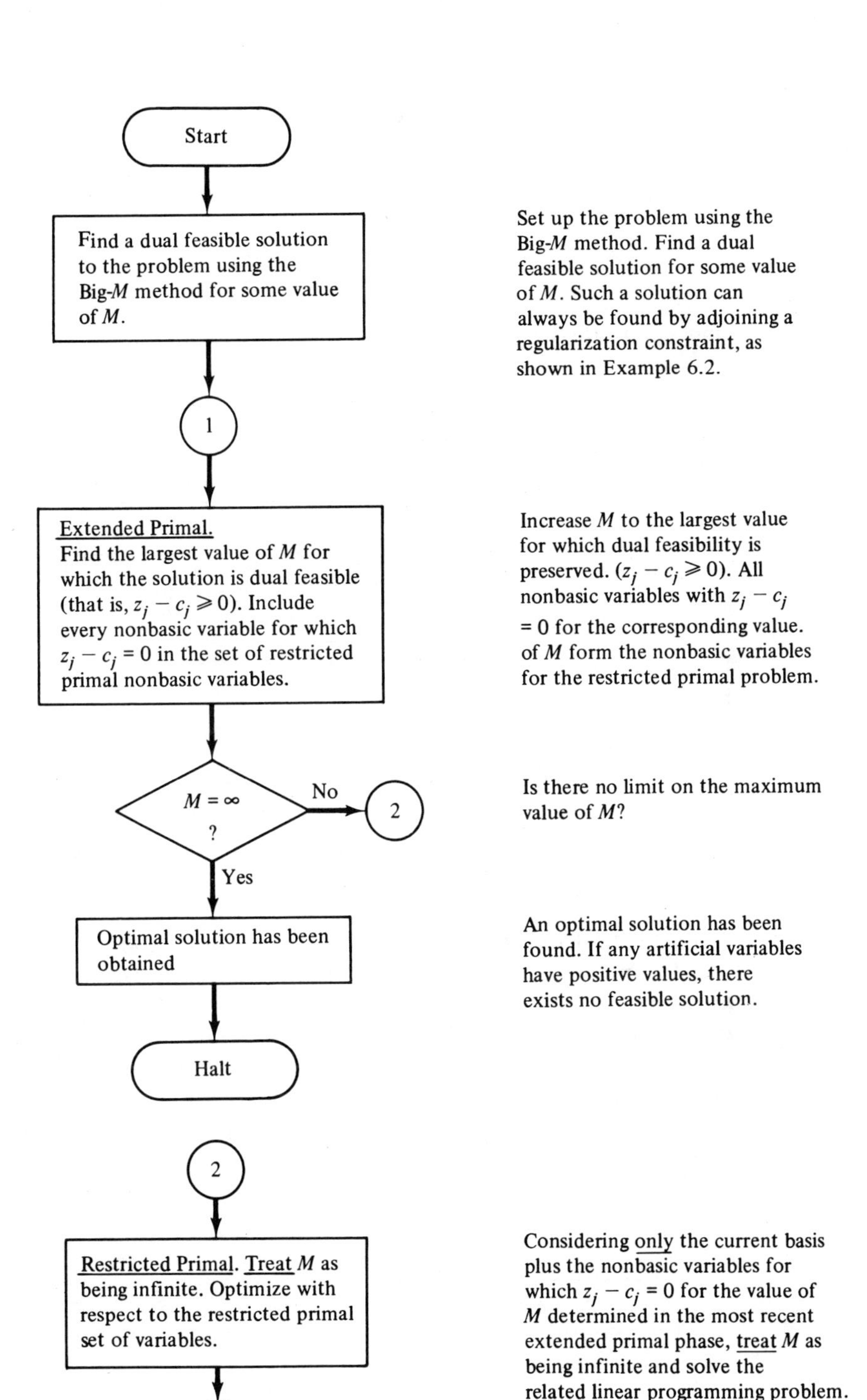


**Figure 6.1**
A Flow Chart of the Dantzig-Ford-Fulkerson Primal-Dual Algorithm

Adjoining slacks and an artificial variable, we have

$$\begin{aligned} \text{Maximize } z = -x_1 - x_2 \qquad\qquad\qquad & - Mx_a \\ \text{subject to:} \quad 3x_1 + 2x_2 + x_3 \qquad\qquad & = 20 \\ x_1 + 2x_2 \qquad - x_4 + \quad x_a & = \ 2 \\ x_1, \ldots, x_4, x_a & \geq \ 0 \end{aligned}$$

Pivoting $x_a$ into the basis yields Tableau 1, the starting tableau.

**Tableau 1**

|  |  | $x_1$ | $x_2$ ↓ | $x_3$ | $x_4$ | $x_a$ |  |
|---|---|---|---|---|---|---|---|
| $z$ | $-2M$ | $1-M$ | $1-2M$ | 0 | $M$ | 0 |  |
| $x_3$ | 20 | 3 | 2 | 1 | 0 | 0 |  |
| $x_a$ | 2 | 1 | [2] | 0 | $-1$ | 1 | → |
|  |  | ▬ |  |  | ▬ |  |  |

*Extended Primal 1*: The problem is dual feasible so long as $z_j - c_j \geq 0$, so for the nonbasic variables we have

$$\begin{aligned} x_1: \ z_1 - c_1 \geq 0 &\Longrightarrow 1 - \ \ M \geq 0 \Longrightarrow M \leq 1 \\ x_2: \ z_2 - c_2 \geq 0 &\Longrightarrow 1 - 2M \geq 0 \Longrightarrow M \leq .5 \\ x_4: \ z_4 - c_4 \geq 0 &\Longrightarrow \qquad\ \ M \geq 0 \Longrightarrow M \geq 0 \end{aligned}$$

Therefore the most restrictive constraint limits $M$ to .5, and accordingly $M$ is set to .5, and $x_2$ is the (only) nonbasic variable for the restricted primal. Then, for the restricted primal, obtained by ignoring $x_1$ and $x_4$ (except for updating purposes), we treat $M$ as infinite and then solve. The bars under certain columns indicate that such corresponding variables are *omitted* from the relevant restricted primal problem.

*Restricted Primal 1:* $x_2$ replaces $x_a$, yielding Tableau 2.

**Tableau 2**

|  |  | $x_1$ | $x_2$ | $x_3$ | $x_4$ | $x_a$ |
|---|---|---|---|---|---|---|
| $z$ | $-1$ | .5 | 0 | 0 | .5 | $-.5+M$ |
| $x_3$ | 18 | 2 | 0 | 1 | 1 | $-1$ |
| $x_2$ | 1 | .5 | 1 | 0 | $-.5$ | .5 |
|  |  | ▬ |  |  | ▬ |  |

The restricted primal is now optimal. We proceed to the extended primal.

*Extended Primal 2:* It can be seen that the problem is dual feasible for *any* value of $M$ greater than or equal to .5. Moreover, no artificial variables are in the basis. Hence, the problem has been solved and the optimal solution found.

If there is no dual feasible solution to the original problem immediately available, then one must be made available. This can always be accomplished by adjoining a regularization constraint. This constraint includes all variables that do *not* form part of the starting basis, and an artificial variable (which is seemingly unnecessary). Designating $c_{\max}$ as the maximum of the values of the objective function coefficients, we set the objective function coefficient of the artificial variable of the regularization constraint to $-M + c_{\max}$. It can easily be shown that, using this scheme, $M = 0$ will always provide a feasible solution to the dual. An example, which is the same problem as shown in Example 4.1, illustrates the method.

EXAMPLE 6.2:

$$\begin{aligned}
\text{Maximize } z &= 2x_1 + 3x_2 \\
\text{subject to:} \quad & x_1 + x_2 + x_4 = 4 \\
& x_1 - x_2 + x_3 - x_5 = 2 \\
& x_1 - x_6 = 1 \\
& x_1 + x_2 + x_3 = 5 \\
& x_1, \ldots, x_6 \geq 0
\end{aligned}$$

Noting that $c_{\max} = 3$, the coefficient of $x_2$ in the objective function, we add the regularization constraint, using $P$ as the limit of the sum of the variables, yielding

$$\begin{aligned}
\text{Maximize } \quad z &= 2x_1 + 3x_2 - Mx_a - Mx_b - Mx_c + (-M + 3)x_d \\
\text{subject to:} \quad & x_1 + x_2 + x_4 = 4 \\
& x_1 - x_2 + x_3 - x_5 + \boxed{x_a} = 2 \\
& x_1 - x_6 + \boxed{x_b} = 1 \\
& x_1 + x_2 + x_3 + \boxed{x_c} = 5 \\
\text{(Regularization constraint)} \quad & x_1 + x_2 + x_3 + x_5 + x_6 + x_7 + \boxed{x_d} = P
\end{aligned}$$

where $x_7$ is the slack on the regularization constraint.

Partial pivots are indicated in the above example problem, as is our practice. Performing the indicated partial pivots to obtain the associated basis yields Tableau 1, the starting basic solution.

**Tableau 1**

| | | $x_1$ | ↓ $x_2$ | $x_3$ | $x_4$ | $x_5$ | $x_6$ | $x_7$ | $x_a$ | $x_b$ | $x_c$ | $x_d$ | |
|---|---|---|---|---|---|---|---|---|---|---|---|---|---|
| $z$ | $-(8+P)M + 3P$ | $-4M+1$ | $-M$ | $-3M+3$ | 0 | 3 | 3 | $-M+3$ | 0 | 0 | 0 | 0 | |
| $x_4$ | 4 | 1 | [1] | 0 | 1 | 0 | 0 | 0 | 0 | 0 | 0 | 0 | → |
| $x_a$ | 2 | 1 | −1 | 1 | 0 | −1 | 0 | 0 | 1 | 0 | 0 | 0 | |
| $x_b$ | 1 | 1 | 0 | 0 | 0 | 0 | −1 | 0 | 0 | 1 | 0 | 0 | |
| $x_c$ | 5 | 1 | 1 | 1 | 0 | 0 | 0 | 0 | 0 | 0 | 1 | 0 | |
| $x_d$ | $P$ | 1 | 1 | 1 | 0 | 1 | 1 | 1 | 0 | 0 | 0 | 1 | |
| | | ▬ | | ▬ | | ▬ | ▬ | ▬ | | | | | |

*Extended Primal 1:* The maximum value of $M$ that preserves dual feasibility is $M = 0$. Hence, $x_2$ is the (only) nonbasic variable for the restricted primal and bars are indicated under all other nonbasic variables in Tableau 1.

*Restricted Primal 1:* $x_2$ replaces $x_4$ in the basis, yielding Tableau 2, which is optimal for the restricted primal (where $M$ is treated as being arbitrarily large with respect to this problem).

**Tableau 2**

| | | ↓ $x_1$ | $x_2$ | $x_3$ | $x_4$ | $x_5$ | $x_6$ | $x_7$ | $x_a$ | $x_b$ | $x_c$ | $x_d$ | |
|---|---|---|---|---|---|---|---|---|---|---|---|---|---|
| $z$ | $-(4+P)M + 3P$ | $-3M+1$ | 0 | $-3M+3$ | $M$ | 3 | 3 | $-M+3$ | 0 | 0 | 0 | 0 | |
| $x_2$ | 4 | 1 | 1 | 0 | 1 | 0 | 0 | 0 | 0 | 0 | 0 | 0 | |
| $x_a$ | 6 | 2 | 0 | 1 | 1 | −1 | 0 | 0 | 1 | 0 | 0 | 0 | |
| $x_b$ | 1 | [1] | 0 | 0 | 0 | 0 | −1 | 0 | 0 | 1 | 0 | 0 | → |
| $x_c$ | 1 | 0 | 0 | 1 | −1 | 0 | 0 | 0 | 0 | 0 | 1 | 0 | |
| $x_d$ | $P-4$ | 0 | 0 | 1 | −1 | 1 | 1 | 1 | 0 | 0 | 0 | 1 | |
| | | | | ▬ | ▬ | ▬ | ▬ | ▬ | | | | | |

*Extended Primal 2:* The maximum value of $M$ that preserves dual feasibility in Tableau 2 is $M = \frac{1}{3}$. For that value, $x_1$ is the (only) nonbasic variable for the restricted primal, and bars are indicated under all other nonbasic variables in Tableau 2.

*Restricted Primal 2:* $x_1$ replaces $x_b$ in the basis, yielding Tableau 3, which is optimal for the restricted primal (for $M$ infinite).

**Tableau 3**

| | | $x_1$ | $x_2$ | $x_3$ ↓ | $x_4$ | $x_5$ | $x_6$ | $x_7$ | $x_a$ | $x_b$ | $x_c$ | $x_d$ | |
|---|---|---|---|---|---|---|---|---|---|---|---|---|---|
| $z$ | $-(1+P)M$ $+3P-1$ | 0 | 0 | $-3M$ $+3$ | $M$ | 3 | $-3M$ $+4$ | $-M$ $+3$ | 0 | $3M$ $-1$ | 0 | 0 | |
| $x_2$ | 3 | 0 | 1 | 0 | 1 | 0 | 1 | 0 | 0 | −1 | 0 | 0 | |
| $x_a$ | 4 | 0 | 0 | 1 | 1 | −1 | 2 | 0 | 1 | −2 | 0 | 0 | |
| $x_1$ | 1 | 1 | 0 | 0 | 0 | 0 | −1 | 0 | 0 | 1 | 0 | 0 | |
| $x_c$ | 1 | 0 | 0 | [1] | −1 | 0 | 0 | 0 | 0 | 0 | 1 | 0 | → |
| $x_d$ | $P-4$ | 0 | 0 | 1 | −1 | 1 | 1 | 1 | 0 | 0 | 0 | 1 | |
| | | | | | ▬ | ▬ | ▬ | ▬ | | | | | |

*Extended Primal 3:* $x_b$ has left the basis and is discarded. The maximum value of $M$ that preserves dual feasibility in Tableau 3 is $M = 1$. For that value, $x_3$ is the (only) nonbasic variable for the restricted primal, and bars are indicated under all other nonbasic variables in Tableau 3.

*Restricted Primal 3:* $x_3$ replaces $x_c$ in the basis, yielding Tableau 4, which is optimal for the restricted primal (for $M$ infinite).

**Tableau 4**

| | | $x_1$ | $x_2$ | $x_3$ | $x_4$ | $x_5$ | $x_6$ ↓ | $x_7$ | $x_a$ | $x_c$ | $x_d$ | |
|---|---|---|---|---|---|---|---|---|---|---|---|---|
| $z$ | $-(P-2)M$ $+3P-4$ | 0 | 0 | 0 | $-2M$ $+3$ | 3 | $-3M$ $+4$ | $-M$ $+3$ | 0 | $3M$ $-3$ | 0 | |
| $x_2$ | 3 | 0 | 1 | 0 | 1 | 0 | 1 | 0 | 0 | 0 | 0 | |
| $x_a$ | 3 | 0 | 0 | 0 | 2 | −1 | [2] | 0 | 1 | −1 | 0 | → |
| $x_1$ | 1 | 1 | 0 | 0 | 0 | 0 | −1 | 0 | 0 | 0 | 0 | |
| $x_3$ | 1 | 0 | 0 | 1 | −1 | 0 | 0 | 0 | 0 | 1 | 0 | |
| $x_d$ | $P-5$ | 0 | 0 | 0 | 0 | 1 | 1 | 1 | 0 | −1 | 1 | |
| | | | | | ▬ | ▬ | | ▬ | | | | |

*Extended Primal 4:* $x_c$ has left the basis and is discarded. The maximum value of $M$ that preserves dual feasibility is $M = \frac{4}{3}$. For that value $x_6$ is the (only) nonbasic variable for the restricted primal, and bars are indicated under all other nonbasic variables in Tableau 4.

*Restricted Primal 4:* $x_6$ enters the basis, replacing $x_a$ and yielding Tableau 5, which is optimal for the restricted primal (for $M$ infinite).

**Tableau 5**

| | | $x_1$ | $x_2$ | $x_3$ | $x_4$ | $x_5$ | $x_6$ | $x_7$ | $x_a$ | $x_d$ |
|---|---|---|---|---|---|---|---|---|---|---|
| $z$ | $-(P-6.5)M + 3P - 10$ | 0 | 0 | 0 | $M-1$ | $-1.5M + 5$ | 0 | $-M + 3$ | $1.5M - 2$ | 0 |
| $x_2$ | 1.5 | 0 | 1 | 0 | 0 | .5 | 0 | 0 | −.5 | 0 |
| $x_6$ | 1.5 | 0 | 0 | 0 | 1 | −.5 | 1 | 0 | .5 | 0 |
| $x_1$ | 2.5 | 1 | 0 | 0 | 1 | −.5 | 0 | 0 | .5 | 0 |
| $x_3$ | 1 | 0 | 0 | 1 | −1 | 0 | 0 | 0 | 0 | 0 |
| $x_d$ | $P - 6.5$ | 0 | 0 | 0 | −1 | 1.5 | 0 | [1] | −.5 | 1 |

*Extended Primal 5:* $x_a$ has left the basis and is discarded. The maximum value of $M$ that preserves dual feasibility is $M = 3$. For that value $x_7$ is the (only) nonbasic variable for the restricted primal, and bars are indicated under all other variables in Tableau 5.

*Restricted Primal 5:* $x_7$ enters the basis, replacing $x_d$ and yielding Tableau 6, which is optimal for the restricted primal (for $M$ infinite).

**Tableau 6**

| | | $x_1$ | $x_2$ | $x_3$ | $x_4$ | $x_5$ | $x_6$ | $x_7$ | $x_d$ |
|---|---|---|---|---|---|---|---|---|---|
| $z$ | 9.5 | 0 | 0 | 0 | 2 | .5 | 0 | 0 | $M - 3$ |
| $x_2$ | 1.5 | 0 | 1 | 0 | 0 | .5 | 0 | 0 | 0 |
| $x_6$ | 1.5 | 0 | 0 | 0 | 1 | −.5 | 1 | 0 | 0 |
| $x_1$ | 2.5 | 1 | 0 | 0 | 1 | −.5 | 0 | 0 | 0 |
| $x_3$ | 1 | 0 | 0 | 1 | −1 | 0 | 0 | 0 | 0 |
| $x_7$ | $P - 6.5$ | 0 | 0 | 0 | −1 | 1.5 | 0 | 1 | 1 |

*Extended Primal 6:* $x_d$ has left the basis and could be discarded. The maximum value of $M$ that preserves dual feasibility is infinite. Hence the problem is solved and the solution is given in Tableau 6. Using the primal-dual algorithm required one iteration more than did the method used in Example 4.1.

The question of computational efficiency is relevant. In Hadley [144], it is reported that Richard Mills, in a master's thesis at the Sloan School of Management, MIT, found inconclusive results in comparing the primal-dual algorithm with the revised simplex method. Each method was superior in different cases.

An additional interpretation of the primal-dual algorithm may be useful. By stating a problem with artificial variables added so that a dual feasible solution is available, the procedure may be initiated. If necessary, a regularization constraint may be added as indicated above. Then a parametric programming method (which is discussed in Chapter 7) is used to find the optimal solution to the augmented problem (containing the artificial variables) as a function of the parameter $M$, as $M$ is increased. The solution for $M$ infinite corresponds to the optimal solution to the original problem.

## 6.2 THE CRISS-CROSS METHOD FOR SOLVING LINEAR PROGRAMMING PROBLEMS

This section describes the criss-cross method developed by Zionts [309]. The ordinary simplex method requires a primal feasible starting solution, and the dual simplex method requires a dual feasible solution. Problems that have neither a primal nor a dual feasible basic solution readily available are normally solved by setting up an altered problem with artificial variables, the optimal solution of which coincides with that of the original problem. Artificial variables, if any, are present in the optimal solution only at zero quantity. Such altered problems normally require considerably more time for solution than do similar sized problems with either a primal or a dual feasible basic solution readily available.

The criss-cross method does not employ an altered problem to find a basic solution that is either primal or dual feasible; if the starting solution is either primal or dual feasible, the method reduces to the primal or dual simplex method, respectively. Otherwise this method alternates between (properly defined) primal and dual iterations until a primal feasible solution, a dual feasible solution, or a primal and dual feasible solution is reached. In the first two cases the primal or the dual simplex method, respectively, is utilized to obtain an optimal solution; in the third case the optimal solution has already been found. If at any point no iteration can be taken in accordance with the prescribed rules, then no finite optimal solution exists.

For reasons of symmetry, the criss-cross method uses the simplex tableau representation omitting the identity matrix; that is, it uses a contracted tableau (see problem 3, Chapter 3). No artificial variables need be added to the problem. Inequality constraints of the form $\sum_j a_{ij}x_j \geq b_i > 0$ are represented as $\sum_j - a_{ij}x_j \leq -b_i$. As part of the algorithm, a variable is made basic in each equality constraint. Similarly, variables unrestricted in sign are made basic and remain basic. (The associated constraint in which such variables are basic may be dropped.)

The problem can then be expressed as in Figure 6.2a, where the basic variables are suppressed. After each iteration (by criteria to be specified), we

PROBLEM a—THE PARTITIONED PROBLEM*

$$\begin{aligned} \text{Maximize} \quad & z = +c_1'x_1 - c_2'x_2 \\ \text{subject to:} \quad & A_{11}x_1 + A_{12}x_2 \leq -b_1 \\ & A_{21}x_1 + A_{22}x_2 \leq b_2 \\ & x_1, x_2 \geq 0 \end{aligned}$$

PROBLEM b—THE PRIMAL PORTION OF THE PROBLEM

$$\begin{aligned} \text{Maximize} \quad & z = +c_1'x_1 - c_2'x_2 \\ \text{subject to:} \quad & A_{21}x_1 + A_{22}x_2 \leq b_2 \\ & x_1, x_2 \geq 0 \end{aligned}$$

PROBLEM c—THE DUAL PORTION OF THE PROBLEM

$$\begin{aligned} \text{Maximize} \quad & z = -c_2'x_2 \\ \text{subject to:} \quad & A_{12}x_2 \leq -b_1 \\ & A_{22}x_2 \leq b_2 \\ & x_2 \geq 0 \end{aligned}$$

* $c_1$ and $x_1$ are $n_1$ by 1, $c_2$ and $x_2$ are $n_2$ by 1; $b_1$ is $m_1$ by 1, $b_2$ is $m_2$ by 1; $A_{11}$ is $m_1$ by $n_1$, $A_{12}$ is $m_1$ by $n_2$, $A_{21}$ is $m_2$ by $n_1$, $A_{22}$ is $m_2$ by $n_2$; $n_1$ is the number of dual infeasibilities; $m_1$ is the number of primal infeasibilities; $m = m_1 + m_2$; $n = n_1 + n_2$; $b_2$ and $c_2$ are nonnegative; and $b_1$ and $c_1$ are strictly positive.

**Figure 6.2**

Various Portions of the Problem Referred to in the Criss-Cross Method

shall partition the updated matrix to appear as in Figure 6.2a, where $x_1$ and $x_2$ represent the nonbasic variables and the unwritten variables (which *appear* to be slack variables but which may or may not be) are basic. (It should be pointed out here that the [unwritten] basic variables are slack variables only with respect to the starting solution, providing that all constraints are inequalities, but need not, in general, be the slack variables of the original problem.) For such a solution, then, the first $m_1$ constraints are violated and the next $m_2$ constraints are satisfied. Similarly, for such a solution, the first $n_1$ nonbasic variables correspond to dual constraints that are violated, and the next $n_2$ nonbasic variables correspond to dual constraints that are satisfied. (The partition changes after each iteration, and so do the numbers $m_1$, $m_2$, $n_1$, and $n_2$ in general.) We shall denote as a primal iteration one that uses the usual simplex method criteria on the problem in Figure 6.2b, but then performs the pivot on the entire problem of Figure 6.2a. We shall denote as a dual iteration one that uses the usual dual simplex method criteria on the problem in Figure 6.2c, but then performs the pivot on the entire problem of Figure 6.2a. In other words, for pivot selection a primal iteration ignores primal infeasible constraints and a dual iteration ignores dual infeasible constraints. The algorithm consists of performing alternate primal and dual iterations. In the event that a (primal or dual) variable would enter the basis in an infinite quantity, and certain no finite optimal solution criteria are *not* satisfied, an alternate iteration of the same kind (primal or dual, as the case may be) is attempted, ignoring the variable that would have entered the basis in an infinite quantity. If no such iteration can be taken, then the iteration of the opposite type (dual or primal, as appropriate) is attempted. If this type of iteration also cannot be taken, then it will be shown that no finite optimal feasible solution exists. If, at any point, a primal or dual feasible solution is reached, then only primal or dual (as appropriate) iterations are utilized to achieve an optimal solution to the problem.

Convergence of this method has not been rigorously proven, although it is believed that a proof of convergence can be constructed. The method has been used to solve thousands of problems in practice. Every problem attempted has either converged to an optimum or given evidence of no finite optimal solution. The following scheme will, however, assure convergence in all cases. If a primal or dual feasible solution has not been reached in some large number of iterations (arbitrarily chosen as $2m + 2n$), a regularization constraint of the form $\sum_j x_j \leq M$, $M$ sufficiently large, is appended to the problem. Then only primal iterations, as defined above, are utilized until the problem becomes dual feasible. Then only dual iterations are utilized until the optimal feasible solution is found, or an unbounded solution to the dual problem has been found. In the latter case no optimal feasible solution to the problem exists. In the former case, the optimal solution has been found, provided that the regularization constraint is *not binding*. If the constraint is binding, no finite optimal feasible solution to the problem exists. The above device has been found to provide a successful way for proving convergence. It is cumbersome, and has not, to the author's knowledge, been implemented in any criss-cross computer programs.

The algorithm is best stated in flow chart form, and is presented in Figure 6.3. The device to insure convergence is not included in the flow chart. (A more detailed flow chart is found in Zionts [309].)

EXAMPLE 6.3:

The following example will be solved using the criss-cross method. Note that the partitioning indicated in Figure 6.2 is *not* actually undertaken after each iteration.

$$\begin{aligned}
\text{Maximize } z = 3x_1 - 4x_2 & \\
\text{subject to: } \quad x_1 + 2x_2 &\geq 2 \\
3x_1 + x_2 &\geq 4 \\
x_1 - x_2 &\leq 1 \\
x_1 + x_2 &\leq 3 \\
x_1, x_2 &\geq 0
\end{aligned}$$

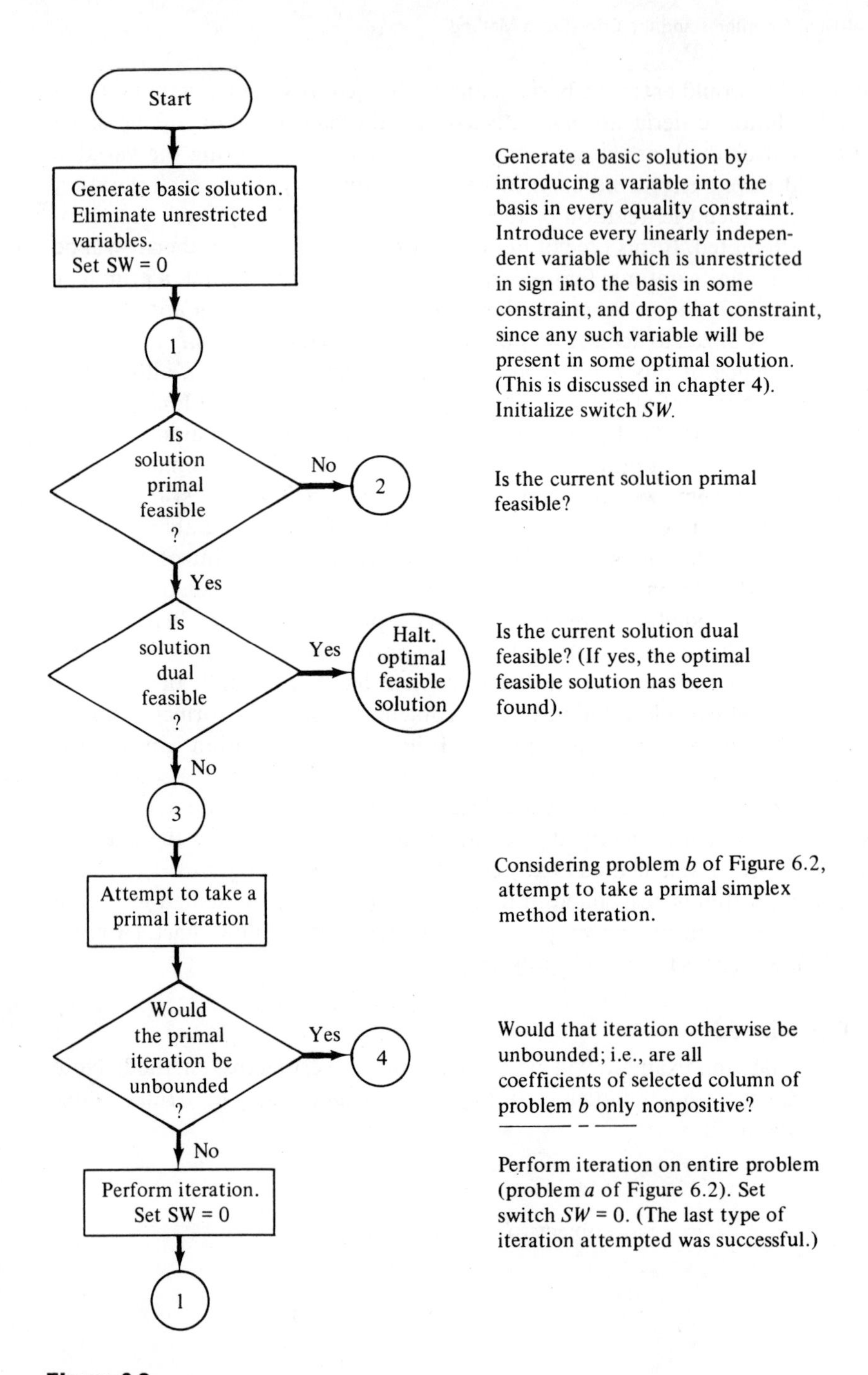


**Figure 6.3**

Flow Chart of the Criss-Cross Method

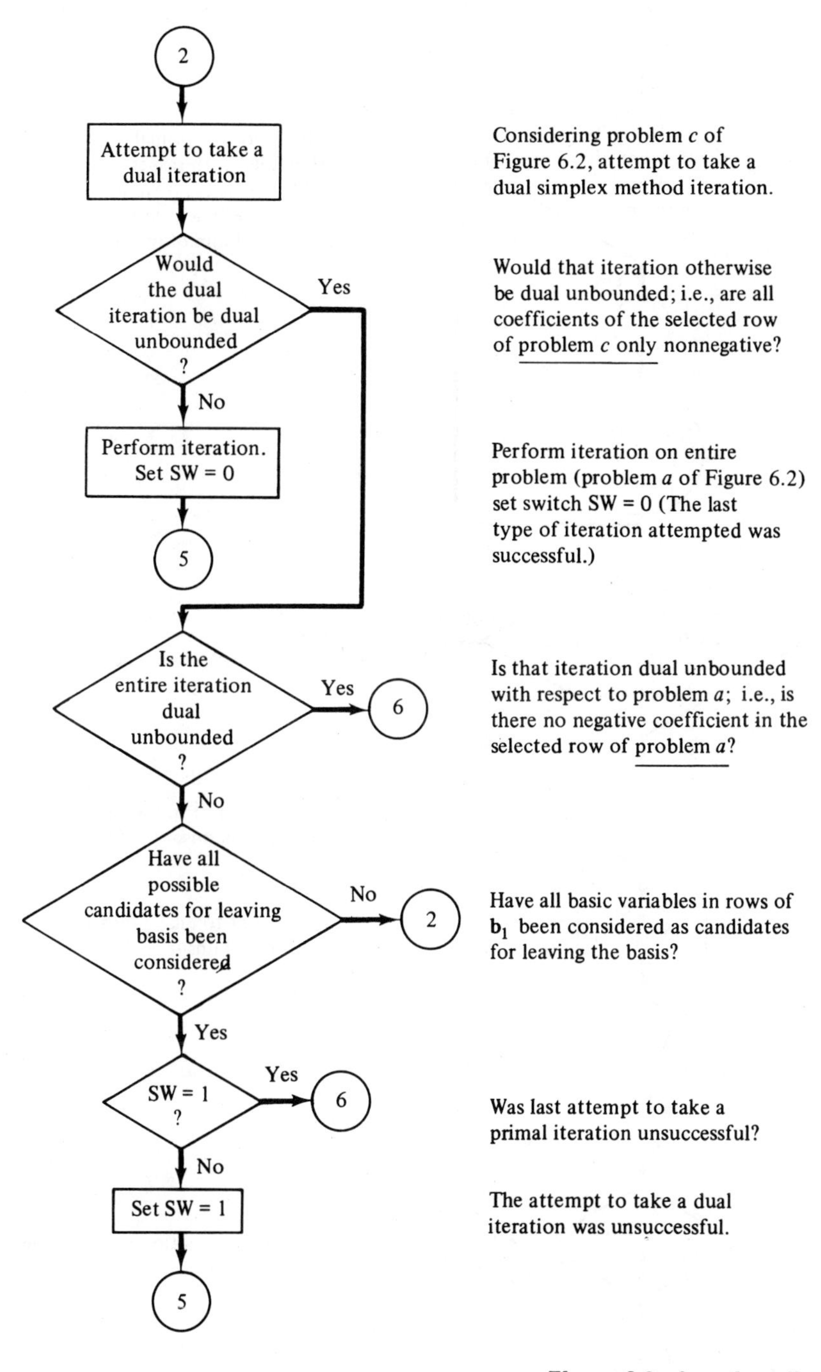


**Figure 6.3 (continued)**

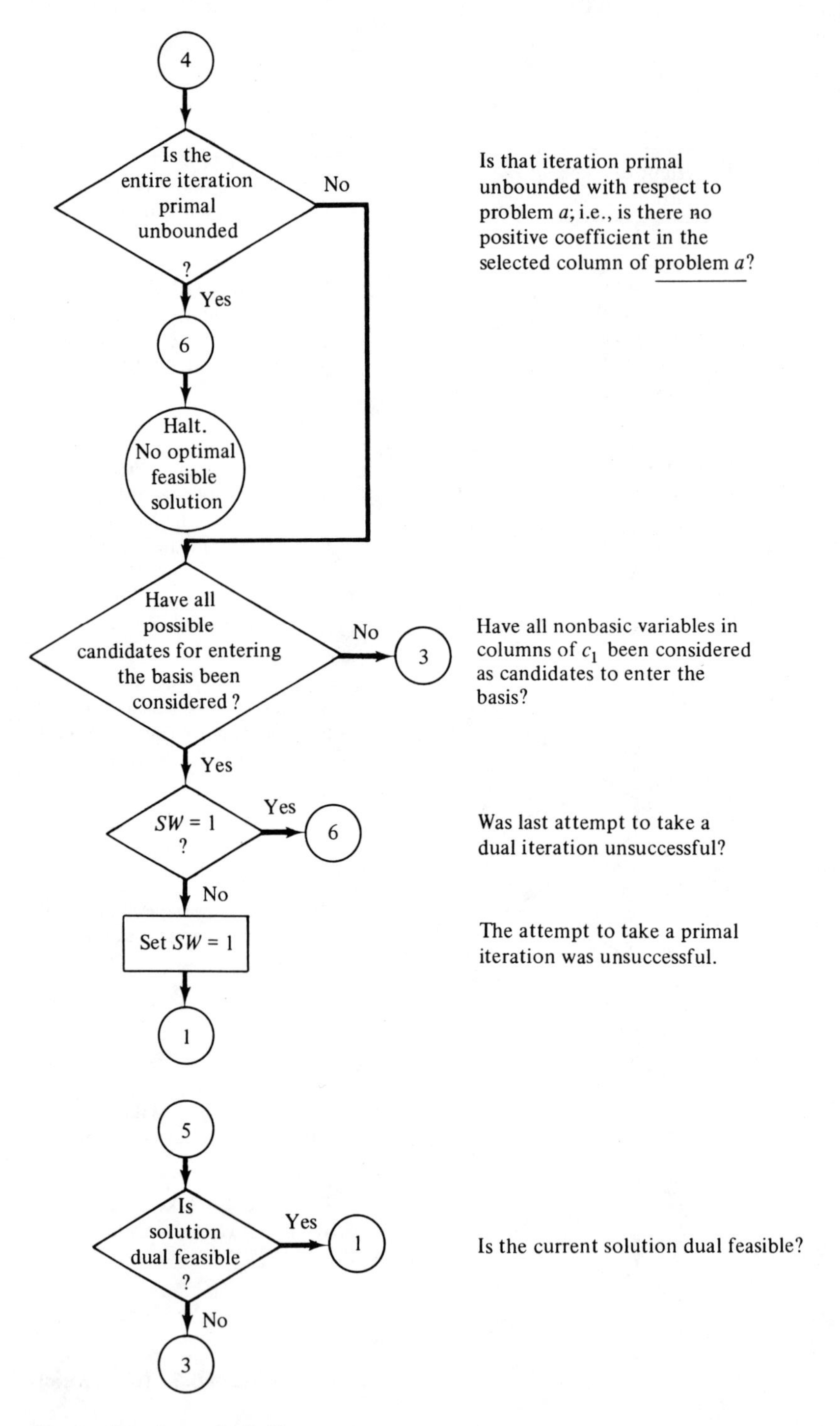


**Figure 6.3 (concluded)**

**Tableau 1**

| | | | $x_1$ | $x_2$ ↓ |
|---|---|---|---|---|
| | $z$ | 0 | $-3$ | 4 |
| initial | $x_3$ | $-2$ | $-1$ | $-2$ |
| slack | $x_4$ | $-4$ | $-3$ | [$-1$] → |
| variables | $x_5$ | 1 | 1 | $-1$ |
| | $x_6$ | 3 | 1 | 1 |

Begin with a dual iteration. The most negative $b_i$ is $-4$, corresponding to $x_4$. Replace $x_4$ by $x_2$, yielding Tableau 2.

**Tableau 2**

| | | $x_1$ ↓ | $x_4$ |
|---|---|---|---|
| | $-16$ | $-15$ | 4 |
| $x_3$ | 6 | [5] | $-2$ → |
| $x_2$ | 4 | 3 | $-1$ |
| $x_5$ | 5 | 4 | $-1$ |
| $x_6$ | $-1$ | $-2$ | 1 |

Utilize a primal iteration. The most negative $z_j - c_j$ is $-15$, corresponding to $x_1$. Replace $x_3$ by $x_1$, which yields Tableau 3.

**Tableau 3**

| | | $x_3$ | $x_4$ ↓ |
|---|---|---|---|
| | 2 | 3 | $-2$ |
| $x_1$ | 1.2 | $\frac{1}{5}$ | $-\frac{2}{5}$ |
| $x_2$ | .4 | $-\frac{3}{5}$ | $\frac{1}{5}$ |
| $x_5$ | .2 | $-\frac{4}{5}$ | [$\frac{3}{5}$] → |
| $x_6$ | 1.4 | $\frac{2}{5}$ | $\frac{1}{5}$ |

The solution is now primal feasible. Therefore, only primal iterations need be utilized. Replace $x_6$ by $x_4$, yielding Tableau 4, which is optimal.

**Tableau 4**

| | | $x_3$ | $x_5$ |
|---|---|---|---|
| | $\frac{8}{3}$ | $\frac{1}{3}$ | $\frac{10}{3}$ |
| $x_1$ | $\frac{4}{3}$ | $-\frac{1}{3}$ | $\frac{2}{3}$ |
| $x_2$ | $\frac{1}{3}$ | $-\frac{1}{3}$ | $-\frac{1}{3}$ |
| $x_4$ | $\frac{1}{3}$ | $-\frac{4}{3}$ | $\frac{5}{3}$ |
| $x_6$ | $\frac{4}{3}$ | $\frac{2}{3}$ | $-\frac{1}{3}$ |

We present another example.

EXAMPLE 6.4:
Example 6.2 will be solved using the criss-cross method.

$$\begin{aligned} \text{Maximize } z &= 2x_1 + 3x_2 \\ \text{subject to:} \quad & x_1 + x_2 \leq 4 \\ & -x_1 + x_2 - x_3 \leq -2 \\ & -x_1 \leq -1 \\ & \boxed{x_1} + x_2 + x_3 = 5 \end{aligned}$$

A variable must be introduced into the basis in the fourth equation. Arbitrarily making $x_1$ basic in that constraint leads to Tableau 1. A more systematic rule, whose efficiency has not yet been explored, is given in the problems.

**Tableau 1**

| | | $x_2$ | $x_3$ ↓ | |
|---|---|---|---|---|
| $z$ | 10 | −1 | 2 | |
| $x_4$ | −1 | 0 | [−1] | → |
| $x_5$ | 3 | 2 | 0 | |
| $x_6$ | 4 | 1 | 1 | |
| $x_1$ | 5 | 1 | 1 | |

Begin with a dual iteration. Choose $x_4$ to leave the basis, replaced by $x_3$, yielding Tableau 2.

**Tableau 2**

| | | ↓ $x_2$ | $x_4$ | |
|---|---|---|---|---|
| $z$ | 8 | −1 | 2 | |
| $x_3$ | 1 | 0 | −1 | |
| $x_5$ | 3 | [2] | 0 | → |
| $x_6$ | 3 | 1 | 1 | |
| $x_1$ | 4 | 1 | 1 | |

Note that the problem is now primal feasible. Replace $x_5$ by $x_2$, yielding Tableau 3, which is optimal.

**Tableau 3**

| | | $x_5$ | $x_4$ |
|---|---|---|---|
| $z$ | 9.5 | .5 | 2 |
| $x_3$ | 1 | 0 | −1 |
| $x_2$ | 1.5 | .5 | 0 |
| $x_6$ | 1.5 | −.5 | 1 |
| $x_1$ | 2.5 | −.5 | 1 |

The criss-cross method required four iterations (including the first arbitrary iteration), as opposed to the simplex method, which required five iterations (Example 4.1), and the primal-dual algorithm, which required six iterations (Example 6.2). Obviously, we cannot conclude anything about efficiency from one problem. Some statements regarding efficiency are given in the next section.

Illustrations of other aspects of the method are given in the problems at the end of the chapter.

## 6.3 THEORETICAL CONSIDERATIONS FOR THE CRISS-CROSS METHOD

We shall now consider some of the theoretical aspects of the criss-cross method.

**THEOREM 6.1**

THE CRISS-CROSS METHOD CONVERGES IN A FINITE NUMBER OF ITERATIONS TO AN OPTIMUM, IF ONE EXISTS.

PROOF:

If the algorithm converges in less than $2m + 2n$ iterations, there is nothing to prove. (Recall that the choice of $2m + 2n$ was arbitrary, but that it should be large, since the insured convergence procedure is cumbersome.) Hence, assume that $2m + 2n$ iterations have been utilized and the solution is neither primal nor dual feasible, for otherwise the problem would converge under either the primal or dual simplex method. Now state the problem so that it appears as in Figure 6.2.b (The problem of Fig. 6.2b is obtained by ignoring any violated inequalities.) Then, if a regularization constraint is added to problem b to assure a finite solution, the finiteness of the simplex method using the perturbation methods can be relied upon to solve for an optimal feasible solution to this newly generated problem in a finite number of iterations. The solution to problem b thereby obtained is clearly dual feasible. When the ignored constraints are reintroduced, however, the dual feasible solution may not be primal feasible. At this point, the dual simplex method using the perturbation methods will guarantee an optimal feasible solution in a finite number of iterations. This completes the proof.

Refer to the problem after $2m + 2n$ iterations as being in the convergence mode, and further define as the primal portion of the convergence mode the iterations taken in the convergence mode prior to a dual feasible solution being obtained. The remaining portion of the convergence mode is defined as the dual portion of the convergence mode.

COROLLARY 1:

If, in the dual portion of the convergence mode, a dual unbounded solution is obtained, no optimal feasible solution to the problem exists. The proof will not be given here, since it follows directly from the dual simplex method as described in Chapter 5.

COROLLARY 2:

If, in the final optimal feasible solution (at the completion of the convergence mode), the regularization constraint is binding, then there is no finite optimal feasible solution to the problem. The proof will not be given here, since it follows directly from the material concerning regularization described in Chapter 4.

COROLLARY 3:

If, at any stage, a nonbasic variable has *all* coefficients nonpositive and is attractive for entry into the basis (i.e., $z_i - c_j < 0$), then there is no feasible solution to the dual problem. The proof follows from writing the dual constraint corresponding to the variable, and noting that no sum of negative quantities can exceed some positive constant.

COROLLARY 4:

If, at any stage, a row has *all* coefficients nonnegative and has a negative $\boldsymbol{b}$ value, then there is no feasible solution to the primal problem. The proof follows from writing the constraint and noting that no sum of positive quantities can be less than some negative quantity.

There is then only one situation in which the algorithm can terminate that has not already been covered either by the existing theory of linear programming or by Theorem 6.1; that is, if the problem is neither primal nor dual feasible and no more iterations can be taken by way of the algorithm. The following theorem provides for such an outcome.

**THEOREM 6.2**

GIVEN A SOLUTION THAT IS NOT BOTH PRIMAL AND DUAL FEASIBLE, THE CONDITION THAT NEITHER A PRIMAL NOR A DUAL ITERATION, AS DESCRIBED IN THE ALGORITHM, CAN BE TAKEN, IS SUFFICIENT FOR NONEXISTENCE OF A FINITE OPTIMAL FEASIBLE SOLUTION TO THE LINEAR PROGRAMMING PROBLEM.

PROOF:

First, in the case of a primal or a dual feasible solution, the inability to undertake further iterations corresponds to an unbounded or infeasible problem solution, respectively. Hence, the proof for a solution that is neither primal nor dual feasible needs to be established. Suppose, therefore, that at a particular (neither primal nor dual feasible) solution to a linear programming problem, neither a primal nor a dual iteration can be taken. Let the problem at that iteration be represented as in Figure 6.2a. In accordance with the above suppositions that no iterations are possible using the algorithm, all the elements of $\boldsymbol{A}_{12}$ must be nonnegative and all the elements of $\boldsymbol{A}_{21}$ must be nonpositive; otherwise a primal or a dual iteration would be possible.

Now assume that an optimal feasible solution to the problem exists; that is, there exist nonnegative values $\boldsymbol{x}_1^*$, $\boldsymbol{x}_2^*$, $\boldsymbol{y}_1^*$, and $\boldsymbol{y}_2^*$, so that

$$\boldsymbol{A}_{11}\boldsymbol{x}_1^* + \boldsymbol{A}_{12}\boldsymbol{x}_2^* \leq -\boldsymbol{b}_1 \tag{6.1}$$

$$\boldsymbol{A}_{21}\boldsymbol{x}_1^* + \boldsymbol{A}_{22}\boldsymbol{x}_2^* \leq \boldsymbol{b}_2 \tag{6.2}$$

$$\boldsymbol{y}_1^{*\prime}\boldsymbol{A}_{11} + \boldsymbol{y}_2^{*\prime}\boldsymbol{A}_{21} \geq \boldsymbol{c}_1' \tag{6.3}$$

$$\boldsymbol{y}_1^{*\prime}\boldsymbol{A}_{12} + \boldsymbol{y}_2^{*\prime}\boldsymbol{A}_{22} \geq -\boldsymbol{c}_2' \tag{6.4}$$

($\boldsymbol{y}_1$ is $m_1$ by 1, and $\boldsymbol{y}_2$ is $m_2$ by 1.) No assumptions are made about the elements of $\boldsymbol{A}_{11}$ or $\boldsymbol{A}_{22}$. Note that (6.1) and (6.2) are the constraints of the primal problem, and that (6.3) and (6.4) are the constraints of the dual problem. $\boldsymbol{x}_1^*$ cannot be zero in all of its elements; if it were, the constraints of (6.1) become $\boldsymbol{A}_{12}\boldsymbol{x}_2^* \leq -\boldsymbol{b}_1$ which could not be satisfied for any $\boldsymbol{x}_2^* \geq \boldsymbol{0}$, because $\boldsymbol{b}_1 > \mathbf{0}$. Hence, at least one element of $\boldsymbol{x}_1^* > \mathbf{0}$. By a similar analysis of (6.3), $\boldsymbol{y}_1^*$ cannot be zero, and therefore must have at least one positive element. Multiplying (6.1) by $\boldsymbol{y}_1^{*\prime}$ and (6.3) by $\boldsymbol{x}_1^*$ yields

$$\boldsymbol{y}_1^{*\prime}\boldsymbol{A}_{11}\boldsymbol{x}_1^* + \boldsymbol{y}_1^{*\prime}\boldsymbol{A}_{12}\boldsymbol{x}_2^* \leq -\boldsymbol{y}_1^*\boldsymbol{b}_1' < 0 \tag{6.5}$$

and

$$\boldsymbol{y}_1^{*\prime}\boldsymbol{A}_{11}\boldsymbol{x}_1^* + \boldsymbol{y}_2^{*\prime}\boldsymbol{A}_{21}\boldsymbol{x}_1^* \geq \boldsymbol{c}_1'\boldsymbol{x}_1^* > 0 \tag{6.6}$$

and the strict inequalities are as indicated for the right-hand sides of (6.5) and (6.6).

Combining these inequalities and simplifying yields

$$\boldsymbol{y}_1^{*\prime}\boldsymbol{A}_{12}\boldsymbol{x}_2^* < 0 < \boldsymbol{y}_2^{*\prime}\boldsymbol{A}_{21}\boldsymbol{x}_1^* \tag{6.7}$$

which is a contradiction because of the nonnegativity constraints on the variables and the conditions that every element of $\boldsymbol{A}_{12}$ is greater than or equal to zero, and every element of $\boldsymbol{A}_{21}$ is less than or equal to zero. This completes the proof.

## Further Research

The convergence of the criss-cross method without the special convergence mode is still an open question worthy of further study. A number of proofs have been attempted, but none have been successful. It is conjectured that a proof can be devised. Perhaps the most promising approach attempted thus far is the following: If the method does not converge, then it must cycle. It is easy to show that two and three iteration cycles are not possible. Then, use induction to show that $n + 1$ iteration cycles are not possible, given that $n$ iteration cycles are not possible.

The problem of degeneracy has not been studied, although perturbation techniques can certainly be utilized. However, intuitively, it seems likely that, because of the alteration of primal and dual iterations, fewer degenerate solutions will be encountered with the criss-cross method as compared to the simplex method.

Empirical tests of the criss-cross method and its comparison with the simplex method have been undertaken [314] utilizing the CDC 6400 computer. The results indicate that the performance of the criss-cross method is significantly better than that of the simplex method, in terms of both time and number of iterations. Omitting problems that are either primal or dual feasible, the least favorable comparison of average times for the criss-cross method is about a 30 per cent reduction in computation time.

The method may readily be extended to a product-form method, but because a dual iteration is about 1.5 times as costly as a primal iteration, some number (greater than one) of primal iterations may be alternated with each dual iteration.

## 6.4 OTHER PRIMAL-DUAL ALGORITHMS

In addition to those presented in this chapter, we shall briefly mention a few other algorithms. The Mutual Primal-Dual Algorithm of Balinski and Gomory [16] employs a hierarchy of subtableaus (alternatively primal and dual feasible). It solves the linear programming problem by using a primal or dual simplex pivot choice rule, as appropriate, on the highest ordered subproblem until degeneracies occur, and then a dual or primal simplex pivot choice rule, as appropriate, on a higher order subproblem until the degeneracies are resolved. Subproblems are redefined as appropriate after each iteration. The procedure is continued until an optimal feasible solution is found, or evidence is generated indicating that no finite optimal feasible solution exists. No experience with the algorithm is cited, but the algorithm is compared with the Dantzig, Ford, and Fulkerson algorithm [66], and it is implied there that the computational advantage resides with the Balinski-Gomory approach. By virtue of the extensive computations that appear to be made within the coefficient matrix, it would seem cumbersome to employ an inverse form of this method.

The Minit Method of Llewellyn [200] and the symmetric method of Talacko and Rockafellar [275] are very similar to each other, as well as to the criss-cross method. The major differences are that, with slight variation, each method chooses either the primal or dual iteration that alters the objective function most (the two methods differ slightly in specification of criteria), whereas in the criss-cross method iterations are alternated using a simplex-like criterion (choosing the most violated constraint to obtain the variable to leave the basis). The characterization of an iteration is also not precisely the same as in the criss-cross method. Finally, convergence is not proven for either method, and a problem that does not converge has been found for the symmetric procedure. (That problem is solved without difficulty by the criss-cross method.)

Other methods have been developed; some are listed in the references. Since, to the author's knowledge, none of them is in widespread use, one can only conclude that if any of them are efficient, they have been overlooked.

### Selected Supplemental References

Section 6.1
[66], [194]
Sections 6.2, 6.3
[309], [314]
Section 6.4
[16], [130], [136], [149], [275]

General
[132], [248]

## 6.5 PROBLEMS

**1.** For the Dantzig-Ford-Fulkerson primal-dual algorithms in the restricted primal phase, why must only variables with $z_j - c_j = 0$ be considered for basis entry?

**2.** Solve the following problem using the criss-cross method. (This illustrates a possible unbounded solution—with respect to Fig. 6.2c only and not for the whole problem—of the dual in which an attempted dual iteration fails.)

$$\begin{aligned} \text{Maximize } z = x_1 + 3x_2 & \\ \text{subject to:} \quad x_1 \qquad & \geq 1 \\ x_1 + x_2 & \leq 2 \\ x_1, x_2 & \geq 0 \end{aligned}$$

**3.** Solve the following problem using the criss-cross method. (This illustrates a possible unbounded solution—with respect to Fig. 6.2b only and not for the whole problem—of the primal in which an attempted primal iteration fails.)

$$\begin{aligned} \text{Maximize } z = x_1 - 2x_2 - 12x_3 & \\ \text{subject to:} \quad x_2 \qquad & \geq 3 \\ -x_1 + x_2 \qquad & \leq 1 \\ -x_1 + .5x_2 + 10x_3 & \geq 6 \\ x_1, x_2, x_3 & \geq 0 \end{aligned}$$

**4.** Solve the following problem using the criss-cross method. (This problem has an infinite solution which is discovered at the first primal iteration.)

$$\begin{aligned} \text{Maximize } z = x_1 - 2x_2 - x_3 & \\ \text{subject to:} \quad x_2 \qquad & \geq 3 \\ x_1 - x_2 \qquad & \geq 1 \\ 6x_2 + 10x_3 & \geq 6 \\ x_1, x_2, x_3 & \geq 0 \end{aligned}$$

**5.** Solve the following problem using the criss-cross method. (For this problem no iterations are possible and the results of Theorem 6.2 are used, even though there is no conclusive unbounded column or infeasible row; hence there is no optimal feasible solution.)

$$\begin{aligned} \text{Maximize } z = x_1 - 3x_2 + x_3 & \\ \text{subject to:}\quad -x_1 \qquad\qquad & \leq 1 \\ -x_1 + 2x_2 + 4x_3 & \leq -2 \\ x_1 + x_2 - x_3 & \leq -1 \\ x_1, x_2, x_3 & \geq 0 \end{aligned}$$

**6.** By dominance considerations, $x_2$ and the first constraint in problem 5 can be ignored. Show that the problem that remains has precisely the same discovery of the infeasible solution, i.e., through using the results of Theorem 6.2.

$$\begin{aligned} \text{Maximize } z = x_1 + x_3 & \\ \text{subject to:}\quad -x_1 + 4x_3 & \leq -2 \\ x_1 - x_3 & \leq -1 \\ x_1, x_3 & \geq 0 \end{aligned}$$

Examine the set of feasible solutions of this problem and its dual.

**7.** Solve problem 2 using the primal-dual algorithm.

**8.** Solve problem 3 using the primal-dual algorithm.

**9.** Solve problem 4 using the primal-dual algorithm.

**10.** Solve problem 5 using the primal-dual algorithm.

**11.** Solve problem 6 using the primal-dual algorithm.

**12.** Develop a flow chart for a product-form criss-cross method.

**13.** Prove that the method for developing a dual feasible starting solution for the primal-dual algorithm, as given in section 6.1, is valid.

**14.** Systematic rules (whose efficiency has not been explored) for introducing a variable into an equality constraint in row $r$ are as follows (using the notation of Fig. 6.2):

a. $b_r$ must be nonnegative. If it is not, multiply the constraint through by $-1$.
b. For $a_{rj} > 0$, and $c_j < 0$, choose a variable for which $|c_j/a_{rj}|$ is maximum. That is the variable which is to enter the basis.
c. If no variable satisfies step b, then for $a_{rj} > 0$, and $c_j \geq 0$, choose a variable for which $c_j/a_{rj}$ is minimized. That is the variable to enter the basis.
d. If no variable satisfies step c, then if $b_r > 0$ (and all $a_{rj} < 0$) there is no solution; otherwise proceed to step e.
e. If all coefficients of the constraint are zero, then the constraint is

vacuous and may be dropped. Otherwise, multiply the equation through by $-1$ and go to step b.

Explain the rationale behind this procedure, and solve Example 6.4 using it.

**15.** Systematic rules (whose efficiency has not been explored) for introducing a variable in column $k$ unrestricted in sign into the basis are as follows (using the notation of Fig. 6.2):

a. $c_k$ must be nonnegative; if it is negative, multiply the column by $-1$.

b. For $a_{ik} > 0$, and $b_i < 0$, choose a row $i$ for which $|b_i/a_{ik}|$ is maximum. That is the row containing the basic variable which is to leave the basis.

c. If no variable satisfies step b, then for $a_{ik} > 0$, and $b_i \geq 0$, choose a row $i$ for which $b_i/a_{ik}$ is minimum. That is the row containing the basic variable which is to leave the basis.

d. If no variable satisfies step c, then if $c_k > 0$ (and all $a_{ik} < 0$) there is no solution; otherwise proceed to step e.

e. If all the coefficients of the variable are zero, the variable is vacuous and may be dropped. Otherwise, multiply the variable by $-1$ and go to step b.

Explain the rationale behind this procedure, and solve Example 4.9 using it.

**16.** Solve problem 15 of Chapter 2 using the primal-dual algorithm and the criss-cross method.

**17.** Solve problem 16 of Chapter 2 using the primal-dual algorithm and the criss-cross method.

**18.** Solve problem 18 of Chapter 2 using the primal-dual algorithm and the criss-cross method.

**19.** Solve problem 5 of Chapter 4 using the primal-dual algorithm and the criss-cross method. In addition, solve it using the criss-cross method involving the procedure proposed in problem 14.

# 7

# Postoptimal Analysis and Parametric Programming

In presenting the various aspects of linear programming, we did not explore the question of the precision of data used in the problem. For example, in the dog food example we assumed that 180,000 pounds of meat were available per month. There are a number of ways in which this figure could have been incorrect. First, 180,000 may be an approximation, and perhaps 180,527.9 is the correct value. Second, some of the meat may have spoiled, and only 176,500 pounds may be available. Alternatively, the contract supplier may be willing to provide more on the same terms as the original at the time of delivery. In any event, depending on which coefficients may be incorrect or altered, it is desired to know in what way, if any, the solution to the original problem is altered. Answers to questions relating to the optimality (and feasibility) of a particular basic solution can be obtained. If the optimal basis for the original problem is not optimal for the current values, a procedure is available for proceeding to a new optimum.

Another reason for exploring changes in problem formulation and how they affect the optimal solution is that the problem data may be best estimates, with the actual values unknown. It may therefore be useful to know how sensitive the optimal solution is to changes in input data.

Still another problem is the question of exploring all optimal solutions to a linear programming problem as a function of some parameter. To continue the above example, suppose that the availability of meat was uncertain, but that the price of meat was fixed at what we had contracted. It would be useful to know the optimal production combinations for *any* amount of meat available, from some lower limit (perhaps zero) to some upper limit (possibly unlimited). Parametric programming, which is essentially a variation of sensitivity analysis, can be used for this purpose.

In this chapter we shall explore the above methods, as well as methods for adding variables and constraints to linear programming problems.

## 7.1 SENSITIVITY ANALYSIS AND CHANGING PROBLEM COEFFICIENTS

We shall develop sensitivity analysis by treating each of three sets of coefficients:

1. The objective function coefficients.
2. The requirements vector entries.
3. The matrix of coefficients.

For each of the above sets of coefficients, we shall explore three questions:

1. Holding all other factors equal, over what range of values of a given coefficient is the original optimal basis still optimal? This analysis is usually referred to as sensitivity analysis.
2. For a given change in a coefficient, or in a number of coefficients, is the original optimal basis optimal? If so, in what way are the optimal objective function values and primal and dual variable solution values altered?
3. If the answer to the first part of question 2 is negative, what is the optimal solution to the problem, having made any necessary changes? A procedure for doing this (that does not require re-solving the problem from the original starting point) is presented. If the changes made to the original problem are relatively minor, then this procedure is generally superior to re-solving the problem completely, starting from the initial solution. If the changes are great enough, it is better to re-solve the problem completely.

### The Objective Function Coefficients

We repeat in Figure 7.1 a simplex tableau (from Fig. 3.1) which gives a partition of basic and nonbasic variables. To examine the effects of changes in objective function coefficients, we add a parameter $\theta$ to each objective function coefficient, one at a time, and find the permissible range of values of $\theta$ for each coefficient so that the optimality and feasibility conditions are fulfilled. That is, the conditions $\boldsymbol{c}'_B\boldsymbol{B}^{-1}\boldsymbol{N} - \boldsymbol{c}'_N \geq \mathbf{0}$ and $\boldsymbol{B}^{-1}\boldsymbol{b} \geq 0$ must still be fulfilled. The former of these two conditions is the only one of the two which contains either $\boldsymbol{c}_N$ or $\boldsymbol{c}_B$.

| | OBJECTIVE FUNCTION VALUE AND VALUE OF THE BASIC VARIABLES | BASIC VARIABLES $x_B$ | NONBASIC VARIABLES $x_N$ |
|---|---|---|---|
| $z$ | $c'_B B^{-1} b$ | $0'$ | $c'_B B^{-1} N - c'_N$ |
| $x_B$ | $B^{-1} b$ | $I$ | $B^{-1} N$ |

**Figure 7.1**

A Simplex Tableau in Matrix Form

Consider first a nonbasic variable $x_{N_j}$. Let $e_j$ be a unit vector whose unit element is in position $j$. The first condition above is satisfied for any value of $c_{N_j} + \theta$ in the range

(7.1) $$-\infty \leq c_{N_j} + \theta \leq c'_B B^{-1} N_j$$

This statement may be proven by noting that the vector of reduced costs becomes

$$c'_B B^{-1} N - c'_N - e'_j \theta$$

which differs from the original only for the $j^{\text{th}}$ nonbasic variable. That element and its restriction is

$$c'_B B^{-1} N_j - c_{N_j} - \theta \geq 0$$

which is satisfied so long as $\theta$ is in the range indicated, given a partition of the basic and nonbasic variables, by expression (7.1).

EXAMPLE 7.1:

Tableau 1 is the optimal tableau for the dog food problem. Over what range of the contribution to profits of unused cereal ($x_3$) is the solution optimal? The contribution to profits of unused cereal is otherwise zero, but we change it to $\theta$. The reduced cost entry for $x_3$ then becomes $.035 - \theta$.

**Tableau 1**

| | | $x_1$ | $x_2$ | $x_3$ | $x_4$ | $x_5$ |
|---|---|---|---|---|---|---|
| $z$ | 71,400 | 0 | 0 | .035 | .35 | 0 |
| $x_5$ | 50,000 | 0 | 0 | .5 | −1 | 1 |
| $x_2$ | 90,000 | 0 | 1 | .75 | −.5 | 0 |
| $x_1$ | 60,000 | 1 | 0 | −.5 | 1 | 0 |

We can see that the present solution will be optimal for any contribution to profits of unused cereal ranging from $-\infty$ to $+.035$. If the contribution to profits of unused cereal exceeds .035, the above solution is no longer optimal. Then $x_3$ will enter the basis, replacing $x_5$, and possibly additional iterations will have to be utilized. Similarly, for nonbasic variable $x_4$ (unused meat), the original optimal basic solution will be optimal so long as the contribution to profit of unused meat is between $-\infty$ and .35. For a contribution of greater than .35 the above solution will no longer be optimal, so $x_4$ will replace $x_1$ and additional iterations may have to be utilized.

For basic variables, an entirely analogous argument holds, except that *all nonbasic reduced costs* must be examined to determine the limits. As in the case of nonbasic variables, replace $\boldsymbol{c}_B$ by $\boldsymbol{c}_B + \theta \boldsymbol{e}_j$ and check the optimality conditions:

$$(\boldsymbol{c}'_B + \theta \boldsymbol{e}'_j)\boldsymbol{B}^{-1}\boldsymbol{N} - \boldsymbol{c}'_N \geq \boldsymbol{0}$$

or

$$\boldsymbol{c}'_B\boldsymbol{B}^{-1}\boldsymbol{N} + \theta \boldsymbol{e}'_j\boldsymbol{B}^{-1}\boldsymbol{N} - \boldsymbol{c}'_N \geq \boldsymbol{0} \tag{7.2}$$

Before proceeding, note that $\boldsymbol{e}'_j\boldsymbol{B}^{-1}$ is simply the $j^{\text{th}}$ row of the inverse, and that $\boldsymbol{e}'_j\boldsymbol{B}^{-1}\boldsymbol{N}$ is the $j^{\text{th}}$ row of the updated coefficients corresponding to the nonbasic variables. Also, $\boldsymbol{c}'_B\boldsymbol{B}^{-1}\boldsymbol{N} - \boldsymbol{c}'_N$ is the vector of reduced costs of nonbasic variables. The procedure then is to solve the linear inequalities in (7.2), one for each nonbasic variable, to find the range of values of $\theta$ (i.e., the most limiting inequalities) for which the solution is optimal. The associated change in objective function value is found by computing the objective function value as a function of $\theta$.

$$(\boldsymbol{c}'_B + \theta \boldsymbol{e}'_j)\boldsymbol{B}^{-1}\boldsymbol{b} = \boldsymbol{c}'_B\boldsymbol{B}^{-1}\boldsymbol{b} + \theta \boldsymbol{e}'_j\boldsymbol{B}^{-1}\boldsymbol{b}$$

Furthermore, $\boldsymbol{e}'_j\boldsymbol{B}^{-1}\boldsymbol{b}$ is just the value in the original optimal basis of the variable basic in row $j$, and hence the associated change in the objective function value is $\theta$ times the number of units of the corresponding basic variable in the original optimal solution.

EXAMPLE 7.2:

Find the ranges in cost, for each basic variable for which the original optimal solution holds, for the dog food example whose original optimal tableau is given in Example 7.1.

For $x_5$, the unused packaging capacity on Frisky Pup dog food, suppose that such capacity has a contribution to profits of $\theta$ per unit. Then the reduced costs of the nonbasic variables would be $\boldsymbol{c}'_B\boldsymbol{B}^{-1}\boldsymbol{N} + \theta \boldsymbol{e}'_j\boldsymbol{B}^{-1}\boldsymbol{N} - \boldsymbol{c}'_N$

and from (7.2) we have

$$(0 \quad .42 \quad .56)\begin{pmatrix} .5 & -1 & 1 \\ .75 & -.5 & 0 \\ -.5 & 1 & 0 \end{pmatrix}\begin{pmatrix} 1 & 0 \\ 0 & 1 \\ 0 & 0 \end{pmatrix}$$

$$+ \theta(1 \quad 0 \quad 0)\begin{pmatrix} .5 & -.1 & 1 \\ .75 & -.5 & 0 \\ -.5 & 1 & 0 \end{pmatrix}\begin{pmatrix} 1 & 0 \\ 0 & 1 \\ 0 & 0 \end{pmatrix} - (0 \quad 0) \geq \mathbf{0}$$

where $c_B' = (0 \quad .42 \quad .56)$, corresponding respectively to the objective function coefficients of $x_5$, $x_2$, and $x_1$, or

$$(.035 \quad .35) + \theta(1 \quad 0 \quad 0)\begin{pmatrix} .5 & -1 \\ .75 & -.5 \\ -.5 & 1 \end{pmatrix} \geq \mathbf{0}$$

or

$$(.035 \quad .35) + \theta(.5 \quad -1) \geq \mathbf{0}$$

As indicated above, we can obtain the inequalities on $\theta$ *directly* from the optimal simplex tableau. The coefficients of $\theta$ are found in the $x_5$ row (the row of the basic variable under consideration) and from the $x_3$ and $x_4$ columns (the nonbasic variables). The constant terms are the reduced costs of $x_3$ and $x_4$ (the nonbasic variables).

$$\text{For } x_3\text{:} \quad .035 + .5\theta \geq 0 \Longrightarrow \theta \geq .07$$
$$\text{For } x_4\text{:} \quad .35 + \theta(-1) \geq 0 \Longrightarrow \theta \leq .35$$

The original solution is optimal, holding other factors equal, for any contribution to profits of unused Frisky Pup packaging capacity from $-\$.07$ to \$.35. For any smaller contribution (that is, more negative than $-.07$), $x_3$ would replace $x_5$ in the basis in the first iteration proceeding to a new optimum. For any contribution larger than .35, $x_4$ would replace $x_1$ in the basis in the first iteration proceeding to a new optimum.

For $x_2$, the number of packages of Husky Hound dog food, suppose that the contribution to overhead and profits were changed from .42 to $.42 + \theta$. Then the reduced costs for the nonbasic variables are, according to (7.2),

$$(.035 \quad .35) + \theta(0 \quad 1 \quad 0)\begin{pmatrix} .5 & -1 \\ .75 & -.5 \\ -.5 & 1 \end{pmatrix} \geq \mathbf{0}$$

or

$$(.035\ .35) + \theta(.75\ -.5) \geq \mathbf{0}$$

or

$$\text{for } x_3\text{: } .035 + .75\theta \geq 0 \Longrightarrow \theta \geq -.0466$$
$$\text{for } x_4\text{: } .35 - .5\theta \geq 0 \Longrightarrow \theta \leq .7$$

so that the original solution is optimal for any contribution to profits for Husky Hound dog food from .42 − .0466 to .42 + .7, or from .3733 to 1.12. If the contributions to profits were below .3733, $x_3$ would replace $x_5$ in the basis in the first iteration proceeding to a new optimum, and if contributions to profits were to exceed 1.12, $x_4$ would replace $x_1$ in the basis in the first iteration proceeding to a new optimum.

For $x_1$, the number of packages of Frisky Pup dog food, suppose that the contribution to overhead and profits were changed from .56 to .56 + $\theta$. Then the reduced costs for the nonbasic variables are, according to (7.2),

$$(.035 \quad .35) + \theta(0 \quad 0 \quad 1)\begin{pmatrix} .5 & -1 \\ .75 & -.5 \\ -.5 & 1 \end{pmatrix} \geq \mathbf{0}$$

or

$$(.035 \quad .35) + \theta(-.5 \quad 1) \geq \mathbf{0}$$

or

$$\text{for } x_3\text{: } .035 - .5\theta \geq 0 \Longrightarrow \theta \leq .07$$
$$\text{for } x_4\text{: } .35 + \theta \geq 0 \Longrightarrow \theta \geq -.35$$

so that the original solution is optimal for any contribution to profits from Frisky Pup dog food from .56 − .35 to .56 + .07, or from .21 to .63. For a contribution above .63, $x_3$ would replace $x_5$ in the basis, and for a contribution below .21, $x_4$ would replace $x_1$ in the basis. Again the above calculations can be simplified by using the tableau to get the coefficients of $\theta$.

To illustrate how a change in basis would be made, consider an example building in part upon Example 7.2.

EXAMPLE 7.3:

Suppose that the contribution to overhead and profits for $x_2$ in the dog food problem were .30. What is the optimal solution to the problem that results?

Using the calculations made in the previous example (since $\theta = -.12$ for $x_2$), the reduced costs become $.035 + .75\,(-.12) = -.055$ and $.35 - .5(-.12) = .41$ for $x_3$ and $x_4$, respectively. Hence $x_3$ enters the basis, replacing $x_5$ and yielding Tableau 3 on p. 62, except for the $z$-row. The reduced costs, using the same analysis as before, become $.42 + 1(-.12) = +.30$ and $-.07 - 1.5(-.12) = .11$ for $x_4$ and $x_5$, respectively. The new solution is therefore optimal. For this problem only one iteration was required proceeding from the previous optimal, whereas in general more or less iterations may be required. Further, given many significant changes, proceeding from a previous optimal solution may not *always* be the best strategy; it may be better to solve the altered problem from the beginning.

Adjustments may be made to more than one cost at a time in a manner precisely analogous to the above procedure, but we shall not present any examples of this until we present an example of changing requirement vector entries as well.

### The Requirement Vector Entries

The analysis for the requirement vector entries is entirely symmetric to that of the objective function coefficients. In fact, it is precisely the same analysis on the dual problem. From earlier analysis and by examination of the tableau of Figure 7.1, changing an entry of $\boldsymbol{b}$ by some amount $\theta$, the original basis will be optimal so long as $\boldsymbol{B}^{-1}\boldsymbol{b} \geq \boldsymbol{0}$. In a manner analogous to that in the previous section, if a component—say in the original constraint $j$—is changed by an amount $\theta$, then the previous optimal basis is optimal so long as the following inequality is fulfilled:

$$\boldsymbol{B}^{-1}(\boldsymbol{b} + \theta \boldsymbol{e}_j) \geq \boldsymbol{0} \tag{7.3}$$

or

$$\boldsymbol{B}^{-1}\boldsymbol{b} + \theta \boldsymbol{B}^{-1}\boldsymbol{e}_j \geq \boldsymbol{0}$$

Observe that $\boldsymbol{B}^{-1}\boldsymbol{e}_j$ is just column $j$ of the inverse or, equivalently, the updated column corresponding to the original basic variable of constraint $j$. The inequalities of (7.3) must be examined in specific cases to determine the limiting values of $\theta$.

EXAMPLE 7.4:

Compute the range of validity for the optimal basis of Example 7.1 for the original $\boldsymbol{b}$ values. First consider cereal (constraint 1) and replace the amount of cereal available, 240,000 pounds, by 240,000 + $\theta$ pounds. Then by (7.3) the basis is optimal if the following inequality is fulfilled:

$$\begin{pmatrix} .5 & -1 & 1 \\ .75 & -.5 & 0 \\ -.5 & 1 & 0 \end{pmatrix} \left[ \begin{pmatrix} 240{,}000 \\ 180{,}000 \\ 110{,}000 \end{pmatrix} + \theta \begin{pmatrix} 1 \\ 0 \\ 0 \end{pmatrix} \right] \geq \boldsymbol{0}$$

or

$$\begin{pmatrix} 50{,}000 \\ 90{,}000 \\ 60{,}000 \end{pmatrix} + \begin{pmatrix} .5 \\ .75 \\ -.5 \end{pmatrix} \theta \geq \mathbf{0}$$

or

$$\begin{aligned} x_5 &= 50{,}000 + .5\ \theta \geq 0 \Longrightarrow \theta \geq -100{,}000 \Longrightarrow \theta \geq -100{,}000 \\ x_2 &= 90{,}000 + .75\theta \geq 0 \Longrightarrow \theta \geq -120{,}000 \\ x_1 &= 60{,}000 - .5\ \theta \geq 0 \Longrightarrow \theta \leq \ \ 120{,}000 \Longrightarrow \theta \leq 120{,}000 \end{aligned} \tag{7.4}$$

so that the original optimal basis is optimal for any value of cereal availability from 240,000 – 100,000 to 240,000 + 120,000, or from 140,000 to 360,000 pounds, and the solution values are as indicated in (7.4). (Only the most restrictive inequalities of expression (7.3) are used: in the case of expression (7.4), the first and third constraints.) If less than 140,000 pounds of cereal were available, then the first iteration to proceed toward a new optimum would be to replace $x_5$ in the basis by $x_4$, and if more than 360,000 pounds of cereal were available, then the first iteration to proceed toward a new optimum would be to replace $x_1$ in the basis by $x_3$.
Observe that the inequalities (7.3) can be written directly from the previous optimal tableau. We shall make use of this in the following examples for the second and third constraints. Were meat availability 180,000 + $\theta$ pounds instead of 180,000 pounds, then the basis is optimal so long as (7.3) is fulfilled, or so long as

$$\begin{pmatrix} x_5 \\ x_2 \\ x_1 \end{pmatrix} = \begin{pmatrix} 50{,}000 \\ 90{,}000 \\ 60{,}000 \end{pmatrix} + \theta \begin{pmatrix} -1 \\ -.5 \\ 1 \end{pmatrix} = \begin{pmatrix} 50{,}000 - \theta \\ 90{,}000 - .5\theta \\ 60{,}000 + \theta \end{pmatrix} \geq \mathbf{0} \tag{7.5}$$

or

$$\begin{aligned} 50{,}000 - \ \ \theta \geq 0 &\Longrightarrow \theta \leq \ \ 50{,}000 \Longrightarrow \theta \leq 50{,}000 \\ 90{,}000 - .5\theta \geq 0 &\Longrightarrow \theta \leq \ 180{,}000 \\ 60{,}000 + \ \ \theta \geq 0 &\Longrightarrow \theta \geq -60{,}000 \Longrightarrow \theta \geq -60{,}000 \end{aligned}$$

The original optimal basis is optimal for any value of meat available from 180,000 – 60,000 to 180,000 + 50,000, or from 120,000 to 230,000 pounds, with changes as indicated in expression (7.5), i.e., $x_2 = 90{,}000 - .5\theta$.
For the third constraint expression (7.3) becomes

$$\begin{pmatrix} x_5 \\ x_2 \\ x_1 \end{pmatrix} = \begin{pmatrix} 50{,}000 \\ 90{,}000 \\ 60{,}000 \end{pmatrix} + \theta \begin{pmatrix} 1 \\ 0 \\ 0 \end{pmatrix} \geq \mathbf{0}$$

which is satisfied as long as $-50{,}000 \leq \theta < \infty$.

The preceding analysis and examples for the requirements vector are analogous to the analysis of the basic variable objective function coefficients. For basic variables we can determine the limiting values of the constraints $x_j \geq 0 + \theta$, as in the analysis of the nonbasic variable objective function coefficients.

EXAMPLE 7.5:

Determine the ranges on the constraints $x_j \geq 0 + \theta$ for basic variables.
For $x_5$, the basis is optimal for $x_5 \geq \theta$ for any $\theta$ between $-\infty$ and 50,000.
For $x_2$, the basis is optimal for $x_2 \geq \theta$ for any $\theta$ between $-\infty$ and 90,000.
For $x_1$, the basis is optimal for $x_1 \geq \theta$ for any $\theta$ between $-\infty$ and 60,000.

Thus far we have been treating lower bound constraints such as $x_j \geq 5$ explicitly, but the above example hints that we do not have to treat them as such. We shall show how to handle them implicitly in Chapter 8.

Before proceeding to the matrix of coefficients, we shall show how to handle more than one change at a time.

EXAMPLE 7.6:

Two unusual months are forecast for Canine Products. In the first month, the figures are to be changed as follows: Because of price changes, the contribution to overhead and profit will be \$.63 for Frisky Pup dog food and \$.49 for Husky Hound. 200,000 pounds of cereal and 150,000 pounds of meat have been contracted for.

For the second month, only 50,000 units of Frisky Pup can be packaged; otherwise the data are the same as in the original example, except that the contribution of Frisky Pup is \$.63 per package.

By using the optimal tableau in Tableau 1, Example 7.1, it can be seen that

$$\boldsymbol{B}^{-1} = \begin{pmatrix} .5 & -1 & 1 \\ .75 & -.5 & 0 \\ -.5 & 1 & 0 \end{pmatrix}$$

and that for

$$\boldsymbol{b} = \begin{pmatrix} 200{,}000 \\ 150{,}000 \\ 110{,}000 \end{pmatrix}$$

$$\boldsymbol{B}^{-1}\boldsymbol{b} = \begin{pmatrix} 60{,}000 \\ 75{,}000 \\ 50{,}000 \end{pmatrix}$$

which is feasible.

Checking the optimality conditions, it is found that $\boldsymbol{c}_B'\boldsymbol{B}^{-1} = (.0 \;\; .49 \;\; .63)\boldsymbol{B}^{-1} = (.0525 \;\; .385 \;\; .0)$, and that $\boldsymbol{c}_B'\boldsymbol{B}^{-1}\boldsymbol{N} - \boldsymbol{c}_N' = (.0525 \;\; .385) \geq \boldsymbol{0}$, so that the optimal solution has the same basis, but with $x_1 = 50{,}000$, $x_2 = 75{,}000$, and $x_5 = 60{,}000$. Profits in this month would be $c'_B B^{-1} b$, or \$68,250. Hence the original optimal basis is correct, and no additional iterations are required.

In the second case we proceed in exactly the same manner, with the indicated changes.

$$B^{-1}b = B^{-1}\begin{pmatrix} 240{,}000 \\ 180{,}000 \\ 50{,}000 \end{pmatrix} = \begin{pmatrix} -10{,}000 \\ 90{,}000 \\ 60{,}000 \end{pmatrix}$$

which is not nonnegative, and

$$c'_B B^{-1} N - c'_N = (0 \quad .42 \quad .63) B^{-1} N - (0 \quad 0) = (0 \quad .42)$$

which is still nonnegative. Additional iterations are carried out using the dual simplex method; the sequence of tableaus is given below.

**Tableau 1**

| | | $x_1$ | $x_2$ | $x_3$ | $x_4$ ↓ | $x_5$ | |
|---|---|---|---|---|---|---|---|
| $z$ | 75,600 | 0 | 0 | 0 | .42 | 0 | |
| $x_5$ | −10,000 | 0 | 0 | .5 | [−1] | 1 | ⟶ |
| $x_2$ | 90,000 | 0 | 1 | .75 | −.5 | 0 | |
| $x_1$ | 60,000 | 1 | 0 | −.5 | 1 | 0 | |

**Tableau 2**

| | | $x_1$ | $x_2$ | $x_3$ | $x_4$ | $x_5$ |
|---|---|---|---|---|---|---|
| $z$ | 71,400 | 0 | 0 | .21 | 0 | .42 |
| $x_4$ | 10,000 | 0 | 0 | −.5 | 1 | −1 |
| $x_2$ | 95,000 | 0 | 1 | .50 | 0 | .5 |
| $x_1$ | 50,000 | 1 | 0 | 0 | 0 | −1 |

Tableau 2 is optimal; it is purely a coincidence that the objective function happens to have the same value as that of the original problem.

### Matrix of Coefficients Entries

We shall briefly consider sensitivity analysis on the matrix of coefficients. Although the sensitivity of the matrix of coefficients to errors is important, the computational aspects are not as easy as the others, and therefore such sensitivity analysis is not as frequently undertaken. For any nonbasic variable, we can compute the entry of the vector $\boldsymbol{c}'_B\boldsymbol{B}^{-1}\boldsymbol{N} - \boldsymbol{c}'_N$ as a function of a parameter $\theta$, and then find the limits of $\theta$ so that the reduced costs are nonpositive. A change in such an element may be carried out by altering the old vector and computing its reduced costs. If the reduced cost of such an altered vector is nonnegative, the original solution is still optimal; otherwise, additional iterations are necessary.

If the element to be altered is contained in a basic vector, the procedure is somewhat different. To find the range, a new inverse must be computed incorporating the parameter. However, the computation of such an inverse requires only one "iteration." The checking of $\boldsymbol{B}^{-1}\boldsymbol{b}$, $\boldsymbol{c}'_B\boldsymbol{B}^{-1}\boldsymbol{N} - \boldsymbol{c}'_N$ for limiting values of $\theta$ must be done, as will be shown in an example. To alter an element in a basic vector, it appears reasonable to append the vector as altered, and then to proceed by designating the corresponding basic variable to leave the basis. The new optimum solution can be found by the criss-cross method (or the simplex method). These points are illustrated in Exampl• 7.7.

EXAMPLE 7.7:

Given the problem below and the corresponding optimal Tableau 1, find the ranges of the matrix elements which have $\theta_i$'s appended (one at a time), and determine the corresponding optimal solutions for each element's range.

$$
\begin{aligned}
\text{Maximize } z = \quad & x_1 - x_4 + .5x_5 \\
\text{subject to:} \quad & (1+\theta_3)x_1 + 2x_4 + (1+\theta_2)x_5 = 4 \\
& x_1 + x_2 + x_4 + (1+\theta_1)x_5 = 5 \\
& x_3 + x_5 = 4 \\
& x_1, \ldots, x_5 \geq 0
\end{aligned}
$$

where $\theta_i$ denotes the element to be varied at step $i$. The optimal solution for all $\theta_i = 0$ is given in Tableau 1. (For now, ignore the appended column.)

**Tableau 1**

| | | $x_1$ | $x_2$ | $x_3$ | $x_4$ | $x_5$ | $x'_1$ |
|---|---|---|---|---|---|---|---|
| $z$ | 4 | 0 | 0 | 0 | 3 | $\frac{1}{2}$ | $\theta_3$ |
| $x_1$ | 4 | 1 | 0 | 0 | 2 | 1 | $1+\theta_3$ |
| $x_2$ | 1 | 0 | 1 | 0 | $-1$ | 0 | $-\theta_3$ |
| $x_3$ | 4 | 0 | 0 | 1 | 0 | 1 | 0 |

*First alteration, $\theta_1$:* The inverse of the optimal basis (for $\theta_3 = 0$) can be shown to be

$$\boldsymbol{B}^{-1} = \begin{pmatrix} 1 & 0 & 0 \\ -1 & 1 & 0 \\ 0 & 0 & 1 \end{pmatrix}$$

and the reduced cost of $x_5$ is

$$\begin{aligned} \boldsymbol{c}'_B \boldsymbol{B}^{-1} \boldsymbol{N}_j - c_{N_j} &= (1 \quad 0 \quad 0) \left[ \begin{pmatrix} 1 & 0 & 0 \\ -1 & 1 & 0 \\ 0 & 0 & 1 \end{pmatrix} \begin{pmatrix} 1 \\ 1 + \theta_1 \\ 1 \end{pmatrix} \right] - .5 \\ &= (1 \quad 0 \quad 0) \begin{pmatrix} 1 \\ \theta_1 \\ 1 \end{pmatrix} - .5 = 1 - .5 = .5 \end{aligned}$$

for all values of $\theta_1$, so that the solution of Tableau 1 is optimal over the range of all values of $\theta_1$.

*Second alteration, $\theta_2$:* Following the first case, the reduced cost of $x_5$ in Tableau 1 becomes

$$(1 \quad 0 \quad 0) \begin{pmatrix} 1 + \theta_2 \\ 1 \\ 1 \end{pmatrix} - .5 = 1 + \theta_2 - .5$$

which is nonnegative for $-.5 \leq \theta_2 \leq \infty$. For $\theta_2 < -.5$, $x_5$ enters the basis, replacing $x_1$.

*Third alteration, $\theta_3$:* For $x_1$ of the altered form, i.e., with $\theta_3$ different from zero, we would have an updated column for $x_1$, in Tableau 1, of the following form:

$$\boldsymbol{B}^{-1} \begin{pmatrix} 1 + \theta_3 \\ 1 \\ 0 \end{pmatrix} = \begin{pmatrix} 1 + \theta_3 \\ -\theta_3 \\ 0 \end{pmatrix}$$

which is the augmented column of Tableau 1. (The $z_j - c_j$ entry for the augmented column is calculated in the standard way.) The augmented column would be correct were $x_1$ nonbasic, but since $x_1$ is basic we must first replace $x_1$ in the basis by $x'_1$. This yields Tableau 2, with

$$\boldsymbol{B}^{-1} = \begin{pmatrix} \dfrac{1}{1 + \theta_3} & 0 & 0 \\ -\dfrac{1}{1 + \theta_3} & 1 & 0 \\ 0 & 0 & 1 \end{pmatrix}$$

In Tableau 2 and what follows we suppress the subscript on $\theta_3$ and instead use $\theta$. (Some care must be exercised when solving the inequalities for $\theta$ because the sign of $\theta$ is not, in general, known.)

**Tableau 2**

| | | $x_1'$ | $x_2$ | $x_3$ | $x_4$ | $x_5$ |
|---|---|---|---|---|---|---|
| $z$ | $4 - (4\theta/(1+\theta))$ | 0 | 0 | 0 | $3 - (2\theta/(1+\theta))$ | $\frac{1}{2} - (\theta/(1+\theta))$ |
| $x_1'$ | $4/(1+\theta)$ | 1 | 0 | 0 | $2/(1+\theta)$ | $1/(1+\theta)$ |
| $x_2$ | $1 + (4\theta/(1+\theta))$ | 0 | 1 | 0 | $-1 + (2\theta/(1+\theta))$ | $\theta/(1+\theta)$ |
| $x_3$ | 4 | 0 | 0 | 1 | 0 | 1 |

Tableau 2 is optimal so long as the solution is primal feasible, i.e.,

$$4/(1+\theta) \geq 0 \Longrightarrow \theta > -1$$

$$1 + (4\theta/(1+\theta)) \geq 0 \Longrightarrow \begin{cases} \theta \geq -.2 \\ \text{or} \\ \theta < -1 \end{cases}$$

$$4 \geq 0$$

and so long as the solution is dual feasible, i.e.,

$$3 - (2\theta/(1+\theta)) \geq 0 \Longrightarrow \begin{cases} \theta > -1 \\ \text{or} \\ \theta \leq -3 \end{cases}$$

$$\tfrac{1}{2} - (\theta/(1+\theta)) \geq 0 \Longrightarrow \begin{cases} \theta > -1 \\ \text{and} \\ \theta \leq 1 \end{cases}$$

or so long as $-.2 \leq \theta \leq 1$, the net effect of all the constraints on $\theta$. For $\theta < -.2$, $x_2$ leaves the basis, replaced by $x_4$. For $\theta > 1$, $x_5$ enters the basis, replacing $x_3$.

## 7.2 PARAMETRIC PROGRAMMING

Parametric programming is a means for finding all possible optimal solutions to a linear programming problem in which one or more coefficients are a linear function of some parameter. All optimal solutions are determined for a specified range of the parameter. Parametric programming is an extension of sensitivity analysis. The usual procedure is to find an optimal solution for the parameter (which we shall continue to call $\theta$) for one of the limits of the range of desired values of that parameter. Then the range of the parameter for which that basic solution is optimal is determined. Then the optimal solution for the parameter in an adjacent range is determined, and the process is repeated until all desired solutions have been found. Although parametric programming can in principle be done on any coefficient or combination of coefficients, two particular types are most often used, primarily because of their relatively easy computations:

1. Making one or more requirements vector entries a linear function of a parameter.
2. Making one or more objective function coefficients a linear function of a parameter.

We shall illustrate one example of each type. In addition, we shall present a parametric analysis of a coefficient of the matrix of coefficients (an analysis which is cumbersome and seldom carried out).

Although the formal analysis is not presented, it follows directly from that of the previous section.

EXAMPLE 7.8: (Parametric right-hand side)

Find the optimal solution to the dog food problem as a function of the availability of meat (the second constraint). First assume that no meat is available. Designating $\theta$ as the amount of meat available, the initial tableau is as in Tableau 1. Assume that the range of $\theta$ to be explored is $0 \leq \theta \leq 1{,}000{,}000$. Initially, we can treat $\theta$ as being zero. Tableau 3 is the optimal solution, and it is optimal so long as $\boldsymbol{B}^{-1}\boldsymbol{b}$, or the solution vector, is nonnegative. The solution vector is nonnegative and hence Tableau 3 is optimal so long as

$$\begin{aligned} 240{,}000 - 2\theta \geq 0 &\Longrightarrow \theta \leq 120{,}000 \\ \theta \geq 0 &\Longrightarrow \theta \geq 0 \end{aligned}$$

and the corresponding solution is plotted in Figure 7.2. The value of additional meat in the range given above is \$.42 per pound. For $\theta > 120{,}000$, $x_3$ is negative and must be removed from the basis (using the dual simplex method). The result is Tableau 4. Tableau 4 is optimal so long as the solution vector ($\boldsymbol{B}^{-1}\boldsymbol{b}$) is nonnegative, or so long as

$$\begin{aligned} \theta - 120{,}000 \geq 0 &\Longrightarrow \theta \geq 120{,}000 \Longrightarrow \theta \geq 120{,}000 \\ 180{,}000 - .5\theta \geq 0 &\Longrightarrow \theta \leq 360{,}000 \\ 230{,}000 - \theta \geq 0 &\Longrightarrow \theta \leq 230{,}000 \Longrightarrow \theta \leq 230{,}000 \end{aligned}$$

or for $120{,}000 \leq \theta \leq 230{,}000$, as plotted in Figure 7.2. Note that the value of additional meat in the above range is \$.35 per pound. For $\theta > 230{,}000$,

$x_5$ is replaced in the basis by $x_4$, yielding Tableau 5, which is optimal for $\theta \geq 230{,}000$. The value of additional meat is zero for such solutions. Note how the diagram in Figure 7.2 can be used. First, for any available amount of meat such as 180,000, the optimal solution can be read from the graph (see the dotted line). Second, the value of the meat is useful in determining how much meat to purchase, assuming that the amount of cereal to be purchased is held at 240,000 pounds. To illustrate the point, suppose that the price of meat were \$.20 per pound. Then 230,000 pounds of meat would be ordered. (Because .20 lies between the shadow price of meat of .35 in Tableau 4 and of .00 in Tableau 5, we use the value of $\theta$ common to both tableaus.) If \$.40 per pound were the price of meat, then only 120,000 pounds would be ordered, and so on. Were the amount of cereal to be purchased a variable as well, a different linear programming problem would have to be solved.

**Tableau 1**

| | | $x_1$ ↓ | $x_2$ | $x_3$ | $x_4$ | $x_5$ | |
|---|---|---|---|---|---|---|---|
| $z$ | 0 | −.56 | −.42 | 0 | 0 | 0 | |
| $x_3$ | 240,000 | 1 | 2 | 1 | 0 | 0 | |
| $x_4$ | $\theta$ | [1.5] | 1 | 0 | 1 | 0 | → |
| $x_5$ | 110,000 | 1 | 0 | 0 | 0 | 1 | |

**Tableau 2**

| | | $x_1$ | $x_2$ ↓ | $x_3$ | $x_4$ | $x_5$ | |
|---|---|---|---|---|---|---|---|
| $z$ | $.387\theta$ | 0 | −.033 | 0 | .387 | 0 | |
| $x_3$ | $240{,}000 - .667\theta$ | 0 | 1.333 | 1 | −.667 | 0 | |
| $x_1$ | $.667\theta$ | 1 | [.667] | 0 | .667 | 0 | → |
| $x_5$ | $110{,}000 - .667\theta$ | 0 | −.667 | 0 | −.667 | 1 | |

**Tableau 3**

| | | $x_1$ ↓ | $x_2$ | $x_3$ | $x_4$ | $x_5$ | |
|---|---|---|---|---|---|---|---|
| $z$ | $.42\theta$ | .07 | 0 | 0 | .42 | 0 | |
| $x_3$ | $240{,}000 - 2\theta$ | [−2] | 0 | 1 | −2 | 0 | → |
| $x_2$ | $\theta$ | 1.5 | 1 | 0 | 1 | 0 | Optimal tableau for |
| $x_5$ | 110,000 | 1 | 0 | 0 | 0 | 1 | $0 \leq \theta \leq 120{,}000$ |

**Tableau 4**

| | | $x_1$ | $x_2$ | $x_3$ | $x_4$ ↓ | $x_5$ | |
|---|---|---|---|---|---|---|---|
| $z$ | $.35\theta + 8{,}400$ | 0 | 0 | .035 | .35 | 0 | Optimal tableau for $120{,}000 \leq \theta \leq 230{,}000$ |
| $x_1$ | $\theta - 120{,}000$ | 1 | 0 | −.5 | 1 | 0 | |
| $x_2$ | $180{,}000 - .5\theta$ | 0 | 1 | .75 | −.5 | 0 | |
| $x_5$ | $230{,}000 - \theta$ | 0 | 0 | .5 | [−1] | 1 | → |

**Tableau 5**

| | | $x_1$ | $x_2$ | $x_3$ | $x_4$ | $x_5$ | |
|---|---|---|---|---|---|---|---|
| $z$ | 88,900 | 0 | 0 | .21 | 0 | .35 | Optimal tableau for $\theta \geq 230{,}000$ |
| $x_1$ | 110,000 | 1 | 0 | 0 | 0 | 1 | |
| $x_2$ | 65,000 | 0 | 1 | .5 | 0 | −.5 | |
| $x_4$ | $\theta - 230{,}000$ | 0 | 0 | −.5 | 1 | −1 | |

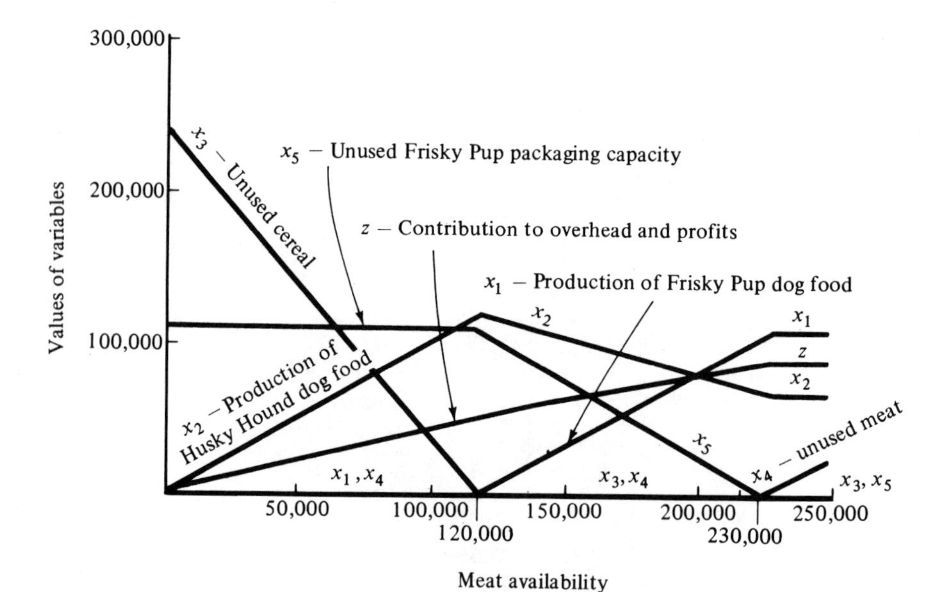


**Figure 7.2**

Parametric Solution to Dog Food Problem with Meat Availability as the Parameter

EXAMPLE 7.9: (Parametric objective function)

Find the optimal solution to the dog food problem as a function of the contribution to overhead and profits of Frisky Pup dog food ($x_1$). Designating $\theta$ as the contribution to overhead and profits of $x_1$, we have Tableau 1 as the starting tableau. For any value of $\theta$, the solution is not optimal, and so an iteration is taken which leads to Tableau 2. Tableau 2 is optimal for any value of $\theta$ in the range $-\infty < \theta < +.21$. See the solution plot in Figure 7.3. For $\theta > .21$, Tableau 2 is not optimal, hence an iteration is

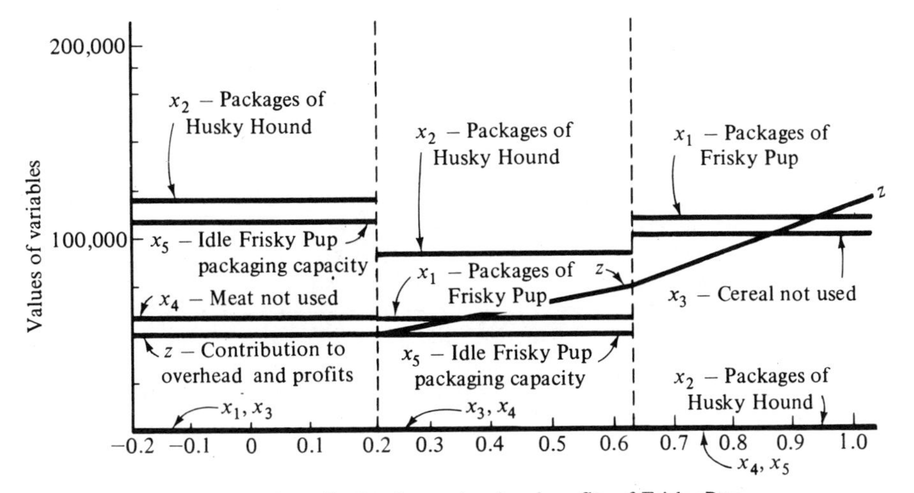


**Figure 7.3**

Parametric Solution to Dog Food Problem with the Contribution to Overhead and Profits of Frisky Pup Dog Food as the Parameter

taken leading to Tableau 3. Tableau 3 is optimal for any value of $\theta$ such that $z_j - c_j \geq 0$, i.e., $.315 - .5\theta \geq 0$ and $\theta - .21 \geq 0$ or $.21 \leq \theta \leq .63$. The associated solutions are plotted in Figure 7.3. For $\theta > .63$, Tableau 3 is not optimal, hence an iteration is taken leading to Tableau 4, which is optimal for all $\theta \geq .63$. The usefulness of this parametric program can be seen in the same way as the earlier one, by giving an optimal strategy as a function of price.

**Tableau 1**

| | | $x_1$ | $x_2$ ↓ | $x_3$ | $x_4$ | $x_5$ | |
|---|---|---|---|---|---|---|---|
| $z$ | 0 | $-\theta$ | $-.42$ | 0 | 0 | 0 | |
| $x_3$ | 240,000 | 1 | [2] | 1 | 0 | 0 | → |
| $x_4$ | 180,000 | 1.5 | 1 | 0 | 1 | 0 | |
| $x_5$ | 110,000 | 1 | 0 | 0 | 0 | 1 | |

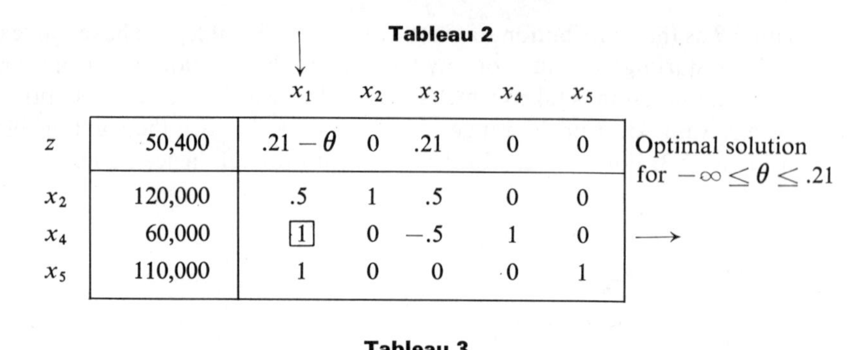

**Tableau 2**

↓ (entering $x_1$)

| | | $x_1$ | $x_2$ | $x_3$ | $x_4$ | $x_5$ | |
|---|---|---|---|---|---|---|---|
| $z$ | 50,400 | $.21-\theta$ | 0 | .21 | 0 | 0 | Optimal solution for $-\infty \leq \theta \leq .21$ |
| $x_2$ | 120,000 | .5 | 1 | .5 | 0 | 0 | |
| $x_4$ | 60,000 | [1] | 0 | $-.5$ | 1 | 0 | ⟶ |
| $x_5$ | 110,000 | 1 | 0 | 0 | 0 | 1 | |

**Tableau 3**

↓ (entering $x_3$)

| | | $x_1$ | $x_2$ | $x_3$ | $x_4$ | $x_5$ | |
|---|---|---|---|---|---|---|---|
| $z$ | 37,800 $+ 60{,}000\theta$ | 0 | 0 | $.315-.5\theta$ | $-.21 + \theta$ | 0 | |
| $x_2$ | 90,000 | 0 | 1 | .75 | $-.5$ | 0 | Optimal solution for $.21 \leq \theta \leq .63$ |
| $x_1$ | 60,000 | 1 | 0 | $-.5$ | 1 | 0 | |
| $x_5$ | 50,000 | 0 | 0 | [.5] | $-1$ | 1 | ⟶ |

**Tableau 4**

| | | $x_1$ | $x_2$ | $x_3$ | $x_4$ | $x_5$ | |
|---|---|---|---|---|---|---|---|
| $z$ | 6,300 $+ 110{,}000\theta$ | 0 | 0 | 0 | .42 | $-.63+\theta$ | |
| $x_2$ | 1,500 | 0 | 1 | 0 | 1 | $-1.5$ | Optimal solution for $.63 \leq \theta \leq \infty$ |
| $x_1$ | 110,000 | 1 | 0 | 0 | 0 | 1 | |
| $x_3$ | 100,000 | 0 | 0 | 1 | $-2$ | 2 | |

EXAMPLE 7.10:

The final example in this section is an example of parametric programming on an element of the matrix of coefficients. Solve the problem of Example 7.7, as a function of the parameter $\theta_3$.

The complete Tableau 1 of Example 7.7, for which $\theta_3 = 0$ is optimal, but with $\theta$ representing $\theta_3$ as indicated, is given in Tableau 1 of the present example. Tableau 1 is obtained using the inverse of the basis or by starting from Tableau 1 of Example 7.7, (where $K = 1/(1+\theta)$). So long as $\theta \neq -1$, in which case one or more coefficients of Tableau 1 are not defined, we may work in terms of $K$. From Tableau 1, it can be seen that the solution is optimal and feasible so long as $4K \geq 0$, $5 - 4K \geq 0$, $1 + 2K \geq 0$, and $-.5 - K \geq 0$, which are true so long as $-.2 \leq \theta \leq 1$. (We repeat that care must be exercised when solving the inequalities for $\theta$ because the sign of $\theta$ is not, in general, known. A straightforward method is to express each solution value or reduced cost as a ratio and separate the regions bounded by values of $\theta$ for which the numerator or denominator vanishes.)

For $\theta > 1$, Tableau 1 is not optimal, since $-.5 + K = -.5 + 1/(1 + \theta) < 0$. Hence, an iteration is taken in which $x_3$ is replaced by $x_5$, yielding Tableau 2, which is optimal for $\theta \geq 1$ (and also for $\theta \leq -3$).

Proceeding in the other direction, for $\theta < -.2$, Tableau 1 is not feasible since $5 - 4K = 5 - 4/(1 + \theta) < 0$. Hence an iteration is taken (shown by the dashed pivot in Tableau 1) in which $x_2$ is replaced by $x_4$, yielding Tableau 3 where $L = 1/(\theta - 1)$, which is optimal (and feasible) for $-.333 \leq \theta \leq -.2$. For $\theta < -.333$, Tableau 3 is not optimal, since $-1.5 - 2L = -1.5 - 2/(\theta - 1) \leq 0$. Hence an iteration is taken in which $x_4$ is replaced by $x_5$, yielding Tableau 4, which is optimal for $-1 \leq \theta \leq -.333$.

**Tableau 1**

| | | $x_1$ | $x_2$ | $x_3$ | $x_4$ | $x_5$ ↓ | |
|---|---|---|---|---|---|---|---|
| $z$ | $4K$ | 0 | 0 | 0 | $1 + 2K$ | $-.5 + K$ | $K = \dfrac{1}{1+\theta}$ |
| $x_1$ | $4K$ | 1 | 0 | 0 | $2K$ | $K$ | Optimal for |
| $x_2$ | $5 - 4K$ | 0 | 1 | 0 | $\boxed{1 - 2K}$ (dashed) | $1 - K$ | $-.2 \leq \theta \leq 1$ |
| $x_3$ | 4 | 0 | 0 | 1 | 0 | $\boxed{1}$ | ⟶ |

**Tableau 2**

| | | $x_1$ | $x_2$ | $x_3$ | $x_4$ | $x_5$ | |
|---|---|---|---|---|---|---|---|
| $z$ | 2 | 0 | 0 | $-K + .5$ | $1 + 2K$ | 0 | $K = \dfrac{1}{1+\theta}$ |
| $x_1$ | 0 | 1 | 0 | $-K$ | $2K$ | 0 | Optimal for |
| $x_2$ | 1 | 0 | 1 | $K - 1$ | $1 - 2K$ | 0 | $\theta \leq -3$ or $\theta \geq 1$ |
| $x_5$ | 4 | 0 | 0 | 1 | 0 | 1 | |

**Tableau 3**

| | | $x_1$ | $x_2$ | $x_3$ | $x_4$ | $x_5$ ↓ | |
|---|---|---|---|---|---|---|---|
| $z$ | $-5 - 12L$ | 0 | $-1 - 4L$ | 0 | 0 | $-1.5 - 2L$ | $L = \dfrac{1}{\theta - 1}$ |
| $x_1$ | $-6L$ | 1 | $-2L$ | 0 | 0 | $-L$ | Optimal for $-.333 \leq \theta \leq -.2$ |
| $x_4$ | $5 + 6L$ | 0 | $1 + 2L$ | 0 | 1 | $\boxed{1 + L}$ | ⟶ |
| $x_3$ | 4 | 0 | 0 | 1 | 0 | 1 | |

**Tableau 4**

| | | $x_1$ | $x_2$ | $x_3$ | $x_4$ ↓ | $x_5$ | |
|---|---|---|---|---|---|---|---|
| $z$ | $2.5 - .5N$ | 0 | $.5 - .5N$ | 0 | $1.5 + .5N$ | 0 | $N = \frac{1}{\theta}$ |
| $x_1$ | $-N$ | 1 | $-N$ | 0 | $N$ | 0 | Optimal for $-1 \leq \theta \leq -.333$ |
| $x_5$ | $5 + N$ | 0 | $1 + N$ | 0 | $1 - N$ | 1 | |
| $x_3$ | $-N - 1$ | 0 | $-1 - N$ | 1 | $\boxed{N - 1}$ | 0 | $\longrightarrow$ |

**Tableau 5**

| | | $x_1$ | $x_2$ ↓ | $x_3$ | $x_4$ | $x_5$ | |
|---|---|---|---|---|---|---|---|
| $z$ | $1 - 4L$ | 0 | $-1 - 4L$ | $1.5 + 2L$ | 0 | 0 | $L = \frac{1}{\theta - 1}$ |
| $x_1$ | $-2L$ | 1 | $-2L$ | $L$ | 0 | 0 | Optimal for $-3 \leq \theta \leq -1$ |
| $x_5$ | 4 | 0 | 0 | 1 | 0 | 1 | |
| $x_4$ | $1 + 2L$ | 0 | $\boxed{1 + 2L}$ | $-1 - L$ | 1 | 0 | $\longrightarrow$ |

For $\theta < -1$, Tableau 4 is not feasible because $-N - 1 = 1/(\theta - 1) < 0$. Hence, an iteration is taken in which $x_3$ is replaced by $x_4$, yielding Tableau 5, which is optimal for $-3 \leq \theta \leq -1$. For $\theta < -3$, Tableau 5 is not optimal since $-1 - 4L = (-3 - \theta)/(\theta - 1) < 0$. Hence an iteration is taken in which $x_4$ (or $x_1$) is replaced by $x_2$, yielding Tableau 2. A plot of the solution values is given in Figure 7.4. The highly nonlinear nature of some of the plots is due to the parameter appearing in the denominator of a term. This method can be implemented on a digital computer using appropriate reciprocal expressions, such as $K$, $L$, and $N$.

Analyses where two or more coefficients vary together as a function of a parameter are also possible, but the computations are cumbersome, except where two or more right-hand side elements are linear functions of the same parameter, or where two or more objective function coefficients are linear functions of the same parameter. The computations for these two cases are relatively straightforward; they are considered in the problems.

## 7.3 ADDING VARIABLES OR CONSTRAINTS

In many programming problems of real situations, it is desired, after obtaining preliminary solutions of the problem, to alter them by adding variables or constraints not considered earlier. The procedure is relatively straightforward, and if there are a modest number of additions, it is usually convenient to use the optimal solution to the preceding problem as a starting point for solving the altered problem.

An intuitive argument will now be made to provide some insight on adding variables and constraints. Adding a variable to a problem provides a new dimension and increases the solution space. All previously feasible solutions are still feasible, and some new possibilities are also available. Hence, the solution to a problem with an added variable can never be worse than the same problem without the variable. Conversely, when adding a constraint, no new options are made available; in fact, some old options may be eliminated. In such a case, therefore, an optimal solution to a linear programming problem after adding a constraint can never be better than the same problem without that constraint.

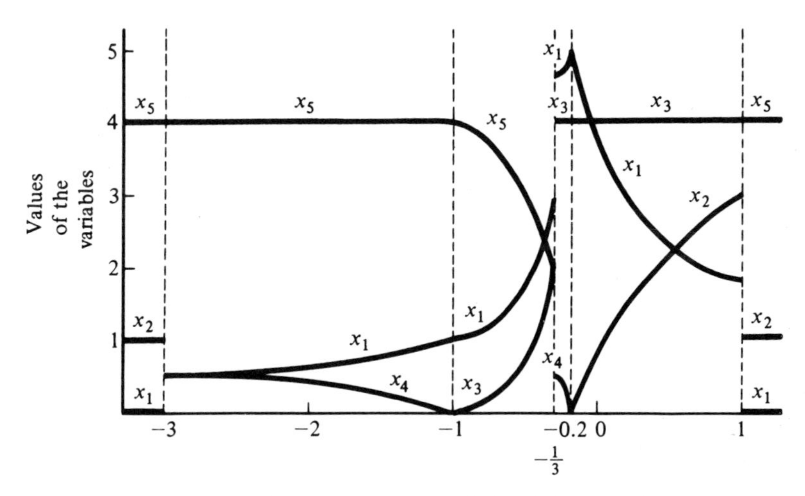


**Figure 7.4**

A Plot of Optimal Solutions of Example 7.10 as a Function of $\theta$*

*Where variable values are not shown, they are zero.

The procedure for adding constraints and/or variables is to test the constraints to see whether or not they are violated, and to test the variables to see whether or not they should enter the basis. If any variables should enter the basis or if any constraints are violated, then the variables and/or constraints are adjoined and the iterations continued until a new optimum has been found. If a variable is added, it must be updated to the current basis using the inverse. If a constraint is added, the old basis plus a slack or artificial variable form a new basis; the steps necessary to expand the basis are only partial iterations. The above points will be illustrated in two examples.

EXAMPLE 7.11:

In the dog food example, whose original optimal solution appears in Example 7.1, a new economy dog food is introduced. It is an all cereal dog food which uses two pounds of cereal per package, one unit of Frisky Pup packaging capacity, and contributes \$.35 per package to overhead and profits. Designating the variable as $x_6$, we can write its coefficients as $(2 \quad 0 \quad 1)'$, and .35 is its contribution to the objective function. First, to check whether or not it is to enter the basis, compute $\mathbf{c}_B'\mathbf{B}^{-1}\mathbf{N}_j - c_{N_j}$ where $\mathbf{c}_B'\mathbf{B}^{-1}$ is the row of evaluators above the original basic variables (in this case $x_3, x_4, x_5$).

$$z_j - c_j = (.035 \quad .35 \quad 0)(2 \quad 0 \quad 1)' - .35 = -.28$$

Hence, it should enter the basis. The updated column is also needed and can be computed using the inverse, which is contained in the position of the original identity matrix.

$$\mathbf{B}^{-1}\mathbf{N}_j = \begin{pmatrix} .5 & -1 & 1 \\ .75 & -.5 & 0 \\ -.5 & 1 & 0 \end{pmatrix} \begin{pmatrix} 2 \\ 0 \\ 1 \end{pmatrix} = \begin{pmatrix} 2 \\ 1.5 \\ -1 \end{pmatrix}$$

Adjoining this column to the originally optimal tableau, we have Tableau 1. Then $x_6$ replaces $x_5$, yielding Tableau 2, which is optimal.

**Tableau 1**

| | | $x_1$ | $x_2$ | $x_3$ | $x_4$ | $x_5$ | $x_6$ ↓ |
|---|---|---|---|---|---|---|---|
| $z$ | 71,400 | 0 | 0 | .035 | .35 | 0 | −.28 |
| $x_5$ | 50,000 | 0 | 0 | .5 | −1 | 1 | [2] → |
| $x_2$ | 90,000 | 0 | 1 | .75 | −.5 | 0 | 1.5 |
| $x_1$ | 60,000 | 1 | 0 | −.5 | 1 | 0 | −1 |

**Tableau 2**

| | | $x_1$ | $x_2$ | $x_3$ | $x_4$ | $x_5$ | $x_6$ |
|---|---|---|---|---|---|---|---|
| $z$ | 78,400 | 0 | 0 | .105 | .21 | .14 | 0 |
| $x_6$ | 25,000 | 0 | 0 | .25 | −.5 | .5 | 1 |
| $x_2$ | 52,500 | 0 | 1 | .375 | .25 | −.75 | 0 |
| $x_1$ | 85,000 | 1 | 0 | −.25 | .5 | .5 | 0 |

EXAMPLE 7.12:

For the original dog food problem (excluding the dog food added in Example 7.11), management has decided to institute a policy requiring that Husky Hound ($x_2$) not exceed Frisky Pup ($x_1$) in production by more than 20,000 packages per month. The constraint stipulates that $x_2 - x_1 \leq 20{,}000$. Denoting the slack as $x_7$, we have an altered Tableau 0, which, through performance of the dashed partial iterations, gives the starting basic solution in Tableau 1. Using the dual simplex method, an optimal solution is reached in one iteration in Tableau 2.

**Tableau 0**

| | | $x_1$ | $x_2$ | $x_3$ | $x_4$ | $x_5$ | $x_7$ |
|---|---|---|---|---|---|---|---|
| $z$ | 71,400 | 0 | 0 | .035 | .35 | 0 | 0 |
| $x_5$ | 50,000 | 0 | 0 | .5 | −1 | 1 | 0 |
| $x_2$ | 90,000 | 0 | [1] | .75 | −.5 | 0 | 0 |
| $x_1$ | 60,000 | [1] | 0 | −.5 | 1 | 0 | 0 |
| $x_7$ | 20,000 | −1 | 1 | 0 | 0 | 0 | 1 |

**Tableau 1**

| | | $x_1$ | $x_2$ | $x_3$ ↓ | $x_4$ | $x_5$ | $x_7$ | |
|---|---|---|---|---|---|---|---|---|
| $z$ | 71,400 | 0 | 0 | .035 | .35 | 0 | 0 | |
| $x_5$ | 50,000 | 0 | 0 | .5 | −1 | 1 | 0 | |
| $x_2$ | 90,000 | 0 | 1 | .75 | −.5 | 0 | 0 | |
| $x_1$ | 60,000 | 1 | 0 | −.5 | 1 | 0 | 0 | |
| $x_7$ | −10,000 | 0 | 0 | [−1.25] | 1.5 | 0 | 1 | → |

**Tableau 2**

| | | $x_1$ | $x_2$ | $x_3$ | $x_4$ | $x_5$ | $x_7$ |
|---|---|---|---|---|---|---|---|
| $z$ | 71,120 | 0 | 0 | 0 | .392 | 0 | .023 |
| $x_5$ | 46,000 | 0 | 0 | 0 | .4 | 1 | .4 |
| $x_2$ | 84,000 | 0 | 1 | 0 | .4 | 0 | .6 |
| $x_1$ | 64,000 | 1 | 0 | 0 | .4 | 0 | −.4 |
| $x_3$ | 8,000 | 0 | 0 | 1 | −1.2 | 0 | −.8 |

## 7.4 SUMMARY

This chapter presented a number of different postoptimal analyses. These include:

1. Sensitivity analysis or exploring the range of a coefficient over which the optimal basis remains optimal, holding all other factors fixed at their original values.
2. Changing one or more problem coefficients, and finding the optimal solution to the altered problem.
3. Parametric programming, or finding optimal solutions of a linear programming problem for a coefficient (or coefficients) varying over some specified range of values.
4. Adding constraints and/or variables to a problem and then solving them by proceeding from the previous optimal solution.

Certain variations of the above methods are implemented in computer programs. They are widely used in practice, and serve as useful accessories in solving linear programming problems.

### Selected Supplemental Reference

Section 7.2
[17]

## 7.5 PROBLEMS

**1.** Show how all the calculations for Examples 7.2 and 7.4 may be simplified as much as possible.

**2.** Determine the number of operations required in general (i.e., as a function of $m$ and $n$) to perform a sensitivity analysis on a problem using the simplex method and using the revised simplex method.

**3.** Show that the sensitivity analysis on the objective function coefficients is the dual to the sensitivity analysis on the requirements vector.

**4.** For each of the following problems, perform a sensitivity analysis on each of the right-hand side elements:

a. Chapter 1, problems 1, 2, 3, 4, 5, 10, 13.
b. Chapter 2, problems 12, 13, 14, 15, 16, 17, 18, 19.
c. Chapter 3, problems 6, 7, 8, 9.
d. Chapter 4, problem 5.

e. Chapter 5, problems 6, 8, 9, 10, 11.
f. Chapter 6, problems 2, 3, 4, 5.

**5.** For each of the problems of problem 4 above, perform a sensitivity analysis on each of the objective function coefficients.

**6.** Develop for each of the methods in section 7.1 the procedure for determining which iteration should be taken when a parameter is altered so that it is no longer contained in its region of optimality.

**7.** Find the optimal solution to the dog food problem as a function of the availability of cereal (the first constraint).

**8.** Find the optimal solution to the dog food problem as a function of the contribution to overhead and profits of Husky Hound dog food ($x_2$).

**9.** For each of the problems of problem 4 above, perform a parametric linear programming solution for each of the right-hand side elements.

**10.** For each of the problems of problem 4 above, perform a parametric linear programming solution for each of the objective function coefficients.

**11.** Show that the primal-dual algorithm of Chapter 6 may be viewed as a parametric program on the objective function of the form

$$\text{Maximize } z = \boldsymbol{c}'\boldsymbol{x} + M\boldsymbol{d}'\boldsymbol{x}$$

where $\boldsymbol{x}$ is the vector of variables (including any artificials) and $\boldsymbol{d}$ has components of $-1$ or 0, corresponding to artificial and nonartificial variables, respectively. The range on $M$ is from some lower bound—usually zero—to infinity.

**12.** Develop the parametric methods for varying two or more right-hand side elements as a function of the same parameter $\theta$. Can this be considered as a special case of the right-hand side parametric programming developed in the text?

**13.** Develop the parametric methods for varying two or more objective function elements as a function of the same parameter $\theta$. Can this be considered as a special case of the objective function parametric programming developed in the text?

**14.** Consider the following problem:

$$\begin{aligned}
\text{Maximize} \quad & (5 + \theta_1)x_1 + (2 + \theta_2)x_2 \\
\text{subject to:} \quad & (1 + \theta_5)x_1 + 2x_2 \leq 12 + \theta_3 \\
& 4x_1 + 3x_2 \leq 24 + \theta_4 \\
& x_1, x_2 \geq 0
\end{aligned}$$

a. Solve the problem for $\theta_1 = \theta_2 = \theta_3 = \theta_4 = \theta_5 = 0$.
b. Solve five parametric programming problems, varying one $\theta_j$ at a time ($j = 1, \ldots, 5$). (Hold all $\theta$'s fixed at zero, except $\theta_j$.)

# 8

# Specially Structured Linear Programming Problems: Bounded Variables, Generalized Upper Bounds, and Decomposition

In this chapter we shall consider methods that take advantage of certain special but common structures in linear programming problems to achieve their computational efficiency. We shall discuss methods that preserve much of the simplex method framework of linear programming, whereas in the next chapter we shall encounter methods which, although developed from the simplex method, are quite different from it. The methods presented in this chapter are of two types—those employing implicitly handled bounds on sets of variables, and a decomposition method which provides a special way of handling matrices of coefficients with block triangular structure.

The methods presented appear to be computationally efficient, and they have been programmed for computer usage.

## 8.1 VARIABLES WITH LOWER BOUNDS

In the formulation of linear programming problems, constraints which specify lower bounds on variables are common. For example:

$$x_3 \geq 6$$

indicates that at least 6 units of a product denoted by $x_3$ must be produced.

By adding a slack to such an inequality we have

$$x_3 - s_1 = 6$$

or

$$x_3 = s_1 + 6 \tag{8.1}$$

So long as the slack $s_1$ is nonnegative, $x_3$ is nonnegative. Therefore, $x_3$ can be eliminated from all other constraints using expression (8.1). Once an optimal solution has been found, the value of $s_1$ will be known and $x_3$ can be solved for using (8.1).

In matrix notation, let the constraint set for a problem be

$$\begin{aligned} \mathbf{Ax} &\le \mathbf{b} \\ \mathbf{x} &\ge \mathbf{h} \end{aligned} \tag{8.2}$$

where $\mathbf{h}$ is an $n$ by 1 vector of constants (it should be noted that it is *not* necessary to assume that $\mathbf{h} \ge \mathbf{0}$). Then

$$\mathbf{x} - \mathbf{s} = \mathbf{h} \tag{8.3}$$

where $\mathbf{s}$ is a vector of slack variables. Solving equation (8.3) for $\mathbf{x}$ yields

$$\mathbf{x} = \mathbf{h} + \mathbf{s} \tag{8.4}$$

which, substituted into (8.2), yields

$$\begin{aligned} \mathbf{As} &\le \mathbf{b} - \mathbf{Ah} \\ \mathbf{s} &\ge \mathbf{0} \end{aligned} \tag{8.5}$$

thereby eliminating explicit representation of all lower bound constraints. Once the solution has been achieved, (8.4) is used to solve for $\mathbf{x}$. Of course, for any component of $\mathbf{h}$—such as $h_j$—equal to zero, $x_j = s_j$. (In such a situation, no lower bound substitution is normally made.)

EXAMPLE 8.1:

Suppose that, for the dog food example, 70,000 packages of Frisky Pup dog food had been contracted for. Then the problem would have been

$$\begin{aligned} \text{Maximize } z = .56x_1 &+ .42x_2 \\ \text{subject to: } \quad x_1 &+ 2x_2 \le 240{,}000 \\ 1.5x_1 &+ x_2 \le 180{,}000 \\ x_1 & \qquad \le 110{,}000 \\ x_1 & \qquad \ge 70{,}000 \\ & x_1, x_2 \ge 0 \end{aligned}$$

Denoting $s_1$ as the slack variable on the lower bound constraint, and making use of the substitution $x_1 = 70{,}000 + s_1$, we have

$$\begin{aligned} \text{Maximize } z &= .56s_1 + .42x_2 + 39{,}200 \\ \text{subject to:} \quad & s_1 + 2x_2 \leq 170{,}000 \\ & 1.5s_1 + x_2 \leq 75{,}000 \\ & s_1 \leq 40{,}000 \\ & s_1, x_2 \geq 0 \end{aligned}$$

The optimal solution to this problem may be seen to be $s_1 = 0$, $x_2 = 75{,}000$, and $z = \$70{,}700$. We may then solve for $x_1$ as $x_1 = 70{,}000 + s_1 = 70{,}000$.

What has been achieved is a way of representing lower bound constraints implicitly.

## 8.2 VARIABLES WITH UPPER BOUNDS— THE METHOD OF UPPER BOUNDS

In a manner completely analogous to that of the last section, a frequently occuring type of constraint is the upper bound on a variable which, for example, might be

$$x_4 \leq 12$$

Let us see if it is possible to make a substitution similar to that of the last section. Adding a slack variable, we have

$$x_4 + s_2 = 12$$

or

$$x_4 = 12 - s_2$$

By substituting for $x_4$ in the above equation, we guarantee that $x_4 \leq 12$. However, we do *not* guarantee that $x_4 \geq 0$ unless we stipulate that

$$x_4 = 12 - s_2 \geq 0$$

or

$$s_2 \leq 12$$

In other words, to make the substitution indicated, the slack variable on an upper bound constraint must have the same upper bound as the variable itself.

We shall henceforth denote the slack of an upper bound constraint as the complement of the variable. We further use the notation that $\bar{x}_j$ is the complement of $x_j$. In addition, the complement of a complement is the original variable. It may appear that nothing has been gained by such a substitution because, after the substitution, the complement is a variable which has an upper bound. The value of the substitution is that either the variable or its complement—but never both—need be represented explicitly. Thus one variable and one constraint may be represented implicitly, rather than explicitly.

We may think of three states with respect to whether or not the variable or its complement are basic, assuming an expanded representation having both variables. The states are:

1. The variable is basic, and the complement is nonbasic.
2. Both the variable and the complement are basic.
3. The variable is nonbasic and the complement is basic.

By virtue of the upper bound constraint, provided that the nonnegativity constraint and the upper bound constraint are enforced within the framework of the algorithm, either the variable or its complement—but not both—need be represented explicitly. But which one? Referring to the three states given above, we can deduce that in state 1 the complement should be explicitly written as nonbasic, and in state 3 the variable should be explicitly written as nonbasic. In state 2, either the variable or its complement may be explicitly written as basic.

How then is the change from variable to complement, or vice versa, to be accomplished? Denoting $u_j$ as the upper bound of the variable $x_j$, we have two substitutions:

(8.6) $$x_j = u_j - \bar{x}_j$$

(8.7) $$\bar{x}_j = u_j - x_j$$

Expression (8.6) is the substitution going from explicit representation of the variable to explicit representation of the complement; expression (8.7) is the reverse substitution.

The simplex iteration, using upper bounds, may be of three types:

1. As an incoming nonbasic variable is increased, some basic variable is decreased to zero. This is the usual type of simplex iteration which we have been using all along.
2. As an incoming nonbasic variable is increased, some basic variable is increased to its upper bound. This type of iteration can be converted into the first type by first using substitution (8.6) or (8.7), as appropriate, to replace the appropriate basic variable by its complement, and then performing a type 1 iteration.
3. As an incoming nonbasic variable is increased, it reaches its upper bound. This type of iteration, which can be accomplished by substituting for the

incoming variable, using expression (8.6) or (8.7), as appropriate, is called an upper bound iteration, and is much simpler than either of the other two types.

Two parts of the simplex method are altered to include the method of upper bounds:

1. Choosing the variable to leave the basis.
2. The iteration.

In Figure 8.1 we present a portion of a flow chart containing the changes that implement the method. By using these charts to supersede those in Figure 3.3, the simplex method incorporating the method of upper bounds is obtained.
We shall provide an example to illustrate the method.

EXAMPLE 8.2:

Noting that the dog food example problem contains an upper bound constraint ($x_1 \leq 110{,}000$), solve the problem using the method of upper bounds. Tableau 1 contains the starting tableau. We have indicated the upper bounds below each variable.

**Tableau 1**

| | | $x_1$ ↓ | $x_2$ | $x_3$ | $x_4$ |
|---|---|---|---|---|---|
| $z$ | 0 | −.56 | −.42 | 0 | 0 |
| $x_3$ | 240,000 | 1 | 2 | 1 | 0 |
| $x_4$ | 180,000 | 1.5 | 1 | 0 | 1 |
| $u_j$ | | 110,000 | ∞ | ∞ | ∞ |

*Iteration 1:* $x_1$ is selected to enter the basis. By reference to the flow chart, $u_1 = 110{,}000$, $f_1 = 120{,}000$, $r_1 = 2$, $f_2 = \infty$, and $r_2 = 0$. The iteration is of type 3: $x_1$ is replaced by its complement $(110{,}000 - \bar{x}_1)$, yielding Tableau 2.

**Tableau 2**

| | | $\bar{x}_1$ | $x_2$ ↓ | $x_3$ | $x_4$ | |
|---|---|---|---|---|---|---|
| $z$ | 61,600 | .56 | −.42 | 0 | 0 | |
| $x_3$ | 130,000 | −1 | 2 | 1 | 0 | |
| $x_4$ | 15,000 | −1.5 | [1] | 0 | 1 | → |
| $u_j$ | | 110,000 | ∞ | ∞ | ∞ | |

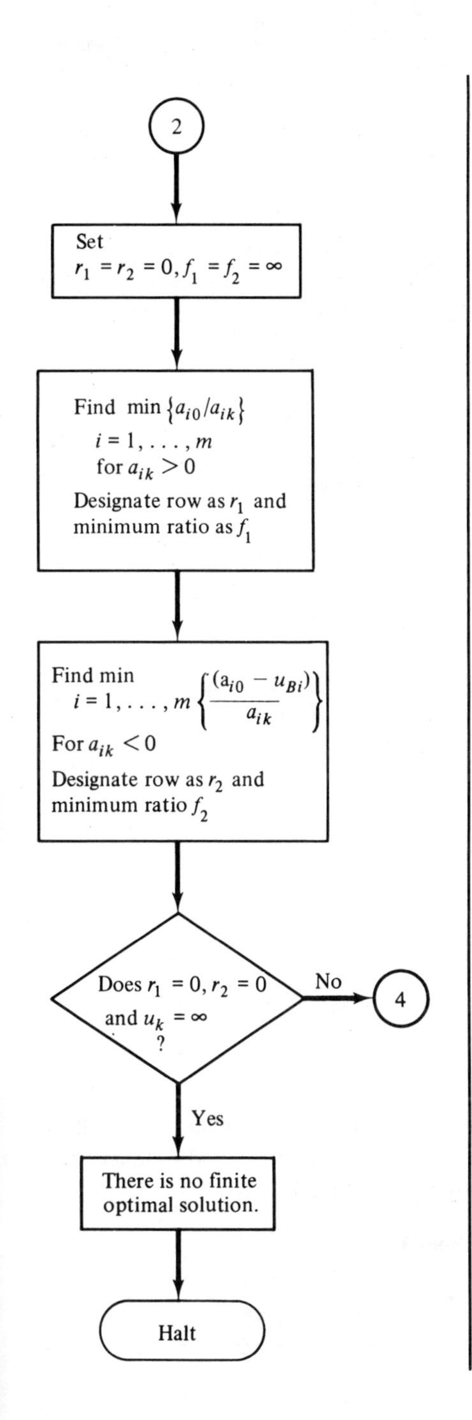


Initialize counters:

$r_1$ – row index of type 1 iteration
$r_2$ – row index of type 2 iteration
$f_1$ – minimum ratio of type 1 iteration
$f_2$ – minimum ratio of type 2 iteration

For positive coefficients in the column of the incoming variable $x_k$, determine which iteration of type 1 is most limiting, i.e., as $x_k$ is increased which basic variable becomes zero first? $f_1$ gives the corresponding value of $x_k$.

For negative coefficients in the column of the incoming variable $x_k$, determine which iteration of type 2 is most limiting, i.e., as $x_k$ is increased which basic variable reaches its upper bound first? $f_2$ gives the corresponding value of $x_k$. ($u_{Bi}$ is the upper bound of the basic variable in row $i$.)

Check to see if solution is unbounded.

No finite optimal solution exists.

**Figure 8.1**

A Flow Chart of the Alterations to the Simplex Method to Implement the Method of Upper Bounds (to supersede indicated portions of Fig. 3.3)

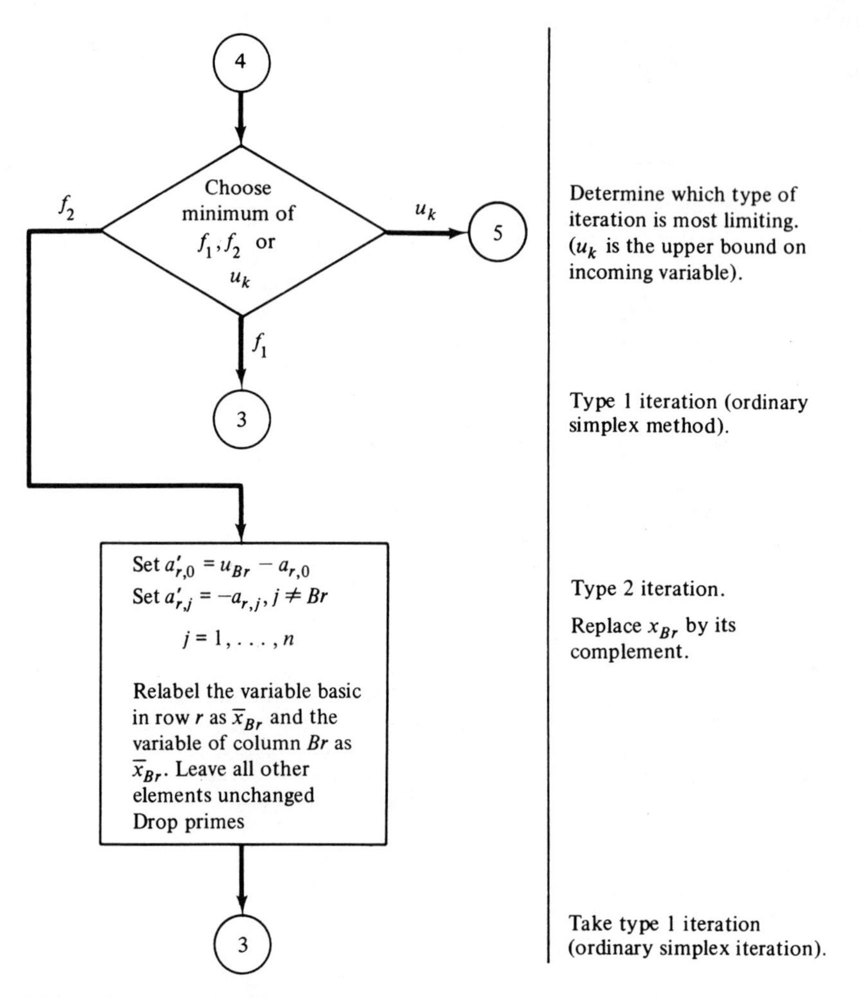


**Figure 8.1 (continued)**

*Iteration 2:* $x_2$ is selected to enter the basis. By reference to the flow chart, $f_1 = 15{,}000$, $r_1 = 2$, $f_2 = \infty$, $u_2 = \infty$, and $r_2 = 0$. Hence, the iteration is of type 1: $x_4$ is replaced by $x_2$, yielding Tableau 3.

**Tableau 3**

| | | ↓ $\bar{x}_1$ | $x_2$ | $x_3$ | $x_4$ | |
|---|---|---|---|---|---|---|
| $z$ | 67,900 | −.07 | 0 | 0 | .42 | |
| $x_3$ | 100,000 | [2] | 0 | 1 | −2 | → |
| $x_2$ | 15,000 | −1.5 | 1 | 0 | 1 | |
| $u_j$ | | 110,000 | ∞ | ∞ | ∞ | |

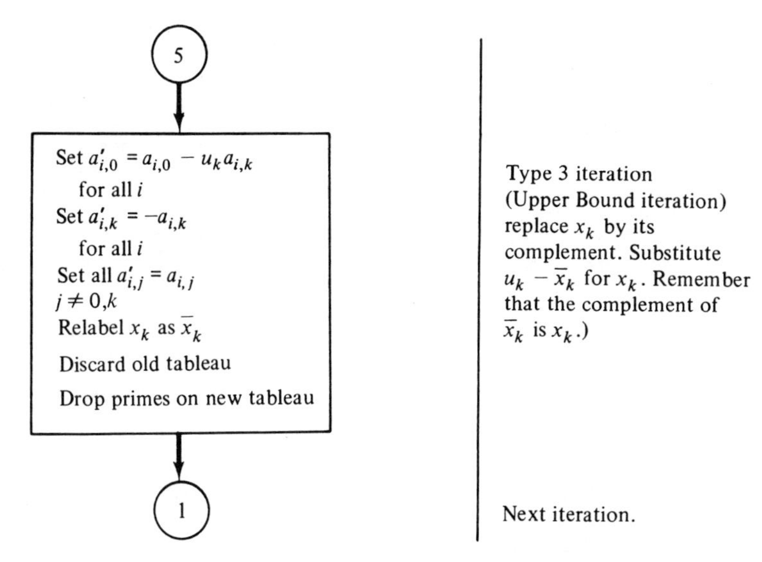


**Figure 8.1 (concluded)**

*Iteration 3:* $\bar{x}_1$ is selected to enter the basis. By reference to the flow chart, $f_1 = 50{,}000$, $r_1 = 1$, $f_2 = \infty$, $r_2 = 0$, and $u_1 = 110{,}000$. Hence, the iteration is of type 1: $x_3$ is replaced by $\bar{x}_1$, yielding Tableau 4, which is optimal. It should be noted that $\bar{x}_1$ is basic at 50,000. Therefore $x_1 = u_1 - \bar{x}_1 = 110{,}000 - 50{,}000$ or 60,000. It is of interest to compare this procedure for solving the problem with that of the simplex method in Example 3.3.

**Tableau 4**

| | | $\bar{x}_1$ | $x_2$ | $x_3$ | $x_4$ ↓ | |
|---|---|---|---|---|---|---|
| $z$ | 71,400 | 0 | 0 | .035 | .35 | |
| $\bar{x}_1$ | 50,000 | 1 | 0 | .5 | [−1] | → |
| $x_2$ | 90,000 | 0 | 1 | .75 | −.5 | |
| $u_j$ | | 110,000 | $\infty$ | $\infty$ | $\infty$ | |

In order *to illustrate an iteration of type* 2, we shall now introduce $x_4$ into the basis in Tableau 4, yielding Tableau 5. In this case $f_1 = \infty$, $r_1 = 0$, $u_4 = \infty$, $f_2 = 60{,}000$, and $r_2 = 1$.
For convenience, we write the intermediate tableau as Tableau 4a, having replaced $\bar{x}_1$ by its complement, $x_1$.
Of course, Tableau 5 is not optimal.
Because of the presence of upper bound constraints in most problems and

**Tableau 4a**

| | | $x_1$ | $x_2$ | $x_3$ | $x_4$ ↓ | |
|---|---|---|---|---|---|---|
| $z$ | 71,400 | 0 | 0 | .035 | .35 | |
| $x_1$ | 60,000 | 1 | 0 | −.5 | [1] | → |
| $x_2$ | 90,000 | 0 | 1 | .75 | −.5 | |
| $u_j$ | | 110,000 | $\infty$ | $\infty$ | $\infty$ | |

**Tableau 5**

| | | $x_1$ | $x_2$ | $x_3$ | $x_4$ |
|---|---|---|---|---|---|
| $z$ | 50,400 | −.35 | 0 | .21 | 0 |
| $x_4$ | 60,000 | 1 | 0 | −.5 | 1 |
| $x_2$ | 120,000 | .5 | 1 | .5 | 0 |
| $u_j$ | | 110,000 | $\infty$ | $\infty$ | $\infty$ |

the efficiency of the method of upper bounds, the method is widely used, and is included in most competitive linear programming computer systems.

## 8.3 GENERALIZED UPPER BOUNDS

(This material is based on the reference by Dantzig and VanSlyke [68] and is used by permission of Academic Press Inc.)

The method of upper bounds has been generalized by G.B. Dantzig and R.M. VanSlyke [68] to handle any problem in which the first $m$ constraints may be of an arbitrary nature, and each variable has at most one nonzero coefficient in the last L constraints. Without loss of generality, we assume that all the nonzero coefficients in the last $L$ constraints are ones. (The transformation is accomplished by scaling, and an example of such a transformation is included in the problems at the end of the chapter.)

The variables may be partitioned into $L + 1$ sets:

$$S_0, S_1, \ldots, S_L$$

The variables in set $S_0$ are constrained only by the first $m$ constraints. The variables in sets $S_i$ ($i \neq 0$) sum to one, as stipulated by constraint $m + i$. (The case for which negative coefficients are permitted in the last $L$ constraints is also examined by Dantzig and VanSlyke; we shall not consider that extension here, but it is included in the problems.) A problem of the appropriate matrix form is given below:

$$
\text{(8.8)}\quad
\begin{array}{llcccccl}
\text{Maximize} & z = & c_0'x_0 & + c_1'x_1 & + c_2'x_2 & + \cdots & + c_L'x_L & \\
\text{subject to:} & & A_0x_0 & + A_1x_1 & + A_2x_2 & + \cdots & + A_Lx_L & = b \\
& & & \mathbf{1}'x_1 & & & & = 1 \\
& & & & \mathbf{1}'x_2 & & & = 1 \\
& & & & & \ddots & & \vdots \\
& & & & & & \mathbf{1}'x_L & = 1 \\
& & & & & & x_0, x_1, \ldots, x_L & \geq \mathbf{0}
\end{array}
$$

where $c_j$ is $n_j$ by 1, $A_j$ is $m$ by $n_j$, $x_j$ is $n_j$ by 1, $b$ is $m$ by 1, and the $\mathbf{1}$'s are column vectors of ones of appropriate order. The method of Dantzig and VanSlyke permits using a working basis of order $m$ by $m$. Where $m$ is small compared to $L$, the computational savings can be appreciable.

By reference to the formulation (8.8), variables in the set $S_0$ have vector components $m + 1$ through $m + L$ equal to zero. Variables in the set $S_j$ $(j \neq 0)$ have component $m + j$ equal to one, with components of $m + k$ $(k \neq j)$ equal to zero.

EXAMPLE 8.3:

Consider the following problem (adapted from Dantzig and VanSlyke [68]):

$$
\begin{array}{ll|cccc|cc|cc|l}
\text{Maximize} & z = x_1 & & & - x_4 & - 2x_5 & & & & & \\
\text{subject to:} & x_1 & & + 2x_3 & + x_4 & & + 5x_6 & + x_7 & - x_8 & - 12x_9 & = 8 \\
& x_1 + & x_2 & - x_3 & + 2x_4 & & + 4x_6 & + 2x_7 & - 3x_8 & + 6x_9 & = 4 \\
& & & & & x_5 & & & & & = 0 \\
\hline
& & x_2 & + x_3 & + x_4 & + x_5 & & & & & = 1 \\
\hline
& & & & & & x_6 & + x_7 & & & = 1 \\
\hline
& & & & & & & & x_8 & + x_9 & = 1
\end{array}
$$

The sets of variables are as partitioned by dashed lines in the problem statement.

$S_0$: $x_1$

$S_1$: $x_2, x_3, x_4, x_5$

$S_2$: $x_6, x_7$

$S_3$: $x_8, x_9$

The various constraint sets are also partitioned in the problem statement. We shall return to this example shortly.

Dantzig and VanSlyke show that at least one variable from each set $S_j$ ($j \neq 0$) must be basic. An outline of the proof is that if none were basic in some set, then the rank of a basis for (8.8) would be less than $m + L$, contrary to fact. Sets of variables which contain two or more basic variables at a particular stage are called *essential sets*, and sets which contain one basic variable at a particular stage are called *inessential sets*. Since variables enter and leave the basis at each iteration, the status of a set of variables—essential or inessential—may change as iterations are undertaken.

Any one *basic* variable from each set $S_j$ ($j \neq 0$) is denoted as the key variable of the set. Let the associated column be called the key column. Basic variables and their associated columns which are not key are called nonkey. To motivate the development of the method of generalized upper bounds we present the method in an exaggerated way: we explicitly make a transformation of the problem which is implicitly made by the method. The problem may be viewed as being transformed by subtracting the key column of a set from every other column of that set. We may view the transformed problem as being of the form (8.8), with unit vectors in place of the sum (**1**) vectors, and the matrices and variables suitably transformed. We shall illustrate the transformation in Example 8.4. In the transformed problem it can be seen that the transformed key variables must equal one (since all other coefficients in the row are zero). Hence, once we have set all the key variables to one, the last $L$ constraints may be eliminated. For the transformed set of $m$ constraints, there is a reduced basis which may be used to solve the larger problem.

EXAMPLE 8.4:

Continuing with the problem of Example 8.3, assume that *somehow* we know that $x_1$, $x_2$, $x_3$, $x_5$, $x_6$, and $x_8$ form a feasible basis. If we did not know a feasible basis, then artificial variables would have to be used to generate a starting basis. Further, we arbitrarily designate $x_2$, $x_6$, and $x_8$ as key variables. (Only the choice of $x_2$ as a key variable is arbitrary; $x_3$ or $x_5$ could have been chosen instead.)

Performing the transformation indicated (subtracting key columns of a set from associated nonkey columns), we have the following problem (as pointed out above, this step is strictly for illustration purposes and is not normally done):

$$\begin{aligned}
\text{Maximize } z &= x_1 - x_4 - 2x_5 \\
\text{subject to: } & x_1 + 2x_3 + x_4 + 5x_6' - 4x_7 - x_8' - 11x_9 = 8 \\
& x_1 + x_2' - 2x_3 + x_4 - x_5 + 4x_6' - 2x_7 - 3x_8' + 9x_9 = 4 \\
& x_5 = 0 \\
& x_2' = 1 \\
& x_6' = 1 \\
& x_8' = 1
\end{aligned}$$

(The primes on the key variables indicate that the variables have been redefined, e.g., $x_2' = x_2 + x_3 + x_4 + x_5$.)
In this form we can eliminate the last three equations by substituting the indicated values of the key variables into the first $m$ equations. (We could also eliminate the third equation because of its peculiar form, but it is included to make the problem less trivial, and therefore is not eliminated.) The reduced form of the example is as follows:

$$\begin{aligned}
\text{Maximize } z &= x_1 - x_4 - 2x_5 \\
\text{subject to: } & x_1 + 2x_3 + x_4 - 4x_7 - 11x_9 = 4 \\
& x_1 - 2x_3 + x_4 - x_5 - 2x_7 + 9x_9 = 2 \\
& x_5 = 0
\end{aligned}$$

The reduced basis form, in terms of the nonkey variables, is as follows:

$$\begin{aligned}
x_1 + 2x_3 \qquad &= 4 \\
x_1 - 2x_3 - x_5 &= 2 \\
x_5 &= 0
\end{aligned}$$

The method of generalized upper bounds does not explicitly perform the subtraction of key variables. To see how the transformation is accomplished, we partition the basis as follows (the vectors of the key variables first), yielding the following:

$$\boldsymbol{B} = \begin{pmatrix} \boldsymbol{Q} & \boldsymbol{F} \\ \boldsymbol{I} & \boldsymbol{G} \end{pmatrix} \tag{8.9}$$

where $\boldsymbol{Q}$ is $m$ by $L$, $\boldsymbol{F}$ is $m$ by $m$, and $\boldsymbol{I}$ is an $L$ by $L$ identity matrix. $\boldsymbol{G}$ is an $L$ by $m$ matrix of zeroes and ones. All entries of $\boldsymbol{G}$ are zero, except that each column vector of $\boldsymbol{G}$ corresponding to a nonkey variable in set $S_j$ has element $j$ equal to 1.

We may write the basic solution to the original problem as

$$\boldsymbol{Bx}_B = \boldsymbol{b} \tag{8.10}$$

We want to transform $\boldsymbol{B}$ into a form whereby the lower right $L$ by $m$ submatrix of the transformation is zero. (Why?) We then transform the above relationship as follows:

$$\boldsymbol{BIx}_B = (\boldsymbol{BT})(\boldsymbol{T}^{-1}\boldsymbol{x}_B) = \boldsymbol{b} \tag{8.11}$$

where $\boldsymbol{T}$ is such that

$$\boldsymbol{BT} = \begin{pmatrix} \boldsymbol{Q} & \boldsymbol{H} \\ \boldsymbol{I} & \boldsymbol{O} \end{pmatrix} \tag{8.12}$$

thereby accomplishing the transformation of $G$ to zero. It can be shown that

$$H = -QG + F \tag{8.13}$$

and that

$$T = \begin{pmatrix} I & -G \\ O & I \end{pmatrix} \quad \text{and} \quad T^{-1} = \begin{pmatrix} I & G \\ O & I \end{pmatrix}$$

Now, let us express $B^{-1}$ in partitioned form

$$B^{-1} = (BTT^{-1})^{-1} = T(BT)^{-1} \tag{8.14}$$

Since

$$(BT)^{-1} = \begin{pmatrix} O & I \\ H^{-1} & -H^{-1}Q \end{pmatrix}$$

we have by equation (8.14)

$$B^{-1} = \begin{pmatrix} I & -G \\ O & I \end{pmatrix} \begin{pmatrix} O & I \\ H^{-1} & -H^{-1}Q \end{pmatrix} = \begin{pmatrix} -GH^{-1} & I + GH^{-1}Q \\ H^{-1} & -H^{-1}Q \end{pmatrix} \tag{8.15}$$

Before shrugging off expression (8.15) as worthless, observe that the only inverse in $B^{-1}$ is of an $m$ by $m$ matrix $H$. Even more significant is the multiplication $B^{-1}a$, where $a$ is an incoming vector which is partitioned so that the last $L$ components have at most one nonzero element, and that element is 1.

Let the first $m$ elements of $a$ be $\alpha$. Then

$$B^{-1}a = B^{-1}\begin{pmatrix} \alpha \\ e_i \end{pmatrix} = \begin{pmatrix} -GH^{-1}\alpha + e_i + GH^{-1}q_i \\ H^{-1}\alpha - H^{-1}q_i \end{pmatrix}$$

or

$$B^{-1}a = \begin{pmatrix} e_i - G(H^{-1}\alpha - H^{-1}q_i) \\ H^{-1}\alpha - H^{-1}q_i \end{pmatrix} \tag{8.16}$$

where $e_i$ is a unit vector of order $L$, $i$ is the index of the set $S_i$ which contains the incoming variable (define $e_0$ as an $L$ order null vector) and $q_i$ is *column* $i$ of $Q$. Because $Q$ is the matrix of coefficients of the first $m$ constraints of the key variables, $q_i$ is a vector consisting of the first $m$ coefficients of the key variable of set $S_i$. (For convenience, define $q_0$ as an $m$ order null vector.)

Therefore, to update an incoming vector, we essentially need multiply $H^{-1}$ times the first $m$ components of the incoming column of the incoming vector and the vector of its key variable, and perform a little matrix addition on the resulting vectors. (Multiplication by the matrix $G$ is equivalent to adding certain elements of the vector together.)

By partitioning $b$ as

$$b = \begin{pmatrix} p \\ 1 \end{pmatrix}$$

where $p$ is the vector of the first $m$ components of $b$, we have*

$$B^{-1}b = B^{-1}\begin{pmatrix} p \\ 1 \end{pmatrix} = \begin{pmatrix} 1 - G(H^{-1}p - H^{-1}Q1) \\ H^{-1}p - H^{-1}Q1 \end{pmatrix} \tag{8.17}$$

$B^{-1}b$ is usually completely updated by the method; accordingly, expression (8.17) is not generally used except for checking accuracy or for matrix inversion.

We now consider the development of the simplex multipliers $c'_B B^{-1}$, which we partition as $(\pi' \; \mu')$, where $\pi$ is the $m$ vector of multipliers for the first $m$ constraints and $\mu$ is the $L$ vector of multipliers for the last $L$ constraints. We further partition $c'_B$ as $c'_B = (c'_{Bk} \; c'_{Bn})$, where $c_{Bk}$ is the $L$ vector of key variables and $c_{Bn}$ is the $m$ vector of nonkey variables. Using expression (8.15), we have

$$(\pi' \quad \mu') = (c'_{Bk} \quad c'_{Bn})B^{-1} = (c'_{Bk} \quad c'_{Bn})\begin{pmatrix} -GH^{-1} & I + GH^{-1}Q \\ H^{-1} & -H^{-1}Q \end{pmatrix}$$

or

$$(\pi' \quad \mu') = (-(c'_{Bk}G - c'_{Bn})H^{-1} \mid c'_{Bk} + (c'_{Bk}G - c'_{Bn})H^{-1}Q) \tag{8.18}$$

Separating expression (8.18), we have

$$\pi' = -(c'_{Bk}G - c'_{Bn})H^{-1} \tag{8.19}$$

and

$$\mu' = c'_{Bk} + (c'_{Bk}G - c'_{Bn})H^{-1}Q \tag{8.20}$$

By substituting expression (8.19) into (8.20) we have

$$\mu' = c'_{Bk} - \pi' Q \tag{8.21}$$

The evaluation of reduced costs for nonbasic variables is made by left-multiplying a nonbasic column by $c'_B B^{-1}$ or $(\pi' \; \mu')$, as given by expressions

* The calculations in expressions (8.16) and (8.17) can be simplified slightly by factoring out $H^{-1}$ where possible.

(8.19) and (8.21). Because the last $L$ elements of the nonbasic column are a unit vector or a null vector, the calculation of $c'_B B^{-1} a_j - c_j$ reduces to

$$c'_B B^{-1} a_j - c_j = \pi' \alpha_j + \mu_i - c_j \tag{8.22}$$

where $S_i$ is the set containing $x_j$. (Define $\mu_0$ as zero.) It is possible to work with only the $m$ elements of the updated reduced right-hand side and make any calculation for the $L$ elements, which have been eliminated, directly from the reduced form, instead of working with the full set of $m + L$ elements of the updated right-hand side. We shall not pursue that strategy here, however.

EXAMPLE 8.5:

Determine the first interation to be taken in the problem of Examples 8.3 and 8.4. Writing the key variables first, we have the complete basis from the original formulation.

$$\begin{aligned}
5x_6 - x_8 + x_1 + 2x_3 \quad &= 8 \\
x_2 + 4x_6 - 3x_8 + x_1 - x_3 \quad &= 4 \\
x_5 &= 0 \\
x_2 \qquad + x_3 + x_5 &= 1 \\
x_6 \qquad &= 1 \\
x_8 \qquad &= 1
\end{aligned}$$

Using the partition of $B$ as in (8.9), we have

$$Q = \begin{pmatrix} 0 & 5 & -1 \\ 1 & 4 & -3 \\ 0 & 0 & 0 \end{pmatrix} \qquad F = \begin{pmatrix} 1 & 2 & 0 \\ 1 & -1 & 0 \\ 0 & 0 & 1 \end{pmatrix} \qquad G = \begin{pmatrix} 0 & 1 & 1 \\ 0 & 0 & 0 \\ 0 & 0 & 0 \end{pmatrix}$$

and

$$H = -QG + F = \begin{pmatrix} 1 & 2 & 0 \\ 1 & -2 & -1 \\ 0 & 0 & 1 \end{pmatrix}$$

$$H^{-1} = \begin{pmatrix} .5 & .5 & .5 \\ .25 & -.25 & -.25 \\ 0 & 0 & 1 \end{pmatrix}$$

We next compute the simplex multipliers $\pi$ and $\mu$, using expressions (8.19) and (8.21). Designating the vectors $c_{Bk}$ and $c_{Bn}$ as

$$c'_{Bk} = (0 \quad 0 \quad 0); \quad c'_{Bn} = (1 \quad 0 \quad -2)$$

we have

$$\boldsymbol{\pi}' = -((0 \quad 0 \quad 0) - (1 \quad 0 \quad -2))\boldsymbol{H}^{-1} = (.5 \quad .5 \quad -1.5)$$
$$\boldsymbol{\mu}' = (0 \quad 0 \quad 0) - \boldsymbol{\pi}'\boldsymbol{Q} = (0 \quad 0 \quad 0) - (.5 \quad 4.5 \quad -2)$$
$$\boldsymbol{\mu}' = (-.5 \quad -4.5 \quad +2)$$

Next we evaluate the nonbasic columns using expression (8.22). The results are

$$\text{for } x_4\text{: } z_4 - c_4 = \boldsymbol{\pi}'\boldsymbol{\alpha}_4 + \mu_1 - c_4 = (.5 \quad .5 \quad -1.5)\begin{pmatrix}1\\2\\0\end{pmatrix} + \mu_1 - c_4$$
$$= 1.5 + (-.5) - (-1) = 2$$
$$\text{for } x_7\text{: } z_7 - c_7 = \boldsymbol{\pi}'\boldsymbol{\alpha}_7 + \mu_2 - c_7 = 1.5 + (-4.5) - 0 = -3$$
$$\text{for } x_9\text{: } z_9 - c_9 = \boldsymbol{\pi}'\boldsymbol{\alpha}_9 + \mu_3 - c_9 = -3 + 2 - 0 = -1$$

Hence, we choose $x_7$ to enter the basis. To obtain the updated coefficients $\boldsymbol{B}^{-1}\boldsymbol{a}_7$ we use expression (8.16). (In this case $i = 2$.)
Hence

$$\boldsymbol{B}^{-1}\boldsymbol{a}_7 = \begin{pmatrix}\boldsymbol{e}_2 - \boldsymbol{G}(\boldsymbol{H}^{-1}\boldsymbol{\alpha}_7 - \boldsymbol{H}^{-1}\boldsymbol{q}_2)\\ \boldsymbol{H}^{-1}\boldsymbol{\alpha}_7 - \boldsymbol{H}^{-1}\boldsymbol{q}_2\end{pmatrix}$$

$$\boldsymbol{H}^{-1}\boldsymbol{\alpha}_7 - \boldsymbol{H}^{-1}\boldsymbol{q}_2 = \begin{pmatrix}1.5\\-.25\\0\end{pmatrix} - \begin{pmatrix}4.5\\.25\\0\end{pmatrix} = \begin{pmatrix}-3\\-.5\\0\end{pmatrix}$$

and

$$\boldsymbol{G}(\boldsymbol{H}^{-1}\boldsymbol{\alpha}_7 - \boldsymbol{H}^{-1}\boldsymbol{q}_2) = \begin{pmatrix}0 & 1 & 1\\0 & 0 & 0\\0 & 0 & 0\end{pmatrix}\begin{pmatrix}-3\\-.5\\0\end{pmatrix} = \begin{pmatrix}-.5\\0\\0\end{pmatrix}$$

Hence

$$\boldsymbol{B}^{-1}\boldsymbol{a}_7 = \begin{pmatrix}\begin{pmatrix}0\\1\\0\end{pmatrix} - \begin{pmatrix}-.5\\0\\0\end{pmatrix}\\ \begin{pmatrix}-3\\-.5\\0\end{pmatrix}\end{pmatrix} = \begin{pmatrix}.5\\1\\0\\-3\\-.5\\0\end{pmatrix}$$

Since

$$\boldsymbol{H}^{-1}\boldsymbol{p} - \boldsymbol{H}^{-1}\boldsymbol{Q}\mathbf{1} = \begin{pmatrix}6\\1\\0\end{pmatrix} - \begin{pmatrix}3\\.5\\0\end{pmatrix} = \begin{pmatrix}3\\.5\\0\end{pmatrix}$$

we have by expression (8.17)

$$
B^{-1}b = \begin{pmatrix} \begin{pmatrix} 1 \\ 1 \\ 1 \end{pmatrix} - \begin{pmatrix} 0 & 1 & 1 \\ 0 & 0 & 0 \\ 0 & 0 & 0 \end{pmatrix} \begin{pmatrix} 3 \\ .5 \\ 0 \end{pmatrix} \\ \begin{pmatrix} 3 \\ .5 \\ 0 \end{pmatrix} \end{pmatrix} = \begin{pmatrix} .5 \\ 1 \\ 1 \\ 3 \\ .5 \\ 0 \end{pmatrix}
$$

which means that $x_2 = .5$, $x_6 = 1$, $x_8 = 1$, $x_1 = 3$, $x_3 = .5$, $x_5 = 0$. There is a tie in selecting a variable to leave the basis between $x_2$ and $x_6$. We shall return to the example shortly.

Given an incoming variable, there are three types of iterations which may occur analogous to those of the method of upper bounds.

1. The outgoing variable is not a key variable. In this case, the iteration is very much like an ordinary revised simplex method iteration, in that the reduced basis changes.
2. The outgoing variable is a key variable in an essential set. This iteration is analogous to an iteration in the method of upper bounds in which a variable enters the basis and a basic variable goes to its upper bound. The procedure is to make another variable key in the set in question. The iteration then becomes one of the first type.
3. The outgoing variable is a key variable in an inessential set. This iteration is analogous to an upper bound iteration as described in the previous section. The reduced basis does not change.

We shall now consider the consequences of each type of iteration. For type 1, $\boldsymbol{Q}$ is not affected; hence, only $\boldsymbol{H}^{-1}$ need be updated along with the nonkey variables, as in the revised simplex method. The matrix $\boldsymbol{G}$ is changed, but does not come into play until the next iteration.

For iterations of type 2, we must first make the outgoing variable (which is an essential key variable) nonkey, and an appropriate basic nonkey variable key. Thus, because we exchange one nonkey variable for another, the reduced basis $\boldsymbol{H}$ is changed in order to perform an iteration. The change is accomplished by multiplying one column of the original reduced basis $\boldsymbol{H}$ by $-1$ and subtracting that column of $\boldsymbol{H}$ from all other nonkey variables of that set. That is equivalent to right-multiplying the reduced basis by a matrix $\boldsymbol{R}$ which is an identity matrix except for one row. That row—the row corresponding to the column of $\boldsymbol{H}$ containing the variable to be made key—has a $-1$ on the diagonal and one or more additional $-1$'s in it, corresponding to columns of $\boldsymbol{H}$ containing other nonkey basic variables of the set. The matrix $\boldsymbol{R}$ is an elementary row matrix (whereas what we had previously defined as elementary matrices are elementary column matrices). Further, it can be shown that

$$\boldsymbol{R}^{-1} = \boldsymbol{R}$$

Hence the inverse of the new reduced basis is obtained by left-multiplying the old inverse by $\boldsymbol{R}^{-1}$ or $\boldsymbol{R}$. Of course, $\boldsymbol{Q}$ is altered by replacing the column of the old key variable with that of the new key variable. Once the transformation has been made, the iteration becomes a type 1 iteration and the procedure described earlier is followed.

For a type 3 iteration, $\boldsymbol{Q}$ changes but $\boldsymbol{H}$ does not, because the outgoing variable is a key variable from an inessential set. In this case the incoming variable is in the same inessential set as the outgoing variable. Hence, the iteration only involves recomputing the associated $\mu_i$ and $\boldsymbol{B}^{-1}\boldsymbol{b}$.

A flow chart of the entire procedure is given in Figure 8.2. Note that neither $\boldsymbol{Q}$ nor $\boldsymbol{G}$ need be maintained explicitly. The columns of $\boldsymbol{Q}$ are available from the original statement of the problem. Further, only the first $m$ constraints of the original problem need be stated explicitly, using a partition of the variables into sets $S_i$.

Example 8.6:

Complete the solution of the problem of Example 8.5 using the method of generalized upper bounds, proceeding from the calculations of Example 8.5. We choose $x_6$ to leave the basis (instead of $x_2$). Since $x_6$ is a nonessential key variable, the iteration is of type 3. We indicate the new $\boldsymbol{Q}$ for convenience:

$$\boldsymbol{Q} = \begin{pmatrix} 0 & 1 & -1 \\ 1 & 2 & -3 \\ 0 & 0 & 0 \end{pmatrix}$$

The values of the variables are altered during the iteration from Tableau 1 to Tableau 2 as follows (the calculations of $\boldsymbol{B}^{-1}\boldsymbol{b}$ and the updated incoming column of $x_7$ in Tableau 1 were obtained in Example 8.5):

**Tableau 1**

| *Basic Variables* | *Incoming Column* $x_7$ | $\boldsymbol{B}^{-1}\boldsymbol{b}$ |
|---|---|---|
| $x_2$ | .5 | .5 |
| $x_6$ | $\boxed{1}$ | 1 |
| $x_8$ | 0 | 1 |
| $x_1$ | −3 | 3 |
| $x_3$ | −.5 | .5 |
| $x_5$ | 0 | 0 |

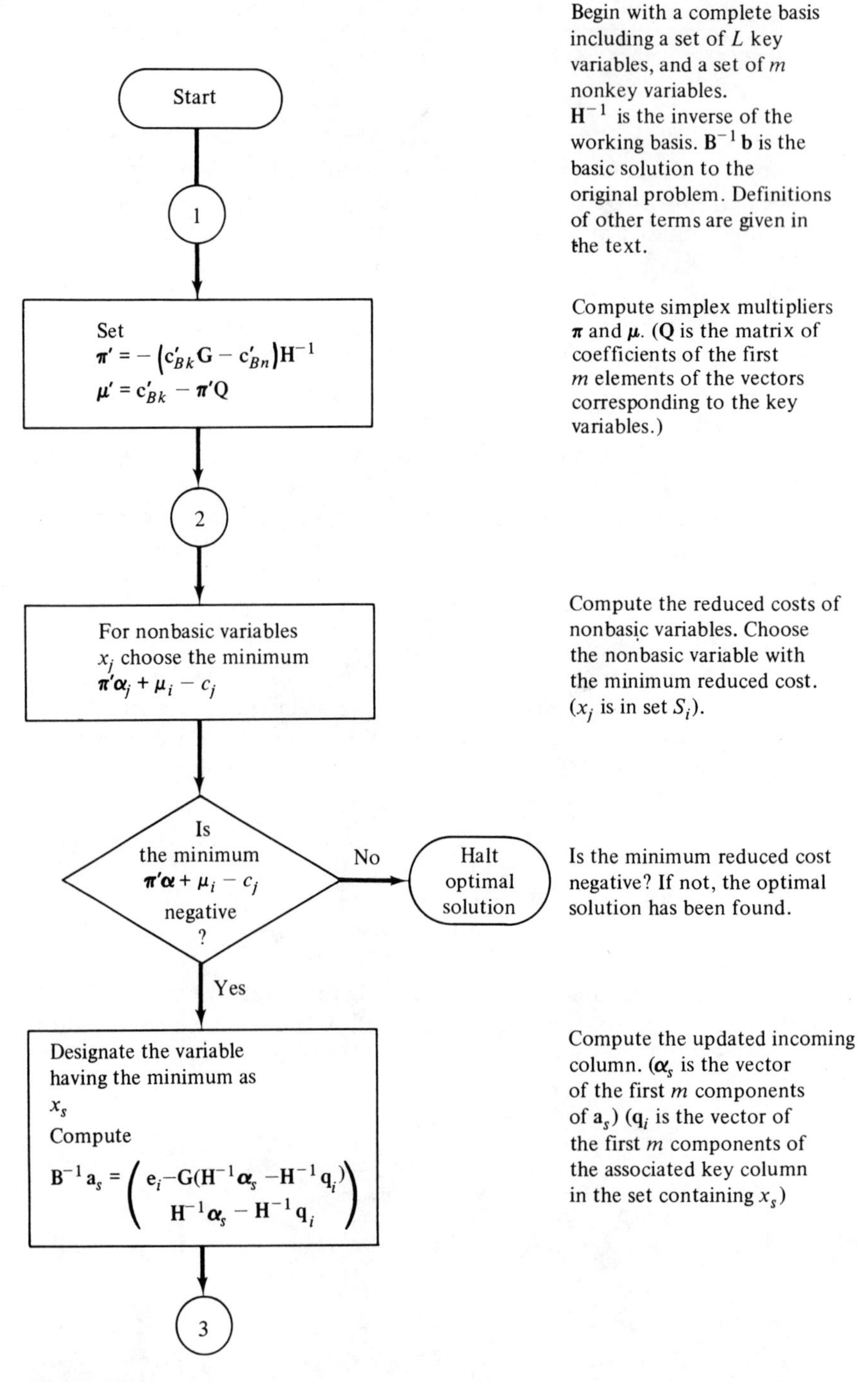


**Figure 8.2**

Flow Chart of Generalized Upper Bound Procedure

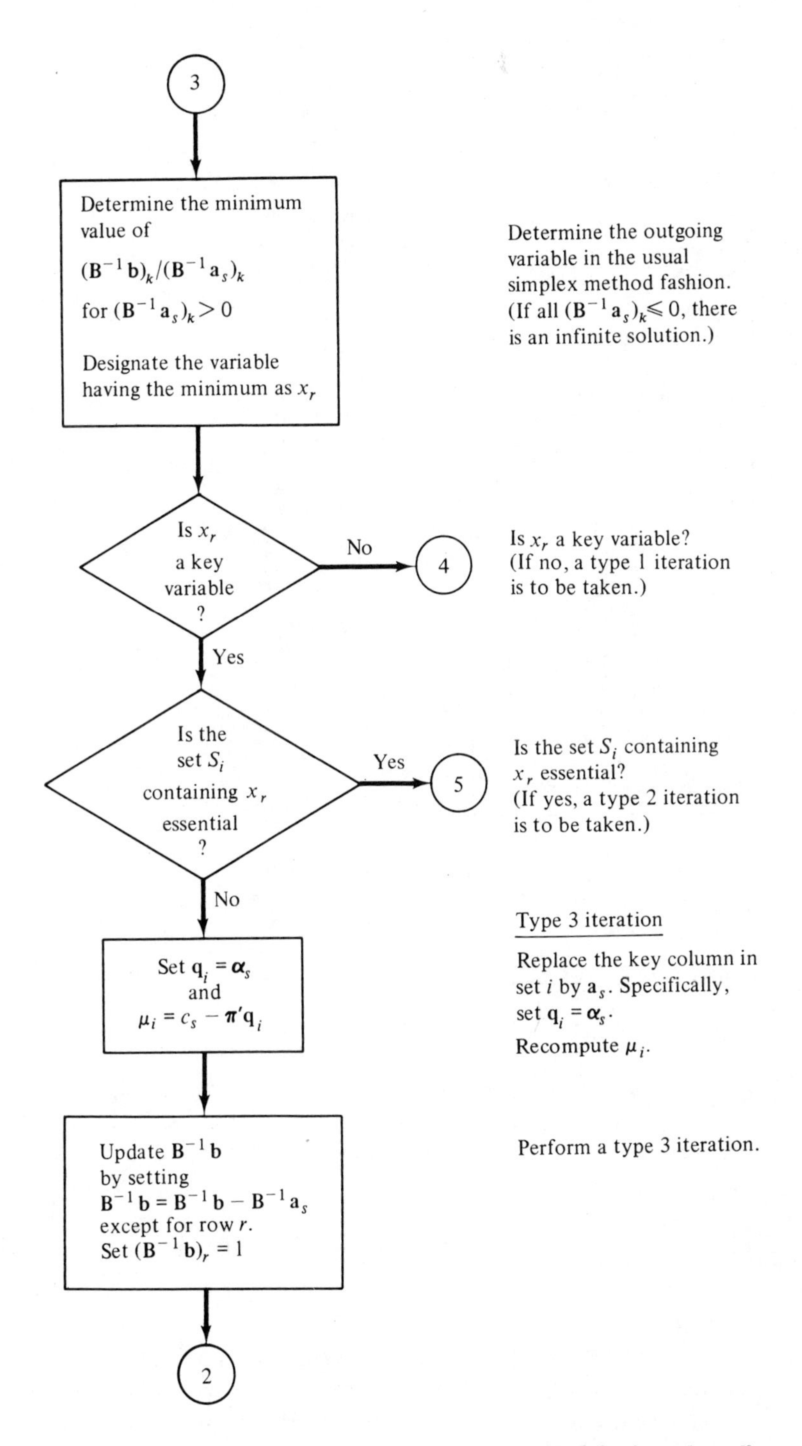


**Figure 8.2 (continued)**

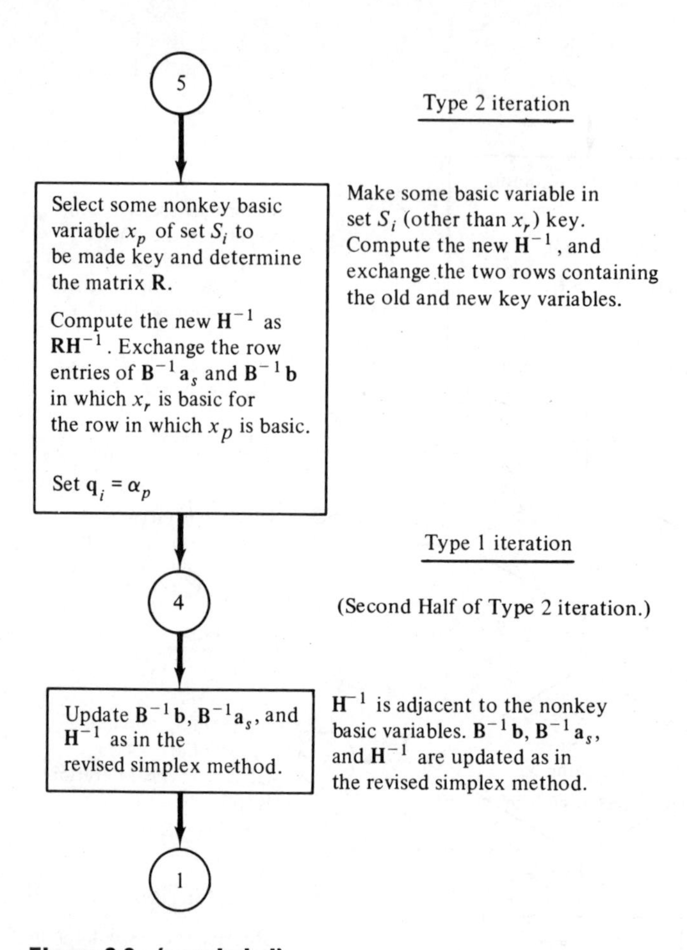


**Figure 8.2 (concluded)**

The new problem solution is $x_2 = 0$, $x_7 = 1$, $x_8 = 1$, $x_1 = 6$, $x_3 = 1$, $x_5 = 0$. (Because the objective function is not automatically computed, it may be convenient to represent it as a variable, as is done by Dantzig and Van-Slyke.) We now recompute $\mu_2$ using expression (8.21):

$$\mu_2 = 0 - (.5 \quad .5 \quad -1.5)\begin{pmatrix} 1 \\ 2 \\ 0 \end{pmatrix} = -1.5$$

The reduced costs for nonbasic variables are the same as for the previous iteration, except for those in constraint set $S_2$. This is because $\boldsymbol{\pi}$ has not changed, and only $\mu_2$ has changed. Hence, the reduced cost of the only nonbasic variable of $S_2$ is for $x_6$:

$$\boldsymbol{\pi}'\boldsymbol{\alpha}_6 + \mu_2 - c_6 = 4.5 - 1.5 - 0 = 3$$

We therefore choose $x_9$ to enter the basis. The representation of the nonbasic column for $x_9$ is achieved using expression (8.16).

$$H^{-1}\boldsymbol{\alpha}_9 = \begin{pmatrix} -3 \\ -4.5 \\ 0 \end{pmatrix} \qquad H^{-1}\boldsymbol{q}_3 = \begin{pmatrix} -2 \\ .5 \\ 0 \end{pmatrix}$$

Hence, we have

$$B^{-1}\boldsymbol{a}_9 = \begin{pmatrix} \begin{pmatrix} 0 \\ 0 \\ 1 \end{pmatrix} - \begin{pmatrix} -5 \\ 0 \\ 0 \end{pmatrix} \\ \begin{pmatrix} -3 \\ -4.5 \\ 0 \end{pmatrix} - \begin{pmatrix} -2 \\ .5 \\ 0 \end{pmatrix} \end{pmatrix} = \begin{pmatrix} 5 \\ 0 \\ 1 \\ -1 \\ -5 \\ 0 \end{pmatrix}$$

which appears in Tableau 2.

**Tableau 2**

| *Basic Variables* | *Incoming Column* $x_9$ | $x_7$ | $B^{-1}b$ |
|---|---|---|---|
| $x_2$ | 5 | 0 | 0 |
| $x_7$ | 0 | 1 | 1 |
| $x_8$ | 1 | 0 | 1 |
| $x_1$ | $-1$ | 0 | 6 |
| $x_3$ | $-5$ | 0 | 1 |
| $x_5$ | 0 | 0 | 0 |

By reference to Tableau 2, $x_2$ is to leave the basis. Because $x_2$ is key in an essential set, we have a type 2 iteration and must first make another variable key in that set. We choose $x_3$, and instead of having as nonkey variables $x_1$, $x_3$, and $x_5$, we want to have $x_1$, $x_2$, and $x_5$. Because $x_3$ and $x_5$ are both in the set $S_1$, we have

$$R = \begin{pmatrix} 1 & 0 & 0 \\ 0 & -1 & -1 \\ 0 & 0 & 1 \end{pmatrix}$$

$$R^{-1}H^{-1} = RH^{-1} = \begin{pmatrix} 1 & 0 & 0 \\ 0 & -1 & -1 \\ 0 & 0 & 1 \end{pmatrix} \begin{pmatrix} .5 & .5 & .5 \\ .25 & -.25 & -.25 \\ 0 & 0 & 1 \end{pmatrix}$$

$$= \begin{pmatrix} .5 & .5 & .5 \\ -.25 & .25 & -.75 \\ 0 & 0 & 1 \end{pmatrix}$$

$\boldsymbol{Q}$ is now altered by replacing the first column $\boldsymbol{a}_2$ by $\boldsymbol{a}_3$,

$$\boldsymbol{Q} = \begin{pmatrix} 2 & 1 & -1 \\ -1 & 2 & -3 \\ 0 & 0 & 0 \end{pmatrix} \quad \text{and} \quad \boldsymbol{G} = \begin{pmatrix} 0 & 1 & 1 \\ 0 & 0 & 0 \\ 0 & 0 & 0 \end{pmatrix}$$

The order of variables is now $x_3, x_7, x_8, x_1, x_2, x_5$, and the remainder of the iteration is a type 1 iteration.

**Tableau 2 (revised)**

| *Basic Variables* | *Old Inverse* | | | *Incoming Column* $x_9$ | $\boldsymbol{B}^{-1}\boldsymbol{b}$ |
|---|---|---|---|---|---|
| $x_3$ | | | | $-5$ | 1 |
| $x_7$ | | | | 0 | 1 |
| $x_8$ | | | | 1 | 1 |
| $x_1$ | .5 | .5 | .5 | $-1$ | 6 |
| $x_2$ | $-.25$ | .25 | $-.75$ | $\boxed{5}$ | 0 |
| $x_5$ | 0 | 0 | 1 | 0 | 0 |

**Tableau 3**

| *Basic Variables* | *New Inverse* | | | $x_9$ | $\boldsymbol{B}^{-1}\boldsymbol{b}$ |
|---|---|---|---|---|---|
| $x_3$ | | | | 0 | 1 |
| $x_7$ | | | | 0 | 1 |
| $x_8$ | | | | 0 | 1 |
| $x_1$ | .45 | .55 | .35 | 0 | 6 |
| $x_9$ | $-.05$ | .05 | $-.15$ | 1 | 0 |
| $x_5$ | 0 | 0 | 1 | 0 | 0 |

We now compute the simplex multipliers $\boldsymbol{\pi}$ and $\boldsymbol{\mu}$. By equations (8.19) and (8.21), we have

$$\begin{aligned} \boldsymbol{\pi}' &= -\left((0 \quad 0 \quad 0)\begin{pmatrix} 0 & 0 & 1 \\ 0 & 0 & 0 \\ 0 & 1 & 0 \end{pmatrix} - (1 \quad 0 \quad -2)\right)\begin{pmatrix} .45 & .55 & .35 \\ -.05 & .05 & -.15 \\ 0 & 0 & 1 \end{pmatrix} \\ &= (.45 \quad .55 \quad -1.65) \end{aligned}$$

$$\boldsymbol{\mu}' = (0 \quad 0 \quad 0) - \boldsymbol{\pi}'\boldsymbol{Q} = -(.45 \quad .55 \quad -1.65)\begin{pmatrix} 2 & 1 & -1 \\ -1 & 2 & -3 \\ 0 & 0 & 0 \end{pmatrix}$$

$$\boldsymbol{\mu}' = (-.35 \quad -1.55 \quad 2.1)$$

Evaluating the reduced costs for nonbasic variables, we have

$$\begin{aligned}
&\text{for } x_2\text{: } \boldsymbol{\pi}'\boldsymbol{\alpha}_2 + \mu_1 - c_2 = .55 - .35 - 0 = .2 \\
&\text{for } x_4\text{: } \boldsymbol{\pi}'\boldsymbol{\alpha}_4 + \mu_1 - c_4 = 1.55 - .35 - (-1) = 2.2 \\
&\text{for } x_6\text{: } \boldsymbol{\pi}'\boldsymbol{\alpha}_6 + \mu_2 - c_6 = 4.45 - 1.55 - 0 = 2.9
\end{aligned}$$

Hence, the solution given in Tableau 3 is optimal.

On the question of computational efficiency, a test problem having $m = 39$, $L = 780$, and 2813 variables was solved, as referenced in Dantzig and VanSlyke [68]. The solution required 15 minutes on an IBM 7094. They estimate the running time for that problem, using a general linear programming code, at 150 minutes.

#### Extensions

Lasdon [188], extending the work of Kaul [173], indicates how generalized upper bounds may be extended to problems having block angular structure. The results appear promising, although no computational experience is reported. Related results are developed by Saigal [245] and Sakarovitch and Saigal [247].

## 8.4 THE DECOMPOSITION PRINCIPLE

In linear programming formulations of decentralized organizations, a frequently occurring model framework is the following:

$$\begin{aligned}
\text{Maximize } z = \; & \boldsymbol{c}_0'\boldsymbol{x}_0 \quad \boldsymbol{c}_1'\boldsymbol{x}_1 + \boldsymbol{c}_2'\boldsymbol{x}_2 + \cdots + \boldsymbol{c}_h'\boldsymbol{x}_h \\
\text{subject to: } \; & \boldsymbol{A}_{00}\boldsymbol{x}_0 + \boldsymbol{A}_{01}\boldsymbol{x}_1 + \boldsymbol{A}_{02}\boldsymbol{x}_2 + \cdots + \boldsymbol{A}_{0h}\boldsymbol{x}_h = \boldsymbol{b}_0 \\
& \boldsymbol{A}_{11}\boldsymbol{x}_1 = \boldsymbol{b}_1 \\
& \boldsymbol{A}_{22}\boldsymbol{x}_2 = \boldsymbol{b}_2 \\
& \quad \vdots \\
& \boldsymbol{A}_{hh}\boldsymbol{x}_h = \boldsymbol{b}_h \\
& \boldsymbol{x}_1, \boldsymbol{x}_2, \cdots, \boldsymbol{x}_h \geq \boldsymbol{0}
\end{aligned} \tag{8.23}$$

where $c_j$ and $x_j$ are $n_j$ by 1, $A_{ij}$ is $m_i$ by $n_j$, and $b_i$ is $m_i$ by 1. We assume that each convex set $A_{ii}x_i = b_i$ is bounded. This may be assured by adjoining a regularization constraint to each such constraint set. The objective function is typically the contribution to corporate profits of the selected set of activities. The first set of $(m_0)$ constraints are the constraints due to centralized resource limitations and other commitments. These constraints thereby include such factors as limited availability of centrally procured raw materials, cash flows, and manpower, as well as constraints due to sales commitments. Each additional set of $(m_i, i \neq 0)$ constraints are constraints local to the $i^{\text{th}}$ location, such as limitations due to plant capacities, maintenance, other manpower constraints, and local resource (e.g., electricity) availability. The variables (vectors $x_1, \ldots, x_h$) are generally final or intermediate products produced at the respective locations. Once set up in the above framework, the model is simply a linear programming problem consisting of $\sum_{i=0}^{h} m_i$ constraints and $\sum_{j=0}^{h} n_j$ variables. Many such problems have been formulated and solved for by a large number of organizations. In general, however, such problems are huge and may not be solvable given the present state of computer programs, or may require an exorbitant amount of time for computation. This helped provide the motivation for development of the decomposition principle of Dantzig and Wolfe [69],

**Table 8.1**

Data for Example Problem

| PLANT 1 | | | |
|---|---|---|---|
| | *Pan 1* | *Pan 2* | *Total Monthly Time Available of Production Facility in Seconds* |
| Contribution to overhead and profits | \$.20 | \$.10 | — |
| Seconds of production facility A time required | 30 | 20 | 1,200,000 |
| Seconds of production facility B time required | 40 | 0 | 800,000 |

| PLANT 2 | | | | |
|---|---|---|---|---|
| | *Pan 3* | *Pan 4* | *Pan 5* | *Total Monthly Time Available of Production Facility in Seconds* |
| Contribution to overhead and profits | \$.12 | \$.15 | \$.10 | — |
| Seconds of production facility A time required | 20 | 20 | 0 | 800,000 |
| Seconds of production facility B time required | 30 | 0 | 20 | 1,000,000 |

which we shall consider now. (The Dantzig-Wolfe decomposition method preceded the generalized upper bounds development and its extension to problems of this nature by Kaul [173] and others mentioned in the previous section.)

An example is useful to introduce the principle. A manufacturer of pots and pans has two plants where the pans are produced. Plant 1 produces two kinds of pans; plant 2 produces three kinds of pans. A schedule of the time required on each production facility, the contribution to overhead and profits of each pan, and the production facility availability are given in Table 8.1. Each pan requires one unit of stainless steel sheet, of which the company has 100,000 units per month. It is assumed that all production can be sold. What is the production plan which maximizes contribution to overhead and profits? The problem corresponding to the data in Table 8.1 is formulated below.

$$
\begin{array}{llllllll}
\text{Maximize } z = & .20x_{11} + & .10x_{12} + & .12x_{21} + & .15x_{22} + & .10x_{23} & & \\
\text{subject to:} & x_{11} + & x_{12} + & x_{21} + & x_{22} + & x_{23} & \leq 10 & \text{(limit on stainless steel sheet)} \\
& 30x_{11} + & 20x_{12} & & & & \leq 120 & \text{(facility A, plant 1)} \\
& 40x_{11} & & & & & \leq 80 & \text{(facility B, plant 1)} \\
& & & 20x_{21} + & 20x_{22} & & \leq 80 & \text{(facility A, plant 2)} \\
& & & 30x_{21} & & + 20x_{23} & \leq 100 & \text{(facility B, plant 2)}
\end{array}
$$

The first subscript on variables refers to the subproblem number, and the second subscript refers to the variable number. For convenience, all production is measured in 10,000 units, and time is measured in 10,000 second units. The above problem has five constraints and ten variables (including slack variables).

Except for one constraint (which in general is a set of $m_0$ constraints), the problem decomposes into subproblems. In other words, were the first $m_0$ constraints not present, the problem could be solved simply by solving $h$ independent linear programming problems.

The decomposition principle consists of solving the overall problem via repeated solution of $h + 1$ linear programming problems—a master problem which will be defined shortly in which prices are set on the resources and/or commitments of the set of common constraints, and $h$ subproblems which are suboptimizations using the latest sets of prices generated by the master problem. A flow chart of the procedure is given in Figure 8.3.

To develop the decomposition principle, we first note that an alternative formulation to the original linear programming problem is to list *all* extreme point solutions to each subproblem. For subproblem $i$ there can be at most $p_i = \binom{n_i}{m_i}$ extreme point solutions. Denote these as $\boldsymbol{x}_i^k$, $k = 1, \ldots, p_i$. Corresponding to each extreme point solution $\boldsymbol{x}_i^k$, the amount of corporate resources and/or commitments used and/or met is $\boldsymbol{A}_{0i}\boldsymbol{x}_i^k$, an $m_0$ by 1 vector. The contribution to the objective function of the extreme point solution is $\boldsymbol{c}_i'\boldsymbol{x}_i^k$, a scalar. Any convex combination of $\boldsymbol{x}_i^k$ is a feasible solution to subproblem $i$. Designating as $y_i^k$ the (scalar) fraction $(0 \leq y_i^k \leq 1)$ of $\boldsymbol{x}_i^k$ in the convex combination, it is seen that $\sum_{k=1}^{p_i} (\boldsymbol{A}_{0i}\boldsymbol{x}_i^k)y_i^k$ is the total usage of resources of that combination, and that $(\boldsymbol{c}_i'\boldsymbol{x}_i^k)y_i^k$ is the contribution to overhead and profits of that combination. The following formulation of the problem can then be made:

$$
\begin{aligned}
&\text{Maximize} && z = \sum_{i=1}^{h} \sum_{k=1}^{p_i} (\boldsymbol{c}_i'\boldsymbol{x}_i^k)y_i^k \\
&\text{subject to:} && \boldsymbol{A}_{00}\boldsymbol{x}_0 + \sum_{i=1}^{h} \sum_{k=1}^{p_i} (\boldsymbol{A}_{0i}\boldsymbol{x}_i^k)y_i^k = \boldsymbol{b}_0 \\
&(8.24) && \sum_{k=1}^{p_i} y_i^k = 1, i = 1, \ldots, h \\
& && y_i^k \geq 0, i = 1, \ldots, h \\
& && k = 1, \ldots, p
\end{aligned}
$$

($\boldsymbol{x}_i^k$ in the above formulation is a constant vector, one of the extreme point solutions to subproblem $i$.)

Compare the above formulation with that of (8.23). The first formulation has $\sum_{i=0}^{h} m_i$ constraints and $\sum_{i=1}^{h} n_i$ variables. Formulation (8.24) has $m_0 + h$ constraints and an upper limit of $n_0 + \sum_{i=1}^{h} \binom{n_i}{m_i}$ variables where

$$\binom{n}{m} = (n!)/((m!)(n-m)!)$$

The number of constraints of formulation (8.24) is generally much less than formulation (8.23), whereas the number of variables of formulation (8.24) is usually much greater than the number of variables of formulation (8.23). In fact, the number of variables in formulation (8.24) is generally tremendous in comparison to formulation (8.23). Even if the revised simplex method were to be used for the second formulation, all variables would have to be listed initially and evaluated between iterations. The amount of work associated with the computational aspects of all the columns makes any net advantage of having fewer constraints doubtful, at least based on the discussion above.

It is possible, however, to solve the second formulation *without explicitly considering* all the variables; that procedure is the Dantzig-Wolfe decomposition principle [69]. The procedure involves solving formulation (8.24) (hereafter called the master problem) using a small subset of variables, then using the

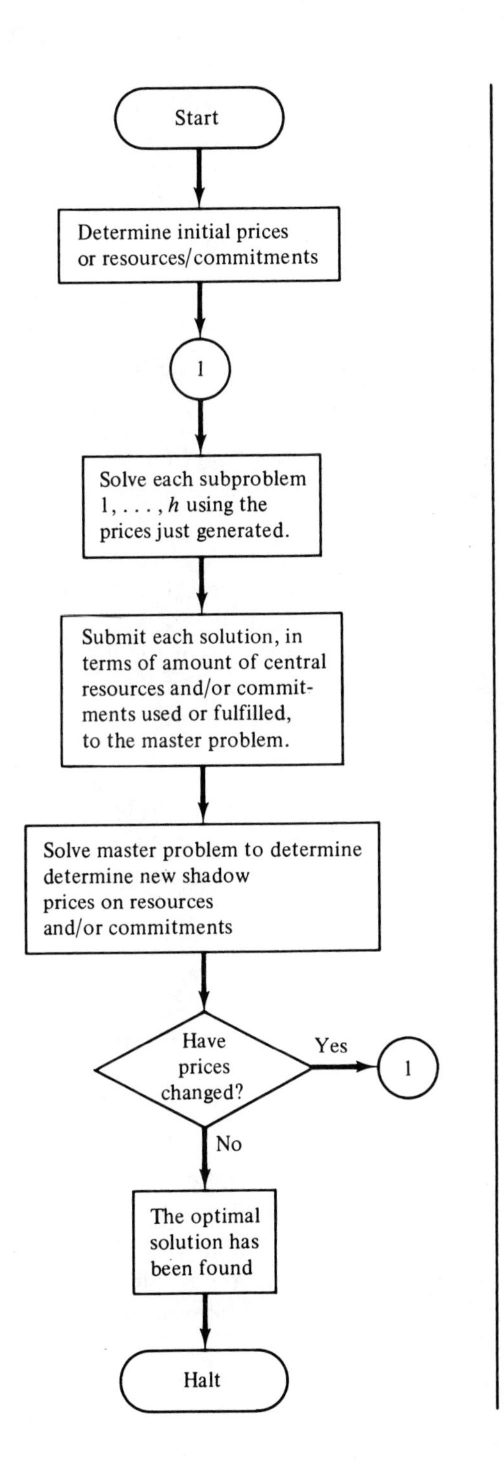


Determine initial prices. Use artificial variables if necessary to obtain a starting feasible basis.

Ignoring the first $m_0$ constraints, solve each of the $h$ subproblems using current prices for interlinking resources and/or commitments.

Express the optimal solution to each subproblem in terms of the amount of interlinking resources and/or commitments used or satisfied respectively and in terms of associated contribution to the objective function.

Re-solve the master problem given the new solutions.

Have the prices changed?

The optimal solution has been found. Express it in terms of the original variables and stop.

**Figure 8.3**

A Simplified Flow Chart of the Decomposition Principle

optimal shadow prices from that solution to solve the subproblems. The subproblem solutions are then expressed in terms of common resources and/or commitments and appended as new variables to the master problem, and the process is repeated. An optimal solution has been found when the master problem solution does not change between successive iterations.

#### An Algorithm for the Decomposition Principle

More formally, an algorithm which is essentially the Dantzig-Wolfe decomposition principle is the following:

1. Assume a starting (basic) solution to formulation (8.24) (if no such solution is available, add artificial variables). Only the basic variables need be stated explicitly in the initial formulation.
2. Solve the problem for both the primal and dual variables. Note that the dual variables on the first $m_0$ constraints are the marginal values of the $m_0$ central resources. Denote the dual variable vectors as $\boldsymbol{\pi}_0$, $\boldsymbol{\pi}_1$, where $\boldsymbol{\pi}_0$ is an $m_0$ by 1 vector that consists of the dual variables for the first $m_0$ constraints of (8.24), and $\boldsymbol{\pi}_1$ is an $h$ by 1 vector that consists of the dual variables for the next $h$ constraints of (8.24), that is, the constraints of the form $\sum_{k=1}^{p_i} y_i^k = 1$. We wish to find the variable $y_i^k$ having the greatest reduced cost (or updated value of $\mathbf{c}_i'\mathbf{x}_i^k$) for each $i$. This can be done by computing $\mathbf{c}_i' - \boldsymbol{\pi}_0'\mathbf{A}_{0i}$ where $\mathbf{A}_{0i}$ and $\mathbf{c}_i$ are from (8.23), and then solving

$$\text{(8.25)} \qquad \begin{aligned} &\text{Maximize } z = (\mathbf{c}_i' - \boldsymbol{\pi}_0'\mathbf{A}_{0i})\mathbf{x}_i \\ &\text{subject to:} \qquad \mathbf{A}_{ii}\mathbf{x}_i = \mathbf{b}_i \\ &\qquad\qquad\qquad\quad \mathbf{x}_i \geq \mathbf{0} \end{aligned}$$

   Note that (8.25) is just the local problem with a charge (possibly negative) for the use of common resources or the satisfaction of common constraints.
3. Solve the $h$ problems of the form (8.25) for $i = 1, \ldots, h$.
4. Compute $\mathbf{A}_{0i}\mathbf{x}_i^k$ and $\mathbf{c}_i'\mathbf{x}_i^k$ for $i = 1, \ldots, h$ for the solutions found in step 3, thereby creating new variables $y_i^k$, $i = 1, \ldots, h$.
5. Determine the objective function coefficients for all the new variables $y_i^k$ (i.e., $\mathbf{c}_i'\mathbf{x}_i - \boldsymbol{\pi}_0'(\mathbf{A}_{0i}\mathbf{x}_i^k) - \boldsymbol{\pi}_1'\mathbf{e}_i$), and see if any are candidates for entry into the basis. If so, update the coefficients of $y_i^k$, append to the current solution, and go to step 2. Otherwise, go to step 6.
6. The optimal solution to (8.24) has been found. Relate back to (8.23).

None of the variables $y_i^k$ need be remembered, except the ones in the basis, to assure convergence. Experience has shown, however, that it is advisable to retain a limited number of them in memory to provide a reasonably uniform convergence to the optimal solution. This is done typically by retaining the ($hr$) most attractive (even though not candidates for entry into the basis) nonbasic variables $y_i^k$, corresponding to the $r$ most attractive nonbasic variables for each of the $h$ subproblems of (8.23), where $r$ is some positive integer.

The example problem will be used first to prepare formulation (8.24), *a step which is not necessary at all*, included here only for clarity of exposition. First we list the vectors of coefficients corresponding to plant 1:

$$\begin{matrix} x_{11} & x_{12} & x_{13} & x_{14} \\ \begin{pmatrix} 30 \\ 40 \end{pmatrix} & \begin{pmatrix} 20 \\ 0 \end{pmatrix} & \begin{pmatrix} 1 \\ 0 \end{pmatrix} & \begin{pmatrix} 0 \\ 1 \end{pmatrix} \end{matrix}$$

The latter two vectors correspond to slack variables. Using the same notation as before, the first subscript indicates the problem number and the second subscript indicates the variable number. Basic feasible solutions to the subproblems of formulation (8.25) are as follows (the corresponding $y$ variables are the fraction of those solutions that appear in formulation (8.24)):

1. $y_1^1$ which corresponds to $x_{11}$ and $x_{12}$ basic ($x_{11} = 2$, $x_{12} = 3$, $x_{13} = 0$, and $x_{14} = 0$), or $\boldsymbol{x}_1^{1\prime} = (2 \quad 3 \quad 0 \quad 0)$ and $\boldsymbol{c}'_1\boldsymbol{x}_1^1 = (.2 \quad .1 \quad 0 \quad 0)\begin{pmatrix} 2 \\ 3 \\ 0 \\ 0 \end{pmatrix} = .7$, and

   $$\boldsymbol{A}_{01}\boldsymbol{x}_1^1 = (1 \quad 1 \quad 0 \quad 0)\begin{pmatrix} 2 \\ 3 \\ 0 \\ 0 \end{pmatrix} = 5.$$

2. $y_1^2$ which corresponds to $x_{11}$ and $x_{13}$ basic, or $\boldsymbol{x}_1^{2\prime} = (2 \quad 0 \quad 60 \quad 0)$ and $\boldsymbol{c}'_1\boldsymbol{x}_1^2 = .4$, and $\boldsymbol{A}_{01}\boldsymbol{x}_1^2 = 2$.
3. $y_1^3$ which corresponds to $x_{12}$ and $x_{14}$ basic, or $\boldsymbol{x}_1^{3\prime} = (0 \quad 6 \quad 0 \quad 80)$ and $\boldsymbol{c}'_1\boldsymbol{x}_1^3 = .6$, and $\boldsymbol{A}_{01}\boldsymbol{x}_1^3 = 6$.
4. $y_1^4$ which corresponds to $x_{13}$ and $x_{14}$ basic, or $\boldsymbol{x}_1^{4\prime} = (0 \quad 0 \quad 120 \quad 80)$ and $\boldsymbol{c}'_1\boldsymbol{x}_1^4 = 0$, and $\boldsymbol{A}_{01}\boldsymbol{x}_1^4 = 0$.

These are the four feasible *extreme point solutions* to subproblem 1.

Next, we list the vectors of coefficients corresponding to plant 2:

$$\begin{matrix} x_{21} & x_{22} & x_{23} & x_{24} & x_{25} \\ \begin{pmatrix} 20 \\ 30 \end{pmatrix} & \begin{pmatrix} 20 \\ 0 \end{pmatrix} & \begin{pmatrix} 0 \\ 20 \end{pmatrix} & \begin{pmatrix} 1 \\ 0 \end{pmatrix} & \begin{pmatrix} 0 \\ 1 \end{pmatrix} \end{matrix}$$

Basic feasible solutions to the second subproblem are as follows (using (8.23)):

1. $y_2^1$ which corresponds to $x_{21}$ and $x_{22}$ basic ($x_{21} = 3.33$, $x_{22} = 0.67$, $x_{23} = 0$,

$x_{24} = 0$, and $x_{25} = 0$), or $x_2^{1\prime} = (3.33 \quad 0.67 \quad 0 \quad 0 \quad 0)$ and $c_2' x_2^1 =$

$$(.12 \quad .15 \quad .10 \quad 0 \quad 0)\begin{pmatrix} 3.33 \\ .67 \\ 0 \\ 0 \\ 0 \end{pmatrix} = .50, \text{ and } A_{02} x_1^2 = (1 \quad 1 \quad 1 \quad 0 \quad 0)\begin{pmatrix} 3.33 \\ .67 \\ 0 \\ 0 \\ 0 \end{pmatrix}$$

$= 4.$

2. $y_2^2$ which corresponds to $x_{21}$ and $x_{24}$ basic, or $x_2^{2\prime} = (3.33 \quad 0 \quad 0 \quad 13.33 \quad 0)$ and $c_2' x_2^2 = .40$, and $A_{02} x_2^2 = 3.33$.
3. $y_2^3$ which corresponds to $x_{22}$ and $x_{23}$ basic, or $x_2^{3\prime} = (0 \quad 4 \quad 5 \quad 0 \quad 0)$ and $c_2' x_2^3 = 1.10$, and $A_{20} x_2^3 = 9$.
4. $y_2^4$ which corresponds to $x_{22}$ and $x_{25}$ basic, or $x_2^{4\prime} = (0 \quad 4 \quad 0 \quad 0 \quad 100)$ and $c_2' x_2^4 = .60$, and $A_{02} x_2^4 = 4$.
5. $y_2^5$ which corresponds to $x_{23}$ and $x_{24}$ basic, or $x_2^{5\prime} = (0 \quad 0 \quad 5 \quad 80 \quad 0)$ and $c_2' x_2^5 = .5$, and $A_{02} x_2^5 = 5$.
6. $y_2^6$ which corresponds to $x_{24}$ and $x_{25}$ basic, or $x_2^{6\prime} = (0 \quad 0 \quad 0 \quad 80 \quad 100)$ and $c_2' x_2^6 = 0$, and $A_{02} x_2^6 = 0$.

Adding the definition that $y_0^1$ is the slack corresponding to unused stainless steel sheet, the formulation in the form of (8.24) is as follows (all variables are nonnegative):

Maximize

$$z = .7y_1^1 + .4y_1^2 + .6y_1^3 + 0y_1^4 + .5y_2^1 + 4y_2^2 + 1.1y_2^3 + .6y_2^4 + .5y_2^5 + 0y_2^6$$

subject to:

$$\begin{aligned} (8.26) \quad y_0^1 + 5y_1^1 + 2y_1^2 + 6y_1^3 + 0y_1^4 + 4y_2^1 + 3.3y_2^2 + 9y_2^3 + 4y_2^4 + 5y_2^5 + 0y_2^6 &= 10 \\ y_1^1 + y_1^2 + y_1^3 + y_1^4 &= 1 \\ y_2^1 + y_2^2 + y_2^3 + y_2^4 + y_2^5 + y_2^6 &= 1 \end{aligned}$$

The example problem has three constraints and eleven variables, as compared to the original formulation with five constraints and ten variables (including slacks). The number of variables in formulation (8.26) is only one larger than the number in the original formulation, because there are relatively few subproblem extreme points for this problem. We could solve (8.26) directly using the simplex method. However, that would not illustrate the decomposition principle, and we shall not solve problem (8.26) directly.

Instead, making use of the fact that $y_0^1$, $y_1^4$, and $y_2^6$ form a starting feasible (slack) basis to the problem, the initial master problem is the following:

$$\begin{aligned} \text{Maximize } z &= 0y_0^1 + 0y_1^4 + 0y_2^6 \\ \text{subject to: } \quad y_0^1 &= 10 \\ y_1^4 &= 1 \\ y_2^6 &= 1 \end{aligned}$$

Were the above feasible solution not available, one or more artificial variables would have to be employed. We shall use the same names for the variables that were used in formulation (8.26). In solving problems using decomposition in general, each variable is named when it is first used (and possibly given a different name in subsequent generation). The labeling that we chose was strictly arbitrary and bears *no intended relationship* with the order in which they are generated.

The (trivial) optimal solution to the original master problem is contained in Tableau 1, ignoring the last two columns. The corresponding values of $\boldsymbol{\pi}_0$ and $\boldsymbol{\pi}_1$ are $(\boldsymbol{\pi}_0' \ \boldsymbol{\pi}_1') = (0 \ \ 0 \ \ 0)$. Each subproblem is solved using $\boldsymbol{\pi}_0 = \mathbf{0}$, which means that the plants are charged nothing for using stainless steel. Hence, the subproblems are, for plant 1:

$$\text{Maximize } z = ((.2 \ \ .1 \ \ 0 \ \ 0) - 0(1 \ \ 1 \ \ 0 \ \ 0))\begin{pmatrix} x_{11} \\ x_{12} \\ x_{13} \\ x_{14} \end{pmatrix}$$

or

$$\begin{aligned} &\text{Maximize } z = .2x_{11} + .1x_{12} \\ &\text{subject to:} \quad 30x_{11} + 20x_{12} + x_{13} = 120 \\ &\qquad\qquad\quad 40x_{11} + x_{14} = 80 \\ &\qquad\qquad\quad x_{11}, x_{12}, x_{13}, x_{14} \geq 0 \end{aligned} \tag{8.27}$$

and for plant 2:

$$\text{Maximize } z = ((.12 \ \ .15 \ \ .10 \ \ 0 \ \ 0) - 0(1 \ \ 1 \ \ 1 \ \ 0 \ \ 0))\begin{pmatrix} x_{21} \\ x_{22} \\ x_{23} \\ x_{24} \\ x_{25} \end{pmatrix}$$

or

$$\begin{aligned} &\text{Maximize } z = .12x_{21} + .15x_{22} + .10x_{23} \\ &\text{subject to:} \quad 20x_{21} + 20x_{22} + x_{24} = 80 \\ &\qquad\qquad\quad 30x_{21} + 20x_{23} + x_{25} = 100 \\ &\qquad\qquad\quad x_{21}, x_{22}, x_{23}, x_{24}, x_{25} \geq 0 \end{aligned} \tag{8.28}$$

The optimal solutions to problems (8.27) and (8.28) are $\boldsymbol{x}_1^{1\prime} = (2 \ \ 3 \ \ 0 \ \ 0)$ and $\boldsymbol{x}_2^{3\prime} = (0 \ \ 4 \ \ 5 \ \ 0 \ \ 0)$. The simplex method is usually used to compute the optimal subproblem solution. The coefficients for the master problem are $\boldsymbol{c}_i'\boldsymbol{x}_i^k$ as the objective function coefficient, and $\boldsymbol{A}_{0i}\boldsymbol{x}_i^k$ as the vector of common resources utilized (for the example, a scalar). For this set of subproblems $\boldsymbol{c}_1'\boldsymbol{x}_1^1 = .7$, $\boldsymbol{A}_{01}\boldsymbol{x}_1^1 = 5$, $\boldsymbol{c}_2'\boldsymbol{x}_2^3 = 1.1$, and $\boldsymbol{A}_{20}\boldsymbol{x}_2^3 = 9$. Accordingly, the updated columns (the original columns multiplied by $\boldsymbol{B}^{-1}$, in this case $\boldsymbol{I}$) of the $y_i^k$ variables are appended to Tableau 1, and the optimal solution appears after three iterations in Tableau 4, ignoring the appended columns. From Tableau 4, $\boldsymbol{\pi}_0$ is seen to be 11/90 (the reduced cost of $y_0^1$), which means that the plants should be charged 11/90 of a dollar for a unit of stainless steel. The objective functions for the subproblems become, for subproblem 1:

$$(\boldsymbol{c}_1' - \boldsymbol{\pi}_0'\boldsymbol{A}_{01}) = (.2 \quad .1 \quad 0 \quad 0) - \frac{11}{90}(1 \quad 1 \quad 0 \quad 0) = \left(\frac{7}{90} \quad -\frac{2}{90} \quad 0 \quad 0\right)$$

and for subproblem 2:

$$(\boldsymbol{c}_2' - \boldsymbol{\pi}_0'\boldsymbol{A}_{02}) = (.12 \quad .15 \quad .10 \quad 0 \quad 0) - \frac{11}{90}(1 \quad 1 \quad 1 \quad 0 \quad 0)$$
$$= \left(\frac{1}{450} \quad \frac{1}{36} \quad -\frac{1}{45} \quad 0 \quad 0\right)$$

**Tableau 1**

| | | $y_0^1$ | $y_1^4$ | $y_2^6$ | $y_1^1$ | $y_2^3$ ↓ |
|---|---|---|---|---|---|---|
| $z$ | 0 | 0 | 0 | 0 | −.7 | −1.1 |
| $y_0^1$ | 10 | 1 | 0 | 0 | 5 | 9 |
| $y_1^4$ | 1 | 0 | 1 | 0 | 1 | 0 |
| $y_2^6$ | 1 | 0 | 0 | 1 | 0 | [1] → |

**Tableau 2**

| | | $y_0^1$ | $y_1^4$ | $y_2^6$ | $y_1^1$ ↓ | $y_2^3$ |
|---|---|---|---|---|---|---|
| $z$ | 1.1 | 0 | 0 | 1.1 | −.7 | 0 |
| $y_0^1$ | 1 | 1 | 0 | −9 | [5] | 0 → |
| $y_1^4$ | 1 | 0 | 1 | 0 | 1 | 0 |
| $y_2^3$ | 1 | 0 | 0 | 1 | 0 | 1 |

**Tableau 3**

| | | $y_0^1$ | $y_1^4$ | $y_2^6$ ↓ | $y_1^1$ | $y_2^3$ |
|---|---|---|---|---|---|---|
| $z$ | 1.24 | .14 | 0 | −.16 | 0 | 0 |
| $y_1^1$ | .2 | .2 | 0 | −1.8 | 1 | 0 |
| $y_1^4$ | .5 | −.2 | 1 | [1.8] | 0 | 0 → |
| $y_2^3$ | .1 | 0 | 0 | 1 | 0 | 1 |

**Tableau 4**

| | | $y_0^1$ | $y_1^4$ | $y_2^6$ | $y_1^1$ | $y_2^3$ | $y_1^2$ | $y_2^4$ ↓ |
|---|---|---|---|---|---|---|---|---|
| $z$ | 1.311 | $\frac{11}{90}$ | $\frac{8}{90}$ | 0 | 0 | 0 | $-\frac{2}{30}$ | $-\frac{10}{90}$ |
| $y_1^1$ | 1 | 0 | 1 | 0 | 1 | 0 | 1 | 0 |
| $y_2^6$ | $\frac{4}{9}$ | $-\frac{1}{9}$ | $\frac{5}{9}$ | 1 | 0 | 0 | $\frac{1}{3}$ | $\boxed{\frac{5}{9}}$ → |
| $y_2^3$ | $\frac{5}{9}$ | $\frac{1}{9}$ | $-\frac{5}{9}$ | 0 | 0 | 1 | $-\frac{1}{3}$ | $\frac{4}{9}$ |

**Tableau 5**

| | | $y_0^1$ | $y_1^4$ | $y_2^6$ | $y_1^1$ | $y_2^3$ | $y_1^2$ | $y_2^4$ |
|---|---|---|---|---|---|---|---|---|
| $z$ | 1.4 | .1 | .2 | .2 | 0 | 0 | 0 | 0 |
| $y_1^1$ | 1 | 0 | 1 | 0 | 1 | 0 | 1 | 0 |
| $y_2^4$ | .8 | −.2 | 1 | 1.8 | 0 | 0 | .6 | 1 |
| $y_2^3$ | .2 | .2 | −1 | −.8 | 0 | 1 | −.6 | 0 |

Each subproblem is then solved using the above objective functions. (Generally speaking, it is efficient to proceed from the most recent optimal solution, as described in Chapter 7). The optimal solutions turn out to be $\boldsymbol{x}_1^{2\prime} = (2 \ \ 0 \ \ 6 \ \ 0)$ for subproblem 1 and $\boldsymbol{x}_2^{4\prime} = (0 \ \ 4 \ \ 0 \ \ 0 \ \ 100)$ for subproblem 2. The coefficients for the master problem are, for the solution to subproblem 1, $\boldsymbol{c}_1'\boldsymbol{x}_1^2 = .4$, $\boldsymbol{A}_{01}\boldsymbol{x}_1^2 = 2$, and for the solution to subproblem 2, $\boldsymbol{c}_2'\boldsymbol{x}_2^4 = .6$ and $\boldsymbol{A}_{02}\boldsymbol{x}_2^4 = 4$. Using these to form the columns of the matrix for the corresponding $y_i^k$'s, they are multiplied by the inverse to obtain the updated columns which appear in Tableau 4. Then $y_2^4$ replaces $y_2^6$ in the basis, yielding Tableau 5, which is optimal. From Tableau 5, it is seen that $\boldsymbol{\pi}_0 = .1$ (the charge for a unit of stainless steel has become \$.10), and hence the objective functions for the subproblems become, for subproblem 1:

$$(\boldsymbol{c}_1' - \boldsymbol{\pi}_0'\boldsymbol{A}_{01}) = (.2 \quad .1 \quad 0 \quad 0) - .1(1 \quad 1 \quad 0 \quad 0) = (.1 \quad 0 \quad 0 \quad 0)$$

and for subproblem 2:

$$\begin{aligned}(\boldsymbol{c}_2' - \boldsymbol{\pi}_0'\boldsymbol{A}_{02}) &= (.12 \quad .15 \quad .10 \quad 0 \quad 0) - .1(1 \quad 1 \quad 1 \quad 0 \quad 0) \\ &= (.02 \quad .05 \quad 0 \quad 0 \quad 0)\end{aligned}$$

The preceding optimal solutions to the subproblems are still optimal; hence, the optimal solution to the overall problem has been achieved. The optimal solution can be read from Tableau 5:

$$y_1^1 = 1, \quad y_2^4 = .8, \quad y_2^3 = .2$$

This means that for subproblem 1 use $\boldsymbol{x}_1^1$ (corresponding to $y_1^1$), which is (2 3 0 0) or 20,000 pans of type 1 and 30,000 pans of type 2 at plant 1. For subproblem 2 use .8 of $x_2^4$, which is .8(0 4 0 0 100), plus .2 of $\boldsymbol{x}_2^3$, which is .2(0 4 5 0 0), or 40,000 pans of type 4 and 10,000 pans of type 5 at plant 2. No pans of type 3 are produced. From Tableau 5 it can be seen that an alternate optimum exists:

$$y_1^2 = 1, \quad y_2^3 = .8, \quad y_2^4 = .2$$

More generally, there are a number of ways to compute the optimal solution once the problem is solved. One method, as given in the example, is to compute the blend of subproblem solutions as $\boldsymbol{x}_j = \sum_k \boldsymbol{x}_j^k y_j^k$. The above procedure is obviously not possible unless *all* of the generated solutions $\boldsymbol{x}_j^k$ have been remembered. An alternate method is to compute the amount of master problem resources and/or commitments to be used and/or satisfied by subproblem $i$, $\sum_k (\boldsymbol{A}_{0i}\boldsymbol{x}_i^k)y_i^k$. Then adjoin appropriate constraints to each subproblem to reflect the resource limitations on that subproblem, and solve each subproblem so altered one more time to determine the optimal solution using the optimal master problem prices.

### Other Considerations

After each subproblem solution, there is available an upper bound for the problem's optimal objective function value. That value is the existing objective function value plus the sum of the reduced costs for each of the newly generated variables, one for each subproblem. This can be proven by writing the objective function for the master problem with all variables present as

$$z = \boldsymbol{c}_B'\boldsymbol{B}^{-1}\boldsymbol{b} + (\boldsymbol{c}_N - \boldsymbol{c}_B\boldsymbol{B}^{-1}\boldsymbol{N})\boldsymbol{y}_N$$

or

$$z = \boldsymbol{c}_B'\boldsymbol{B}^{-1}\boldsymbol{b} + (\boldsymbol{c}_{N_1} - \boldsymbol{c}_B\boldsymbol{B}^{-1}\boldsymbol{N}_1)\boldsymbol{y}_{N_1} + \cdots + (\boldsymbol{c}_{N_h} - \boldsymbol{c}_B\boldsymbol{B}^{-1}\boldsymbol{N}_h)\boldsymbol{y}_{N_h}$$

where the $h$ partitions correspond to the $h$ subproblems. By choosing the maximum reduced cost for each subproblem (which is the solution of the subproblem during a subproblem solution), we may write

$$z \leq \boldsymbol{c}_B'\boldsymbol{B}^{-1}\boldsymbol{b} + \sum_{i=1}^{h} \max_j (\boldsymbol{e}_j(\boldsymbol{c}_{N_i} - \boldsymbol{c}_B\boldsymbol{B}^{-1}\boldsymbol{N}_i))$$

or

$$z \leq \boldsymbol{c}_B'\boldsymbol{B}^{-1}\boldsymbol{b} - \sum_{i=1}^{h} \min_j (\boldsymbol{e}_j(\boldsymbol{c}_B\boldsymbol{B}^{-1}\boldsymbol{N}_i - \boldsymbol{c}_{N_i}))$$

To illustrate this computation in the example in Tableau 1, an upper bound is seen to be $0 + 1.1 + .7 = 1.8$, or \$18,000. In Tableau 4 an upper bound is seen to be 1.311 + 2/30 + 10/9, or 1.4888 ($14,888). Finally, in Tableau 5, since the minimum reduced cost for both problems is zero, an upper bound is seen to be 1.4 + 0 + 0, or 1.4 ($14,000). It is useful to make the calculation at each stage because the timing of the computation does not relate systematically to the tightness of the bounds.

### Use of the Decomposition Principle for Solving Staircase-Type Problems

We shall briefly consider application of the decomposition principle to multistage or staircase-type problems, as proposed by Dantzig [62]. A typical application for such a program is a time-phased production-inventory system in which goods produced in one period may be stored for sale in a future period. The form for such a problem is of the following type:

$$
\begin{array}{lllllll}
\text{Maximize } z & & & & & & \\
\text{subject to:} & \boldsymbol{A}_1\boldsymbol{x}_1 & & & & = & \boldsymbol{b}_1 \\
& \boldsymbol{D}_1\boldsymbol{x}_1 & + \boldsymbol{A}_2\boldsymbol{x}_2 & & & = & \boldsymbol{b}_2 \\
& & \boldsymbol{D}_2\boldsymbol{x}_2 & + \boldsymbol{A}_3\boldsymbol{x}_3 & & = & \boldsymbol{b}_3 \\
\boldsymbol{A}_0 z & & & + \boldsymbol{D}_3\boldsymbol{x}_3 & + \boldsymbol{A}_4\boldsymbol{x}_4 & = & \boldsymbol{b}_4
\end{array}
$$

By way of explanation, the variables in the vector $\boldsymbol{x}_j$ are the goods produced or inventoried during period $j$, for consumption during period $j$ or a future period. The matrix $\boldsymbol{A}_j$ reflects the consumption of period $j$'s resources and/or the satisfaction of period $j$'s constraints due to $\boldsymbol{x}_j$. (The matrix $\boldsymbol{A}_0$ [a column vector] reflects the effect on the objective function of all the constraints whereby the period costs have been accumulated in the last set of constraints.) The matrix $\boldsymbol{D}_j$ reflects the effect on the resources and other constraints in period $j + 1$ due to $\boldsymbol{x}_j$. It is possible to increase the detail of the staircase representation to handle inventory variables and so on, but we shall not do that here. By making subprograms out of every other stage, we have

$$
\begin{array}{llllll}
\text{Maximize} & z & & & & \\
\text{subject to:} & \boldsymbol{A}_0 z & & + \boldsymbol{D}_3\boldsymbol{x}_3 + \boldsymbol{A}_4\boldsymbol{x}_4 & = & \boldsymbol{b}_4 \\
& & \boldsymbol{D}_1\boldsymbol{x}_1 + \boldsymbol{A}_2\boldsymbol{x}_2 & & = & \boldsymbol{b}_2 \\
\text{Subprogram 1} & & \boldsymbol{A}_1\boldsymbol{x}_1 & & = & \boldsymbol{b}_1 \\
\text{Subprogram 2} & & & \boldsymbol{D}_2\boldsymbol{x}_2 + \boldsymbol{A}_3\boldsymbol{x}_3 & = & \boldsymbol{b}_3
\end{array}
$$

which is in the form (8.23).

The subprograms may be solved as before. Instead of solving the master problem as we did before, however, we can decompose the master problem into a subprogram and a so-called "second level" master program. For a larger number of periods, the master problem may be partitioned into a number of subproblems and a higher level master problem, which is then partitioned into a number of subproblems and a higher level master problem, and so on.

Thus we see that the decomposition principle can be used for a more general class of problems. Unfortunately, no computational experience is reported.

### Primal Decomposition Algorithms

A number of papers have been written on so-called "primal decomposition algorithms" (see, for example, Zschau [316]). Such algorithms solve the same block angular type of problem as does the Dantzig-Wolfe decomposition method, but they operate in a reverse fashion. In the Dantzig-Wolfe decomposition framework, resource shadow prices are determined in the master problem phase, and optimal allocations, given those prices, are made in the subproblems. The resources used in those allocations are then used in the master problem to determine new resource shadow prices, and so on, until the prices remain unchanged. In the primal decomposition, allocation of resources is determined in the master problem, and use of those resources and associated local shadow prices are determined in the subproblem. The proposals are then submitted to the master problem, where a new allocation of resources is made based upon price changes , and so on.

No computational experience using the method is reported.

### Other Algorithms

There are many other special algorithms for special linear programming problems, some of which are presented in Chapter 9. These include the transportation method and the out-of-kilter method. Where certain decomposition-type problems have substructures of such problem types as are presented in Chapter 9, the decomposition approach is *particularly* useful. A number of such algorithms have been proposed—they are commonly referred to as column-generation procedures. See, for example, Gilmore and Gomory [107] and Rao and Zionts [230].

### Selected Supplemental References

Section 8.2
[65]
Section 8.3
[68], [138], [150], [173], [247]
Section 8.4
[9], [69], [189], [217], [230], [244], [245], [293], [316]

General

[27], [29], [30], [31], [32], [33], [100], [151], [188]

## 8.5 PROBLEMS

**1.** Devise a scheme in the method of upper bounds for taking an iteration of type 2 directly, without first replacing a basic variable by its complement.

**2.** Solve the following problem using the method of upper bounds:

$$\begin{aligned} \text{Maximize } z &= 5x_1 + 3x_2 + 2x_3 \\ \text{subject to:} \quad & x_1 + 2x_2 + 1.5x_3 \leq 9 \\ & 3x_1 + x_2 + 2x_3 \leq 10 \\ & 0 \leq x_1 \leq 3 \\ & 0 \leq x_2 \leq 3 \\ & 0 \leq x_3 \leq 3 \end{aligned}$$

**3.** Solve the following problem using the method of upper bounds:

$$\begin{aligned} \text{Maximize } z &= x_1 + 7x_2 + 3x_3 + 2x_4 \\ \text{subject to:} \quad & 3x_1 + 6x_2 + x_3 + x_4 \leq 12 \\ & 2x_1 + 3x_2 - x_3 + 3x_4 \leq 10 \\ & 0 \leq x_1 \leq 3 \\ & 0 \leq x_2 \leq 3 \\ & 0 \leq x_3 \leq 4 \\ & 0 \leq x_4 \leq 4 \end{aligned}$$

**4.** Solve the following problem using the method of upper bounds:

$$\begin{aligned} \text{Maximize } z &= 5x_1 + 8x_2 + 3x_3 + 7x_4 \\ \text{subject to:} \quad & 8x_1 + 7x_2 + 7x_3 + 2x_4 \leq 38 \\ & 6x_1 + 5x_2 - 7x_3 + 6x_4 \leq 45 \\ & 0 \leq x_1 \leq 2 \\ & 0 \leq x_2 \leq 4 \\ & 1 \leq x_3 \leq 5 \\ & 0 \leq x_4 \leq 4 \end{aligned}$$

**5.** Solve the following problem by the method of generalized upper bounds:

$$\begin{aligned}
\text{Maximize } z = \; & 6x_1 + 2x_2 + 7x_3 + 3x_4 + 8x_5 + 6x_6 \\
\text{subject to:} \quad & 3x_1 + 3x_2 + 9x_3 + 2x_4 + 3x_5 - 7x_6 \leq 8 \\
& 4x_1 - 7x_2 + 2x_3 + 4x_4 + 8x_5 + 6x_6 \leq 6 \\
& x_1 + x_2 \geq 1 \\
& x_3 + x_4 \geq 1 \\
& x_5 + x_6 \geq 1 \\
& x_1, \ldots, x_6 \geq 0
\end{aligned}$$

**6.** Convert the following problem to the form amenable to generalized upper bounds given in expression (8.8):

$$\begin{aligned}
\text{Maximize } z = \; & 3x_1 + 2x_2 + 2x_3 + 2x_4 + x_5 + x_6 \\
\text{subject to:} \quad & x_1 + 1.5x_2 + x_3 - 2x_4 + 4x_5 + 2x_6 \leq 4 \\
& 3x_1 + 2x_2 - x_3 + x_4 + 1.5x_5 - x_6 \leq 6 \\
& 2x_1 + 3x_2 + x_3 \leq 4 \\
& 2x_4 + 3x_5 \leq 6 \\
& x_1, \ldots, x_6 \geq 0
\end{aligned}$$

Hint: Set $x_1 = 2x'_1$, $x_2 = 4/3x'_2$, $x_3 = 4x'_3$. Divide the third constraint by 4. Then set $x_4 = 3x'_4$, $x_5 = 2x'_5$, and divide the fourth constraint by 6. (This process is known as scaling the problem to conform to some set form. Scaling with a somewhat different objective is discussed in Chapter 11.)

**7.** Solve the scaled problem 6 by the method of generalized upper bounds.

**8.** For the method of generalized upper bounds, prove that $\boldsymbol{H}$ is nonsingular.

**9.** For the method of generalized upper bounds, explain why, in iteration type 3, the incoming and outgoing variables must be in the same inessential set.

**10.** For the method of generalized upper bounds, prove that the number of sets of partial sum equations containing two or more basic variables is at most $m$.

**11.** For the method of generalized upper bounds, develop the scheme for negative coefficients in the last $L$ constraints. (Hint: A variable with a positive coefficient must still be key, and the key columns must be added to negative columns of the associated set, not subtracted from them.)

**12.** Find the optimal solution to the example of section 8.4, using the results of Tableau 5 and the alternate procedure described on page 210.

**13.** Suppose that we do not require that each convex set $\boldsymbol{A}_{ii}\boldsymbol{x}_i = \boldsymbol{b}_i$ is bounded. Then indicate the changes that must be made to the decomposition algorithm if a subproblem solution turns out to be unbounded. (Hint: The common constraints still limit the solution; hence, the unbounded ray of the subproblem may be used in place of the subproblem extreme point.)

**14.** Solve the following problem using the decomposition algorithm:

$$\begin{aligned}
\text{Maximize } z = \; & 3x_1 + 2x_2 + 2x_3 + 3x_4 + x_5 \\
\text{subject to:} \quad & 2x_1 + 3x_2 + 2x_3 + 2x_4 + 3x_5 \leq 7 \\
& x_1 + 3x_2 \leq 4 \\
& 2x_1 + x_2 \leq 5 \\
& 3x_3 + 2x_4 + 2x_5 \leq 4 \\
& 2x_3 + 4x_4 + x_5 \leq 5 \\
& 3x_3 + x_4 + 3x_5 \leq 3 \\
& x_1, x_2, x_3, x_4, x_5 \geq 0
\end{aligned}$$

**15.** Solve the following problem using the decomposition algorithm:

$$\begin{aligned}
\text{Maximize } z = \; & x_1 + 2x_2 + 2x_3 + 5x_4 + 3x_5 + x_6 \\
\text{subject to:} \quad & 4x_1 + 5x_2 + 3x_3 + 4x_4 + 4x_5 + 2x_6 \leq 26 \\
& x_1 + 4x_2 + 2x_3 + 4x_4 + 4x_5 + 5x_6 \leq 34 \\
& 2x_1 + 2x_2 + 4x_3 \leq 10 \\
& 5x_1 + x_2 + 2x_3 \leq 15 \\
& 2x_4 + 5x_5 + 5x_6 \leq 14 \\
& x_4 + 4x_5 + x_6 \leq 18 \\
& x_1, \ldots, x_6 \geq 0
\end{aligned}$$

# 9

# Specially Structured Linear Programming Problems: Network Flow Methods

In this chapter we shall introduce additional methods that take advantage of a special problem structure different from those structures amenable to methods presented in Chapter 8. The methods presented treat problems whose variables may be represented as flows of materials. We shall focus on the transportation problem (including the assignment problem), network flow methods, and the shortest route problem. In many cases our approach will be to formulate the problem as a linear programming problem, explore the ramifications of the formulation, and use that to lead into the special method. It will be useful for the reader to keep this procedure in mind while studying this chapter.

## 9.1 THE TRANSPORTATION PROBLEM AS A LINEAR PROGRAMMING PROBLEM

We shall introduce the transportation problem with an example.

EXAMPLE 9.1:

The Inland Central Railroad has available 11 flat cars at Akron, New York, and 13 flat cars at Butler, Pennsylvania. It requires 6 flat cars in Cleveland, Ohio, 4 in Detroit, Michigan, and 14 in Erie, Pennsylvania. Direct shipping costs, which are based on various processing costs and have little relationship to distances, are as follows:

| FROM: \ TO: | Cleveland | Detroit | Erie |
|---|---|---|---|
| Akron | \$60 | \$40 | \$30 |
| Butler | \$20 | \$30 | \$50 |

What is the minimum cost scheme of supplying the needed cars?

The problem is a linear programming problem. By designating $x_{AC}$ as the number of cars shipped from Akron to Cleveland, and other variables accordingly, we have the following formulation (equalities in the constraints are present because the total available is equal to the total required):

$$\text{Minimize } z = 60x_{AC} + 40x_{AD} + 30x_{AE} + 20x_{BC} + 30x_{BD} + 50x_{BE}$$

or

$$\text{Maximize } z = -60x_{AC} - 40x_{AD} - 30x_{AE} - 20x_{BC} - 30x_{BD} - 50x_{BE}$$

subject to:

$$\begin{aligned}
x_{AC} + x_{AD} + x_{AE} &= 11 \\
x_{BC} + x_{BD} + x_{BE} &= 13 \\
x_{AC} + x_{BC} &= 6 \\
x_{AD} + x_{BD} &= 4 \\
x_{AE} + x_{BE} &= 14 \\
x_{AC}, x_{AD}, x_{AE}, x_{BC}, x_{BD}, x_{BE} &\geq 0
\end{aligned}$$

Any one constraint of the formulation can be shown to be redundant; we choose to drop the first constraint (which equals the third constraint plus the fourth plus the fifth minus the second). Further, we somehow know that $x_{AE}$, $x_{AC}$, $x_{AD}$, and $x_{BE}$ form a feasible basis which is exhibited in Tableau 1. (A method for finding such a starting basis will be developed; for now we note that an artificial variable method *could have been used* to generate a feasible basis.) Utilizing the solution in Tableau 1 and proceeding via the simplex method, two simplex iterations are needed to reach the optimal solution of Tableau 3.

**Tableau 1**

| | | $x_{AC}$ | $x_{AD}$ | $x_{AE}$ | $x_{BC}$ ↓ | $x_{BD}$ | $x_{BE}$ | |
|---|---|---|---|---|---|---|---|---|
| $z$ | −1200 | 0 | 0 | 0 | −60 | −30 | 0 | |
| $x_{AE}$ | 1 | 0 | 0 | 1 | −1 | −1 | 0 | |
| $x_{AC}$ | 6 | 1 | 0 | 0 | [1] | 0 | 0 | ⟶ |
| $x_{AD}$ | 4 | 0 | 1 | 0 | 0 | 1 | 0 | |
| $x_{BE}$ | 13 | 0 | 0 | 0 | 1 | 1 | 1 | |

**Tableau 2**

| | | $x_{AC}$ | $x_{AD}$ | $x_{AE}$ | $x_{BC}$ | $x_{BD}$ ↓ | $x_{BE}$ | |
|---|---|---|---|---|---|---|---|---|
| $z$ | $-840$ | 60 | 0 | 0 | 0 | $-30$ | 0 | |
| $x_{AE}$ | 7 | 1 | 0 | 1 | 0 | $-1$ | 0 | |
| $x_{BC}$ | 6 | 1 | 0 | 0 | 1 | 0 | 0 | |
| $x_{AD}$ | 4 | 0 | 1 | 0 | 0 | $\boxed{1}$ | 0 | → |
| $x_{BE}$ | 7 | $-1$ | 0 | 0 | 0 | 1 | 1 | |

**Tableau 3**

| | | $x_{AC}$ | $x_{AD}$ | $x_{AE}$ | $x_{BC}$ | $x_{BD}$ | $x_{BE}$ |
|---|---|---|---|---|---|---|---|
| $z$ | $-720$ | 60 | 30 | 0 | 0 | 0 | 0 |
| $x_{AE}$ | 11 | 1 | 1 | 1 | 0 | 0 | 0 |
| $x_{BC}$ | 6 | 1 | 0 | 0 | 1 | 0 | 0 |
| $x_{BD}$ | 4 | 0 | 1 | 0 | 0 | 1 | 0 |
| $x_{BE}$ | 3 | $-1$ | $-1$ | 0 | 0 | 0 | 1 |

The optimal solution is to ship 11 cars from Akron to Erie, 6 from Butler to Cleveland, 4 from Butler to Detroit, and 3 from Butler to Erie. The total cost of shipping is $720.

More formally, the transportation problem is a linear programming problem with a framework in which there are $m$ sources of a commodity, each source having an amount $a_i$ for shipment, and $n$ destinations for that commodity, each requiring an amount $b_j$. It is convenient, but not necessary, to assume that

$$\sum_{i=1}^{m} a_i = \sum_{j=1}^{n} b_j \tag{9.1}$$

The cost of shipping a unit of the commodity from *source i* to *destination j* is $c_{ij}$, and it is desired to find the minimum cost shipping schedule. Stating the problem as a linear programming problem, we have the following formulation:

Minimize $z = c_{11}x_{11} + c_{12}x_{12} + \cdots + c_{1n}x_{1n} + \cdots + c_{m1}x_{m1} + c_{m2}x_{m2} + \cdots + c_{mn}x_{mn}$

subject to:

$$\begin{array}{lllllllll}
x_{11} + & x_{12} + \cdots + & x_{1n} & & & & & & = a_1 \\
 & & & \ddots & & & & & \vdots \\
 & & & & x_{m1} + & x_{m2} + \cdots + & & x_{mn} & = a_m \\
x_{11} + & & \cdots & + & x_{m1} & & & & = b_1 \\
 & x_{12} + & & \cdots & + & x_{m2} & & & = b_2 \\
 & \ddots & & & \vdots & & \ddots & & \vdots \\
 & & x_{1n} + & & \cdots & & + & x_{mn} & = b_n
\end{array} \tag{9.2}$$

To avoid degeneracy, it is additionally useful to assume that no partial sum of $a$'s equals no partial sum of $b$'s. A procedure to overcome this limitation will be given in section 9.2. The method developed allows for degeneracy, although it is not known whether or not a degenerate transportation problem can cycle.

Given the assumption (9.1), *any one* equation of (9.2) is redundant, since it can be generated by a linear combination of $m + n - 1$ other equations. Accordingly, any one constraint of (9.2) may be dropped. For example, the first equation is equal to the sum of the last $n$ equations less the sum of the equations 2 through $m$. Further, every variable in the problem is a source-destination pair representing the number of units transported via that route. Therefore, the problem representation in Figure 9.1 may also be used. Before proceeding, it is useful to assert that a basis for a transportation problem

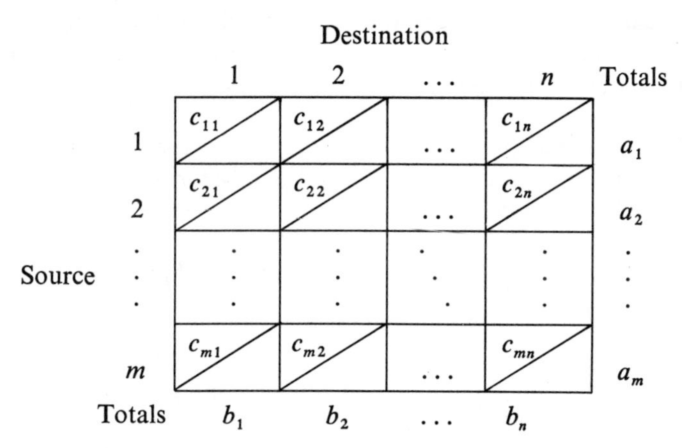


**Figure 9.1**

An Alternate Problem Representation for the Transportation Problem

consists of $m + n - 1$ linearly independent vectors. It can be shown that the basis consists of at most $m + n - 1$ linearly independent vectors by noting that there are at most $m + n - 1$ linearly independent equations in (9.2). It can then be shown that the rank of the matrix of coefficients of the linear programming problem (9.2) is exactly $m + n - 1$ by demonstrating that an inverse to a square submatrix of order $m + n - 1$ exists. For example, delete the first equation of (9.2). Then take the vectors that form the coefficients of $x_{11}, x_{12}, \ldots, x_{1n}, x_{21}, x_{31}, \ldots, x_{m1}$. They form a matrix (dimensions of which are indicated)

$$\boldsymbol{P} = \left(\begin{array}{c|c} \boldsymbol{0}_{m-1 \times n} & \boldsymbol{I}_{m-1 \times m-1} \\ \hline \multirow{2}{*}{$\boldsymbol{I}_{n \times n}$} & \boldsymbol{I}_{1 \times m-1} \\ \cline{2-2} & \boldsymbol{0}_{n-1 \times m-1} \end{array}\right)$$

whose inverse is

$$\boldsymbol{P}^{-1} = \left(\begin{array}{c|c} -\boldsymbol{1}_{1 \times m-1} & \multirow{2}{*}{$\boldsymbol{I}_{n \times n}$} \\ \cline{1-1} \boldsymbol{0}_{n-1 \times m-1} & \\ \hline \boldsymbol{I}_{m-1 \times m-1} & \boldsymbol{0}_{m-1 \times n} \end{array}\right)$$

Hence the rank of the bases of a transportation problem is exactly $m + n - 1$. An extension of the above demonstration can be used to prove that all basis inverses contain only the entries 0, 1, $-1$.

A few results are needed to proceed. First, it can be shown (e.g., Dantzig [62]) that all bases of transportation problems are *triangular*.

**Definition 9.1**

*A square matrix is triangular if, after suitable rearrangement of rows and columns, all coefficients below the principal diagonal are zero.*

REMARK:

A triangular system of equations can be solved recursively by solving the last equation in one variable and then proceeding to the next to last equation, which (after substituting in the solutions already found) is an equation in one variable, and so on.

Consider next the dual problem to (9.2). Designating the dual variables as $u_i, i = 1, \ldots, m$ for the sources and $v_j, j = 1, \ldots, n$ for the destination, we have the following formulation:

$$\begin{array}{llllllll}
\text{Maximize } z = & a_1u_1 + a_2u_2 + \cdots + a_mu_m + b_1v_1 + b_2v_2 + \cdots + b_nv_n & & & & & & \\
\text{subject to:} & u_1 & & & + \; v_1 & & & \leq c_{11} \\
& u_1 & & & & + \; v_2 & & \leq c_{12} \\
& \vdots & & & & & \ddots & \vdots \\
& u_1 & & & & & + \; v_n & \leq c_{1n} \\
& & u_2 & & + \; v_1 & & & \leq c_{21} \\
& & u_2 & & & + \; v_2 & & \leq c_{22} \\
(9.3) & & \vdots & & & & \ddots & \vdots \\
& & u_2 & & & & + \; v_n & \leq c_{2n} \\
& & & \ddots & & & \ddots & \vdots \\
& & & u_m & + \; v_1 & & & \leq c_{m1} \\
& & & u_m & & + \; v_2 & & \leq c_{m2} \\
& & & \vdots & & & \ddots & \vdots \\
& & & u_m & & & + \; v_n & \leq c_{mn} \\
& & & & & & & u_i, v_j \text{ unrestricted in sign}
\end{array}$$

Since any one constraint of (9.2) is redundant, any one dual variable of (9.3) may be chosen to take on some arbitrary value. Further, by duality, for every basic variable in the primal problem (9.2) we have a constraint of the dual problem (9.3) satisfied as an equality. In addition, since our matrix of coefficients for the equality constraints of (9.3) is just the transpose of the corresponding basis of (9.2), the basis for the dual problem is triangular as well!

The above discussion gives us a very powerful result: If we have a basic solution in the framework of Figure 9.1, then

1. We may choose the value of any one dual variable (either a $u_i$ or a $v_j$) arbitrarily.
2. We may compute all the others by virtue of the triangularity of the basis, that is, solve expressions $u_i + v_j = c_{ij}$ for every basic variable using a value of $u_i$ or $v_j$ that is known.
3. We may compute the reduced costs for all nonbasic variables as

$$c_{ij} - u_i - v_j$$

4. We may check to see whether or not the solution is optimal. If it is, then all reduced costs for nonbasic variables are nonnegative

$$c_{ij} - u_i - v_j \geq 0$$

(Remember that this is a minimization problem.)

If there is a negative reduced cost, that is, a $c_{ij} - u_i - v_j < 0$, then the corresponding variable should be made basic. Accordingly, that variable should be expressed in terms of basic variables and the change of basis effected. Now that we have explored certain aspects of the simplex method solution to a transportation problem, we are in a position to develop the special algorithm.

## 9.2 A COMPUTATIONAL ALGORITHM FOR THE TRANSPORTATION PROBLEM

The first step in solving the transportation problem using the algorithm partially implied by the simplex method is to determine a starting basic feasible solution. There are two methods that are commonly used:

1. The Northwest corner method.
2. The Vogel Approximation method [232].

The Northwest corner method is almost as simple as its name (we could alternatively use a Northeast corner method, and so on). To obtain a starting solution, use the following steps:

1. Select the Northwest corner cell of the problem. Set the shipments in that cell to the lesser of source availability and destination requirements. If it is zero, denote the cell as a *basic* variable at zero.
2. Decrement amount in cell from source availability and destination requirements, and delete whichever row or column is zero. In case of a tie, the solution is degenerate; delete only the row or the column, but not both.
3. When all supplies or demands have been exhausted or filled, respectively, stop; otherwise go to step 1.

For example problem 9.1, the Northwest corner method is given in Figure 9.2. Nothing inherent in the Northwest corner method chooses a starting solution on the basis of source-destination costs, so the starting solution obtained using this method may or may not be good, depending on the way the cells happen to be ordered.

On the other hand, the Vogel Approximation method [232] tends to choose a good starting solution. Utilize the following steps:

1. For each row and column, compute the difference between the minimum cost element and the second lowest cost element.
2. Choose the row or column with the maximum difference. Assign as much as possible to the minimum cost entry cell in that row or column (even zero in the case of degeneracy). Decrement the source and destination by the amount assigned. Delete whichever source or destination is then zero, but not both. (Only in the degenerate case can both be zero.)
3. If there is only one row or one column left, make the remaining assignments with one basic element in each cell of the row or column, respectively, and stop. Otherwise go to step 1.

Figure 9.3 illustrates use of the Vogel Approximation method on example problem 9.1. For this example, the method gives the optimal solution, but the Vogel Approximation method does not yield an optimal solution in general.

Before we introduce the transportation method based on the simplex method, it is important to note that, given a basic feasible solution, any incoming variable to the basis must displace the same number of equivalent shipments to preserve balance. Thus, if $x_{ij}$ is to enter the basis (at some level), we must decrease flow from source $i$ to some destination $k$, increase flow from some source $h$ to destination $k$, and so on, and finally decrease flow from some source $p$ to destination $j$, preserving the balance. This may be observed by incrementing any nonbasic variable in a (transportation) simplex tableau. Consider Tableau

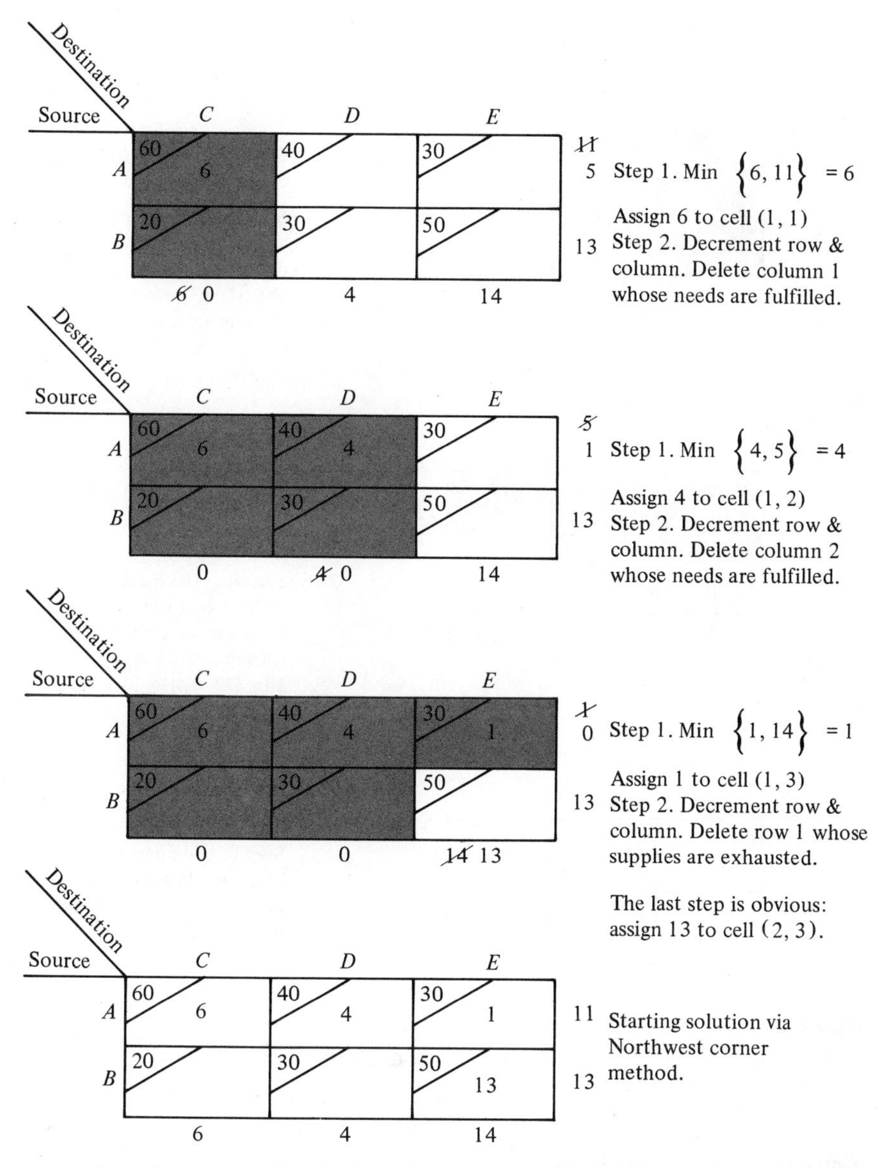


**Figure 9.2**

Illustration of the Northwest Corner Method

1 in Example 9.1. As $x_{BC}$ increases, $x_{BE}$ decreases (a + 1 coefficient in the fourth component of the column for $x_{BC}$), $x_{AE}$ increases, and $x_{AC}$ decreases.

A flow chart presenting the implied method is given in Figure 9.4. This method is equivalent to the simplex method, the main differences being that

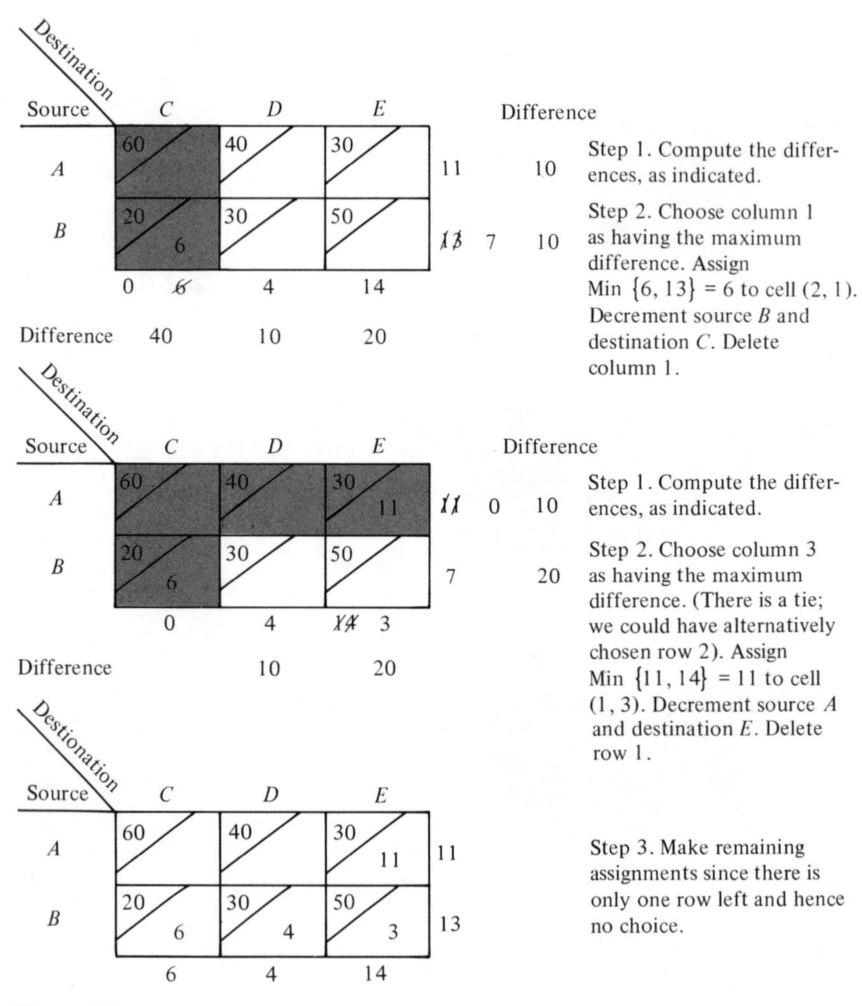


**Figure 9.3**

Illustration of the Vogel Approximation Method for Getting a Starting Solution to the Transportation Problem

a more economical problem representation can be used, and the computations are simpler.

EXAMPLE 9.1 (continued):

The solution steps for the problem of Example 9.1, using the starting solution obtained by the Northwest corner method, are given in Figure 9.5. Starting with Tableau 1, we have the following steps:

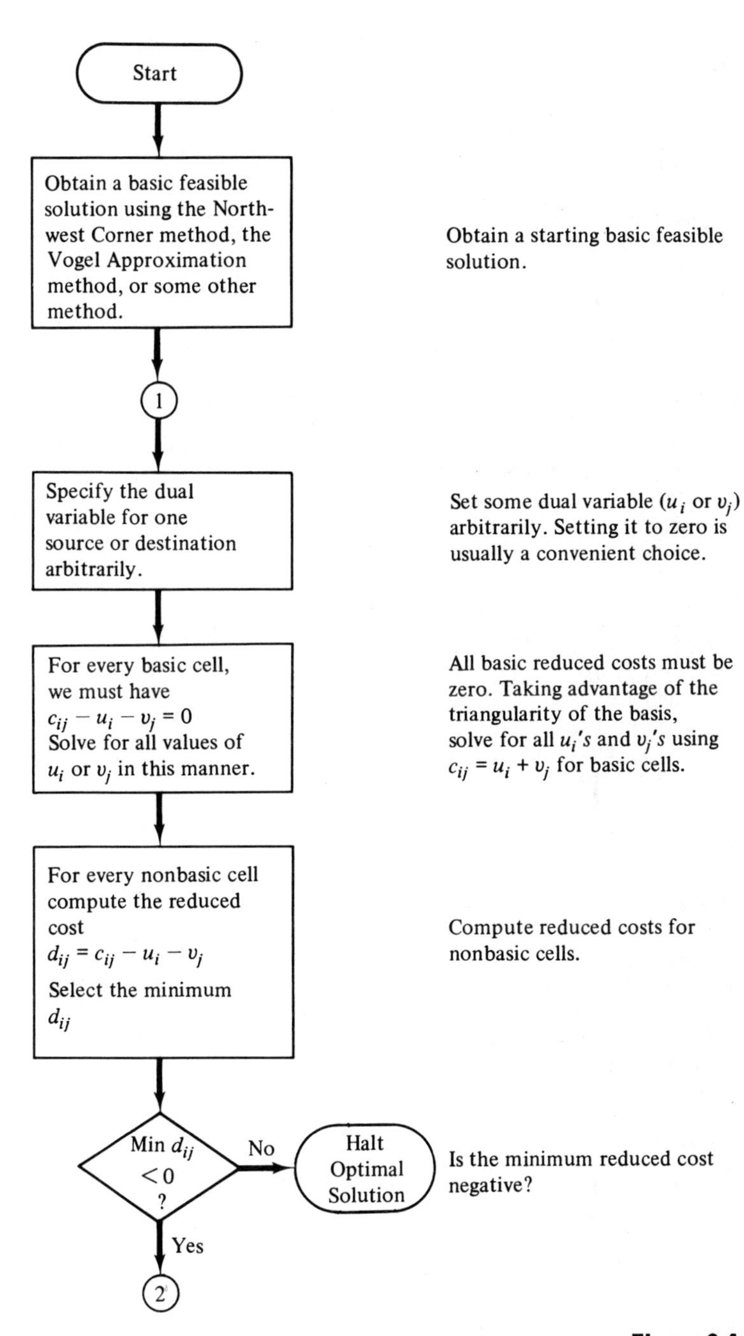


**Figure 9.4**

Flow Chart for the Transportation Method

Designate the variable corresponding to the most negative $d_{ij}$ to enter the basis.

Choose the variable having the most negative reduced cost to enter the basis.

Find a series of alternating vertical and horizontal jumps, analogous to Rook's moves in chess, that start and end on the incoming cell with all intermediate hops on basic cells. No basic cell may be landed upon more than once.

Starting with the cell of the incoming variable, find the unique set of basic cells which are necessary to balance the flow.

Label the incoming cell with a + sign, and alternately label basic cells along the Rook's move with – and + signs.

Express the incoming variable in terms of basic variables. Determine the unique set of adjustments necessary for balance for the incoming variable to be increased.

Choose the smallest basic entry having a minus sign. (If there is a tie, there is degeneracy; break the tie arbitrarily). Add the minimum quantity just determined to the cells with + signs and subtract it from cells with minus signs. A cell whose entry becomes zero is now nonbasic. Drop all signs, $d_{ij}$, $u_i$, and $v_j$'s.

Revise the solution to the new basic solution by increasing the amount of flow in the entering cell and balancing flow accordingly.

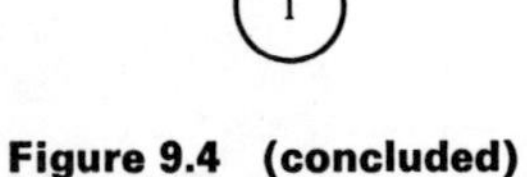


**Figure 9.4 (concluded)**

**Tableau 1**

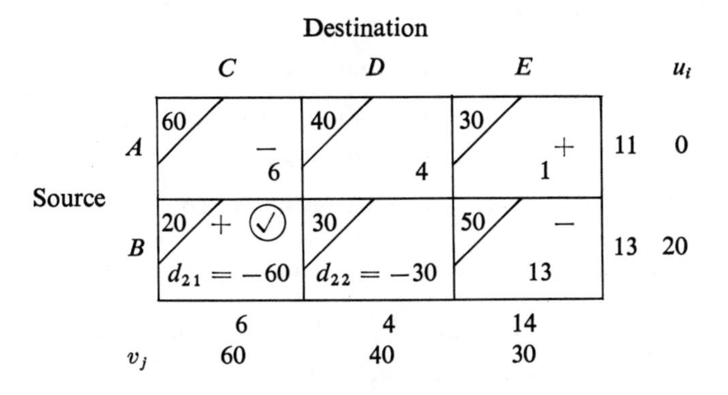

Destination

| Source | $C$ | $D$ | $E$ | | $u_i$ |
|---|---|---|---|---|---|
| $A$ | 60 − 6 | 40 4 | 30 + 1 | 11 | 0 |
| $B$ | 20 + ✓ $d_{21} = -60$ | 30 $d_{22} = -30$ | 50 − 13 | 13 | 20 |
| | 6 | 4 | 14 | | |
| $v_j$ | 60 | 40 | 30 | | |

**Tableau 2**

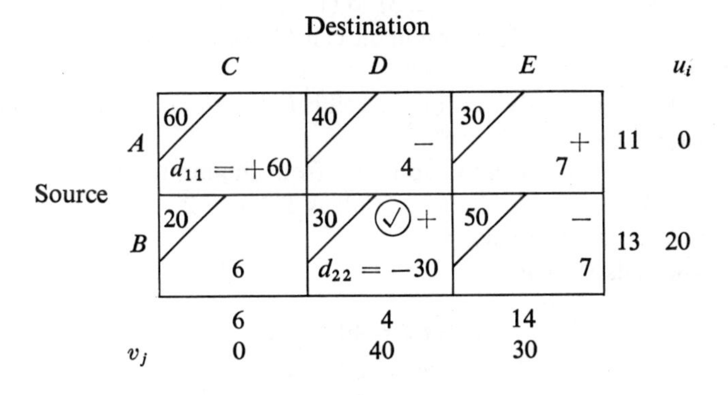

Destination

| Source | $C$ | $D$ | $E$ | | $u_i$ |
|---|---|---|---|---|---|
| $A$ | 60 $d_{11} = +60$ | 40 − 4 | 30 + 7 | 11 | 0 |
| $B$ | 20 6 | 30 ✓ + $d_{22} = -30$ | 50 − 7 | 13 | 20 |
| | 6 | 4 | 14 | | |
| $v_j$ | 0 | 40 | 30 | | |

**Tableau 3**

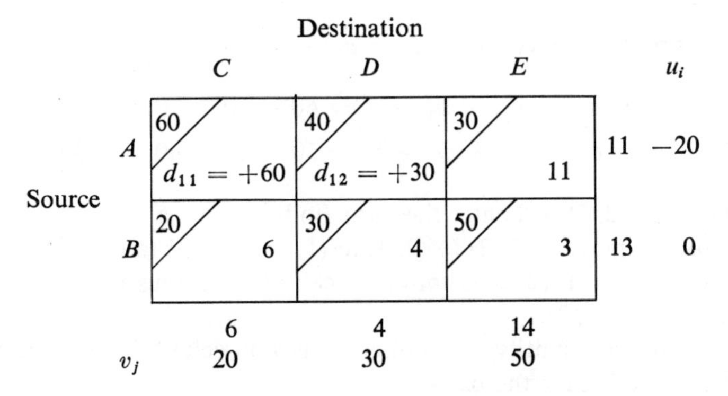

Destination

| Source | $C$ | $D$ | $E$ | | $u_i$ |
|---|---|---|---|---|---|
| $A$ | 60 $d_{11} = +60$ | 40 $d_{12} = +30$ | 30 11 | 11 | −20 |
| $B$ | 20 6 | 30 4 | 50 3 | 13 | 0 |
| | 6 | 4 | 14 | | |
| $v_j$ | 20 | 30 | 50 | | |

**Figure 9.5**

Sequence of Iterations for the Transportation Problem Using the Starting Solution Obtained with the Northwest Corner Method

Choose $u_1 = 0$ (arbitrarily).

For basic cells, $u_i + v_j = c_{ij}$

$$u_1 + v_1 = 60 \Longrightarrow v_1 = 60$$
$$u_1 + v_2 = 40 \Longrightarrow v_2 = 40$$
$$u_1 + v_3 = 30 \Longrightarrow v_3 = 30$$
$$u_2 + v_3 = 50 \Longrightarrow u_2 = 20$$

For nonbasic cells, compute $d_{ij} = c_{ij} - u_i - v_j$

$$d_{21} = c_{21} - u_2 - v_1 = 20 - 20 - 60 = -60$$
$$d_{22} = c_{22} - u_2 - v_2 = 30 - 20 - 40 = -30$$

Hence cell (2, 1) will enter the basis.
The rook's move is (2, 1) to (2, 3) to (1, 3) to (1, 1) to (2, 1).
Put a plus in cell (2, 1), a minus in cell (2, 3), a plus in cell (1, 3), and a minus in cell (1, 1).
The minimum quantity in a minus cell is 6, in cell (1, 1). Since it is unique, cell (1, 1) will leave the basis.
Add 6 to the plus cells and subtract 6 from the minus cells.
Drop (1, 1) from the basis. The result is Tableau 2.
Choose $u_1 = 0$ arbitrarily.
For basic cells $u_i + v_j = c_{ij}$

$$u_1 + v_2 = 40 \Longrightarrow v_2 = 40$$
$$u_1 + v_3 = 30 \Longrightarrow v_3 = 30$$
$$u_2 + v_3 = 50 \Longrightarrow u_2 = 20$$
$$u_2 + v_1 = 20 \Longrightarrow v_1 = 0$$

For nonbasic cells, compute $d_{ij} = c_{ij} - u_i - v_j$

$$d_{11} = c_{11} - u_1 - v_1 = 60 - 0 - 0 = 60$$
$$d_{22} = c_{22} - u_2 - v_2 = 30 - 20 - 40 = -30$$

Hence, cell (2, 2) will enter the basis (indicated by a check $\checkmark$).
The rook's move is (2, 2) to (2, 3) to (1, 3) to (1, 2) to (2, 2).
Put a plus in cell (2, 2), a minus in cell (2, 3), a plus in cell (1, 3), and a minus in cell (1, 2).
The minimum quantity in a minus cell is 4, in cell (2, 1). Since it is unique, cell (2, 1) will leave the basis.
Add 4 to the plus cells and subtract 4 from the minus cells.
Drop cell (2, 1) from the basis. The result is Tableau 3.
Choose $u_2 = 0$ arbitrarily.

$$u_2 + v_1 = 20 \Longrightarrow v_1 = 20$$
$$u_2 + v_2 = 30 \Longrightarrow v_2 = 30$$
$$u_2 + v_3 = 50 \Longrightarrow v_3 = 50$$
$$u_1 + v_3 = 30 \Longrightarrow u_1 = -20$$

For nonbasic cells, compute $d_{ij} = c_{ij} - u_i - v_j$

$$d_{11} = c_{11} - u_1 - v_1 = 60 - (-20) - 20 = 60$$
$$d_{12} = c_{12} - u_1 - v_2 = 40 - (-20) - 30 = 30$$

Hence the solution is optimal. Notice that this solution parallels precisely the earlier solution using the simplex method. A second example is given below.

EXAMPLE 9.2:

An automobile manufacturer assembles automobiles at three plants and produces various parts and subassemblies at four plants. One subassembly is required in each automobile. Table 9.1 gives a one week schedule of the plants, as well as the variable costs (manufacturing and transportation) of supplying this subassembly to the assembly plants. How should the subassemblies be assigned for the given week so that the total cost to the producer is minimized?

**Table 9.1**

Cost and Availability of Assembly and Subassembly Capacity for Example 9.2

| | | ASSEMBLY PLANT 1 | ASSEMBLY PLANT 2 | ASSEMBLY PLANT 3 | CAPACITY OF SUBASSEMBLY SUPPLIER (100's) |
|---|---|---|---|---|---|
| Supplier plant of subassembly | 1 | \$60 | \$30 | \$20 | 44 |
| | 2 | 20 | 30 | 30 | 55 |
| | 3 | 70 | 70 | 80 | 65 |
| | 4 | 30 | 60 | 50 | 20 |
| Auto assembly schedule (100's) | | 50 | 50 | 80 | |

This problem may be viewed as a transportation problem, with the subassembly suppliers as the sources and the auto assembly plants as the destinations, except that $\sum a_i = 184 > \sum b_j = 180$. By adding a dummy destination which requires 4 units from any plant (at zero cost), the necessary form is achieved. The steps necessary for solution of this problem are carried out in Figure 9.6, beginning with a Northwest corner method solution in Tableau 1. $x_{13}$ enters the basis, replacing $x_{11}$ and yielding Tableau 2. Next $x_{34}$ enters the basis at 4, replacing $x_{44}$ and yielding Tableau 3.

**Tableau 1**

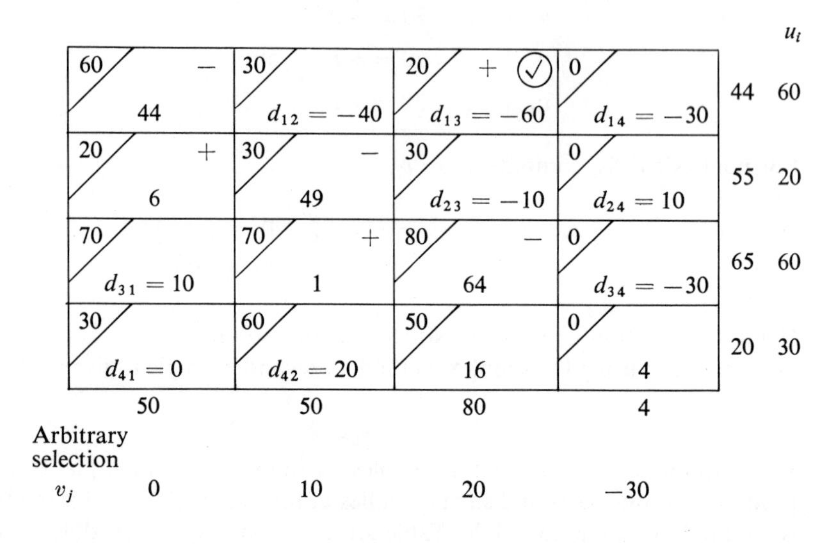

| | | | | | $u_i$ |
|---|---|---|---|---|---|
| 60 − / 44 | 30 / $d_{12} = -40$ | 20 + ✓ / $d_{13} = -60$ | 0 / $d_{14} = -30$ | 44 | 60 |
| 20 + / 6 | 30 − / 49 | 30 / $d_{23} = -10$ | 0 / $d_{24} = 10$ | 55 | 20 |
| 70 / $d_{31} = 10$ | 70 + / 1 | 80 − / 64 | 0 / $d_{34} = -30$ | 65 | 60 |
| 30 / $d_{41} = 0$ | 60 / $d_{42} = 20$ | 50 / 16 | 0 / 4 | 20 | 30 |
| 50 | 50 | 80 | 4 | | |
| Arbitrary selection | | | | | |
| $v_j$: 0 | 10 | 20 | −30 | | |

**Tableau 2**

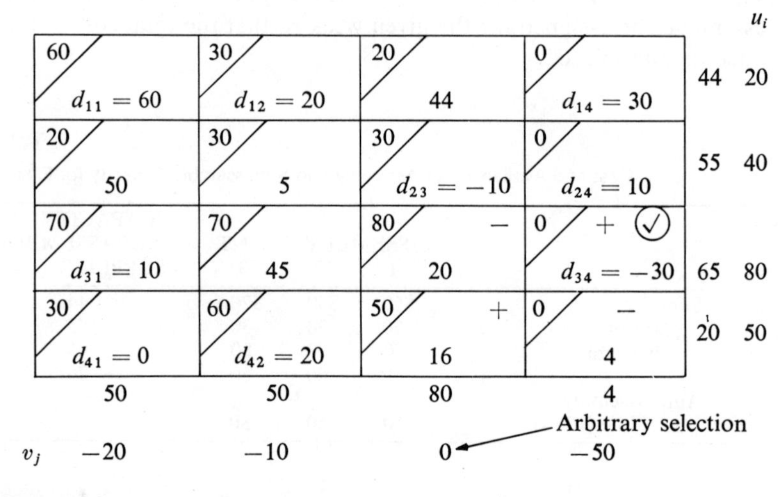

| | | | | | $u_i$ |
|---|---|---|---|---|---|
| 60 / $d_{11} = 60$ | 30 / $d_{12} = 20$ | 20 / 44 | 0 / $d_{14} = 30$ | 44 | 20 |
| 20 / 50 | 30 / 5 | 30 / $d_{23} = -10$ | 0 / $d_{24} = 10$ | 55 | 40 |
| 70 / $d_{31} = 10$ | 70 / 45 | 80 − / 20 | 0 + ✓ / $d_{34} = -30$ | 65 | 80 |
| 30 / $d_{41} = 0$ | 60 / $d_{42} = 20$ | 50 + / 16 | 0 − / 4 | 20 | 50 |
| 50 | 50 | 80 | 4 | | |
| $v_j$: −20 | −10 | 0 ← Arbitrary selection | −50 | | |

**Figure 9.6**

Sequence of Iterations for Example 9.2

Next $x_{23}$ enters the basis at 5, replacing $x_{22}$ and yielding Tableau 4. Next $x_{41}$ enters the basis at 20, replacing $x_{43}$ and yielding Tableau 5, which is optimal. Notice that subassembly plant 3 produces at 400 less than capacity, since it supplies 4 ($\times$ 100) units to the dummy automobile assembly plant.

**Tableau 3**

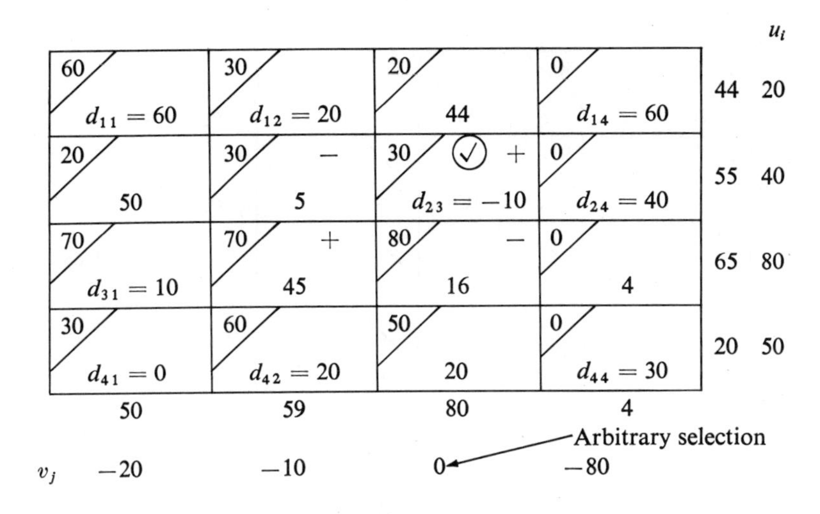

| | | | | | $u_i$ |
|---|---|---|---|---|---|
| 60 / $d_{11} = 60$ | 30 / $d_{12} = 20$ | 20 / 44 | 0 / $d_{14} = 60$ | 44 | 20 |
| 20 / 50 | 30 − / 5 | 30 (✓) + / $d_{23} = -10$ | 0 / $d_{24} = 40$ | 55 | 40 |
| 70 / $d_{31} = 10$ | 70 + / 45 | 80 − / 16 | 0 / 4 | 65 | 80 |
| 30 / $d_{41} = 0$ | 60 / $d_{42} = 20$ | 50 / 20 | 0 / $d_{44} = 30$ | 20 | 50 |
| 50 | 59 | 80 | 4 | | |
| $v_j$: −20 | −10 | 0 ← Arbitrary selection | −80 | | |

**Tableau 4**

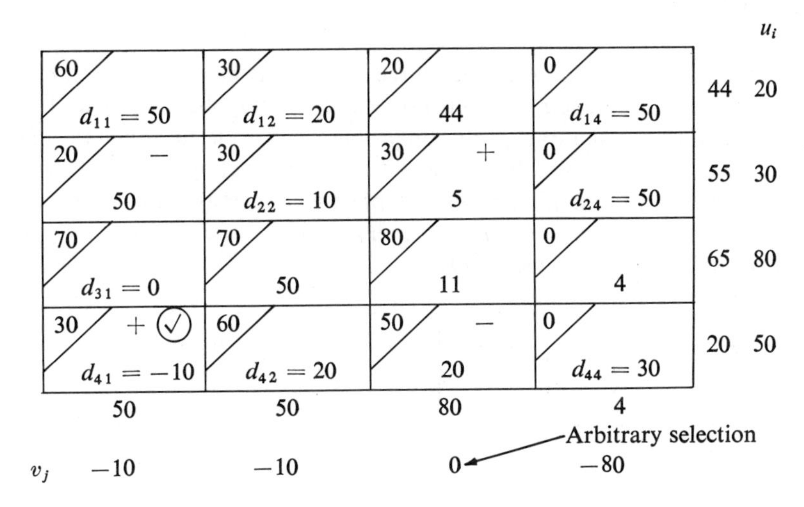

| | | | | | $u_i$ |
|---|---|---|---|---|---|
| 60 / $d_{11} = 50$ | 30 / $d_{12} = 20$ | 20 / 44 | 0 / $d_{14} = 50$ | 44 | 20 |
| 20 − / 50 | 30 / $d_{22} = 10$ | 30 + / 5 | 0 / $d_{24} = 50$ | 55 | 30 |
| 70 / $d_{31} = 0$ | 70 / 50 | 80 / 11 | 0 / 4 | 65 | 80 |
| 30 + (✓) / $d_{41} = -10$ | 60 / $d_{42} = 20$ | 50 − / 20 | 0 / $d_{44} = 30$ | 20 | 50 |
| 50 | 50 | 80 | 4 | | |
| $v_j$: −10 | −10 | 0 ← Arbitrary selection | −80 | | |

**Figure 9.6 (continued)**

In these examples we used the Northwest corner method to obtain a starting solution, primarily in order to illustrate iterations. It is usually better to use the Vogel Approximation method, since the generated solution it obtains usually requires fewer iterations to produce an optimum than that obtained by the Northwest corner method.

One potential difficulty with the transportation method is the problem of

**Tableau 5**

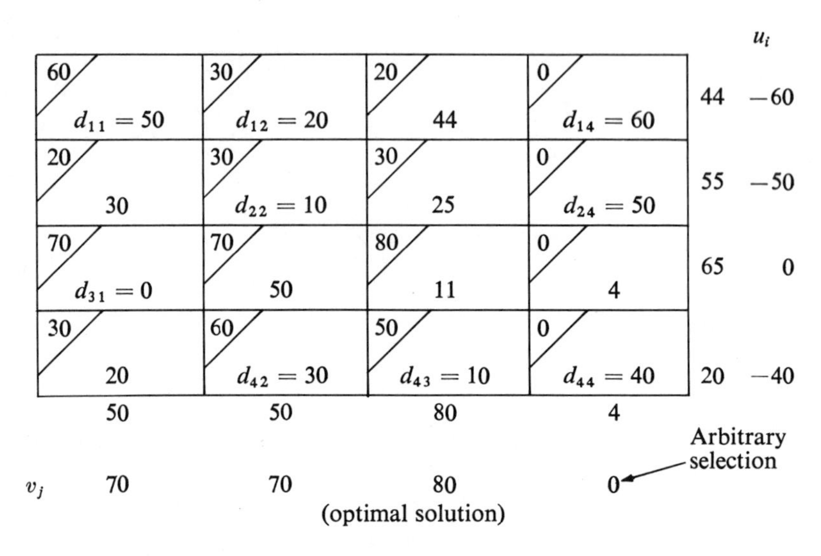

| | | | | | $u_i$ |
|---|---|---|---|---|---|
| 60 / $d_{11} = 50$ | 30 / $d_{12} = 20$ | 20 / 44 | 0 / $d_{14} = 60$ | 44 | $-60$ |
| 20 / 30 | 30 / $d_{22} = 10$ | 30 / 25 | 0 / $d_{24} = 50$ | 55 | $-50$ |
| 70 / $d_{31} = 0$ | 70 / 50 | 80 / 11 | 0 / 4 | 65 | 0 |
| 30 / 20 | 60 / $d_{42} = 30$ | 50 / $d_{43} = 10$ | 0 / $d_{44} = 40$ | 20 | $-40$ |
| 50 | 50 | 80 | 4 | | |
| $v_j$ 70 | 70 | 80 | 0 ← Arbitrary selection | | |

(optimal solution)

**Figure 9.6 (concluded)**

degeneracy. If any partial sums of source availability equal any partial sums of destination requirements, a degenerate solution may be obtained. It is not known whether or not a transportation problem can cycle, although the methods presented earlier allow for degeneracy. To prevent cycling so long as all of the availabilities and demands are strictly positive, we may add $\epsilon$, where $\epsilon$ is a sufficiently small nonzero element, to every demand, and then add $n\epsilon$ to one of the availabilities, as depicted below:

| | | | | |
|---|---|---|---|---|
| $x_{11}$ | $x_{12}$ | $\cdots$ | $x_{1n}$ | $a_1$ |
| $x_{21}$ | $x_{22}$ | $\cdots$ | $x_{2n}$ | $a_2$ |
| $\vdots$ | $\vdots$ | $\ddots$ | $\vdots$ | $\vdots$ |
| $x_{m-1,1}$ | $x_{m-1,2}$ | $\cdots$ | $x_{m-1,n}$ | $a_{m-1}$ |
| $x_{m1}$ | $x_{m2}$ | $\cdots$ | $x_{mn}$ | $a_m + n\epsilon$ |
| $b_1 + \epsilon$ | $b_2 + \epsilon$ | $\cdots$ | $b_n + \epsilon$ | |

This device will insure that no degenerate solution will occur. For $\epsilon$ small enough, the optimal solution to the problem containing the $\epsilon$'s will be the optimal solution to the original problem.

## 9.3 THE PRIMAL-DUAL METHOD FOR SOLVING TRANSPORTATION PROBLEMS

The primal-dual method for solving transportation problems is the predecessor of the Dantzig, Ford, and Fulkerson primal-dual algorithm which was described in Chapter 6. It is analogous to that method because both methods minimize the sum of the artificial variables in the primal problem while preserving the conditions of complementary slackness. We shall not develop the analogy here, primarily because later we shall derive an analogy for a more general class of problems. Instead, we shall describe the primal-dual algorithm and illustrate it by means of an example. The method proceeds by maintaining a feasible solution to the dual problem without necessarily satisfying the primal restrictions. The values of the dual variables are adjusted, preserving dual feasibility but moving toward primal feasibility. Once primal feasibility has been achieved, the optimal solution is attained. A flow chart of the method is given in Figure 9.7.

An important benefit of this method is that degeneracy causes no problems, which is not true of the earlier method.

Some simplifications can be made, but we shall not pursue them here. An example is presented to illustrate the method.

EXAMPLE 9.3:

Solve the problem whose source and destination requirements and costs are as given below, using the primal-dual method for solving the transportation problem.

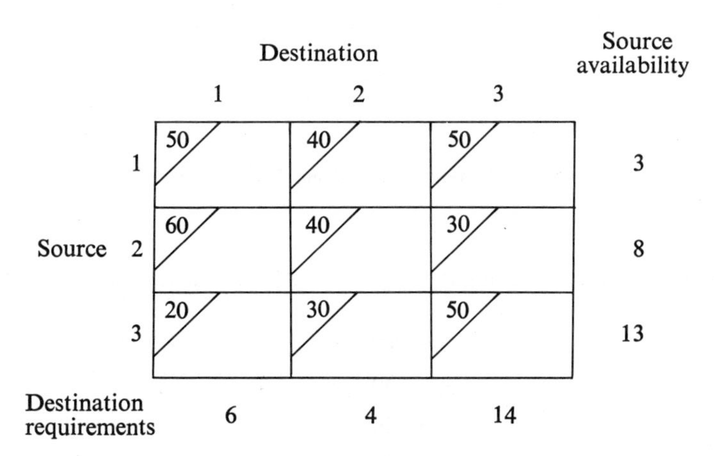

| | Destination 1 | Destination 2 | Destination 3 | Source availability |
|---|---|---|---|---|
| Source 1 | 50 | 40 | 50 | 3 |
| Source 2 | 60 | 40 | 30 | 8 |
| Source 3 | 20 | 30 | 50 | 13 |
| Destination requirements | 6 | 4 | 14 | |

For convenience we shall use tableaus of $c_{ij} - u_i - v_j$ as well as tableaus of feasible and infeasible cells.

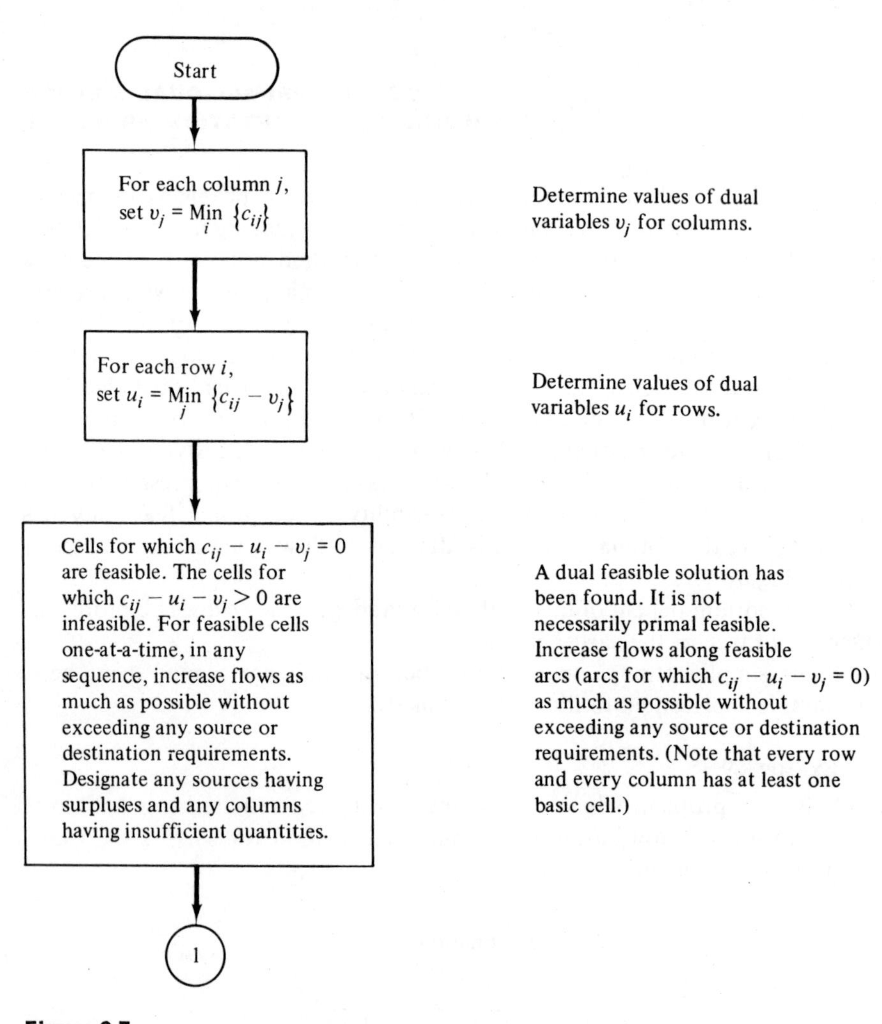


**Figure 9.7**

Flow Chart of the Primal-Dual Method for Solving Transportation Problems

First, we determine initial values of column dual variables $v_j$.

$$v_1 = 20$$
$$v_2 = 30 \qquad (= \operatorname{Min}_i \{c_{ij}\})$$
$$v_3 = 30$$

Next, we determine initial values of row dual variables $u_i$.

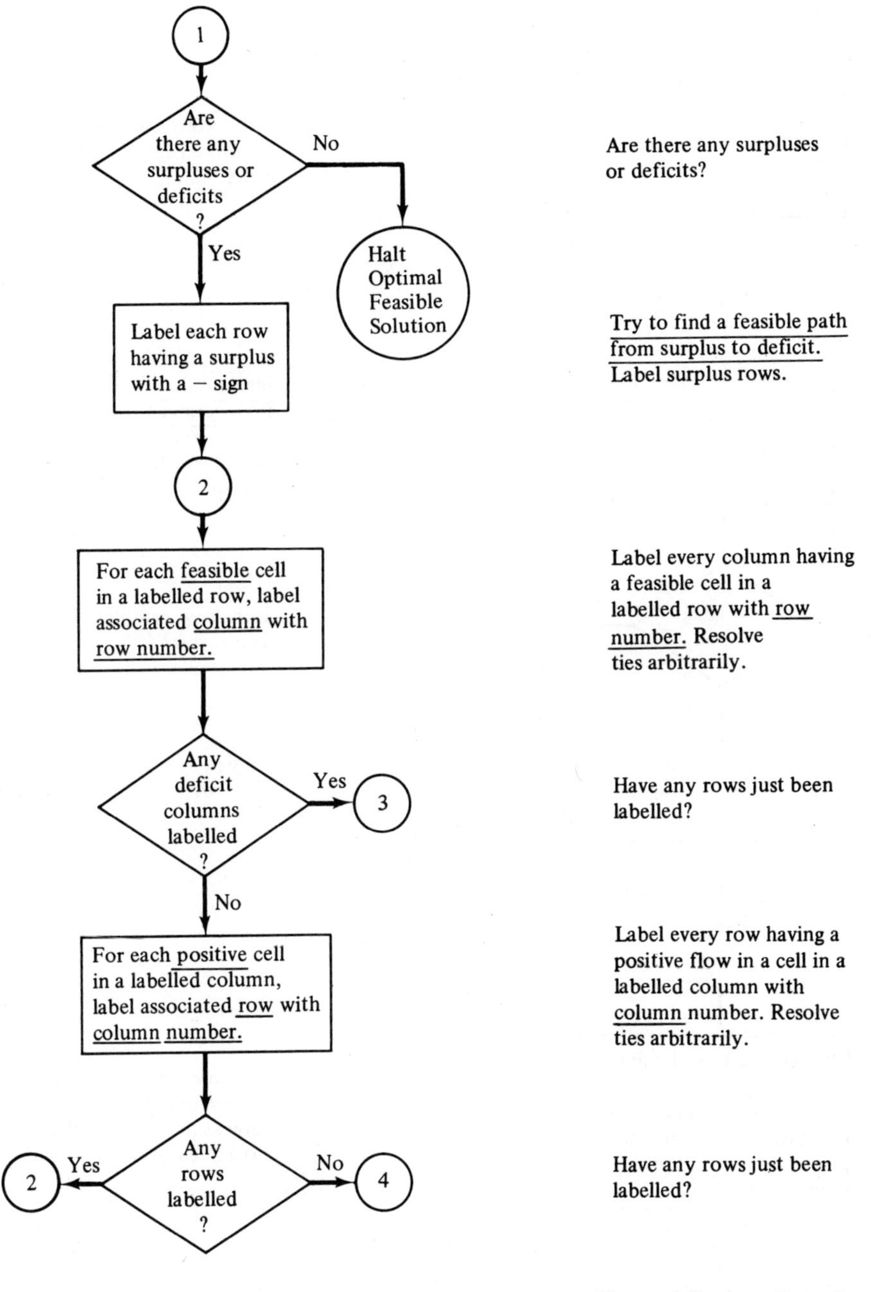


**Figure 9.7 (continued)**

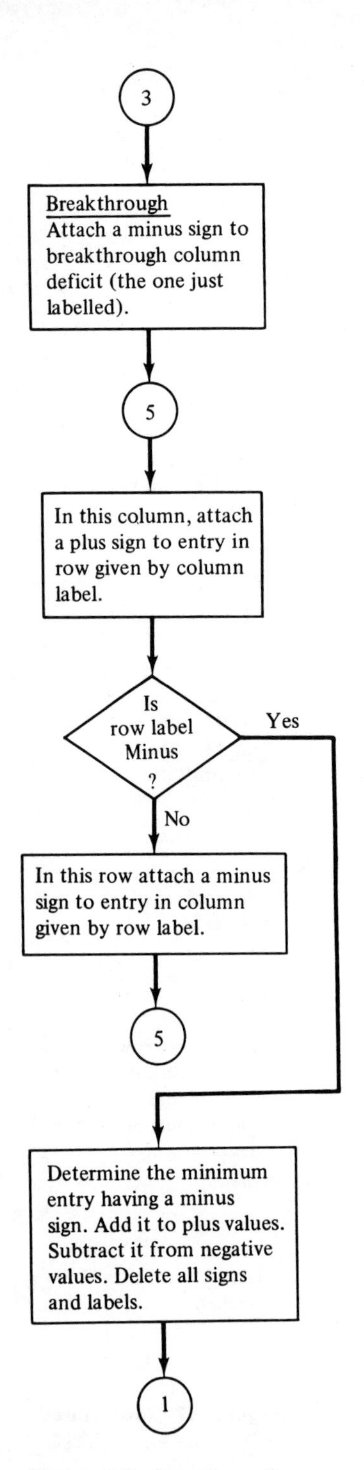


A breakthrough has been achieved. Label column deficit minus.

Identify the feasible path from surplus to deficit. Attach to entry (in row given by column label) a plus sign.

Is the row a surplus source row?

Attach to entry (in column given by row label) a minus sign.

Increase flow from surplus source to deficit destination as much as possible without violating feasibility.

**Figure 9.7 (continued)**

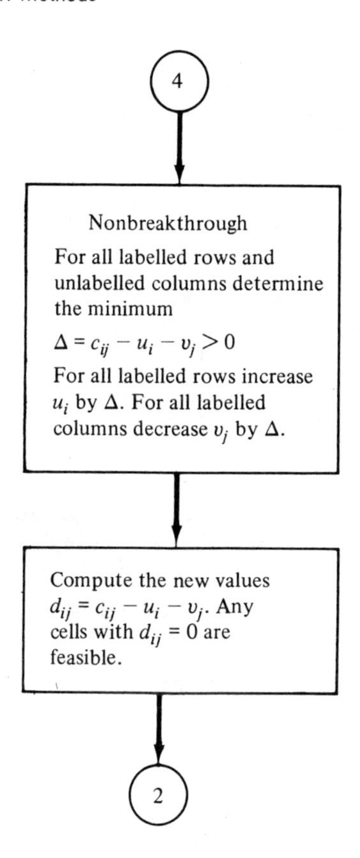


A nonbreakthrough has occured. It is not possible to find a path from surplus to deficit without changing dual variables. Preserve dual feasibility by holding $c_{ij} - u_i - v_j = 0$ for all positive cells (labelled rows and labelled columns) or unlabelled rows and unlabelled columns. At least one previously infeasible cell in a labelled row will now be feasible, and usually, but not always, at least one previously feasible, but zero, cell in a labelled row will be infeasible.

**Figure 9.7 (concluded)**

$$u_1 = 10$$
$$u_2 = 0 \qquad (= \operatorname*{Min}_j \{c_{ij} - v_j\})$$
$$u_3 = 0$$

The tableau of reduced costs $d_{ij} = c_{ij} - u_i - v_j$ together with the $u_i$ and $v_j$ values is as follows:

| | | | | $u_i$ |
|---|---|---|---|---|
| | 20 | 0 | 10 | 10 |
| | 40 | 10 | 0 | 0 |
| | 0 | 0 | 20 | 0 |
| $v_j$ | 20 | 30 | 30 | |

The cells with zero entries are feasible. Accordingly, for feasible cells, we increase flows as much as possible. The results are given below with feasible cells circled. Amounts shipped as well as surpluses and deficits are indicated.

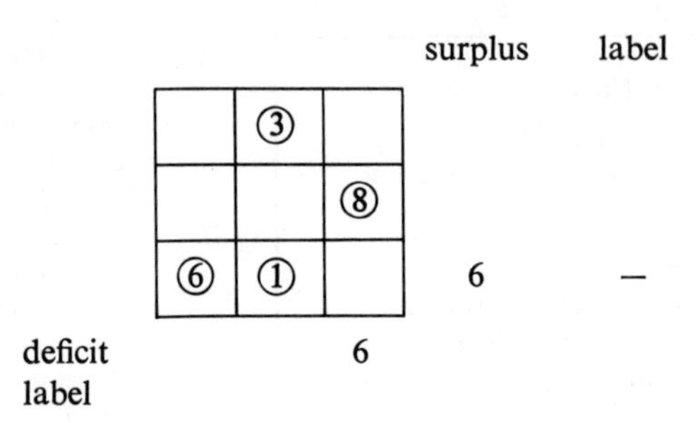


Label the surplus row, row 3, minus, as indicated in the tableau above. Label columns 1 and 2 with the number 3, corresponding to the labeled row in which its feasible cell is contained, as below. Then label row 1 from column 2, since there is a positive flow in cell (1, 2), as below.

| | | | | surplus | label |
|---|---|---|---|---|---|
| | | ③ | | | 2 |
| | | | ⑧ | | |
| | ⑥ | ① | | 6 | — |
| deficit | | | 6 | | |
| label | 3 | 3 | | | |

No new rows or columns can be labeled, resulting in a nonbreakthrough.

$$\text{Labeled rows} = \{1, 3\}$$

$$\text{Unlabeled column} = \{3\}$$

Determine

$$\Delta = \text{Min}\,\{d_{ij} = c_{ij} - u_i - v_j\}_{j=3}^{i=1,3}$$

$$\Delta = 10(= d_{13})$$

Then set

$$u_1' = u_1 + 10 = 20, \quad u_3' = u_3 + 10 = 10$$

$$v_1' = v_1 - 10 = 10, \quad v_2' = v_2 - 10 = 20$$

with the other $u_i$ and $v_j$ values remaining unchanged. The new tableau of reduced costs $d_{ij}$ and the $u_i$ and $v_j$ then is as follows:

| | | | | $u_i$ |
|---|---|---|---|---|
| | 20 | 0 | 0 | 20 |
| | 50 | 20 | 0 | 0 |
| | 0 | 0 | 10 | 10 |
| $v_j$ | 10 | 20 | 30 | |

Cell (1, 3) is now feasible, in addition to the other feasible cells. Preserving the earlier labels, we have the following tableau in which we have labeled column 3 from row 1, and thus we have a breakthrough.

| | | | | surplus | label |
|---|---|---|---|---|---|
| | | (3)− | (+) | | 2 |
| | | | (8) | | |
| | (6) | (1)+ | | 6− | — |
| deficit | | | 6− | | |
| label | 3 | 3 | 1 | | |

The steps follow: Label the breakthrough deficit (in column 3) minus. Attach a plus sign to entry (1, 3) (deficit column with row indicated by deficit column label). Attach a minus sign to entry (1, 2) (row 1 with column given by row 1's label). Attach a plus sign to entry (3, 2) (column 2 with row given by column 2's label). Because row 3's label is minus, we label the corresponding surplus minus. The smallest entry with a minus sign is 3. Add and subtract 3 as indicated by the signs, and drop all signs and labels. The solution is not yet feasible and is given below.

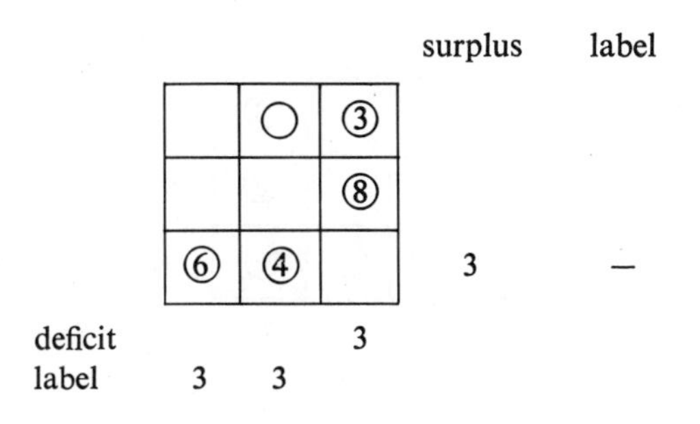

Attach a minus to the surplus row and then label columns 1 and 2 from row 3 as in the above tableau. No new columns or rows can be labeled; hence we have a nonbreakthrough.

$$\text{Labeled row} = \{3\}$$
$$\text{Unlabeled column} = \{3\}$$

Determine

$$\Delta = \text{Min}\,\{d_{ij} = c_{ij} - u_i - v_j\}\begin{matrix} i = 3 \\ j = 3 \end{matrix}$$

or $\Delta = 10\ (= d_{33})$. Then

$$u'_3 = u_3 + 10 = 20$$
$$v'_1 = v_1 - 10 = 0$$
$$v'_2 = v_2 - 10 = 10$$

Then new tableau of reduced costs is then

| | | | $u_i$ |
|---|---|---|---|
| 30 | 10 | 0 | 20 |
| 60 | 30 | 0 | 0 |
| 0 | 0 | 0 | 20 |
| 0 | 10 | 30 | $v_j$ |

Cell (3, 3) is now feasible, and cell (1, 2) is no longer feasible. Preserving the labels, we have the following tableau in which we have labeled column 3 from row 3, resulting in a breakthrough. Attaching pluses and minuses as before, we see that the smallest entry with a minus sign is 3.

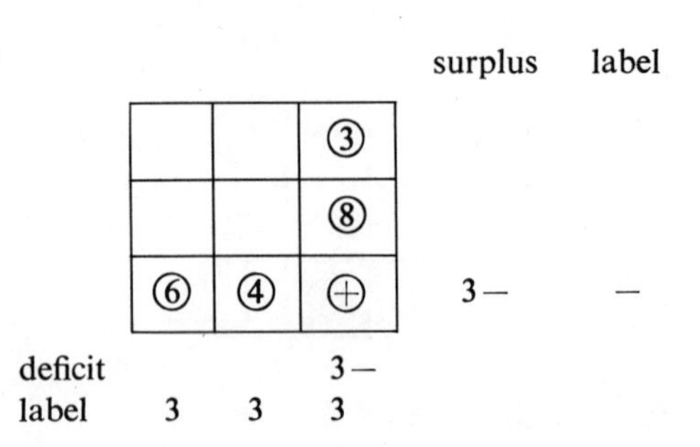

Adding and subtracting 3 as indicated, we have the following tableau which has no surpluses or deficits and hence is an optimal feasible solution.

| | | To destination | | |
|---|---|---|---|---|
| | | 1 | 2 | 3 |
| From | 1 | | | ③ |
| Source | 2 | | | ⑧ |
| | 3 | ⑥ | ④ | ③ |

Because of the space devoted to this example, the method appears cumbersome. In fact, it is not; it is quite efficient in comparison to other methods.

## 9.4 OTHER ASPECTS OF THE TRANSPORTATION PROBLEM

There are other aspects of the transportation problem worthy of mention. If all the availabilities and demands are integer, then the optimal solution will be integer. This can be proven by making use of the triangularity of the basis, as well as the fact that the columns of the basis contain only $+1$ and 0. It can be shown that the uniquely determined variable of a triangular basis is integer. By reducing the problem size to account for the variable whose value is known, the process can be repeated to show that the entire solution is integer.

A special case of the transportation problem is the assignment problem in which $a_i = b_j = 1$ for all $i$ and $j$, and $m = n$. Such a problem might represent the assignment of $n$ men to $n$ jobs, where each man-job combination has a certain measure of effectiveness and the overall sum is to be maximized subject to the requirement that each job be filled and each man be engaged. Because of the extreme degeneracy present in the assignment problem (every solution is badly degenerate), the primal-dual method is particularly useful.

Another problem analogous to the transportation problem—in that it has a special method for its solution—is the generalized transportation problem. The generalized transportation problem has the same coefficient format as the transportation problem, namely (9.2), except that the coefficients are only required to be positive and the equalities may be inequalities in either direction. As in the case of generalized upper bounds, we may assume without loss of generality that either the source or the destination constraints (but not both) have coefficients of $+1$.

Still another problem amenable to this form of solution is the capacitated transportation problem, a transportation problem with capacity limitations on cells. We shall not go into the methods, but will note that they are developed in essentially the same way as the transportation method treated here. Questions concerning certain of these methods are included in the problems at the end of the chapter. Finally, there are various other transportation models with inequality constraints, transshipment provisions, and so on, and these are developed in much the same way. Dantzig [62] provides a comprehensive treatment of some of these methods, and numerous references to other methods are presented in the bibliography. All of the problems discussed thus far in this chapter may be considered as network flow problems, and some of the solution methods can be evolved from network flow solution methods.

## 9.5 NETWORK FLOWS—AN INTRODUCTION

A large class of linear programming problems, including the transportation problem, may be cast in terms of flows in networks. In addition to this apparent convenience in viewing such problems, there is also some value in the computational methods or algorithms that come from this analysis. Before proceeding with it, we need some definitions from graph theory.

**Definition 9.2**
*A* linear graph *or* network *consists of* nodes *or* points *each connected to one or more other nodes by* arcs *or* links. *An arc is also referred to as an edge.*

Example of a linear graph:

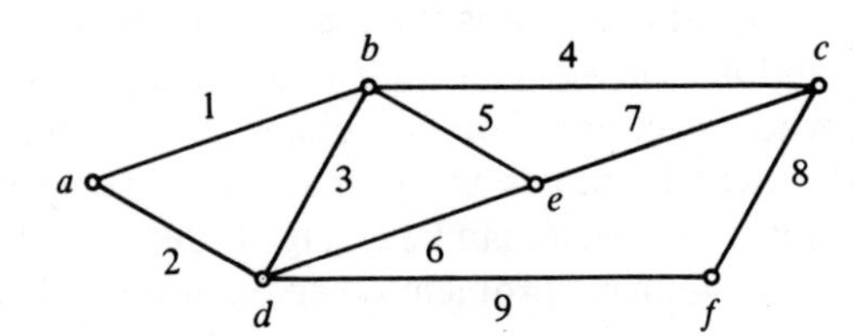


Nodes: $a, b, c, d, e, f$. Arcs: 1, 2, 3, 4, 5, 6, 7, 8, 9.

**Definition 9.3**
*A* directed network *is a network in which flow along an arc may only be in one direction.*

Obviously, any network may be represented as a directed network by replacing an arc on which two-way flow is permitted with two arcs going in opposite directions. Therefore we shall restrict our attention to directed networks, without loss of generality.

**Definition 9.4**

*A* bipartite graph *is a directed graph in which the nodes are divided into two subsets, with all arcs of the graph leading from nodes of one subset to nodes of the other.*

A transportation problem's graph is an example of a bipartite graph, since all flows are from source nodes to destination nodes.

**Definition 9.5**

*A* path *or a* chain *is an ordered set of arcs that connects two nodes by means of intermediate nodes, each of which is on exactly two arcs in the chain.*

An example of a chain in the above graph is the set of arcs 1, 5, and 7, which connects nodes $a$ and $c$ by means of nodes $b$ and $e$.

**Definition 9.6**

*A* connected graph *is a graph in which a path exists between any pair of nodes.* The graph presented above is connected.

**Definition 9.7**

*A* loop *is a chain which connects a node to itself.*

Arcs 1, 5, 7, 8, 9, and 2 in the above graph form a loop connecting node $a$ to itself (or any node in the chain to itself).

**Definition 9.8**

*A* tree *is a connected graph which contains no loops.*

Examples of trees in the above graph include arcs 1, 3, 4, 6, 8 and arcs 2, 3, 4, 5, 8. The set of arcs 1, 2, 3, 4, 7, 8 contains a loop and hence cannot be a tree; the set of arcs 1, 3, 7, 8 does not form a tree, since the graph defined by the set of arcs is not connected. It can be proven that a tree which has $n$ nodes has $n - 1$ arcs, and that there are at least two ends of a tree.

It will sometimes be convenient to refer to a directed network via an incidence matrix. The rows of an incidence matrix correspond to nodes of the network, and the columns of an incidence matrix correspond to arcs of the network. An arc corresponds to possible flow from the node indicated by a $+1$ entry to the node indicated by a $-1$ entry. For example, the directed network of Figure 9.8 has the incidence matrix given in Figure 9.8.

We shall consider two problems that may be viewed as network flow problems. The first is the problem of maximizing the flow through a network from a source of flow to a sink. The second is the problem of maximizing the value of flow through a network. It will be shown that the first problem is a special case

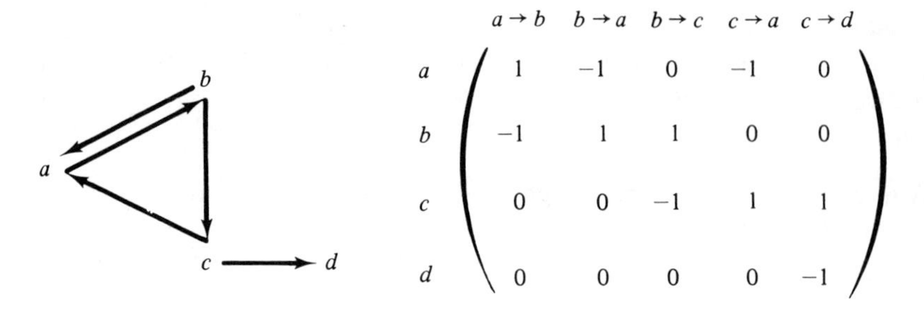


**Figure 9.8**

A Directed Graph and Its Incidence Matrix

of the second, and that all problems of this chapter can be expressed as problems of the second type.

### 9.6 MAXIMAL FLOW THROUGH A NETWORK

An important network problem is determination of the maximal flow between two points in a network. To provide a motivation for this problem, consider the following.

A natural gas producer has a pipeline network as shown in Figure 9.9.

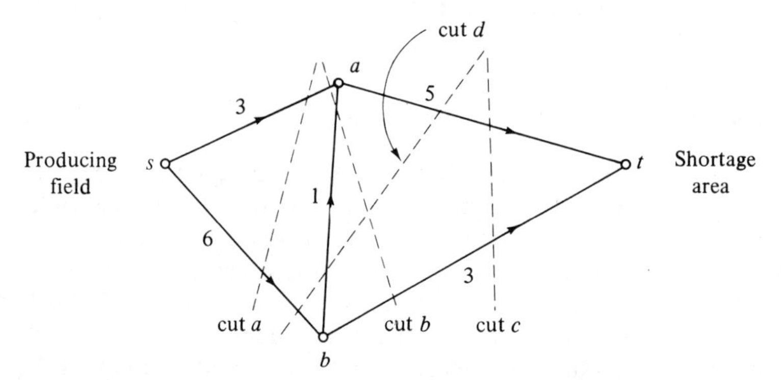


**Figure 9.9**

The Network of a Gas Pipeline

(The capacities of the pipelines indicated on the arcs are in hundred millions of cubic feet per day.) A shortage has occurred at point $t$, and it is desired to ship as much natural gas as possible from the producing field to point $t$. Therefore, the problem is to find the maximal capacity of the network between $s$ and $t$ so that the maximum amount can be moved to point $t$.

More formally, the problem to be considered is that of maximizing the flow from a node $s$ (called the source) to a node $t$ (called the sink) subject to limitations on the capacity of the arcs. Given an incidence matrix representing the network, we add an arc that goes from sink $t$ to source $s$, the purpose of this arc being to insure that the sink and source nodes also conserve flow. The problem can then be represented as a linear programming problem by maximizing the flow from sink $t$ to source $s$, subject to the requirements that the product of the incidence matrix times a vector of flows on arcs be zero (the conservation of flow equations), and that the nonnegativity and upper bound constraints on the flows on the arcs be satisfied. The linear programming problem which can be solved to obtain the maximal flow through the network of the example is the following:

$$
\begin{array}{lrrrrrrll}
 & t\to s & s\to a & s\to b & a\to t & b\to a & b\to t & & \\
\text{Maximize } z = x_0 & & & & & & & & \textit{nodes} \\
\text{subject to:} & x_0 & & & -x_3 & & -x_5 & = 0 & t \\
 & -x_0 & +x_1 & +x_2 & & & & = 0 & s \\
 & & -x_1 & & +x_3 & -x_4 & & = 0 & a \\
 & & & -x_2 & & +x_4 & +x_5 & = 0 & b \\
 & & & & & & & & \textit{arc capacities} \\
 & & x_1 & & & & & \le 3 & s \longrightarrow a \\
 & & & x_2 & & & & \le 6 & s \longrightarrow b \\
 & & & & x_3 & & & \le 5 & a \longrightarrow t \\
 & & & & & x_4 & & \le 1 & b \longrightarrow a \\
 & & & & & & x_5 & \le 3 & b \longrightarrow t \\
 & & & & & & x_0, x_1, x_2, x_3, x_4, x_5 & \ge 0 &
\end{array}
$$

where $x_0$ represents the flow from sink $t$ to source $s$ (which must equal the flow through the network from $s$ to $t$), $x_1$ represents the flow from $s$ to $a$, $x_2$: $s$ to $b$, $x_3$: $a$ to $t$, $x_4$: $b$ to $a$, and $x_5$: $b$ to $t$. The first four equations represent conservation of flow at nodes $t$, $s$, $a$, and $b$, respectively, and the last five constraints are the limits of flow on the arcs. The optimal solution to the problem may be found, using the simplex method, to be $x_0 = 7$, $x_1 = 3$, $x_2 = 4$, $x_3 = 4$, $x_4 = 1$, and $x_5 = 3$, with a corresponding maximum flow of 7.

Let us now formulate the dual of the maximal flow problem of the example.

$$
\begin{array}{lrrrrrrrrrl}
\text{Minimize } z = & & & & & 3\mu_1 & +6\mu_2 & +5\mu_3 & +\mu_4 & +3\mu_5 & \\
\text{subject to:} & \pi_t & -\pi_s & & & & & & & & \ge 1 \\
 & & \pi_s & -\pi_a & & +\mu_1 & & & & & \ge 0 \\
 & & \pi_s & & -\pi_b & & +\mu_2 & & & & \ge 0 \\
 & -\pi_t & & +\pi_a & & & & +\mu_3 & & & \ge 0 \\
 & & & -\pi_a & +\pi_b & & & & +\mu_4 & & \ge 0 \\
 & -\pi_t & & & +\pi_b & & & & & +\mu_5 & \ge 0
\end{array}
$$

$$\pi_t, \pi_s, \pi_a, \pi_b \text{ unrestricted in sign, } \mu_1, \mu_2, \mu_3, \mu_4, \mu_5 \ge 0$$

An optimal solution to the dual is $\pi_t = 1$, $\pi_s = 0$, $\pi_a = 1$, $\pi_b = 0$, $\mu_1 = 1$, $\mu_2 = 0$, $\mu_3 = 0$, $\mu_4 = 1$, and $\mu_5 = 1$. In general, it can be shown that the optimal solution to the dual will have $\pi_t = 1$ and $\pi_s = 0$. All nodes $w$ whose dual values $\pi_w$ are 1 form a tree that includes the sink $t$, and all nodes $v$ whose dual values $\pi_v$ are 0 form a tree that includes the source $s$. By way of proof, were this not true, one of the constraints of the dual problem would be violated. It can also be shown in general that only the arcs which connect the two trees (the $s$ tree to the $t$ tree) will have their arc values ($\mu$'s) equal to unity. The value of the sum of the capacities of the arcs connecting the source tree to the sink tree is therefore a minimum, in general. We have proved, in an informal manner, a fundamental theorem of Ford and Fulkerson [95], which we shall state after the following definitions:

**Definition 9.9** (Cut)

*A* cut *is a minimal set of arcs which, if deleted, divides the network into two disjoint subnetworks—one containing s, the other containing t.*

**Definition 9.10** (Cut capacity)

*The capacity of a cut is defined as the forward capacity (from s to t) of the bipartite graph made up of the arcs of the cut.*

**THEOREM 9.1** (Max-flow-min-cut theorem) [95]

FOR ANY NETWORK THE MAXIMAL FLOW VALUE FROM $s$ TO $t$ IS EQUAL TO THE MINIMAL CUT CAPACITY OF ALL CUTS SEPARATING $s$ AND $t$.

Theorem 9.1 says that the minimum cut capacity in a network is equal to the maximum flow through the network. To illustrate this, for the example given above there are four possible cuts:

1. $s \longrightarrow a$, $s \longrightarrow b$, with a cut capacity of 9.
2. $s \longrightarrow a$, $b \longrightarrow a$, $b \longrightarrow t$, with a cut capacity of 7.
3. $a \longrightarrow t$, $b \longrightarrow t$, with a cut capacity of 8.
4. $s \longrightarrow b$, $b \longrightarrow a$, $a \longrightarrow t$, with a cut capacity of 11.

The cuts are shown in Figure 9.9. We observe that the minimal cut through the network is 7, as is the maximal flow from $s$ to $t$.

To emphasize that the positive flow from source subnetwork to sink subnetwork determines the value of the cut, suppose arc $ba$ is reversed, i.e., suppose that the flow is from $a$ to $b$ and a maximum of 1 unit (100 million cubic feet per day) can flow. Then the minimum cut (and of course the maximum flow) is 6 units.

In the primal form of the current formulation, variables correspond to flows on arcs, and constraints (excluding the upper bounds on arc flows) represent

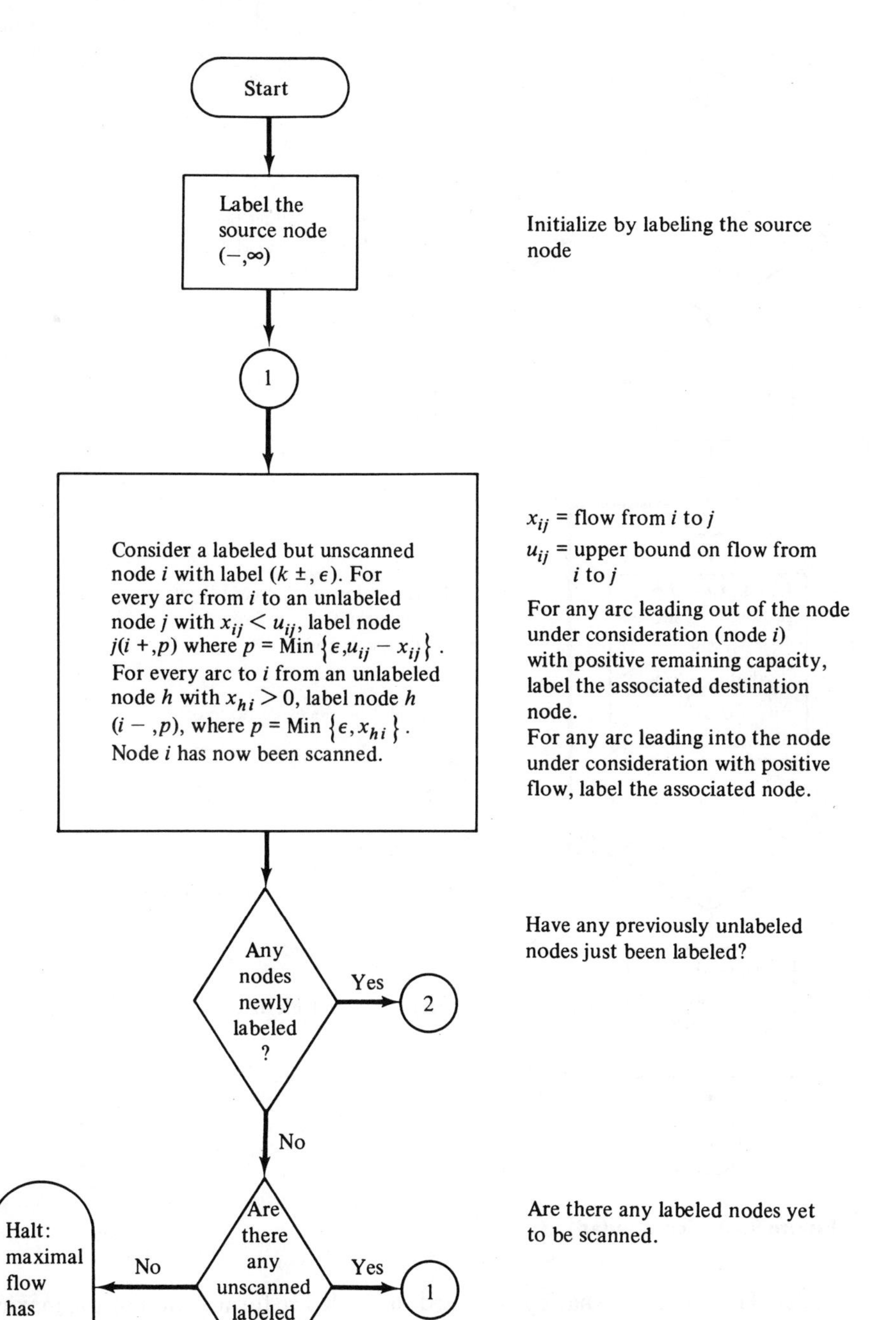


**Figure 9.10**

Flow Chart for Finding the Maximal Flow Through a Network

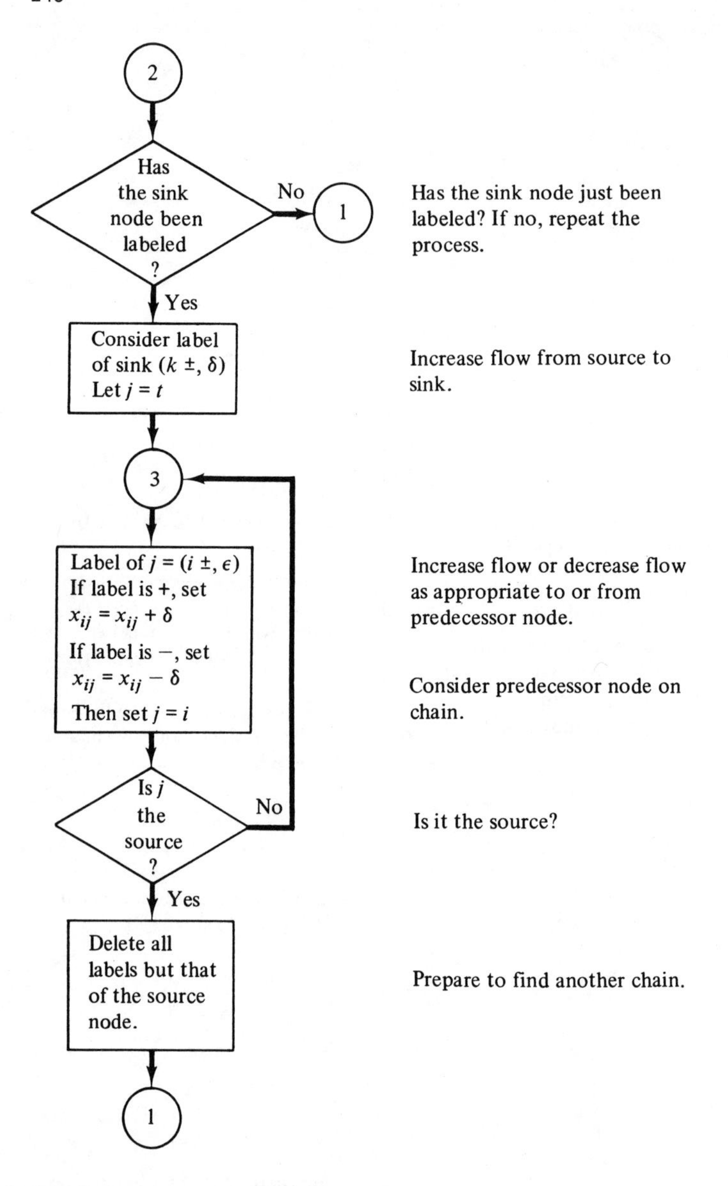


**Figure 9.10 (concluded)**

nodes. This approach has been named the node-arc formulation of the maximal flow problem. In an alternate linear programming formulation, called the arc-chain formulation, a variable represents flow along a chain from source to sink, and all such variables are explicitly stated. The constraints are the capacities of the arcs. The objective is to maximize the total flow from the source or the total flow to the sink. Using the simplex method to solve the arc-chain formulation corresponds to successively introducing chains into the basis until a maximal flow is found. This approach leads to a method for solving the maximal flow problem, a flow chart of which is given in Figure 9.10.

The method consists of finding a chain (of a slightly more general form than that in the arc-chain formulation) from source to sink having positive capacity, and then increasing flow along that chain. The procedure is continued until no chain having positive flow from source to sink can be found. Each node is given a label having two components: the first is its immediate predecessor along a unique chain from the source; the second, the net forward capacity from the source to the node. The procedure begins by labeling the source. Then all arcs leading from the source are used to label any unlabeled nodes. At the completion of the process, the source node is said to have been scanned. The process is repeated until all labeled nodes are scanned, or until the sink has been labeled. If the sink cannot be labeled, then the maximal flow is the sum of all chain flows found thus far; otherwise, the chain used to label the sink is used to update the flow patterns found thus far. The labels are then dropped and the labeling process repeated. By requiring integral capacities on each arc (this is no practical limitation), the procedure can easily be shown to be finite. We now consider an example.

EXAMPLE 9.4:

Find the maximal flow from $s$ to $t$ in the graph below, using the method given by flow chart in Figure 9.10.

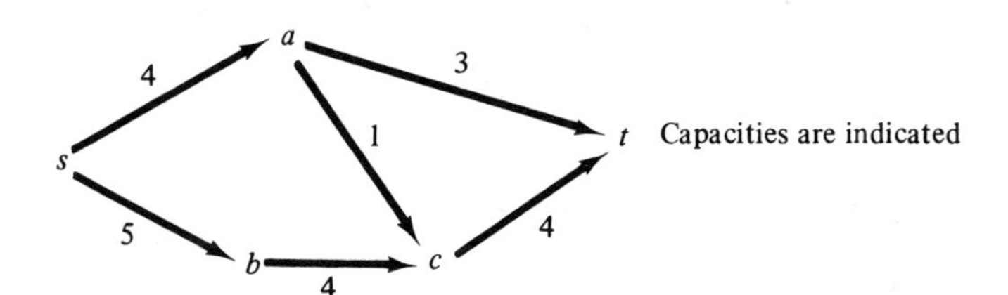


Label $s$ $(-, \infty)$. The only labeled and unscanned node is $s$. Thus we label $a$ $(s+, 4)$ and $b$ $(s+, 5)$, completing the scanning of node $s$. Now both nodes $a$ and $b$ are labeled but unscanned. We arbitrarily choose node $a$ and label $t$ $(a+, 3)$ and $c$ $(a+, 1)$. Since $t$ has been labeled, we work back from $t$ using $\delta = 3$ and increasing flow to $t$ from $a$, and to $a$ from $s$ by 3. The network, with the flow in parentheses, is given below:

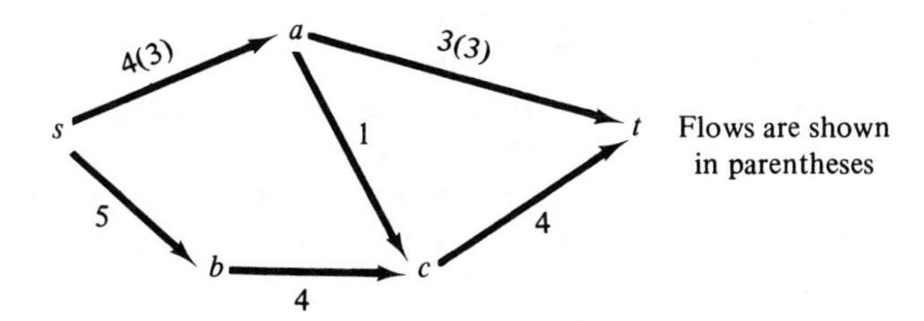

$s$ is labeled $(-, \infty)$. Label $a$ $(s+, 1)$. Label $b$ $(s+, 5)$. Label $c$ $(a+, 1)$. (We label $c$ $(a+, 1)$ because node $a$ was arbitrarily selected for scanning instead of node $b$. Had node $b$ been selected for scanning, $c$ would have been labeled $(b+, 4)$.) Label $t$ $(c+, 1)$. One unit of flow is augmented from $s$ to $a$ to $c$ to $t$.

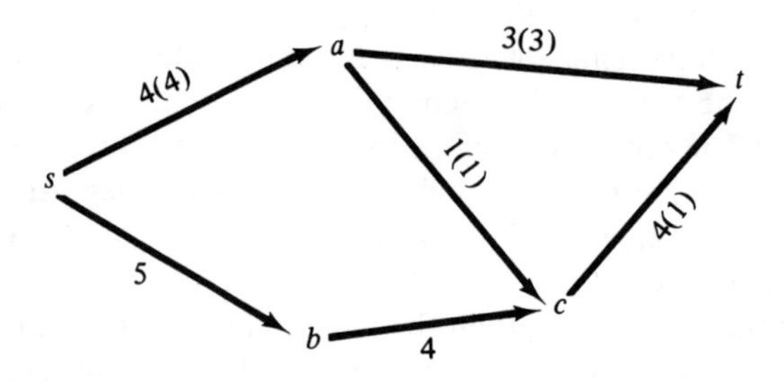


$s$ is labeled $(-, \infty)$. Label $b$ $(s+, 5)$. Label $c$ $(b+, 4)$. Label $a$ $(c-, 1)$. Label $t$ $(c+, 3)$. Three units of flow are augmented from $s$ to $b$ to $c$ to $t$.

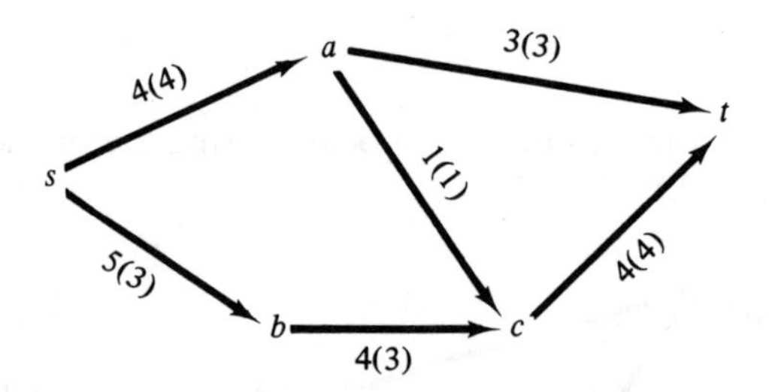


$s$ is labeled $(-, \infty)$. Label $b$ $(s+, 2)$. Label $c$ $(b+, 1)$. Label $a$ $(c-, 1)$. Since no further nodes can be labeled, the flow of 7 from source $s$ to sink $t$ shown above is maximal. (Since flows are indicated in parentheses, we need only compute the total flow from $s$—to $a$ and to $b$—or the total flow into $t$.)

## 9.7 MAXIMUM VALUE OF FLOW THROUGH A NETWORK

All of the problems solved by methods presented in this chapter can be formulated as problems of maximizing or minimizing the value of a flow through a network, given specified maximum and minimum flow limits on arcs. Moreover, the general problem can be solved by a method which builds upon and implies some of the methods described in this chapter. The maximal value of flow problem is sufficiently general, and its solution method sufficiently powerful, for us to devote three sections to it. In this section we describe the problem, using an example, and then state the problem formally and derive the optimality conditions using linear programming. In section 9.8 we present an algorithm and solve an example, and in section 9.9 we consider certain other aspects of the maximum value of flow problem.

To motivate the development, the following problem is an example of a maximal value of flow problem.

EXAMPLE 9.5:

A truck company has an opportunity to contract to haul between two and five loads of a commodity from Atlantic City, N.J., to Bethlehem, Pa. (Two loads are already under contract.) There is a further opportunity to haul a load from Bethlehem to Coatesville, Pa., an opportunity to haul up to two loads from Bethlehem to Dover, Delaware, and an opportunity to haul up to ten loads from Atlantic City to Coatesville. Net revenues of loaded trips (positive entries) and costs of empty trips (negative entries) are given in Table 9.2. Recognizing that only trips involving completed round trips can reflect appropriate profits, for which trips should the company contract, and what is a corresponding feasible routing schedule?

**Table 9.2**

| | TO: | A | B | C | D |
|---|---|---|---|---|---|
| FROM: | A | — | \$500* | \$350 | −\$400 |
| | B | −\$200 | — | 350 | 450 |
| | C | −400 | −300 | — | −300 |
| | D | −400 | −350 | −300 | — |

* Net revenues (negative figures correspond to net costs of empty trips) where A represents Atlantic City, and so on.

The maximum value of flow problem can be formulated as a linear programming problem using a node-arc formulation. The arc-chain formulation could also be used, but the node-arc formulation is better for our purposes. The linear programming formulation, in general, is as follows:

$$\text{Maximize } z = \sum_i \sum_j c_{ij} x_{ij}$$

$$\text{subject to:} \quad \sum_i x_{ij} - \sum_k x_{jk} = 0 \quad j = 1, \ldots, n \text{ where } k \text{ and } i \text{ are nodes incident to } j$$

$$h_{ij} \leq x_{ij} \leq u_{ij}$$

Variable $x_{ij}$ is defined only for nodes $i$ and $j$ if there exists an arc from $i$ to $j$. A linear programming formulation of example problem 9.5 appears in Figure 9.11, and can, of course, be solved using linear programming.

In order to gain some insight into the problem, we write the dual formulation. (Writing the dual appears useful in general for gaining insight into a prob-

Maximize $z$ =

$$5x_{AB} + 3.5x_{AC} - 4x_{AD} - 2x_{BA} + 3.5x_{BC} + 4.5x_{BD} - 4x_{CA} - 3x_{CB} - 3x_{CD} - 4x_{DA} - 3.5x_{DB} - 3x_{DC}$$

subject to:

$$\begin{aligned}
-x_{AB} - x_{AC} - x_{AD} + x_{BA} + x_{CA} + x_{DA} &= 0\\
+x_{AB} - x_{BA} - x_{BC} - x_{BD} + x_{CB} + x_{DB} &= 0\\
+x_{AC} + x_{BC} - x_{CA} - x_{CB} - x_{CD} + x_{DC} &= 0\\
+x_{AD} + x_{BD} + x_{CD} - x_{DA} - x_{DB} - x_{DC} &= 0
\end{aligned}$$

$$2 \leqslant x_{AB} \leqslant 5, \quad 0 \leqslant x_{AC} \leqslant 10, \quad 0 \leqslant x_{BC} \leqslant 1, \quad 0 \leqslant x_{BD} \leqslant 2 \quad \text{all other } x_{ij} \geqslant 0$$

$A$ – Atlantic City
$B$ – Bethlehem
$C$ – Coatesville
$D$ – Dover
$x_{AB}$ – shipment from $A$ to $B$

The objective function has been divided by 100 for convenience.

**Figure 9.11**

The Linear Programming Formulation of the Trucking Problem of Example 9.5.

lem.) Ignoring the bound constraints, we have the following constraints:

$$-\pi_i + \pi_j \geq c_{ij} \text{ for all pairs } i, j \text{ corresponding to arcs}$$
$$\pi_i, \pi_j \text{ unrestricted in sign}$$

The above can be written as $\pi_j \geq c_{ij} + \pi_i$. We know from complementary slackness considerations that if the constraint holds as a strict greater-than inequality, that is, $\pi_j > c_{ij} + \pi_i$, then $x_{ij}$ must be at its lower bound, i.e., $x_{ij} = h_{ij}$. It can similarly be shown that if $\pi_j < c_{ij} + \pi_i$, then $x_{ij}$ must be at its upper bound, i.e., $x_{ij} = u_{ij}$. Also, from complementary slackness considerations for $x_{ij}$ to equal neither $h_{ij}$ or $u_{ij}$, the following expression must hold:

$$\pi_j = c_{ij} + \pi_i$$

The above three conditions are the optimality criteria for the general network flow problem. To aid intuition $\pi_j$ may be thought of as a deposit, or node use tax, paid for a truck entering node $j$. However, the tax is refunded when a truck leaves a node. Since the flow is ultimately balanced, the total of the node use taxes over the network is zero. Nonetheless, the difference in the node use taxes between two nodes gives us the cost in terms of tax differentials in going from one node to another. Thus, to evaluate traversing an arc from node $i$ to node $j$, we compare the tax paid at node $j$, $\pi_j$, with the sum of the profit of moving from $i$ to $j$ and the tax rebate at node $i$, $c_{ij} + \pi_i$. If the former is larger, we ship as little as possible. If the latter is larger, we ship as much as possible. If the two are equal, we are indifferent to how much is shipped, so long as the amount shipped is within the specified limits. (A degenerate possibility is that $x_{ij} = h_{ij}$ or $u_{ij}$ and $\pi_j = c_{ij} + \pi_i$.)

EXAMPLE 9.5 (continued):

We now reconsider example 9.5, the optimal solution of which is given below. We then examine all arcs to show that the optimality conditions hold. A network representation of arcs indicating *positive flows only* for the optimal solution is given below.

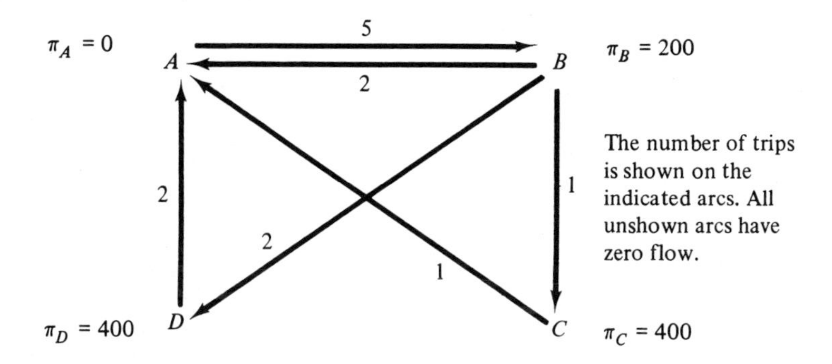


To show optimality, check the arc conditions.

| Arc $ij$ | $\pi_j \overset{?}{\gtreqless} \pi_i + c_{ij}$ | *Associated Condition* |
|---|---|---|
| $AB$ | $200 < 0 + 500$ | $x_{AB} = 5$, upper bound |
| $AC$ | $400 > 0 + 350$ | $x_{AC} = 0$, lower bound |
| $AD$ | $400 > 0 - 400$ | $x_{AD} = 0$, lower bound |
| $BA$ | $0 = 200 - 200$ | $x_{BA} = 2$, basic |
| $BC$ | $400 < 200 + 350$ | $x_{BC} = 1$, upper bound |
| $BD$ | $400 < 200 + 450$ | $x_{BD} = 2$, upper bound |
| $CA$ | $0 = 400 - 400$ | $x_{CA} = 1$, basic |
| $CB$ | $200 > 400 - 300$ | $x_{CB} = 0$, lower bound |
| $CD$ | $400 > 400 - 300$ | $x_{CD} = 0$, lower bound |
| $DA$ | $0 = 400 - 400$ | $x_{DA} = 2$, basic |
| $DB$ | $200 > 400 - 350$ | $x_{DB} = 0$, lower bound |
| $DC$ | $400 > 400 - 300$ | $x_{DC} = 0$, lower bound |

We have shown that the solution satisfies the conditions of optimality; hence the optimum is confirmed. There are three (basic) variables between upper and lower bounds. Since one node constraint is redundant, there should be one less variable at other than lower or upper bounds than the number of nodes (and there are).

In general, the linear programming problem solution will not assure that the solution forms a connected graph. For example, shipments may be required from Los Angeles to San Francisco, and from Philadelphia to New York. The optimal solution will probably consist of two separate, disconnected networks of round trips between the two pairs of cities, and such a solution is not feasible unless there are separate fleets of trucks. Network flow methods have not yet been developed to handle such problems. In many instances, however, the solution does fit, or an approximation is made utilizing the disconnected solutions.

## 9.8 SOLVING THE MAXIMAL VALUE OF FLOW PROBLEM—THE OUT-OF-KILTER METHOD

Thus far we have presented the maximal value of flow problem and stated that every problem considered in this chapter is a special case of that problem. (The formulation of each such problem as a maximal value of flow problem is asked in more detail in the problems at the end of the chapter.) We have not yet considered how to solve the maximal value of flow problem. In this section the out-of-kilter method of Ford and Fulkerson [95], which can be derived from the simplex method, is shown to solve the problem. If the optimality conditions of the previous section are fulfilled by a flow, then the problem has been solved and the arcs of the network are said to be in-kilter. If any of the optimality conditions are violated or if any feasibility conditions ($h_{ij} \leq x_{ij} \leq u_{ij}$) are violated, an arc is out-of-kilter. The out-of-kilter method begins with any balanced flow (a flow which satisfies the node equations), then focuses upon and corrects out-of-kilter arcs, one at a time.

We now state the optimality conditions:

$$\left.\begin{array}{l} 1.\ \pi_j > \pi_i + c_{ij} \Longrightarrow x_{ij} = h_{ij} \\ 2.\ \pi_j < \pi_i + c_{ij} \Longrightarrow x_{ij} = u_{ij} \\ 3.\ h_{ij} < x_{ij} < u_{ij} \Longrightarrow \pi_j = \pi_i + c_{ij} \end{array}\right\} \text{optimality conditions satisfied; arcs in-kilter}$$

There are 6 possible ways in which an arc can be out-of-kilter:

$$\left.\begin{array}{lll} 1.\ \pi_j < \pi_i + c_{ij} & \text{and} & x_{ij} < u_{ij} \\ 2.\ \pi_j > \pi_i + c_{ij} & \text{and} & x_{ij} > h_{ij} \end{array}\right\} \text{optimality conditions violated}$$

$$\left.\begin{array}{lll} 3.\ \pi_j > \pi_i + c_{ij} & \text{and} & x_{ij} < h_{ij} \\ 4.\ \pi_j = \pi_i + c_{ij} & \text{and} & x_{ij} < h_{ij} \\ 5.\ \pi_j < \pi_i + c_{ij} & \text{and} & x_{ij} > u_{ij} \\ 6.\ \pi_j = \pi_i + c_{ij} & \text{and} & x_{ij} > u_{ij} \end{array}\right\} \text{feasibility conditions violated}$$

Optimality conditions one and two also include two feasibility conditions, namely:

for 1: $\pi_j < \pi_i + c_{ij}$ and $x_{ij} < h_{ij}$

for 2: $\pi_j > \pi_i + c_{ij}$ and $x_{ij} > u_{ij}$

If a starting feasible solution to the problem is at hand, only the first two out-of-kilter conditions can occur. If a basic feasible solution is at hand, then it can be shown that the out-of-kilter method corresponds precisely to the simplex method except for determination of the variable to enter the basis. Very often, of course, a feasible solution to a flow problem is not easily obtained, and the initial conditions are selected as a zero flow or some other balanced flow.

To apply the out-of-kilter method, we must first attach to each out-of-kilter condition a measure of how much the arc is "out-of-kilter." The measure used is called the kilter number, and the corresponding calculations for each of the six conditions above are as follows:

$$1.\quad k_{ij} = (\pi_i + c_{ij} - \pi_j)(u_{ij} - x_{ij})$$
$$2.\quad k_{ij} = (\pi_j - \pi_i - c_{ij})(x_{ij} - h_{ij})$$
$$3, 4.\quad k_{ij} = h_{ij} - x_{ij}$$
$$5, 6.\quad k_{ij} = x_{ij} - u_{ij}$$

For conditions 1 and 2 the kilter numbers are equal to the reduced costs, or marginal change in the objective function, times the amount of change in flow necessary to bring the arc into kilter. The kilter numbers for conditions 3 through 6 are the amount by which the appropriate constraint is violated, which is equivalent to the dual reduced cost or the criteria for entry for the dual simplex method (or the criss-cross method dual iteration).

The out-of-kilter method proceeds by choosing from the out-of-kilter arcs the one with the largest kilter number. Suppose that the arc selected goes from node $i$ to node $j$, and that an increase in flow on that arc is desired to make the arc in-kilter. We seek to find a path of positive capacity from node $j$ to $i$ (as in the maximal flow procedure), *without increasing* (possibly decreasing) *any* kilter numbers. (If we wish to decrease the flow on the arc from node $i$ to $j$, we seek an alternate path from node $i$ to $j$.) An attempt is made to increase or decrease the flow on an out-of-kilter arc by finding a loop that is not unprofitable which contains the arc and on which flow can be altered in the desired direction. This is done using a node labeling procedure in which arcs to unlabeled nodes are added if they are not unprofitable; then the node incident to each added arc is labeled. If a loop is found, a *breakthrough* is said to occur, and the flow is altered. If a loop is not found, a nonbreakthrough has occurred; then the node values (the $\pi$'s) of certain nodes (the ones not yet labeled) are decreased just enough to permit the labeling of at least one more node. Decreasing the node values in this way assures that all in-kilter arcs remain in-kilter, and the kilter numbers of one or more out-of-kilter arcs are decreased. The kilter number of the out-of-kilter arc on which attention is focused is decreased at this stage. In the event of a nonbreakthrough once the dual variables are altered, either the original arc becomes in-kilter or at least one new node can be labeled; otherwise there is no feasible solution to the problem. Once all arcs are in-kilter, that solution is optimal.

A detailed flow chart of the procedure is given in Figure 9.12.

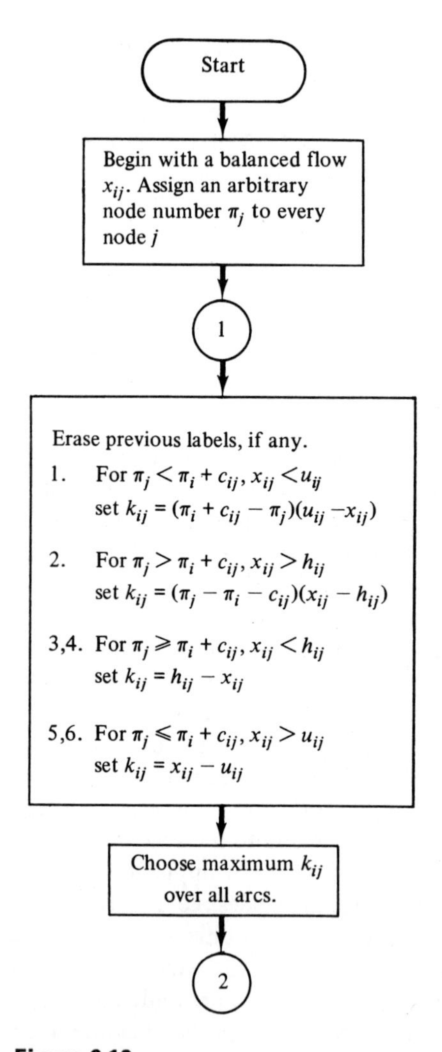


Initialize

Find a balanced flow so that the net accumulation at any node is zero.

Erase all previous labels.
Establish kilter number $k_{ij}$ for out-of-kilter arcs. In-kilter arcs have zero kilter numbers.
$c_{ij}$ = value of unit flow along arc $ij$
On arc $ij$ for out-of-kilter node type indicates we want to do as follows:

Type 1, increase flow to $u_{ij}$;

Type 2, decrease flow to $h_{ij}$;

Type 3, increase flow to $h_{ij}$;

Type 4, increase flow to at least $h_{ij}$ but not more than $u_{ij}$;

Type 5, decrease flow to $u_{ij}$;

Type 6, decrease flow to at most $u_{ij}$ but not less than $h_{ij}$.

Determine arc with highest kilter number.

**Figure 9.12**

Flow Chart of the Out-of-Kilter Procedure

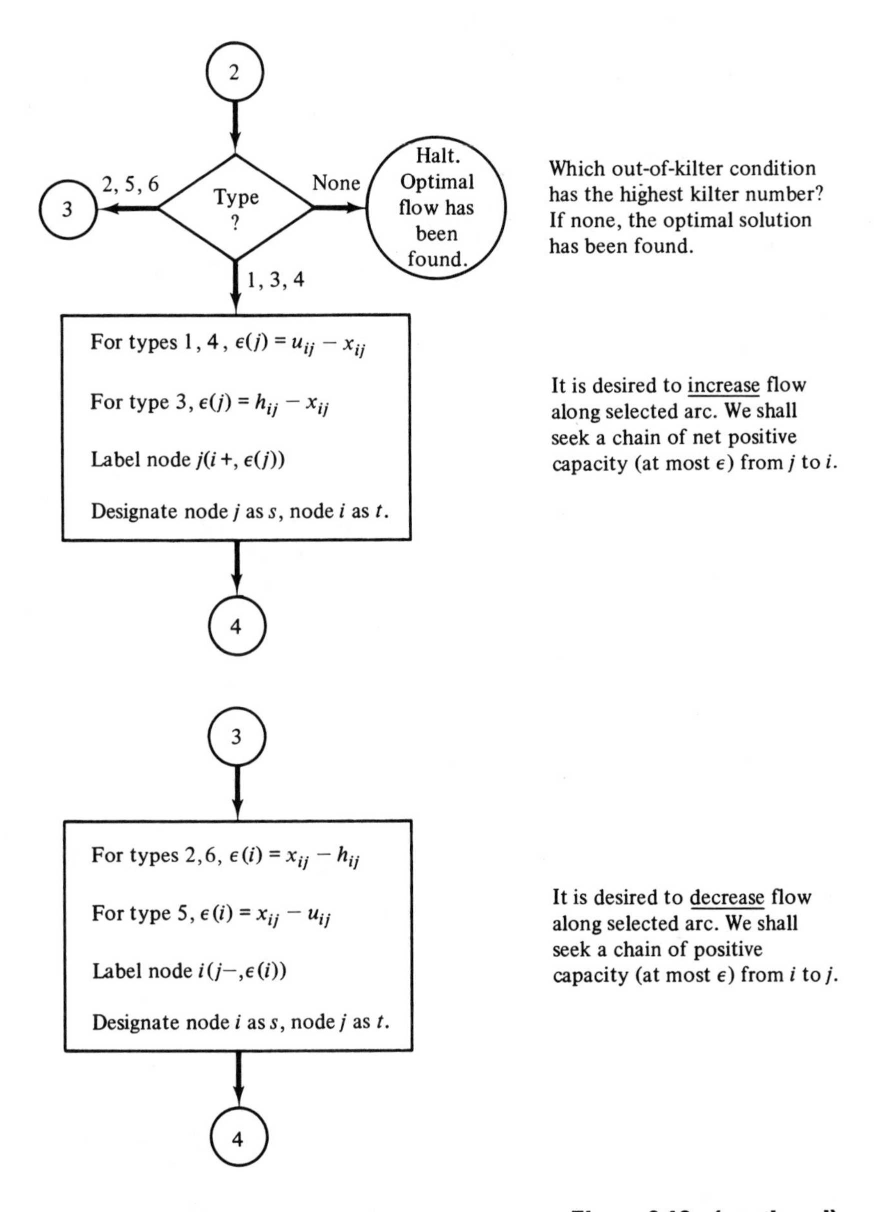


**Figure 9.12 (continued)**

We shall outline a proof that the out-of-kilter procedure converges to an optimum, if one exists, or that otherwise it determines that no feasible solution exists.

Following the out-of-kilter procedure using integral (or rational) maximum and minimum flows on arcs leads to an optimum if one exists. Whenever

For any node $i$ labeled but not scanned and adjacent unlabeled node $j$

1. For $\pi_j \leqslant \pi_i + c_{ij}$ and $x_{ij} < u_{ij}$ label $j$ $(i+, \epsilon(j))$ where $\epsilon(j) = \text{Min}\ \{\epsilon(i), u_{ij} - x_{ij}\}$
2. For $\pi_j > \pi_i + c_{ij}$ and $x_{ij} < h_{ij}$ label $j(i+, \epsilon(j))$ where $\epsilon(j) = \text{Min}\ \{\epsilon(i), h_{ij} - x_{ij}\}$
3. For $\pi_i \geqslant \pi_j + c_{ji}$ and $x_{ji} > h_{ji}$ label $j(i-, \epsilon(j))$ where $\epsilon(j) = \text{Min}\ \{\epsilon(i), x_{ji} - h_{ji}\}$
4. For $\pi_i < \pi_j + c_{ji}$, and $x_{ji} > u_{ji}$ label $j$ $(i -, \epsilon(j))$ where $\epsilon(j) = \text{Min}\ \{\epsilon(i), x_{ji} - u_{ji}\}$

Node $i$ has now been scanned.

Scan a labeled but previously unscanned arc.

Label any adjacent unlabeled nodes that can be labeled so as to increase or decrease flow without increasing and possibly decreasing kilter numbers. Specifically, test 1 indicates that we can increase the flow along out-of-kilter arcs type 1 and 4 and in-kilter arcs of type 3.

Test 2 indicates that we can increase the flow along out-of-kilter arcs of type 3.

Test 3 indicates that we can decrease the flow along out-of-kilter arcs of type 2 and 6 and in-kilter arcs of type 3.

Test 4 indicates that we can decrease the flow along out-of-kilter arcs of type 5.

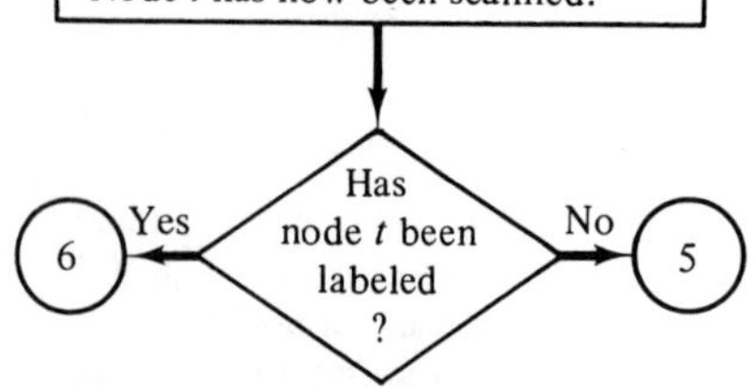


Do we have a breakthrough?

**Figure 9.12 (continued)**

a breakthrough is achieved, at least one kilter number is decreased by a positive integer. In the case of a nonbreakthrough, at least one node is labeled for each nonbreakthrough, and hence there is a finite upper limit on how many nonbreakthroughs can occur between breakthroughs, and on how many breakthroughs can possibly occur.

If a nonbreakthrough cannot be resolved by changing node numbers, then no feasible solution exists. That the converse is true can be demonstrated by using the finiteness of the procedure to show that an arc which cannot be made feasible will eventually have the highest kilter number. Since it is not possible to drive that kilter number to zero, then some unlabeled nodes will have their node numbers decremented to $-\infty$, which indicates an infeasible condition. An unbounded condition can only occur if a loop or cycle of the network has

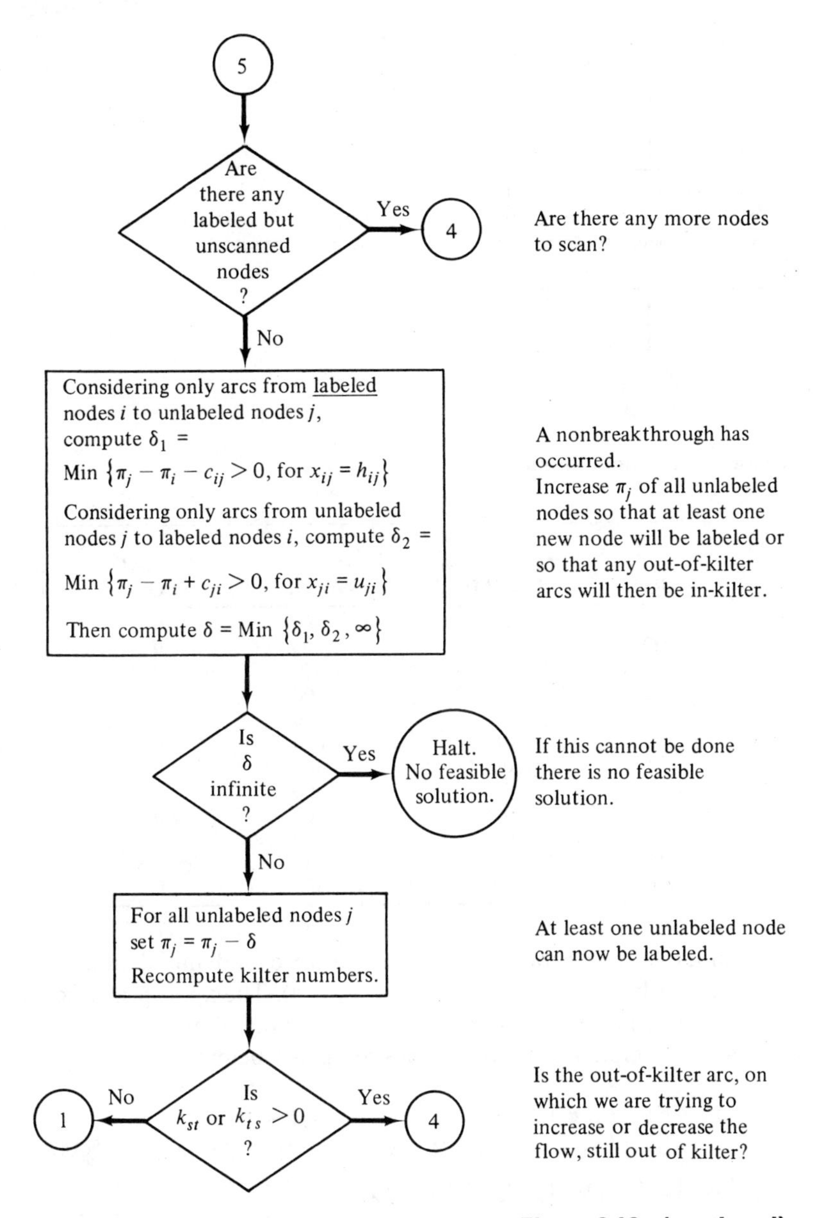


**Figure 9.12 (continued)**

a positive objective function value (i.e., the loop is profitable), and *no arc* of the loop has a finite capacity.

EXAMPLE 9.6:

Solve Example 9.5 using the out-of-kilter method. The arcs and their profit values are given below (ignore the last two columns for the moment).

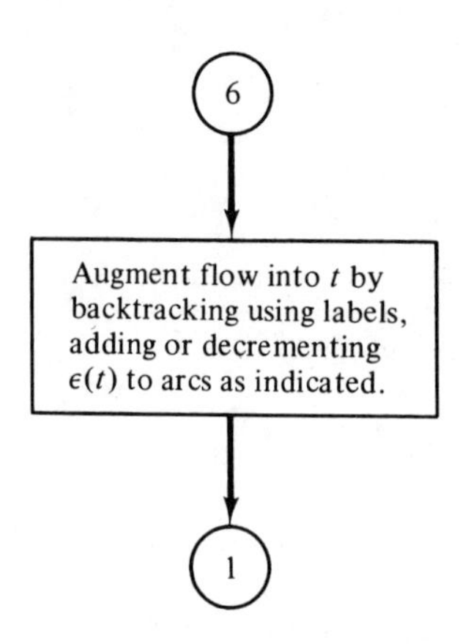


Augment flow as determined by breakthrough.

**Figure 9.12 (concluded)**

| ARC | MINIMUM/ MAXIMUM | VALUE | INITIAL FLOW | INITIAL KILTER NUMBER |
|---|---|---|---|---|
| $AB$ | 2/5 | 500 | 0 | 2500 |
| $AC$ | 0/10 | 350 | 0 | 3500 |
| $AD$ | 0/∞ | −400 | 0 | 0 |
| $BA$ | 0/∞ | −200 | 0 | 0 |
| $BC$ | 0/1 | 350 | 0 | 350 |
| $BD$ | 0/2 | 450 | 0 | 900 |
| $CA$ | 0/∞ | −400 | 0 | 0 |
| $CB$ | 0/∞ | −300 | 0 | 0 |
| $CD$ | 0/∞ | −300 | 0 | 0 |
| $DA$ | 0/∞ | 400 | 0 | 0 |
| $DB$ | 0/∞ | −350 | 0 | 0 |
| $DC$ | 0/∞ | −300 | 0 | 0 |

Initial flow is arbitrarily selected as zero and initial node numbers are arbitrarily set as $\pi_A = \pi_B = \pi_C = \pi_D = 0$. The appropriate flow and kilter numbers are given in the last two columns above.
A few sample kilter calculations are given below:

Arc $AB$: Type 1 $\pi_B < \pi_A + c_{AB}$, $x_{AB} < u_{AB}$;
$$k_{AB} = (\pi_A + c_{AB} - \pi_B)(u_{AB} - x_{AB}) = 500(5) = 2500$$

Arc $AC$: Type 1 $\pi_C < \pi_A + c_{AC}$, $x_{AC} < u_{AC}$;
therefore $k_{AC} = 350(10) = 3500$

Arc $AD$: In-kilter, $\pi_D > \pi_A + c_{AD}$, $x_{AD} = 0$, lower bound

All arcs out-of-kilter for this example are of type 1. Note that if we always start with zero flow and the initial $\pi$'s are zero, all out-of-kilter arcs are of type 1. We select the out-of-kilter arc having the maximum kilter number, in this case arc $AC$. Then we label node $C\,(A+, 10)$ and attempt to increase

flow from $A$ to $C$. Although the three arcs emanating from $C$ are in-kilter, they are all in optimality condition number one, and the flow along all three arcs must remain at their lower bound, zero. Since no node can be labeled from node $C$, we have a nonbreakthrough.
To resolve the nonbreakthrough we consider all arcs from labeled nodes $i$ to unlabeled nodes $j$, in this case arcs $CA$, $CB$, $CD$. Since $x_{ij} = h_{ij}$ in each case, we compute $\pi_j - \pi_i - c_{ij}$ and choose the minimum.

$$\begin{aligned} CA&: \quad \pi_A - \pi_C - c_{CA} = 400 \\ CB&: \quad \pi_B - \pi_C - c_{CB} = 300 \\ CD&: \quad \pi_D - \pi_C - c_{CD} = 300 \end{aligned}$$

Set $\delta = 300$; hence $\pi_A = \pi_B = \pi_D = -300$, and $\pi_C = 0$.
The kilter numbers that change and their new values are:

$$\begin{aligned} k_{AC} &= 500 \\ k_{BC} &= 50 \end{aligned}$$

Since $k_{AC} > 0$, we continue. We had earlier labeled $C\ (A +, 10)$. Now we label $B\ (C +, 10)$ and $D\ (C +, 10)$, since for arcs $CB$ and $CD$, $\pi_B = \pi_C + c_{CB}$ and $\pi_D = \pi_C + c_{CD}$. No other nodes can be labeled; hence we again have a nonbreakthrough. We now consider arcs $BA$, $CA$, and $DA$.

$$\begin{aligned} BA&: \quad \pi_A - \pi_B - c_{BA} = 200 \\ CA&: \quad \pi_A - \pi_C - c_{CA} = 100 \\ DA&: \quad \pi_A - \pi_D - c_{DA} = 400 \end{aligned}$$

Set $\delta = 100$; hence $\pi_A = -400$, $\pi_B = \pi_D = -300$, and $\pi_C = 0$.
The kilter numbers that change and their new values are:

$$\begin{aligned} k_{AB} &= 2000 \\ k_{AC} &= 0 \end{aligned}$$

Since $k_{AC}$ is now 0, we erase all previous labels. Then we choose another arc, $AB$, on which to increase flow. We label node $B\ (A +, 5)$. We can now label $C\ (B +, 1)$, $D\ (B +, 2)$, and then label node $A\ (C +, 1)$, which is a breakthrough. We therefore increase flow 1 unit from $A$ to $B$, $B$ to $C$, and $C$ to $A$. Thus our current solution is as follows:

$$x_{AB} = 1, \quad x_{BC} = 1, \quad x_{CA} = 1$$

and the out-of-kilter arcs are as follows (arc $BC$ is now in-kilter):

$$\begin{aligned} k_{AB} &= 1600 \\ k_{BD} &= 900 \end{aligned}$$

We erase old labels, then select arc $AB$ and label $B(A+, 4)$. We then label $D\,(B+, 2)$. No further nodes can be labeled and we have a nonbreakthrough. Hence, we consider arcs $BA$, $DA$, and $DC$.

$$BA: \quad \pi_A - \pi_B - c_{BA} = 100$$
$$DA: \quad \pi_A - \pi_D - c_{DA} = 300$$
$$DC: \quad \pi_C - \pi_D - c_{DC} = 600$$

Therefore, we set $\delta = 100$; hence $\pi_A = -500$, $\pi_B = \pi_D = -300$, and $\pi_C = -100$. The kilter number that changes and its new value is $k_{AB} = 1200$. (Because $AB$ is still out-of-kilter, we do not erase any labels.) We can now label $A\ (B+, 4)$ which is a breakthrough. We therefore increase flow 4 units from $A$ to $B$ and from $B$ to $A$.

$$x_{AB} = 5$$
$$x_{BA} = 4$$
$$x_{BC} = 1$$
$$x_{CA} = 1$$

and our only out-of-kilter arc $BD$ has kilter number $k_{BD} = 900$. Hence we select arc $BD$ and label $D\,(B+, 2)$. No other nodes can be labeled and we have a nonbreakthrough. We consider arcs $DA$, $DB$, and $DC$.

$$DA: \quad \pi_A - \pi_D - c_{DA} = 200$$
$$DB: \quad \pi_B - \pi_D - c_{DB} = 350$$
$$DC: \quad \pi_C - \pi_D - c_{DC} = 500$$

Therefore, $\delta = 200$, and $\pi_A = -700$, $\pi_B = -500$, and $\pi_C = \pi_D = -300$. Now $k_{BD} = 500$, and we continue by labeling $A\ (D+, 2)$ and $B\ (A-, 2)$ which is a breakthrough. ($B$ is labeled using test 3 in the block following ④ in the flow chart of Figure 9.12.) We therefore increase flow 2 units from $B$ to $D$ and from $D$ to $A$, and decrease flow 2 units from $B$ to $A$. At this point all kilter numbers are zero. The solution is as follows:

$$x_{AB} = 5$$
$$x_{BA} = 2$$
$$x_{BC} = 1$$
$$x_{BD} = 2$$
$$x_{CA} = 1$$
$$x_{DA} = 2$$

which is the optimal (the value of the flow is 2150) and is the same as in the previous section. If we add 700 to every node number $\pi_k$ (which does not alter any kilter condition), we obtain the same node numbers as given in the solution to Example 9.5.

## 9.9 OTHER ASPECTS OF THE MAXIMAL VALUE OF FLOW PROBLEM

In this section we shall consider certain miscellaneous aspects of the maximal value of flow problem. One frequent restriction of network flow problems is that some nodes have capacity limitations. Such a restriction can be handled easily by partitioning the node into two nodes—an entry node and an exit node with an arc between them which has the desired capacity on it. The representation is as given in Figure 9.13.

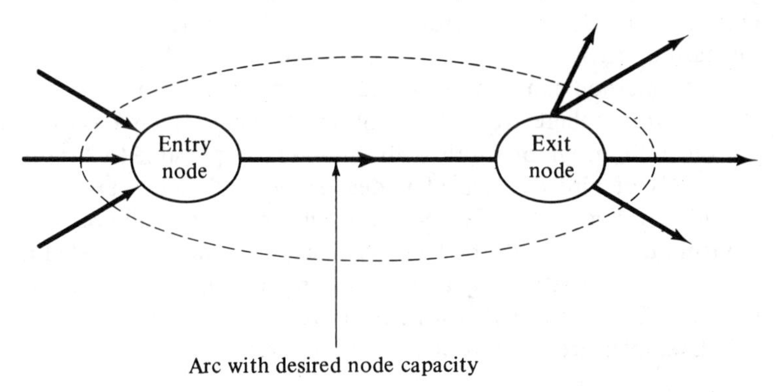


**Figure 9.13**
Representation of a Node with Capacity

In solving network flow problems it is desirable to begin with a feasible solution if it is easily obtainable—if possible, a basic feasible solution. This is most obvious for the transportation problem: the standard method presented in section 9.2 begins with a basic primal feasible solution, while the primal-dual method of section 9.3 begins with a generated dual feasible solution. Otherwise, both methods are essentially the same as the out-of-kilter method. To appreciate the value of such feasible starting solutions, it is suggested that the reader solve a transportation problem in each of the three ways.

An important network flow problem involves multicommodities shipped on the same network in which shipments of different goods do not cancel each other out. Such problems have been studied in some detail, and publications describing some of them include Ford and Fulkerson [95] and Hu [158].

## 9.10 THE SHORTEST PATH BETWEEN TWO NODES OF A NETWORK

An important problem is to find the shortest path between two nodes of a network. The shortest path problem can, of course, represent times or costs instead of distances. It can be formulated and solved as a linear program or as an out-of-kilter problem requiring exactly one unit of flow between the two nodes and minimizing the cost of flow. A method given by Dantzig [62] can be derived from the out-of-kilter method, assuming that all arcs have nonnegative costs. First the source is labeled zero. Then label the unlabeled node for which the sum of an adjacent node label plus the cost along the arc from that node is minimal. Repeat the latter step until the sink node has been labeled. Then the shortest path is found by backtracking along a unique path from the source. A detailed flow chart is found in Figure 9.14. The procedure is readily extended to finding the shortest routes to all nodes from a source, and to finding the shortest routes from all nodes to a sink. Proof of the procedure follows from consideration of shortest routes from the source to different nodes in monotonically increasing order.

The method can also be extended to the case in which some costs are negative, providing there are no loops of negative cost (such a loop would be analogous to a "fountain of youth"). In this case it is possible to "improve" or reduce the labels of already labeled nodes. Hence, all nodes except the source are initially given $\pi$ values of $\infty$. Starting with the source, all nodes on arcs emanating from the source may be labeled. Then, each newly relabeled node should be scanned to see whether any arcs emanating from it can be relabeled. The process is repeated until all relabeled nodes have been scanned.

Examples are given to illustrate the methods.

EXAMPLE 9.7:

Determine the shortest route from $a$ to $e$ in Figure 9.15, ignoring the dashed path from $c$ to $b$.

1. Label $a(0, -)$.
2. Label $b(3, a)$.
3. Label $c(5, a)$.
4. Label $e(7, b)$.

The route is to $e$ from $i(e) = b$ from $i(c) = a$ at a cost of 7.

EXAMPLE 9.8:

Determine the shortest route from $a$ to $e$ in Figure 9.15, replacing the arc from $c$ to $b$ by the dashed arc.

Initial labels: $a(0, -)$; $b(\infty, -)$; $c(\infty, -)$; $d(\infty, -)$; $e(\infty, -)$.

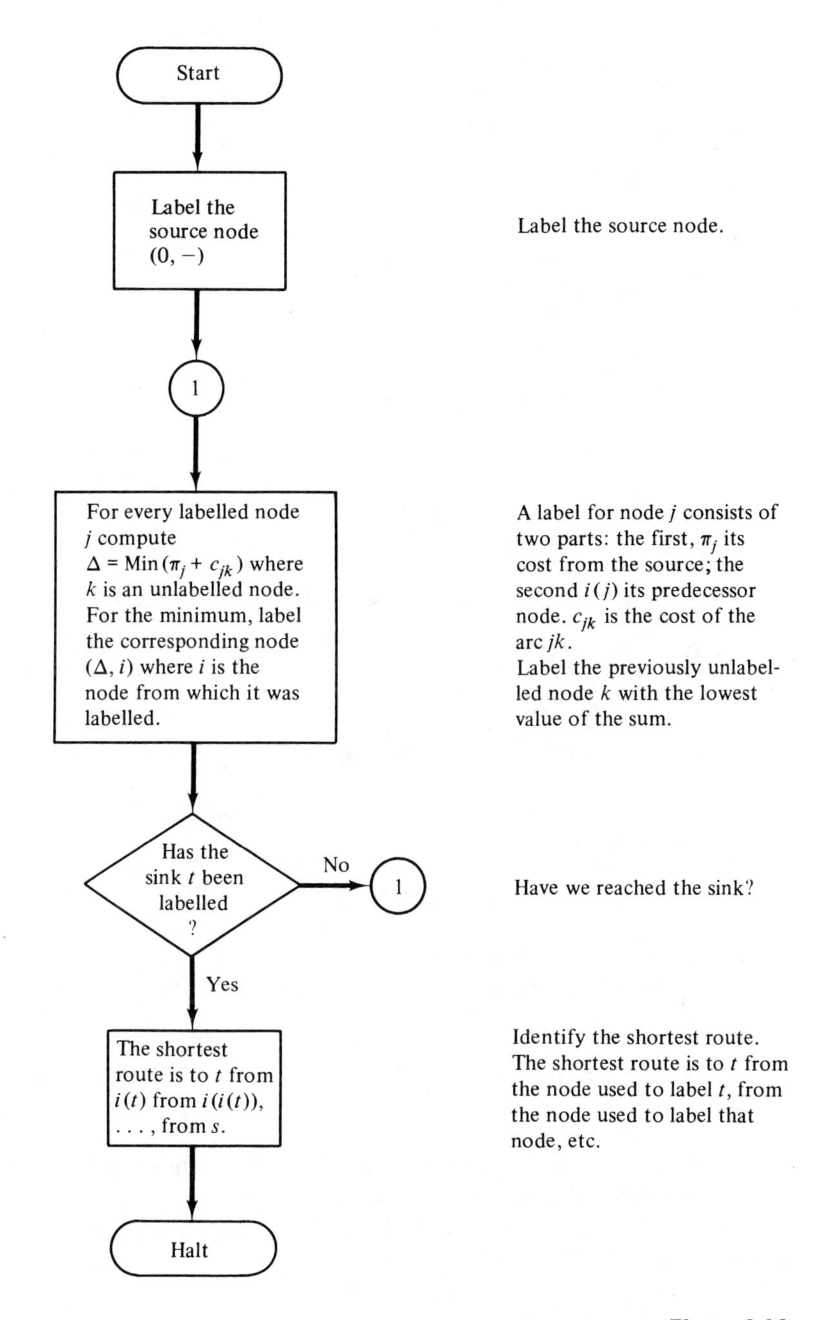


**Figure 9.14**

Flow Chart of a Shortest Route Algorithm

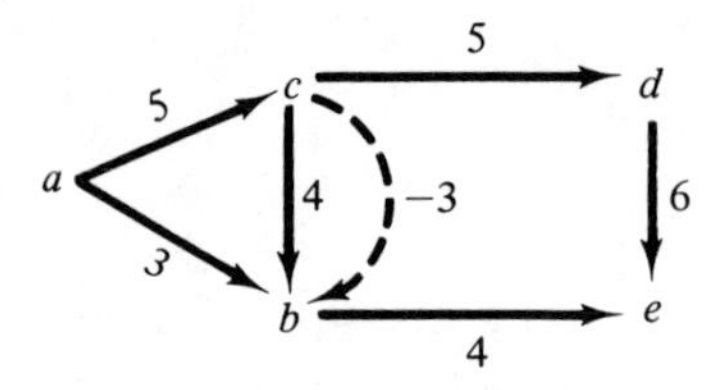


**Figure 9.15**

Network of Examples 9.7 and 9.8

1. Nodes yet to scan: $a$.
   Relabel $b$ and $c$ from $a$:

   Labels: $a(0, -)$; $b(3, a)$; $c(5, a)$; $d(\infty, -)$; $e(\infty, -)$

2. Nodes yet to scan: $b$, $c$.
   Relabel $e$ from $b$:

   Labels: $a(0, -)$; $b(3, a)$; $c(5, a)$; $d(\infty, -)$; $e(7, b)$

3. Nodes yet to scan: $c$, $e$.
   Relabel $b$ and $d$ from $c$:

   Labels: $a(0, -)$; $b(2, c)$; $c(5, a)$; $d(10, c)$; $e(7, b)$

4. Nodes yet to scan: $b$, $d$, $e$.
   Relabel $e$ from $b$:

   Labels: $a(0, -)$; $b(2, c)$; $c(5, a)$; $d(10, c)$; $e(6, b)$

5. Nodes yet to scan: $d$, $e$.
   Scanning produces no improvement; hence we scan the nodes and then stop.

The shortest route, of length 6, is to $e$ from $i(e) = b$ from $i(b) = c$ from $i(c) = a$.

Shortest route methods have been refined to a high degree of efficiency. For further information about some of the efficient schemes, the reader is referred to Hu [158] and Farbey, Land, and Murchland [90], among other works in the field.

**Selected Supplemental References**

Section 9.3
[215]

Section 9.4
[12], [23], [51], [52], [53], [83], [99], [139], [186], [204], [243], [272], [273]
Section 9.5
[157], [216]
Sections 9.6, 9.7, 9.8
[160], [175], [230]
Section 9.9
[58], [236], [237], [238], [239], [240], [283]
Section 9.10
[80], [90], [93], [159], [162], [212], [246], [249], [305], [306]
General
[3], [6], [14], [22], [54], [76], [86], [95], [109], [113], [156], [158], [165], [166], [169], [197], [202], [234], [254]

## 9.11 PROBLEMS

**1.** Prove that all transportation bases are triangular. (Hint: Use the compact representation of Fig. 9.1 and argue that there must be at least one row or column with only one basic cell. Delete that row or column, and proceed inductively.)

**2.** Prove that a loop cannot exist in a transportation problem basis. (Hint: Show that a "basis" corresponding to a loop does not possess an inverse since its columns are not linearly independent.)

**3.** Prove that every transportation problem inverse has only entries of 1, $-1$, and 0.

**4.** Show why no basis of a transportation problem can have a complete loop among basic variables using rook's moves. That is, there cannot be shipments from points $i$ to $j$, $k$ to $j$, $k$ to $m$, . . . , $q$ to $p$, $i$ to $p$ in a basic solution. (Hint: Consider the linear independence of these vectors in the original framework (9.2)).

**5.** Show that, for the transportation method, the direction of the rook's move does not affect the location of the $+$ signs and $-$ signs.

**6.** A method proposed by Charnes and Cooper [47] to solve the transportation problem is called the stepping-stone method. It is identical to the one given in the chapter except for the evaluation of nonbasic cells. In the stepping-stone method a cell is evaluated by being treated as if it were to enter the basis. Attach to the pluses and minuses the associated costs and sum them. A negative sum indicates that a reduction in cost may be achieved by introducing the variable into the basis. Show that the evaluation is equivalent to the $c_{ij} - u_i - v_j$ presented here.

**7.** Develop a method for starting the transportation problem solution with a feasible nonbasic solution. (Hint: Use a method which designates any nonzero cell in a loop to be decreased or increased.)

**8.** Solve Example 9.2 using the Vogel approximation method for a start.

**9.** Solve Example 9.2 using a Northeast corner method to start.

**10.** Show that a Southeast corner method starting solution for a transportation problem is identical to the Northwest corner method.

**11.** Prove that the epsilon method introduced at the end of section 9.2 assures nondegeneracy in the transportation method.

**12.** Show why any one dual variable of the transportation problem can be set to some arbitrary value.

**13.** Develop a method for treating upper and lower bounds implicitly in transportation problems in the framework of the transportation method. The problem is called the capacitated transportation problem. (Hint: Use the out-of-kilter method to help you.)

**14.** Explain the reasoning underlying Vogel's approximation method.

**15.** Solve the following transportation problems by the transportation method, and by the primal-dual method. (Cost entries are indicated in the tables.*)

a.

| | | | | | | Supply |
|---|---|---|---|---|---|---|
| | 4 | 2 | 4 | 2 | 3 | 16 |
| | 4 | 2 | 5 | 5 | 4 | 75 |
| | 11 | 2 | 1 | 2 | 3 | 57 |
| | 5 | 4 | 3 | 3 | 2 | 9 |
| Demand | 48 | 50 | 10 | 38 | 11 | |

b.

| | | | | | | | Supply |
|---|---|---|---|---|---|---|---|
| | 5 | 3 | 5 | 3 | 2 | 5 | 76 |
| | 3 | 2 | 5 | 3 | 2 | 3 | 38 |
| | 3 | 3 | 2 | 2 | 1 | 5 | 27 |
| Demand | 29 | 15 | 26 | 35 | 12 | 24 | |

c.

| | | | | Supply |
|---|---|---|---|---|
| | 9 | 4 | 9 | 6 |
| | 6 | 7 | 8 | 26 |
| | 8 | 5 | 5 | 24 |
| | 3 | 4 | 3 | 9 |
| Demand | 21 | 32 | 12 | |

* Note: The sum of supplies does not necessarily equal the sum of demands.

d.

| | | | | Supply |
|---|---|---|---|---|
| 30 | 14 | 40 | 18 | 100 |
| 20 | 35 | 25 | 30 | 136 |
| 50 | 10 | 30 | 22 | 90 |
| 25 | 20 | 8 | 28 | 154 |
| 10 | 40 | 15 | 35 | 70 |
| Demand 85 | 125 | 190 | 95 | |

e. (Degenerate)

| | | | | Supply |
|---|---|---|---|---|
| 8 | 8 | 5 | 2 | 50 |
| 10 | 5 | 10 | 10 | 80 |
| 7 | 10 | 10 | 8 | 150 |
| 5 | 10 | 10 | 6 | 90 |
| Demand 130 | 90 | 50 | 100 | |

**16.** A woodworking shop has the following orders for cabinets:

| | CABINETS | | |
|---|---|---|---|
| MONTH | *A* | *B* | *C* |
| January | 12 | 22 | 7 |
| February | 22 | 22 | 13 |
| March | 22 | 33 | 27 |

Cabinet *A* requires 6 hours to make, cabinet *B* requires 8 hours to make, and cabinet *C* requires 5 hours to make. There are two workers employed in the shop, and together they work 320 hours per month. In addition, they can work an additional 100 hours per month on overtime if necessary. Contributions to overhead and profits for each cabinet (exclusive of labor and materials, and depending on how long in advance the cabinets are made and must be stored) are given in the table below.

| | *A* | *B* | *C* |
|---|---|---|---|
| Made during regular time for shipment same month | $40 | $50 | $75 |
| Made during overtime for shipment same month | 31 | 38 | 60 |
| Made during regular time for shipment next month | 35 | 45 | 70 |
| Made during overtime for shipment next month | 26 | 33 | 55 |
| Made during regular time for shipment two months hence | 32 | 42 | 67 |
| Made during overtime for shipment two months hence | 23 | 30 | 52 |

Formulate and solve the problem as a transportation problem. (Hint: The problem will be in the proper form except for the values of the coefficients.

Scale the problem, i.e., multiply rows and columns by nonzero constants so that all coefficients are +1.)

**17.** Associated Industries has six different packaging lines. Each can be used to package a product, but the cost of packaging varies according to the product line combination. For the next week, six products are to be packaged, one per line, and the relative variable costs per hour per product line combination are given below. (All lines operate at the same rate, but labor costs vary.) What is the product line combination that minimizes variable hourly costs?

Costs in dollars per hour per combination

| | PRODUCT | | | | | |
|---|---|---|---|---|---|---|
| LINE | *A* | *B* | *C* | *D* | *E* | *F* |
| 1 | 88 | 61 | 84 | 84 | 82 | 93 |
| 2 | 85 | 61 | 58 | 90 | 64 | 52 |
| 3 | 97 | 64 | 88 | 83 | 79 | 96 |
| 4 | 56 | 90 | 90 | 61 | 71 | 88 |
| 5 | 63 | 65 | 63 | 96 | 69 | 73 |
| 6 | 70 | 56 | 93 | 54 | 58 | 95 |

**18.** Derive the primal-dual transportation algorithm from the Dantzig-Ford-Fulkerson primal-dual linear programming algorithm.

**19.** Solve the problem of Example 9.2 using the primal-dual method.

**20.** The production planning supervisor of the Rush Engineering Works, Inc., has to schedule the following four jobs for the quarter ending next March on the three grinding machines in the machine shop:

| JOB NO. | NO. OF PIECES | EQUIVALENT STANDARD MACHINE HOURS |
|---|---|---|
| A | 600 | 450 |
| B | 500 | 500 |
| C | 450 | 600 |
| D | 400 | 800 |

At present the machine shop works one shift of eight hours, five days a week, and the capacity of the three grinding machines, after taking efficiency into account (efficiency depends on age of machine and skill of operator), is as follows:

| GRINDING MACHINE | ACTUAL HOURS FOR THE QUARTER | CAPACITY IN STANDARD HOURS |
|---|---|---|
| 1 | 480 | 450 |
| 2 | 480 | 480 |
| 3 | 480 | 550 |

Since grinding machine 3 requires a specially trained skilled operator, it is not possible to work more than one shift on that machine. However, it is possible to work overtime on that machine up to a maximum of 25 per cent capacity. It is possible to work *one extra shift* on either of the other two machines.

The standard variable costs of grinding are as follows:

| | Grinding Machine | | | | | |
|---|---|---|---|---|---|---|
| | 1 | | 2 | | 3 | |
| JOB NO. | NORMAL | O.T. | NORMAL | O.T. | NORMAL | O.T. |
| A | $0.80 | $0.85 | $0.90 | $1.00 | $1.10 | $1.15 |
| B | 0.90 | 1.00 | 0.95 | 1.05 | 1.05 | 1.15 |
| C | 1.00 | 1.10 | 1.00 | 1.10 | 1.05 | 1.20 |
| D | 1.20 | 1.35 | 1.25 | 1.40 | 1.30 | 1.50 |

Formulate the above problem as a transportation problem and solve it.

**21.** Prove that the graph associated with a transportation problem basis is a tree. (Hint: That it has no loops is asked in problem 2. Hence, prove the graph is connected.)

**22.** Prove that a tree having $n$ nodes has $n - 1$ arcs. (Hint: Use induction.)

**23.** Prove that a tree has at least two ends.

**24.** Set up and solve the problem of section 9.6 using an arc-chain formulation.

**25.** Show that the maximal flow problem is a special case of the maximum value of flow problem. (Hint: All arcs have zero value except for the arc from *sink* to *source*, which has a value of 1.)

**26.** Find the maximal flow from $s$ to $t$ in the following networks, treating the labels on the arcs as capacities.

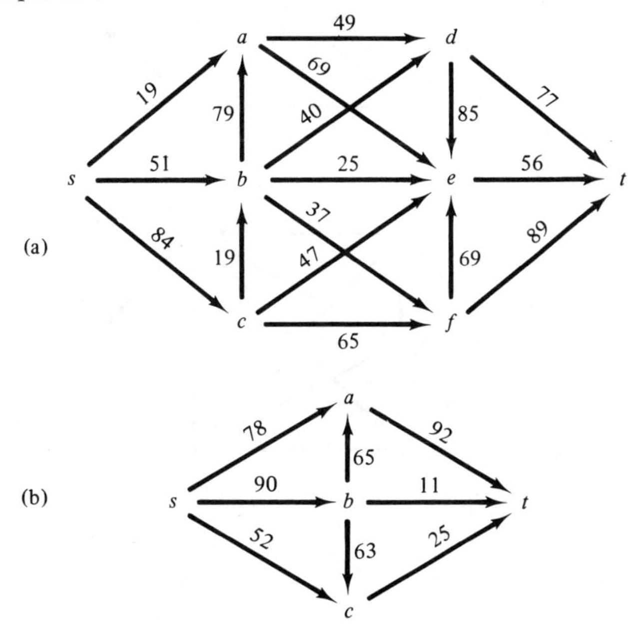

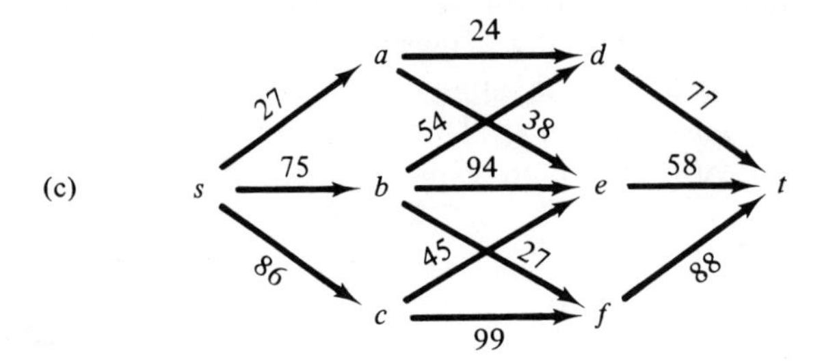

(c)
s
a
b
c
d
e
f
t
27
75
86
24
54
38
94
45
27
99
77
58
88


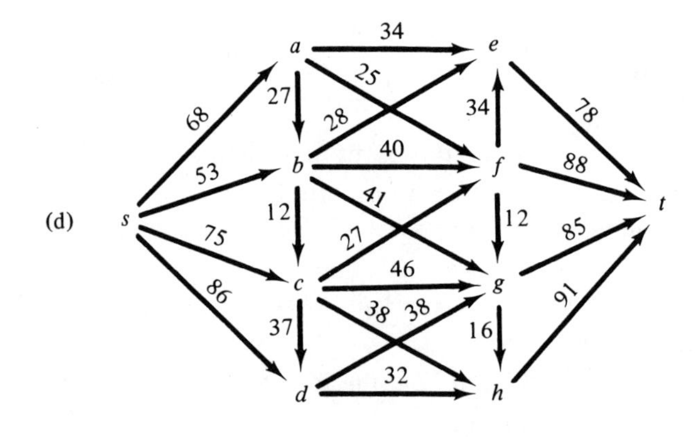

(d)
s
a
b
c
d
e
f
g
h
t
68
53
75
86
34
25
27
28
40
41
12
27
46
38
38
37
32
34
12
16
78
88
85
91


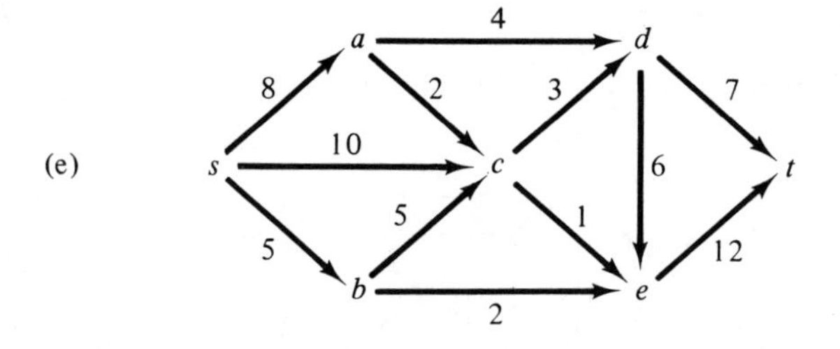

(e)
s
a
b
c
d
e
t
8
10
5
4
2
5
2
3
1
6
7
12


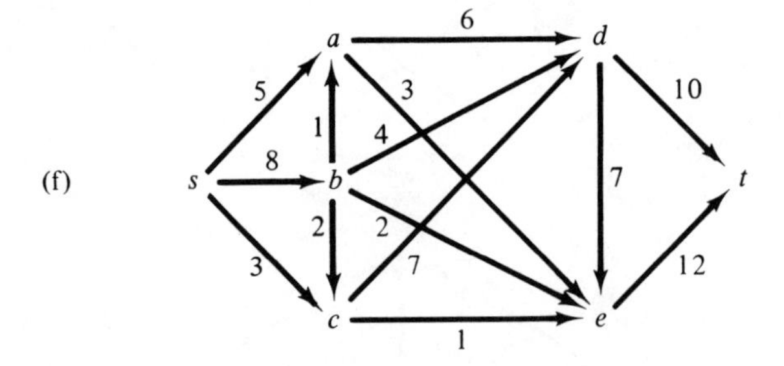

(f)
s
a
b
c
d
e
t
5
8
3
6
1
3
4
2
2
7
1
7
10
12

**27.** Show that the transportation method is a special case of the out-of-kilter method, given a starting primal feasible basic solution.

**28.** Show that the primal-dual transportation method is a special case of the out-of-kilter method, given a starting dual feasible solution.

**29.** Show that the method for solving the maximal flow problem is a special case of the out-of-kilter method.

**30.** The Eastern Steamship Company has a contract to haul the following commodities on the Great Lakes:

5 loads of coal from Conneaut, Ohio, to Port Huron, Michigan
8 loads of limestone from Bay City, Michigan, to Cleveland, Ohio
10 loads of iron ore from Duluth, Minnesota, to Cleveland, Ohio

The variable costs of each trip are as follows (assume that empty trips cost the same as loaded trips):

| | Bay City | Cleveland | Conneaut | Duluth | Port Huron |
|---|---|---|---|---|---|
| Bay City | 0 | $15,000 | $15,000 | $20,000 | $ 6,000 |
| Cleveland | $15,000 | 0 | 500 | 35,000 | 12,000 |
| Conneaut | 15,000 | 500 | 0 | 35,000 | 12,000 |
| Duluth | 20,000 | 35,000 | 35,000 | 0 | 25,000 |
| Port Huron | 6,000 | 12,000 | 12,000 | 25,000 | 0 |

a. What are the minimum cost routings for the steamship to make?
b. Discuss the implications for the company's bidding strategy with respect to the possibility of hauling goods on routes (or nearby routes) that it would otherwise travel empty.

**31.** Find the shortest route from $s$ to $t$ for each of the problems of 26, treating the labels on the arcs as distances.

**32.** Develop a method for determining the longest path between two points in a network. This problem is equivalent to finding the critical path in a network of jobs where the arcs correspond to jobs which must be carried out. Further, no job can be started until all predecessor jobs have been completed. The labels on the arcs are the time necessary to complete the job. (Hint: The method is very much like the shortest route method, except for two changes in labeling procedure:

a. *All* predecessor nodes must be labeled before a node is labeled.
b. The *largest* number is used to label the node.)

Critical path scheduling methods are used widely in scheduling and coordinating such complex projects as large construction jobs.

**33.** Treating the entries along the arcs as time required for job completion, and further stipulating that all jobs terminating at a node must be complete before any jobs originating at that node are begun, find the critical path for each of the networks in problem 26, using the method developed in problem 32.

**34.** Show how the shortest route algorithm can be derived from the linear programming formulation. (Hint: Formulate the problem of finding the minimum value of a unit of flow from source to sink.)

**35.** Formulate the shortest route problem as a maximal value of flow problem.

# 10

# Game Theory and Related Topics

## 10.1 INTRODUCTION

In their excellent *Games and Decisions* [203], Luce and Raiffa summarize a definition of game theory as "a model for situations of conflict among several people, in which two principal modes of resolution are collusion and conciliation." The principal element of games as we shall consider them here is conflict. Examples of such games are:

1. Parlor games such as matching coins, tick-tack-toe, checkers, chess, and bridge.
2. Economic conflict or competition.
3. Military conflict or war.

Generally, in game situations there are one or more "persons" (each team is considered a "person"; e.g., in the card game of bridge there are four players but only two "persons"), and different persons have different objectives. Each person makes a decision (also called action or strategy) concerning a situation that he faces, in which his ultimate intention is to obtain the "best possible" outcome for himself in some sense. That decision could be the only decision of the game or one in a network of decisions. Such decisions might include:

1. The opening move in chess.
2. Where to put the first "X" in tick-tack-toe.

3. Whether or not (and if so, how much) a company should raise prices in response to a competitor's price increase.
4. Where and when to respond to a military attack.

Games are categorized in terms of the following characteristics:

1. Types of outcomes:
   a) Certain—The outcome is precisely defined, given the actions taken.
   b) Chance—Probabilities of different outcomes are known, given the actions taken.
   c) Uncertain—Only the different possible outcomes (but not their probabilities) are known, given the actions taken.
2. Number of persons:
   a) One person—Such games are called games against nature. If the strategy of nature is probabilistic, such games are called decision problems. If nature's strategy is certain, the game is trivial, and if nature's strategy is uncertain, we may regard the game as two-person if we attribute some perversity to nature.
   b) Two persons.
   c) $n$ persons ($n$ greater than 2).
3. Nature of payoffs:
   a) Zero-sum—The sum of all winnings is zero.
   b) Constant-sum—The sum of all winnings is constant, but not zero.
   c) Nonzero nonconstant sum—There is no systematic relationship between payoffs to different persons.
4. Nature of information:
   a) Perfect information—Complete and unambiguous knowledge of all previous moves.
   b) Imperfect information.

In this chapter we shall consider certain aspects of game theory, particularly as they relate to linear programming. Our attention will be focused primarily on the area of greatest development: two-person zero-sum games. The original development and structuring of the theory is by von Neumann and Morgenstern [290]. Our development is based on that of Luce and Raiffa [203].

## 10.2 TWO-PERSON ZERO-SUM GAMES

Two-person zero-sum games are the most highly developed in that the theory of such games is known. Simply stated, two-person zero-sum games are situations in which each of two persons must choose one of his available actions (which may be finite or infinite in number). Once both persons have made their decisions, the decisions are announced and a payoff table (known in advance to both players) is used to determine the payoff from one player to another.

Consider as an example the following game:

Game 1

| | | Y STRATEGIES | | |
|---|---|---|---|---|
| | | *C* | *D* | *E* |
| X STRATEGIES | *A* | 2 | [1] | 2 |
| | *B* | −1 | 0 | 3 |

The entries are the payoffs from $Y$ to $X$.

Player $X$ may choose strategy $A$ or $B$; player $Y$ may choose strategy $C$, $D$, or $E$. Their choices are announced simultaneously, so that neither player has the advantage of knowing the other's choice. The payoff is to be made as indicated in the game matrix (in dollars) from player $Y$ to player $X$, depending on which strategy each chooses. Suppose, for example, that player $X$ chooses strategy $A$ and player $Y$ chooses strategy $C$ (each not knowing the other's choice). Then player $Y$ must pay player $X$ two dollars. The problem to be solved is determination of an optimal strategy for each player. (We postpone for now a discussion of the objective function for each player.)

The above formulation of a game is said to be in normal form (the payoffs are in a table), as opposed to extensive or tree form, given below.

There are two ways of writing the extensive form of the above game. They differ in whose decision *appears* to be made first. In Figure 10.1a, $X$ chooses $A$ or $B$ and then $Y$, not knowing $X$'s decision, chooses $C$, $D$, or $E$. In Figure 10.1b, player $Y$ chooses $C$, $D$, or $E$ and then $X$, not knowing $Y$'s decision, chooses $A$ or $B$. The realized payoff is that occurring at the end of the appropriate branch. The dashed lines indicate that the decision points contained within are indistinguishable to the player making the decision; e.g., in Figure 10.1a, player $Y$ does not know $X$'s choice when choosing among alternatives $C$, $D$, or $E$.

We assume that, in the neighborhood of amounts of payments, utilities of both persons are approximately linear in wealth, so that it will be adequate for our purposes to have cash values reflect preferences. In other words, each dollar received by a player has the same incremental value to that player.

Let us now consider what $X$ would do *if* he knew $Y$'s strategy in advance:

> If $Y$ chooses $C$, then $X$ would choose $A$ with payoff 2;
> If $Y$ chooses $D$, then $X$ would choose $A$ with payoff 1;
> If $Y$ chooses $E$, then $X$ would choose $B$ with payoff 3.

Similarly, we consider what $Y$ would do if he knew $X$'s strategy in advance:

> If $X$ chooses $A$, then $Y$ would choose $D$ with payoff to $X$ of 1;
> If $X$ chooses $B$, then $Y$ would choose $C$ with payoff to $X$ of −1;

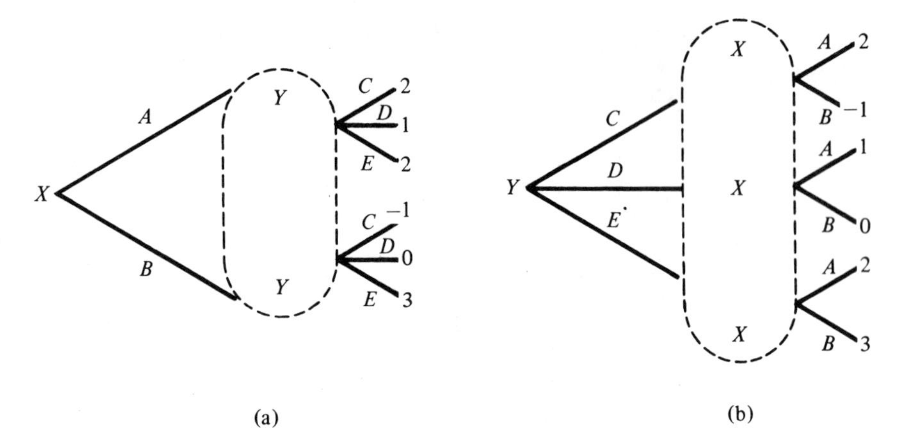


**Figure 10.1**

The Extensive Form of Game 1. All Payoffs Are from *Y* to *X*.

(Remember that all payoffs are from *Y* to *X*.) The analysis does not appear to be of much help, since neither player knows the other's strategy in advance. We shall now consider how to use the above analysis.

Suppose that, in choosing his action, each player acts as if his opponent knows his strategy. Such a supposition leads to very conservative choices. Thus, when choosing a strategy, *X* assumes that *Y* will somehow know to choose a strategy that minimizes the payoff to *X*. We have just considered how *Y* would act if he knew *X*'s strategy. Therefore, *X* would choose *A*, thereby assuring that his payoff is $1 (i.e., assuming that *Y* chooses *D*). Using the same reasoning for *Y*, *Y* chooses *D*, assuming that *X* will then maximize the payoff from *Y* to *X* (i.e., assuming that *X* chooses *A*).

Further, if *Y* chooses *D* and *X* chooses *A*, neither person is better off by changing his action, given that the other holds his original choice. Such a pair of actions *A* and *D* is called an equilibrium pair. It gives the best possible guarantee to each person, assuming that the other wishes to maximize the payoff he receives (or minimize what he pays). The solution we have developed yields an optimal strategy that each person can choose, given that his opponent is also using an optimal strategy. Such an equilibrium solution is called a *saddle point* of a game. We have indicated the saddle point of Game 1 by a square around the appropriate entry in the table.

We now examine how we have determined the optimal strategies in Game 1:

1. For *X*, we consider the minimum (or worst) payoff corresponding to each action, and choose the action having the *maxi*mum *min*ima (best of the worst). *X*'s strategy is said to be *maximin*.
2. For *Y*, we consider the maximum (or worst) payoff corresponding to each action and choose the action having the *mini*mum *max*ima (again, best of the worst). *Y*'s strategy is said to be a *minimax* strategy.

The strategy need not be unique. For example, suppose $Y$ has an added strategy $F$. Game 2 is Game 1 with strategy $F$ added.

Game 2

| | | $Y$ $C$ | $D$ | $E$ | $F$ |
|---|---|---|---|---|---|
| $X$ | $A$ | 2 | 1 | 2 | 1 |
| | $B$ | −1 | 0 | 3 | −2 |

The entries are the payoffs from $Y$ to $X$.

Using the same reasoning as above, $X$ would choose strategy $A$ and $Y$ would choose either strategy $D$ or $F$. Both pairs ($AD$ and $AF$) are in equilibrium.

It would appear necessary only to solve for each person's best of his worst in order to find his optimal strategy and his equilibrium pair. However, as Game 3's payoff table shows, such an equilibrium pair does not always exist.

Game 3

| | | $Y$ $C$ | $D$ | $E$ | $G$ |
|---|---|---|---|---|---|
| $X$ | $A$ | 2 | 1 | 2 | 0 |
| | $B$ | −1 | 0 | 3 | 2 |

The entries are the payoffs from $Y$ to $X$.

This game differs from Game 2 in that one of $Y$'s strategies has been altered. If we follow the above rules for determining a solution, $X$ would choose $A$ (minimum payoff 0) and $Y$ would choose $D$ (maximum payoff to $X$ of 1). However, that pair of actions is not in equilibrium, because $Y$ will be better off (assuming $X$ does not change) if he switches to $G$, and then $X$ might want to switch to $B$ (assuming $Y$ holds $G$), and so on. We therefore classify two-person zero-sum games into those having an equilibrium pair, more properly an *optimal pure strategy*, and those not having an equilibrium pair, which, as we shall see, have *optimal mixed strategies*.

Looking at Game 3, it would appear that some kind of switching of strategy is called for. Any systematic switching by $X$ (e.g., alternating between $A$ and $B$) is inappropriate, because if the systematic switching is detected by $Y$, then $Y$ would choose $G$ when $X$ chooses $A$, and $C$ when $X$ chooses $B$. A similar argument holds for $Y$. Therefore, any variation in choice of acts must have some *randomization* associated with it. Suppose person $X$ has a (possibly biased) coin which he flips to determine his choice. One side is labeled $A$, the other is labeled $B$. Let $p_A$ be the probability of choosing $A$, and $p_B$ the probability of choosing $B$. The expected payoffs for such a random strategy using the coin are as follows (depending on $Y$'s choice):

| $C$ | $D$ | $E$ | $G$ |
|---|---|---|---|
| $2p_A - p_B$ | $p_A$ | $2p_A + 3p_B$ | $2p_B$ |

Obviously person $X$ wants the *smallest expected possible payoff* from $Y$ to be as *large* as possible. By trial and error, we see that for $p_A = \frac{2}{3}$, $p_B = \frac{1}{3}$, we have the smallest expected payoff as large as possible ($\frac{2}{3}$). For convenience, we write the expected payoffs for this strategy as below:

| $C$ | $D$ | $E$ | $G$ |
|---|---|---|---|
| 1 | $\frac{2}{3}$ | $\frac{7}{3}$ | $\frac{2}{3}$ |

Similarly, $Y$ wants to randomize among his alternatives so as to have the *largest entry* as *small* as possible. Designating $q_C$ as the probability that $Y$ assigns to choosing action $C$, and so on, we see that the payoff from $Y$ to $X$ for a random strategy for $Y$ will be:

$$A: \quad 2q_C + q_D + 2q_E$$
$$B: \quad -q_C + 3q_E + 2q_G$$

By trial and error we find that $q_C = 0$, $q_D = \frac{2}{3}$, $q_E = 0$, and $q_G = \frac{1}{3}$ gives a payoff of $\frac{2}{3}$ for either action of $X$. No other set of probabilities for $Y$ gives a smaller maximum. Thus, suitably augmenting the original payoff table as below, we see that explicitly adding the appropriate mixed strategies leads to a *constructed* equilibrium pair, $H$ and $K$. $X$'s optimal strategy, $K$, is to randomly choose $A$ $\frac{2}{3}$ of the time and $B$ $\frac{1}{3}$ of the time. Similarly, $Y$'s optimal strategy, $H$, is to randomly choose $D$ $\frac{2}{3}$ of the time and $G$ $\frac{1}{3}$ of the time. Such a strategy is called an optimal mixed strategy or a constructed equilibrium pair.

| | | $Y$ | | | | $H$ |
|---|---|---|---|---|---|---|
| | | $C$ | $D$ | $E$ | $G$ | $(\frac{2}{3}D + \frac{1}{3}G)$ |
| $X$ | $A$ | 2 | 1 | 2 | 0 | $\frac{2}{3}$ |
| | $B$ | $-1$ | 0 | 3 | 2 | $\frac{2}{3}$ |
| | $K$ $(\frac{2}{3}A + \frac{1}{3}B)$ | 1 | $\frac{2}{3}$ | $\frac{7}{3}$ | $\frac{2}{3}$ | $\frac{2}{3}$ |

Let us now express algebraically what we did above. For person $X$, we wanted the minimum payoff as large as possible. Designating the expected minimum payoff as $u$, person $X$ wishes to solve the following linear programming problem:

$$\text{Maximize } u$$

$$\text{(10.1)} \qquad \text{subject to: } \begin{aligned} 2p_A - p_B &\geq u \\ p_A &\geq u \\ 2p_A + 3p_B &\geq u \\ 2p_B &\geq u \\ p_A + p_B &= 1 \\ p_A, p_B &\geq 0 \end{aligned}$$

It may be convenient to rearrange the equations to have all variables to the left of the equalities and inequalities, thereby preserving the usual linear programming format, but we shall not perform the rearrangement here.

The first four constraints of (10.1) indicate that the expected payoff for *any* action of $Y$ is greater than or equal to $u$, the expected payoff to person $X$. The last equation requires that the probabilities sum to unity. Of course, the probabilities, $p_A$ and $p_B$, must be nonnegative, but the expected payoff of the game, $u$, need not be nonnegative.

Similarly, person $Y$ wishes to solve the following linear programming problem:

$$(10.2)\qquad \begin{aligned} &\text{Minimize } v \\ &\text{subject to:} \quad \begin{aligned} 2q_C + q_D + 2q_E \qquad\quad &\leq v \\ -q_C \qquad + 3q_E + 2q_G &\leq v \\ q_C + q_D + q_E + q_G &= 1 \\ q_C, q_D, q_E, q_G &\geq 0 \end{aligned} \end{aligned}$$

As has been stated above, $p_A = \frac{2}{3}$, $p_B = \frac{1}{3}$, $u = \frac{2}{3}$, and $q_C = 0$, $q_D = \frac{2}{3}$, $q_E = 0$, $q_G = \frac{1}{3}$, $v = \frac{2}{3}$ are the optimal solutions to problems (10.1) and (10.2), respectively. It should not be surprising that problems (10.1) and (10.2) are dual, and a proof of this can be developed in general. (The proof is asked in the problems at the end of the chapter.)

We are now in a position to formalize the above development. Suppose an individual $X$ is to play a game in which his payoff table is $A$, an $m$ by $n$ matrix. The rows correspond to his actions, the columns correspond to his opponent's actions, and the entries in the matrix are the payoffs from the opponent to him. The problem of determining his optimal strategy can be expressed as a linear programming problem:

$$(10.3)\qquad \begin{aligned} &\text{Maximize } u \\ &\text{subject to:} \quad \begin{aligned} \mathbf{A}'\mathbf{x} - \mathbf{1}u &\geq \mathbf{0} \\ \mathbf{1}'\mathbf{x} \qquad &= 1 \\ \mathbf{x} &\geq \mathbf{0} \end{aligned} \\ &\qquad\qquad u \text{ unrestricted in sign} \end{aligned}$$

where $\mathbf{x}$ is an $m$ by 1 vector giving the probabilities of choosing the available actions. The opponent's optimal strategy is the solution to the linear programming problem dual to (10.3) and, of course, may be found directly from the solution to (10.3).

At this point it is useful to state the above in a formal theorem fundamental to game theory.

**THEOREM 10.1** (Minimax)

EACH OF THE FOLLOWING CONDITIONS OF TWO-PERSON ZERO-SUM GAMES IMPLIES THE OTHER TWO:

1. A PAIR OF (PURE OR MIXED) EQUILIBRIUM STRATEGIES EXISTS.
2. THE VALUE OF A MAXIMIN STRATEGY OF A GAME FOR ONE PERSON EQUALS THE NEGATIVE VALUE OF A MINIMAX STRATEGY FOR THE OTHER.
3. NO STRATEGY CAN BE BETTER THAN THE MAXIMIN FOR ONE PERSON OR THE MINIMAX FOR THE OTHER PERSON.

The theorem can be proved by using a linear programming formulation of a game and making use of duality considerations.

Obviously, every two-person zero-sum game may be formulated as a linear programming problem. Less obvious is the result that there is a two-person zero-sum game corresponding to every linear programming problem. The result is not of great value for our purposes, and we shall not consider it further here. The interested reader is referred to Gass [98] or Luce and Raiffa [203] (Appendix 5) for additional information.

We now give a few definitions from game theory.

**Definition 10.1** (Value of a game)

*The value of a two-person zero-sum game is the expected payoff to one person of an equilibrium pair solution allowing both mixed and pure strategies.*

**Definition 10.2** (Fair game)

*A two-person zero-sum game is said to be fair if the value of the game is zero.*

An important concept in game theory is the concept of *dominance.*

**Definition 10.3** (Dominance)

*One strategy is said to dominate another if all outcomes for the first strategy are at least as good as the corresponding outcomes for the second strategy.*

Strategies or actions which are not active in an optimal solution can sometimes be shown to be dominated (either absolutely or conditionally). To illustrate dominance for $Y$ in Game 1, strategy $E$ is dominated by strategy $C$ and by strategy $D$. Therefore, $Y$ will never want to choose $E$. Then, conditionally (with $E$ deleted), person $X$ finds that his strategy $A$ dominates $B$. Finally, for $Y$ (with $B$ deleted), $D$ dominates $C$. Hence, the unique pure strategy is indicated. The above procedure can often (but not always) be employed to determine inactive strategies for both optimal pure and mixed strategy problems.

Sometimes mixed or convex dominance concepts (i.e., a blend of two or more strategies dominating some other strategy) may be employed. An example of a game which has convex dominance is Game 4. Strategy $B$ is dominated by $.5A + .5C$, for example, but by neither $A$ nor $C$ separately.

Game 4

| | | $Y$ | |
|---|---|---|---|
| | | $D$ | $E$ |
| | $A$ | 2 | 0 |
| $X$ | $B$ | 1 | 1 |
| | $C$ | 0 | 3 |

The entries are the payoffs from $Y$ to $X$.

The above development assumes a similar optimizing behavior on the part of both persons. If one person chooses a nonoptimal strategy for himself, then he will usually be at a disadvantage, even if his opponent continues to use his original optimal strategy. The opponent, upon detecting the nonoptimal strategy, will usually be able to do even better by taking into consideration the first person's nonoptimal strategy (assuming that the first person does not then change his strategy).

## 10.3 OTHER TWO-PERSON GAMES

In addition to two-person zero-sum games, there are other two-person games, and the theory for such games has been reasonably well developed. Mathematical programming is of some value in solving such problems. Many two-person nonzero-sum games, however, possess equilibrium strategies (pure or mixed) which are dominated by cooperative nonequilibrium strategies, and a double-cross behavior is encouraged. Cooperation (and usually binding cooperation) is necessary to prevent a double-cross.

One example of such a game is the Prisoner's Dilemma, in which two suspects have been arrested for a crime. The evidence is flimsy, and if neither confesses they will both get light sentences. If one confesses and the other does not, the confessor gets a very light sentence and the nonconfessor gets a very heavy sentence. If both confess, they both receive moderate sentences. Thus we have a payoff matrix as in Game 5.

Game 5

| | | SUSPECT $Y$ | |
|---|---|---|---|
| | | DON'T CONFESS | CONFESS |
| SUSPECT $X$ | DON'T CONFESS | Light sentence for both | Very heavy sentence for $X$<br>Very light sentence for $Y$ |
| | CONFESS | Very light sentence for $X$<br>Very heavy sentence for $Y$ | Moderate sentence for both |

It is difficult to attach meaningful numbers to the different outcomes, but in Game 6 we shall attach "reasonable" numbers to reflect the different payoffs of Game 5.

Game 6

| | | SUSPECT $Y$ | |
|---|---|---|---|
| | | DON'T CONFESS | CONFESS |
| SUSPECT $X$ | DON'T CONFESS | $(-5, -5)$ | $(-100, -1)$ |
| | CONFESS | $(-1, -100)$ | $(-20, -20)$ |

The first entry in each cell is the payoff to $X$; the second entry is the payoff to $Y$.

An equilibrium pair for either Game 5 or Game 6 is easily seen to be confession by both suspects, but a better solution for both players is that neither suspect confesses. This is not an equilibrium solution, because both players are encouraged to double-cross the other and confess. This example is intended to give a hint of some of the problems of two-person nonzero-sum games.

Mathematical programming can be useful in providing a handle for certain $n$-person games, but, except for special cases, a general theory for games other than two-person games is not very well developed. (There are some instances of applications of mathematical programming to $n$-person games. See, for example, Charnes and Cooper [47], pp. 785–798, and Contini and Zionts [57].) In fact, Professors Luce and Raiffa [203] (p. 155) state that ". . . the general theory (of $n$-person games) . . . is . . . very different from the two-person theory and . . . less satisfactory." The principal difference is in the formation and breakdown of coalitions, a situation which can occur in two-person games only when cooperation is permitted. The two-person case of course, permits only one coalition, whereas in $n$-person games the number of possible coalitions increases greatly as $n$ increases. We shall not consider $n$-person games further here.

## 10.4 DECISION THEORY AND CONSTRAINED GAMES

Problems of a decision theoretic nature may be viewed as two-person zero-sum games in which your opponent's strategy is known in advance to the extent that you know your opponent's conditional probabilities of taking a particular action for every action that you have available. (They may also be viewed as one-person games with nature's strategy probabilistic.) In such a case, it can be shown that an optimal pure strategy always exists. Such decision situations may seem trivial, but things become complex in such situations because of the multistage nature of many decision problems, as well as the large number of actions or strategies a person may have.

We may thus consider decision theoretic situations as game situations against a benign opponent whose strategy is fixed (usually mixed) and known, possibly dependent upon our actions. Two-person zero-sum game situations, by contrast, may be viewed as games in hostile environments in which an opponent is out to "beat us." We may construct a continuum between the two, and indeed this is the topic of constrained games explored in some detail by Charnes and Cooper [47]. Constrained games are games in which some information about one's opponent's strategy is known, usually in the form of bounds on or other constraints upon his probabilities. Such bounds or constraints appear irrational for one's opponent because they can only worsen the value of the game to him. Such constraints probably arise from (possibly irrational) policy decisions or legal restrictions on a player. The advantage of having such information is that an optimal strategy can be found (in the sense previously used) *conditional upon* the *additional information.*

The procedure is to adjoin the restriction on probabilities to one's opponent's linear programming problem, then find the dual linear programming problem corresponding to it (one's own altered problem), and then solve that problem. As an example, suppose for Game 3 we somehow know that $Y$ must have $q_E \geq .25$. The corresponding "best" game strategy for $Y$ cannot be better than the unconstrained optimum. We could solve for $Y$'s strategy by solving (10.2), having appended the constraint $q_E \geq .25$.

The result is $q_D = .5833$, $q_E = .25$, $q_G = .1667$, and $v = 1.0833$. However, we are not interested in $Y$'s strategy; we are interested in $X$'s strategy. Therefore, (10.1) suitably altered becomes

$$\begin{aligned}
\text{Maximize } u' &= && + .25w + u \\
\text{subject to:} \quad & 2p_A - p_B && \geq u \\
& p_A && \geq u \\
& 2p_A + 3p_B - w && \geq u \\
& 2p_B && \geq u \\
& p_A + p_B && = 1
\end{aligned}$$

The solution is again $p_A = \frac{2}{3}$, $p_B = \frac{1}{3}$, with $u' = 1.0833$. Note that $u'$ (not $u$) is the expected payoff. In this case $X$'s strategy is not altered, although the expected value of the game is changed.

To consider another example, suppose that $q_D \leq .1$ and that $q_E \geq 0$ again. We would re-solve (10.2) with $q_D \leq .1$ adjoined. The solution is $q_C = .34$, $q_D = .1$, and $q_G = .56$, with $v = .78$. For $X$ the altered problem is

$$\begin{aligned}
\text{Maximize } u'' &= && - .1w' + u \\
\text{subject to:} \quad & 2p_A - p_B && \geq u \\
& p_A + w' && \geq u \\
& 2p_A + 3p_B && \geq u \\
& 2p_B && \geq u \\
& p_A + p_B && = 1 \\
& && p_A, p_B \geq 0
\end{aligned}$$

and the solution is $p_B = \frac{2}{5}$, $p_A = \frac{3}{5}$, $u = \frac{4}{5}$, and $u'' = u - .1w' = .78$, the value of the game.

There are, of course, many other aspects of decision theory, such as revising probabilities based on experience, choice of experiments and sample size, and so forth. We shall not consider such topics here. Instead, we refer the reader to books on statistical decision theory, such as Pratt, Raiffa, and Schlaifer [227].

## 10.5 CONCLUSION

In this chapter we have introduced game theory, indicated how it interacts with linear programming, and presented some fundamental results. Our presentation is by no means complete, since our purpose is limited to the relationship with linear programming.

Some comments concerning practical applications of game theory are in order. Most real game situations are sufficiently complex to make the theory of two-person zero-sum games of little help for anything but simple parlor games.

J.C.G. Boot [36, 37] has described an application of a two-person zero-sum game to a buyer and seller estimating independently the per cent of iron content of a bulk iron ore shipment, and specifying their values for valuation and pricing purposes. If they are sufficiently close, the estimates are averaged. If they are not sufficiently close, an independent umpire is called upon to estimate the ore content. The estimate closer to that of the umpire, the buyer's or seller's, is used for pricing purposes. An optimal strategy is developed using two-person zero-sum game theory, but has apparently not been applied in practice. In addition, the game is not strictly zero-sum because of the payment made to the umpire for his services.

Progress on some more realistic two-person and $n$-person nonconstant-sum games continues to be made. (Certain bargaining and negotiation schemes have been extensively studied, for example.) Further discussion of such games is beyond the scope of this book, and the interested reader is encouraged to refer to some works on game theory and related topics, such as Dresher [79], Isaacs [164], Luce and Raiffa [203], Owen [220], and von Neumann and Morgenstern [290].

### Selected Supplemental References

Section 10.3
[37]
Section 10.4
[222]
General
[56], [77], [164], [203], [220], [252], [255], [263], [290]

## 10.6 PROBLEMS

**1.** Tim and Jim play a game of matching pennies. They each independently decide whether to show heads or tails. If the coins match, Tim wins them, if they don't, Jim wins. Verify the following payoff matrix and determine the optimal strategy, by linear programming or by trial and error:

| | | JIM | |
|---|---|---|---|
| | | HEADS | TAILS |
| TIM | HEADS | 1 | −1 |
| | TAILS | −1 | 1 |

**2.** The game of paper, rock, and scissors is played by two people simultaneously showing a hand outstretched, a fist, or two fingers corresponding respectively to the articles above. The winner is determined by the following rules:

- **a.** Paper covers rock.
- **b.** Rock breaks scissors.
- **c.** Scissors cuts paper.

The winner wins a unit from the loser; in case of a tie there is no exchange. Determine the optimal strategy for both players. Suppose one player announces that he will never play scissors. What is the optimal strategy for the other player?

**3.** Find the optimal strategy for the following matrix games:

| | | $Y$ | |
|---|---|---|---|
| | | $D$ | $E$ |
| $X$ | $A$ | 2 | −1 |
| | $B$ | 1 | 0 |
| | $C$ | −2 | −1 |

| | | $Y$ | |
|---|---|---|---|
| | | $E$ | $F$ |
| $X$ | $A$ | 1 | −3 |
| | $B$ | 0 | −1 |
| | $C$ | −1 | 3 |
| | $D$ | −3 | 1 |

**4.** Develop a graphical method for solving $m$ by 2 or 2 by $n$ games. (Hint: Plot the strategies for the person having 2 actions in one dimension, with payoffs in a second dimension.)

**5.** Use the method of problem 4 to solve the first four games in the chapter, as well as the games of problems 1 and 3.

**6.** For an arbitrary two-person zero-sum game, prove that the problems corresponding to each opponent's game solution are dual to each other. (Hint: Write out the two linear programming problems in general, and show that they are dual.)

**7.** Prove Theorem 10.1.

**8.** A game which has a skew symmetric matrix $A = -A'$ is said to be symmetric.

**a.** Give some examples of symmetric games.

**b.** Prove that a symmetric game has value zero.

**9.** Which of the games in the chapter, including the problems, are symmetric? Which are fair games?

**10.** Show that the optimal solution to Game 3 (in particular that $Y$ will never choose action C) cannot be deduced using only convex dominance considerations.

**11.** Given the following two-person, nonzero-sum game in which two players $X$ and $Y$ cannot communicate with each other, what signaling strategy might one player try in order to encourage a cooperative solution in repeated plays of the game?

| | | $Y$: $C$ | $D$ |
|---|---|---|---|
| $X$ | $A$ | (100, 50) | (−100, −100) |
| | $B$ | (−100, −100) | (50, 100) |

The first figure in parentheses is the payoff to $X$; the second is the payoff to $Y$.

**12.** Show for a two-person zero-sum game that if an opponent's strategy (pure or mixed) is known and fixed regardless of what the player does, then there exists an optimal *pure* strategy for the player. (Hint: Write the linear programming formulation of the game for the player taking into consideration his opponent's strategy.) Then, find the optimal pure strategy for $X$ in Game 3, assuming that $Y$'s strategy remains as $q_C = q_D = q_E = q_G = \frac{1}{4}$.

**13.** Show that a dominated strategy for one player in a game corresponds to an extraneous variable in his linear programming formulation of the game (one which need not be basic in an optimal solution).

**14.** Show that a dominated strategy for a player's opponent in a game corresponds to a constraint which may be omitted in the linear programming formulation of the game for the player.

# 11

# Applying Linear Programming to Problems

## 11.1 INTRODUCTION

Our approach thus far has been primarily to introduce various linear programming methods, explore their theoretical and computational bases, and present examples of how they work. In this chapter we shall address the question of application, consider types of applications, and attempt to provide an introduction to the art of formulating and solving linear programming problems.

We believe that linear programming methods are applicable in solving various real problems. We do not mean to imply that all of the methods presented are necessarily useful in a practical sense; some may only be of theoretical interest, or useful in advancing research on more practical algorithms. In this chapter we indicate some of the problems associated with solving real problems using linear programming methods (when such methods are applicable), techniques of formulating problems, gathering data, solving the problem, and interpreting solutions. In addition we present a few applications of linear programming to real problems.

## 11.2 APPLICATIONS OF LINEAR PROGRAMMING

There have been many different applications of linear programming to real problems. (A recent bibliography of such applications appears in Gass

[98]). Instead of trying to survey them, we describe what seem to be the four types of structure that appear in these applications. Our structure types are not necessarily unique. They do appear to be collectively exhaustive, because all problems can be decomposed into structures of the four types. The structure types may not be mutually exclusive (some parts of a problem may be categorized as having one or more of them).

The four types of problem structure are as follows:

1. Resource allocation structure.
2. Blending structure.
3. Cutting stock structure.
4. Flow structure.

We shall now consider the four types of structures further by referring to simple problems containing each type.

### Resource Allocation Structure

Resource allocation structure is a structure in which it is desirable to determine the most profitable way of allocating available resources (e.g., raw materials and/or processing facilities) to the production of different products. In this formulation the variables are the products and the constraints limit the resources available. The products have per unit profits associated with them. The dog food problem introduced in Chapter 1 is an example of a resource allocation problem. Most simple production planning problems can be viewed as resource allocation problems.

### Blending Structure

In blending structure it is desired to determine the least cost blend having certain specified characteristics. Variables are input materials to be used in a specific product. Constraints specify certain maximum, minimum, or constrained characteristics of the blend desired. One of the earliest blending applications was the diet problem in which the lowest cost daily diet was desired, with limitations on vitamins, calories, protein, carbohydrates, and so on. Other applications have been to the blending of gasolines, the burdening of blast furnaces in iron and steel making, and the blending of animal feed-mix, fertilizer, and peanut butter. An example of a problem with blending structure follows.

EXAMPLE 11.1:

A candy shop has available various quantities of nuts, as follows:

100 pounds of peanuts
30 pounds of cashews
50 pounds of hazelnuts

Peanuts sell for \$ .80 a pound. The Bridge Club mix consists of at least 20 per cent cashews and not more than 50 per cent peanuts, and sells for \$1.20 a pound. The Deluxe mix consists of at least 30 per cent cashews and not more than 30 per cent peanuts, and sells for \$1.80 a pound. Cashews are sold for \$2.20 a pound. The shop likes to have the blended mixes available in advance. Assuming they can sell all that is available, given the nut availability, how should they mix the nuts to maximize sales receipts?

The variables (measured in pounds) are as follows:

$w_p$ peanuts to be sold as peanuts
$x_c$ cashews to be sold as cashews
$y_p$ peanuts to be mixed as Bridge Club mix
$y_c$ cashews to be mixed as Bridge Club mix
$y_h$ hazelnuts to be mixed as Bridge Club mix
$z_p$ peanuts to be mixed as Deluxe mix
$z_c$ cashews to be mixed as Deluxe mix
$z_h$ hazelnuts to be mixed as Deluxe mix

The problem is then to

$$\text{Maximize } z = .8w_p + 1.2y_p + 1.8z_p + 2.2x_c + 1.2y_c + 1.8z_c + 1.2y_h + 1.8z_h$$

subject to:

$$w_p + y_p + z_p \leq 100 \quad \text{Peanut availability}$$

$$x_c + y_c + z_c \leq 30 \quad \text{Cashew availability}$$

$$y_h + z_h \leq 50 \quad \text{Hazel nut availability}$$

$$.5y_p - .5y_c - .5y_h \leq 0 \quad \text{upper limit on peanuts in Bridge Mix}$$

$$.2y_p - .8y_c + .2y_h \leq 0 \quad \text{lower limit on cashews in Bridge Mix}$$

$$.7z_p - .3z_c - .3z_h \leq 0 \quad \text{upper limit on peanuts in Deluxe Mix}$$

$$.3z_p - .7z_c + .3z_h \leq 0 \quad \text{lower limit on cashews in Deluxe Mix}$$

$$w_p, x_c, y_p, y_c, y_h, z_p, z_c, z_h \geq 0$$

To illustrate the development of the limit constraints, consider the lower limit on cashews in the Bridge Mix, of which cashews must make up at least 20 per cent. This yields the following constraint, which is equivalent to that given above.

$$y_c \geq .2(y_p + y_c + y_h)$$

Strictly speaking, the last four constraints are blending structure constraints, whereas the first three constraints may be viewed as resource allocation (or, alternatively, flow) constraints.

### Cutting Stock Structure

A structure somewhere between resource allocation and blending structure is the cutting stock structure. A problem that contains the cutting stock structure is the cutting stock problem, which occurs in various forms in such industries as paper, glass, steel, and aluminum. The problem has many variations. One example is the combining of orders for rolls of paper ordered in different widths. The rolls are to be cut in an optimal manner from the standard widths available. For this problem the variables are the units of produced widths to be cut in a particular way (e.g., a 20 foot width cut into two 8 foot widths and a 3 foot width, with one foot wasted) and constraints that specify the minimum number of each ordered size which must be produced. Costs include the cost of the paper and the cost of making the cuts. An example of a cutting stock problem follows.

EXAMPLE 11.2:

A lumber yard stocks 2″ by 4″ beams in 3 lengths: 8 feet, 14 feet, and 16 feet. The beams are sold by the foot and no charge is made for cuts. The yard has an order for the following lengths:

| | |
|---|---|
| 80 | 12 foot lengths |
| 60 | 10 foot lengths |
| 200 | 8 foot lengths |
| 100 | 4 foot lengths |

The cost of the 2 by 4's to the lumber yard is \$.30 per 8 foot length, \$.60 per 14 foot length, and \$.70 per 16 foot length. Cutting costs can be assumed to be zero. Assuming that the lumber yard has enough of each of the three lengths in stock, what is the minimum cost method of filling the order? Let

$x_1$ be the number of 8 foot lengths sold uncut
$x_2$ be the number of 8 foot lengths cut into two 4 foot lengths
$y_1$ be the number of 14 foot lengths cut into 10 foot and 4 foot lengths
$y_2$ be the number of 14 foot lengths cut into 12 foot lengths
$y_3$ be the number of 14 foot lengths cut into 10 foot lengths
$w_1$ be the number of 16 foot lengths cut into 12 foot and 4 foot lengths

Other possible cutting combinations, such as cutting a 16 foot length into two 8 foot lengths, can be shown to be unprofitable. The linear programming formulation is then

$$\begin{array}{llllllll}
\text{Minimize } z = 30x_1 + & 30x_2 + & 60y_1 + & 60y_2 + & 60y_3 + & 70w_1 & & \\
\text{subject to:} & 2x_2 + & y_1 & & + & w_1 & \geq 100 & \text{(4 ft. lengths)} \\
x_1 & & & & & & \geq 200 & \text{(8 ft. lengths)} \\
& & y_1 & & + y_3 & & \geq 60 & \text{(10 ft. lengths)} \\
& & & y_2 & + & w_1 & \geq 80 & \text{(12 ft. lengths)} \\
& & & & x_1, x_2, y_1, y_2, y_3, & w_1 & \geq 0 &
\end{array}$$

This problem is, strictly speaking, an integer programming problem because the values of the variables must be integral. In this particular case the solution to the linear programming problem turns out to be integer, fortunately.

### Flow Structure

Flow structure appears in most problems, and, of course, in network problems. Flow problems which lead to special methods have been discussed extensively in Chapter 9. These problems generally involve finding the maximal value of flow through a network. Applications of network flows have been made in scheduling transportation systems (e.g., airline operations), as well as flows of goods over time, as in inventory models. Examples of flow structure are given in Chapter 9.

As mentioned above, many problems require more than one type of structure. An example is a multiperiod production and inventory problem, in which each period might consist of a resource allocation or blending problem (possibly both), and the periods would be linked together by forward flows through time.

## 11.3 FORMULATING THE PROBLEM

In formulating a problem, it must be remembered that the problem solution is primary, and the solution technique secondary. Moreover, *linear programming is not a panacea.* It is a mathematical tool which either does or does not fit or approximate a situation. An example of an ordinary tool is a hammer. A hammer, a versatile tool, can be used to hammer nails. But it can also be used to hammer screws, holes (using nails), cuts (again using nails), and bolts. It is obvious that the above tasks are more efficiently done by a screwdriver, drill, saw, and wrench, respectively. However, it may be that for certain tasks there is no appropriate tool, and hence what we have is the best (e.g., a hammer or, in our case, linear programming) available—assuming, of course, that it is somehow applicable.

The first step in problem formulation is problem definition—the system must be defined and the objectives determined. This involves many pitfalls, because the system could be defined either too narrowly or too broadly. Ideally, optimal problem size has been achieved when the marginal cost of increasing the size and scope of the system included in the study just equals the marginal benefit obtained. Practically speaking, these necessary conditions of optimal problem size are useless, except in the crude sense that the model builder "feels" that the "right" problem size has been obtained.

What I try to do in formulating a problem is to begin with as coarse a problem description and as tight a set of system bounds as is tolerable. Then I stingily add what seems absolutely necessary to the problem formulation until I am satisfied with it. This sometimes proves inadequate, and the problem formulation must be altered in a subsequent step.

Objectives must be defined. There is usually a main objective, such as profit maximization, but there are sometimes other conflicting objectives as well. The usual procedure in such cases is to maximize or minimize what seems to be the most important of the objectives subject to constraints specifying minimal levels of the other objectives. Then, when the solution is obtained, it is evaluated by management. The implicit tradeoff between the optimized objective and objectives indicated by constraints is given by the dual variables of the constraints. The levels of such constraints can be adjusted, based on management considerations, until an appropriate optimum is achieved.

A simple example of a problem with conflicting objectives is production planning for a firm with a severe cash constraint. Although the firm wishes to maximize the contribution to overhead and profits obtained by the production plan, it must also conserve on cash expended. Thus, a cash constraint should be adjoined, but, strictly speaking, the amount is usually flexible. It may be useful, therefore, to solve the problem as a parametric linear programming problem in which the cash available is a parameter. The firm can then determine the cash it wants to commit to the plan by making use of the associated shadow price of cash information and the parametric solution.

Another important aspect of the problem formulation stage is that, in computing net benefit, the cost of the solution must be subtracted from any benefits obtained in solving the problem. It is obviously not worth spending $100,000 to save $10,000! Although initially the cost of the solution and the possible benefit value can only be estimated, such estimates must be used in deciding upon the solution method to be used.

Once a problem has been formulated, there are some basic questions which must be addressed:

1. Is this a problem of constrained optimization?
2. Can the relationships be reasonably well approximated by linear functions?

If both answers are yes, then linear programming is a candidate for solving the problem. It should be compared with any other methods which seem appropriate, and the best chosen. (Usually, when both answers are *clearly* affirmative, linear programming is the *only* method that is sensible, but this may not always be the case.)

Certain nonlinearities can be approximated in linear programming. Without loss of generality, the objective function may be assumed to be linear. (An objective function can always be made linear by transferring a nonlinear function to the constraint set. For example, maximizing $f(x)$ subject to some restrictions is equivalent to maximizing $y$ subject to $y - f(x) \leq 0$ and the other restrictions.) Any nonlinearities which lead to a convex set of solutions (maximizing a concave function or minimizing a convex function over a convex set) may be approximated using linear programming. Nonlinearities which lead to a convex set of solutions correspond to situations in which marginal returns to scale are decreasing. More intuitively, per unit costs of goods purchased or produced may increase, and per unit receipts for goods sold may decrease, as the number of units of the good purchased or sold, respectively, increases. Piecewise linear representation of such conditions may be precise if the conditions are piecewise linear, or may be made as accurate as desired if approximated by adding the necessary constraints and variables. If the constraints and objective function are, in addition, separable, whereby every nonlinear function $f(x_1, \ldots, x_n)$ may be written as $g_1(x_1) + g_2(x_2) + \cdots + g_n(x_n)$, then separable programming may be used. An example of a separable programming problem is given in the problems at the end of the chapter.

Nonconvex nonlinearities may be represented using integer programming to ensure an absolute optimum and are discussed further in Chapter 12.

If linear programming is not applicable, whatever approach is needed should be pursued. Here we shall assume that linear programming is appropriate. The next step is to formulate a mock-up of the problem—without data. The form of each type of constraint (e.g., capacities of production units of a given kind or minimal level of production for products) is indicated, as are the categories of variables which appear with nonzero coefficients. Then the number of constraints and number of variables is determined. If the problem is small (less than 50 constraints and 50 nonslack variables *excluding* bounds on variables), proceed to formulate the problem directly. If the problem is moderate (up to 200 to 300 constraints) or large (300 constraints or more), it is usually advisable to first formulate and solve an appropriate subproblem. (The definitions of size are the author's and are *not* universally accepted.)

Generally, there are two ways of defining such a subproblem:

1. Where the formulation is of a multifacility or multiperiod problem, focus on one (or a few) facility(ies) or time period(s).
2. Where the formulation includes many products and associated constraints, eliminate all but a relatively small number of big selling products.

The purpose of first solving a reduced version of a problem is to find inconsistencies in formulation and to use the associated solution as a benchmark for assessing more accurately the reality, cost, and benefit of the study at an early stage. This may provide a useful in-process evaluation of a large project, and will usually be of value in solving the larger problem.

## 11.4 DATA COLLECTION

In spite of the vast management information systems currently in operation, seldom does a problem arise for which all data required is available. In addition, there are usually different definitions of certain terms for different purposes (such as profits for tax purposes as opposed to profits for shareholder reporting purposes versus profits for decision-making purposes). Even when the data is available, there may be errors in the collection or transmission process.

In most studies where data is not available, collecting data over an extended period of time is usually not a feasible alternative. Some data collection may be feasible, but the cost of data collection—including any time delays in the availability of the solution—is part of the cost of problem solution. When some data is not available and data collection is not feasible, whatever is available must be used, and the rest estimated or guessed. After such problems have been solved, sensitivity analysis is especially important in determining the sensitivity of the solution to changes in crudely estimated data, and subsequent refined estimates of data may be utilized. In addition, various parametric analyses of the coefficients may be employed (see Chap. 7).

It is generally useful to record the source of data, particularly with respect to such policy decisions as the minimal production levels of different products. One of the useful results of a linear programming solution is the implicit cost of such policies, as given by the dual variables on the constraint equations. Policies are sometimes changed on the basis of such considerations. In addition, there will occasionally be conflicting constraints in a formulation as well as errors in formulation, and it is important to know the source of the conflicts and/or errors in order to help resolve and correct them.

Because all relationships of the problem are linear, all coefficients must reflect *true variable* costs, times, and so on. There is a tendency to use standard costs or other cost accounting information in determining coefficients, particularly because standards are readily available in many organizations. However, standards include allocations of fixed costs, maintenance, and other factors, and may also be used to determine compensation to certain employee groups. If the latter is true, the standards may be set partly by bargaining and may therefore be biased. Both of these factors—the presence of fixed costs and the possibly biased nature of standards—must be considered and eliminated to the extent possible. In the coefficient data collection phase, it is usually easier to separate variable factors from fixed factors than to eliminate any bias in the standards, unless records on the bias are maintained.

We shall now consider some other aspects of the separation of variable and fixed factors. For cost or profit, we must have incremental (i.e., variable) cost or profit (more properly, contribution to overhead and profits). (We formulated the original problem in Chapter 1 in a manner to bring out the distinction among variable costs, fixed costs, and sunk costs, and their relevance in linear programming.) For processing times of a product on a production unit, we want to include only direct processing time. Indirect processing time, such as maintenance, should be subtracted from total time available to obtain the net production time available.

If fixed costs are an important consideration, a mixed integer approach may be appropriate, but, as we shall see in future chapters, depending on the size or complexity of the problem, it may or may not be solvable. If there are a small number of fixed cost processes, then all possible combinations may be run using linear programming. For example, if there are two fixed cost facilities $A$ and $B$, we may run four linear programming problems corresponding to the possible combinations of such facilities (both $A$ and $B$ operating, only $A$ operating, only $B$ operating, neither $A$ nor $B$ operating). One or more of these possibilities may often be excluded by common sense considerations (e.g., precisely one facility may be required to produce the production required). Another possibility (which is nonoptimal in general) includes ignoring fixed costs for solution purposes and then adding them on, making intuitive alterations to the solution. (Additional runs are often made on the same basis, altering the problem based on previous solutions.)

## 11.5 SETTING UP THE PROBLEM FOR SOLUTION

Virtually all real linear programming problems are solved by computer. (I had an experience a few years ago in which I actually solved a two-variable two-inequality *real* problem graphically and not by computer! Such experiences seem to be the exception rather than the rule.) Programs to solve linear programming problems are available for most computers. The computer programs are very highly developed, and the upper limit to the size problem (without resorting to decomposition and related methods) that can be solved is very large—on the order of 16,000 constraints if the problem is well behaved. There is virtually no upper limit to the number of variables in a problem.

For small problems, the input data cards need only be punched and verified and the program run. The initial input should be checked to assure that the problem is correctly input, i.e., there are no errors in coding and inputing data.

For larger problems, because of the increased probability of errors and solution difficulties, it may be useful to undertake the following activities:

1. Using a computer program to set up the problem.
2. Scaling matrices.
3. Eliminating easily detected redundant equations.

We postpone consideration of the first point until the next section, but we now consider the other two techniques.

In order to reduce round-off error buildup in most computer programs (word lengths are finite and hence calculations are truncated), it is desirable that the determinant of every linear programming basis be as close as possible to one in absolute value. This can be achieved in some cases in the formulation stage, or by scaling the problem after formulation so that nonzero coefficients are close to one. This is accomplished by multiplying rows and columns through by positive constants (for convenience, usually powers of 10). Such multiplication corresponds to changing the units of slacks and variables, respectively. A relatively computationally inexpensive way of scaling a problem automatically is to compute, for each constraint (equality or inequality), the average of the powers of ten which, when multiplied by each nonzero coefficient in that constraint, would place its absolute value between .5 and 5. The average power of ten for that constraint is then rounded to the nearest integer. Then ten to the resulting power is used as a scaling factor to multiply that constraint. The process is sequentially applied to all rows and then all columns, and repeated until there is no change in the scaling multipliers. We illustrate the above scaling procedure on the following constraint:

$$35.4x_1 - 142x_2 + 12x_3 \geq 17$$

$10^{-1}$ $10^{-2}$ $10^{-1}$ $10^{-1}$ Multiplier (power of ten) to convert absolute value of coefficient to between .5 and 5

$$\text{Average power of ten} = \frac{-1 \quad -2 \quad -1 \quad -1}{4} = -1.25 \approx -1$$

Multiplying the constraint by $10^{-1}$ gives us

$$3.54x_1 - 14.2x_2 + 1.2x_3 \geq 1.7$$

Columns are scaled in precisely the same way. (See Fulkerson and Wolfe [96] for a somewhat different method of scaling a problem.)

Many formulators of problems find it convenient to include definitional equations (as defined in Chap. 4, problem 12). However, such equations, in addition to increasing size, make the problems more difficult to solve (see Zionts [313], for example). Where possible, such constraints should be eliminated, either in the formulation stage, in the computer set-up stage, or by the computer program in setting up the computer run.

## 11.6 USE OF THE COMPUTER

As indicated in the previous section, most problems are solved by computer. The capabilities of the current state of linear programming computer programs are such that problem size limitation is virtually no restriction. In this section we shall consider certain salient features of many computer programs.

Input to the programs is usually accomplished by declaring row, and sometimes column, names (usually in five-to-eight digit alphameric mnemonic user-chosen codes), the name of the objective function, and whether the objective is to maximize or minimize. For most programs only the nonzero coefficients of the problem, identified by row and column names, need be input. In addition, identification of upper and lower bounds for use in upper bound routines is indicated where applicable.

Many programs have automatic scaling of linear programming matrices as an option. Other options are the ability to designate certain variables as being basic or nonbasic in an optimal solution, regardless of the usual linear programming considerations, such as primal and dual feasibility. This permits the user to ignore certain variables and constraints if desired, with only a minor alteration of the input. Some programs have options to scan for and eliminate nonbinding constraints and extraneous variables. Many programs have built-in error checking of the optimal solution, sensitivity analysis, and postoptimal analysis capabilities, including parametric programming. In addition, many programs have good separable programming capabilities. (See problem 7 at the end of the chapter.)

Because of the volume of data that must be generated for large problems, many large linear programming computer programming systems have matrix generation capabilities. This includes a set of special control instructions which permits writing programs that set up special tables or matrices for use in formulating a problem.

These tables are then used to set up parts of the problem matrix by transferring the same coefficients to different parts of the matrix, or by performing particular arithmetic operations upon the tables and then creating portions of the matrix. Use of a matrix generation program is usually worthwhile because most large problems have portions of the matrix computed by calculations from tabular information. Hence, correct input of a problem reduces to correct input to the tables (which are usually orders of magnitude smaller than the problem of concern), and correct generation of coefficients from the tables.

Because of the size of large linear programming matrices, it is usually impractical to check them equation by equation, although column and row listings of nonzero coefficients are usually useful. An aid with which many programming systems are equipped is called snapshot, picture, photo, map, or some other similar name. It produces a computer representation of the matrix in which zeroes, signs, and relative magnitudes of matrix coefficients (using one or two digits per entry) are presented, along with appropriate row and column names. Gross errors (and most errors are gross) can be located by scanning such representations. Even though this device achieves a handsome reduction in size compared to an equation-by-equation listing of all coefficients, it can nonetheless be enormous: a representation of a complete 500 by 5000 matrix covers most of the wall space of a 9 by 12 foot office! A listing of the input is useful in completing the correction of an error.

Many large linear programming systems include a report generating package which permits a special management-oriented report to be generated when the problem is solved. Such programs permit the formatted output of a report in virtually any manner desired, with any selected linear programming results exhibited. In addition, calculations may be made using the solution values, and the results output. Normally, report generator programs are coded in a special user-oriented computer language to output the desired reports in a form suitable for use by management where the results are to be used repeatedly on a routine basis.

## 11.7 MODEL VERIFICATION AND DEBUGGING

The first run of a newly formulated problem will almost certainly have many bugs in it. Consequently, the solution may be infeasible, unbounded, or otherwise nonsensical.

In part, this is one good reason for first solving a small prototype of a large problem, and then increasing the size. Most programs give an indication of where the error occurred, such information typically including the name of the variable which became unbounded or the names of the constraints which could not be satisfied, as well as the current values of primal and dual variables, and the objective function. When an error in a run occurs, it is important to check the visual representation of the problem as well as a column-and/or row-wise listing or nonzero elements in order to pinpoint the difficulties.

At various points in the solution process, current bases are remembered. This serves a number of functions, one of which is restarting the debug or solution phase without again solving the problem from the beginning. Once errors have been found and corrected, and provided that the changes are not *too* great, the solution may be started by inverting the corrected matrix directly to a remembered basis. If the changes are too great, this procedure will not be worthwhile. Feel for such decisions is gained with experience.

Another reason to remember bases is for restarting long linear programming solutions in case of computer errors, power failures, and so on. Using the product-form revised simplex method (as do most large programs), periodic reinversions of the current basis are utilized for two reasons:

1. To keep iteration times down.
2. To reduce round-off errors.

When performing reinversions, it is generally easy to output the information necessary for inversion in case an inversion to that basis is desired.

After a sequence of such error hunting and correcting sessions, the problem will appear to solve satisfactorily. The next question is whether the portrayal of the real situation is accurate. A common technique here is to further constrain the problem to a series of known conditions (such as the way the facility is normally operated) and check whether the known conditions and the further constrained solution coincide. If they don't, the formulation must be checked again. Of course, one is never completely sure of an accurate portrayal of a problem, but the above procedure is usually helpful. Although it is not always possible to so constrain a solution, some benchmark runs can usually be made to check the accuracy of the representation.

Once the problem appears to be a reasonable representation of the real situation, we solve the problem of concern and obtain the resulting solution.

## 11.8 SOLUTION INTERPRETATION AND POSTOPTIMAL ANALYSIS

The interpretation of a linear programming solution seems to be relatively straightforward. The optimal primal solution gives the level of activity of each variable at the optimum, and the optimal dual solution gives the optimal shadow prices. There is usually not much difficulty interpreting what that solution means, except that the reason *why* the solution is optimal may not be clear. The answer lies in the constraints, and the dual solution should be examined to identify "tight" constraints, that is, constraints whose shadow prices are relatively large. To elaborate, each constraint represents a restriction. Some restrictions are caused by physical limitations, such as processing hours on a production facility. For such restrictions, the shadow prices give information concerning the marginal value of added capacity *assuming all other factors remain the same.* Such information is a first step in considering expansion of facilities. Constraints based on judgment, such as those limiting the amount of a product which can be sold, have a slightly different interpretation. *We do not mean to question the judgment of setting such limits*, although that can sometimes be a useful exercise. What is useful is the marginal value of relaxing such constraints, for example, the sales constraints. Even though the constraint may not be relaxable under the assumed conditions, it may be possible to relax it *at a price*. Thus, for example, a promotional device such as a special advertising campaign may make it possible to increase sales. Obviously, if the value of increased sales exceeds the cost of the advertising campaign, the advertising campaign should be undertaken. Similar analyses may be employed on other constraints, with attendant benefits.

Postoptimal analyses based on the above discussion may be carried out sequentially. If it is possible to identify which constraints are relaxable systematically, the problem may be altered to allow for this directly. An example is a product whose upper sales limit has been specified as 1000. Suppose that on further study it is determined that sales could be increased by 200 units by an expenditure of \$2000. (The effect of the expenditure is assumed divisible and linear in an advertising campaign.) Denoting the original product as $x_1$, and assuming its profit to be \$50, we have, as *part of the originally formulated problem*, the following expressions:

$$\begin{aligned} &\text{Maximize} \quad \cdots + 50x_1 + \cdots \\ &\text{subject to:} \quad x_1 \leq 1000 \end{aligned}$$

Define $x_2$ as the number of *units* of sales to be achieved by advertising expenditure. For example, if $x_2 = 0$, no expenditure in advertising is made; if $x_2 = 200$, \$2000 is expended in advertising. We may now alter the above portion of the problem to appear as follows:

$$\begin{aligned} &\text{Maximize} \quad \cdots + 50x_1 - 10x_2 + \cdots \\ &\text{subject to:} \quad x_1 - x_2 \leq 1000 \\ &\qquad\qquad\qquad\quad x_2 \leq 200 \end{aligned}$$

The objective function coefficient of $x_2$ is $-10$ because the cost of advertising to sell each additional unit of $x_1$ (above 1000) is \$2000/200, or \$10. The net effect is that the first 1000 units of the product sold contribute \$50 per unit to profit, but the next 200 units contribute \$50 minus the \$10 advertising expenditure per unit. Alternatively, we might have assumed that the advertising effort was not divisible. In that case we would have to compare the optimal solution to the original problem, with the optimal solution to the original problem having relaxed the sales constraint to 1200 units and having imposed a cost of \$2000. (Alternatively, we might have used mixed integer programming.)

Other additional runs may be made, making analogous alterations to the problem.

## 11.9 CASE HISTORY I: A MEDIA SELECTION PROBLEM

A director of advertising at a large metalworking company read an article about using linear programming to allocate advertising expenditures in optimizing the selection of advertising media for a campaign. The formulation seemed relatively straightforward. A variable represented the number of page insertions (or fractions thereof) to be made in a particular publication. Tabular data was obtained from advertising agencies on the breakdown of the readership (approximate number of readers per issue) in a number of dimensions:

1. Geographically, by state.
2. Profession.
3. Nature of employer's business.

Using the kinds of considerations he had used previously, that is, allocating the advertising over the groups in some reasonable fashion to find a low cost selection of media, he proposed a linear programming problem based on the following considerations:

1. He had a preconceived geographic profile of readership that he wanted to reach.
2. He wanted to reach at least certain numbers of readers of a given profession.
3. He wanted to reach at least a certain number of potential customers for the product being promoted.
4. He wanted to limit the number of insertions per publication.
5. He wanted to stay within his $25,000 budget.
6. He wanted to minimize the cost of the advertising campaign.

The problem was set up accordingly, and the constraints, paralleling the above, were as follows:

1. For each state (e.g., New York, California) relevant to the campaign, the minimum number of readers that must be reached, as well as the maximum number of readers that should be reached, was specified. Thus, each state important to the profile had two inequalities.
2. For each profession relevant to the campaign, there was an inequality stipulating the minimum number of readers.
3. For each industry relevant to the campaign, there was an inequality stipulating the minimum number of readers.
4. For each publication, an upper bound constraint expressed the maximum number of insertions that could appear.

5. A budget constraint limiting the total expenditure to a maximum of $25,000 was included.
6. The objective function was to minimize total expenditures for the campaign.

Data on coverage was made available from the advertising agency, and the profiles on target coverage were based on the advertising director's judgment. After the cards were prepared and the problem verified (the problem was relatively small, on the order of 25 constraints and 100 variables), the problem was run; the first trial gave no feasible solution. It appeared as if the budget constraint was causing some difficulty, so it was subsequently decided to drop the constraint because it was "parallel" to the objective function (the objective being to minimize the amount spent). If the minimal amount spent is less than $25,000, then obviously the optimal solution satisfies the budget constraint; otherwise it does not. *The point here is that a constraint on the objective function (which is used for optimization) does not provide any benefits. In fact, it usually causes confusion.*

With the constraint deleted, the problem was re-solved and the optimal solution required spending $90,000, almost four times the budgeted amount! (Can you spot the difficulty and the reason for this answer before reading on?) The advertising director was hardly happy about this development, because he knew he could find a good selection of publications without linear programming. With all the analysis, he had a ridiculous solution!

By looking at the shadow prices of the optimal solution, or—since we do not have them here—using common sense, it appears questionable to limit the *maximum amount of readership* in different states. The minimum remains important, but for this problem the maximum limitation amounts to avoiding excess coverage. One should not be concerned about *excess coverage*, because limiting it can only be costly. The main concern is in *meeting minimal coverage at lowest cost*. What had happened in the solution is that, in order to satisfy the maximum restrictions for such populous states as New York and California, normally used publications (efficient in terms of cost per reader) could not be utilized. Hence, a selection of regional publications (less efficient in terms of cost per reader), which had limited readership in populous states, were included at an attendant higher cost. After eliminating the upper limits on state readership constraints, the optimal solution had a cost of about $18,500.

The point is that all constraints in linear programming must be satisfied at whatever deterioration necessary to the objective function. Because constraints can only worsen solutions, only those constraints which are *absolutely necessary* should be employed. A corollary to this is that inequality constraints—assuming they are appropriate in the problem context—should be preferred to equality constraints. For example, if production of 100 units in a period must at least be met, the constraint $x \geq 100$ is preferred to the constraint $x = 100$. The exception is if, for some reason, production must precisely equal 100, or, more generally, if there is a *cogent* reason for the equality.

## 11.10 CASE HISTORY II: A PRODUCTION ALLOCATION PROBLEM*

This case study is an illustration of the application of linear programming to determining the optimal allocation of customer orders to producing facilities. The products of this example are fabricated aluminum products, but the approach and methodology are sufficiently general that the method may be applied to similar problems in many different industries.

### Background

Two major types of products are produced at the facilities under study. The first group, Group A, is produced from a primary production unit, then inspected, packed, and shipped to customers. The second group of products, Group B, must be processed further through a second group of equipment before it is inspected, packed, and shipped to customers. Within Group A are 30 individual products, each with a different processing time and different cost. Within Group B are 81 different products, each with different processing times for both primary and secondary equipment and, therefore, different costs. A schematic of the product flow is given in Figure 11.1.

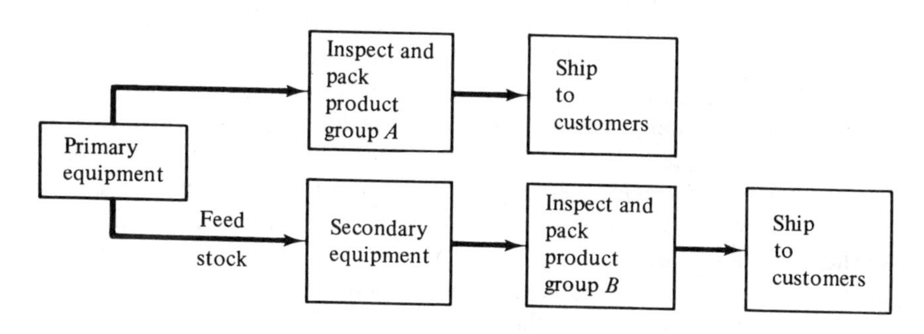


**Figure 11.1**
Schematic of Product Flow

At present, two plants, one located in the Middle Atlantic states and one located in the Southeast, can produce both product Groups A and B. A third, more efficient facility located in the Southwest can produce only a limited number of Group A products. However, with a relatively modest investment in equipment, it can be altered to produce feedstock for the other two mills.

* This section was written by J. Lee O'Nan of the Aluminum Company of America, Pittsburgh, Pennsylvania.

The Manufacturing Manager for both groups of products asked the Corporate Operations Research group if they could help explore the following questions:

1. How should customer orders for the products be allocated to the plants as they are currently configured? Such an allocation should take into account the following characteristics peculiar to each plant:
   a. Production limitations in terms of both equipment types and capacities.
   b. Production costs.
   c. Transportation costs from plant to customer.
   d. Raw material costs—different primarily because of transportation costs to the plants from the raw material source.
2. How much would total cost be lowered via installation of additional equipment? Present equipment limits the Southwest plant to the production of certain Group A products. Yet it is known that this facility is much more efficient than the others, and would therefore have significantly lower operating costs for the rest of the Group A products. On the other hand, the Southwest plant, being much farther from most of the major markets, would have higher transportation costs.

**Problem Formulation**

The formulation phase of a problem is extremely crucial to the success of any operations research study. Although the operations researcher does most of the work, the study should be a joint effort involving both the client (or his staff) and the operations researcher. If the problem is hard to define, or if the operations researcher is not able to grasp it in quantitative terms, the problem formulation phase can be a lengthy affair. Only when the client (the Manufacturing Manager in this case study) and the operations researcher agree that the problem has been completely and correctly formulated are you ready to go on to the next step. Unfortunately, many operations research studies are gathering dust because inadequate problem formulation or misunderstanding of the problem led to a good solution of the wrong problem.

As you will see, the type of problem described above is easily defined as a linear programming model if the relationships are reasonably linear. At the very detailed level, this is not the case. There are production set-up costs for each specification produced that are a function of the specification that was most previously produced. In addition, costs for a given item vary because of the machine used to produce it, the person operating the equipment, and the quantity produced. However, we are not focusing on the detailed scheduling problem. Ours is an overall planning problem. If there were strong evidence that the scheduling practices were poor, our initial recommendation might have been a study directed at improving operating practices. Since the latter seemed sound, we focused on the planning problem. Given this framework, the accountants and industrial engineers, after some study, felt that the production costs were in fact approximately linearly related to volume for a given specification. There was general agreement that overhead costs were not linear, but were step functions. However, since overhead costs were relatively small in comparison to total costs, they were assumed to be linear. If these costs became important to the final results, they would have to be appropriately incorporated in the model.

Although it is usually clear that production costs must be included in a model of this type, it is not at all clear whether inbound and outbound transportation costs should be included. If they are included, the problem becomes much larger, which means more data must be developed, including projected demand by geographical area as well as the transportation costs themselves. However, if transportation costs are not included, then all production of a product will usually be at one plant regardless of where it is to be shipped. In our particular problem, outbound transportation costs for a given plant could easily offset the production cost advantage of one plant over another. Thus the location of customers, plants, and transportation costs had to be considered in this particular model. In order to hold down the size of the model, we defined each state as a geographic region, and chose a single point within the state as an assumed location of all customer demand in that state. This level of abstraction allowed us to include the outbound transportation costs without getting into great detail.

Given the fact that each specification had different facility requirements and, therefore, different production costs, as well as different transportation costs, the decision becomes: At which plant should a given specification that will be shipped to each geographical area of the country be produced? The variables and facilities are defined in terms of

1. Product specification ($i$).
2. Plant ($j$).
3. Geographical area of shipment ($k$).

and the cost for each variable is by specification ($i$) and in terms of

1. Inbound raw material cost to plant $j$.
2. Production cost at primary facility at plant $j$.
3. Production cost at secondary facility at plant $j$.
4. Outbound transportation cost from plant $j$ to state $k$.

In this relatively simple model, the following constraints were considered:

1. Customer demand by specification for each geographical area:

$$\sum_{j=1}^{n} x_{ijk} \leq d_{ik} \quad i = 1, \ldots, m; k = 1, \ldots, q$$

where $x_{ijk}$ is the amount (measured in thousands of pounds) of product $i$ produced at plant $j$ for shipment to state $k$, $d_{ik}$ is the demand for product $i$ in state $k$, $m$ is the number of products, $n$ is the number of plants, and $q$ is the number of states. (Does the above expression assume that every product demanded must be produced? Is every product necessarily profitable?)

2. Utilization of the two major production facilities: Primary equipment constraints:

$$\sum_{i=1}^{m} \sum_{k=1}^{q} a_{ij} x_{ijk} \leq b_j \quad j = 1, \ldots, n$$

Secondary equipment constraints:

$$\sum_{i=1}^{m} \sum_{k=1}^{q} f_{ij} x_{ijk} \leq g_j \quad j = 1, \ldots, n$$

where $a_{ij}$ is the number of primary equipment production hours required per thousand pounds of product $i$ produced at plant $j$, $b_j$ is the number of primary production hours available at plant $j$, $f_{ij}$ is the number of secondary equipment production hours required per thousand pounds of product $i$ produced at plant $j$, and $g_j$ is the number of secondary production hours available at plant $j$.

3. Nonnegativity constraints

$$x_{ijk} \geq 0 \quad \begin{array}{l} i = 1, \ldots, m \\ j = 1, \ldots, n \\ k = 1, \ldots, q \end{array}$$

We now consider formulation of the objective function. We had a choice of minimizing costs or maximizing profits. The latter approach was chosen for a number of reasons, the most important of which was to reflect the effects of prices. Such a model is more useful, given price as an input. The price is reflected in shadow price information, the main implication of which is that we can determine for which products in which states we should try to generate additional sales. On the other hand, a cost minimization formulation takes demand as given, without any reflection of price upon profits. (See question 9c. at the end of the chapter). Thus, the objective function selected was the following:

$$\text{Maximize } z = \sum_{i=1}^{m} \sum_{j=1}^{n} \sum_{k=1}^{q} r_{ijk} x_{ijk}$$

where $r_{ijk}$ represents the net profits (revenues less all costs) per thousand pounds of product $i$ produced at plant $j$ for shipment to state $k$.

### Data Requirements

For this problem, much data had to be collected. This step takes considerable time and incurs considerable expense in most operations research studies. Thus, during the problem definition stage (or model design stage), the important variables should be clearly pinpointed so that extra expense is not incurred because unnecessary data is collected, or time is not lost because data for certain key variables is not collected initially and has to be gathered later.

For this problem, the questions asked led to collection of the following information:

1. Expected demand: Historical shipments in terms of volume and unit revenue broken down by specification and state for the previous year were supplied by Management Information System group. The Sales Department used this data, along with their knowledge of the market, to estimate the year's demand by specification and state.
2. Equipment capacities and their location: Working with the plants, the Manufacturing Manager's staff estimated the total production hours available on each piece of equipment, excluding all downtime.
3. Production times: The Manufacturing Manager's office outlined the equipment and processing time per 1000 pounds that would be required to produce each specification.
4. Production costs: The Accounting Department provided production cost estimates for both present facilities and potential new facilities. These were reviewed with the Manufacturing Manager's staff.
5. Outbound transportation costs: The Traffic Department estimated the cost of shipping the finished product from each plant to each of the geographical areas mentioned in 1 above.
6. Inbound raw material transportation costs: The Traffic Department estimated the transportation costs of shipping raw materials from our suppliers to each of our plants.

### Construction of the Model

We planned to use a standard linear programming code, supplied by the computer manufacturer, to solve the problem. However, before we could use the code, the information gathered above had to be restructured into the proper matrix format. For this project, a special computer program (a Matrix Generator) was written to transform the raw data into the proper linear programming matrix.

The number of possible variables for Group A products could have been as large as 4500 (i.e., 30 [products] $\times$ 3 [plants] $\times$ 50 [states or geographical locations]). However, since there was no demand for certain products in certain states, 300 variables described all the $x_{ijk}$ for Group A. Similarly, there were 900 variables for Group B products, so the initial model contained 1200 variables (excluding slack variables). The problem had 555 constraints: 550 rows relating to customer demand requirements and 5 rows relating to production capacity.

#### Verification of the Model

In this particular study, the first computer result appeared reasonable—all products were produced and the low cost facilities were operating at or near capacity. A more detailed analysis revealed the computer output from the linear programming code (in terms of primal and dual variables) was almost impossible for the Manufacturing Manager's staff to use. It became clear that we needed a report writer program which would take the results and transform them into a format that was easy for the Manufacturing and Sales people to use. Figure 11.2 shows a reproduction of one of the program outputs in a form that the managers could easily read. We have also organized the input data into a special tabulation so that it can be easily reviewed by management as necessary.

As soon as the results were compiled in a form that was easily understood, they were presented to the Manufacturing and Product Sales Managers and their staffs, as well as to the plant managers at each of the locations.

On those specifications where the managers raised questions, revenues, costs, and equipment utilizations were reviewed with the staff groups, any errors in the input data corrected, and the studies rerun. We then asked the managers to review the results and note any areas that looked unusual. After the third cycle, essentially all the input data problems were eliminated, and a number of specifications were allocated in a different manner than they had been in the past. These shifts in allocation of production represented the profit improvement for which we were searching. However, before actual allocation practices were revised, these changes were reviewed by the Sales Department in light of certain other factors:

1. Would the delivery times associated with the changes be acceptable to customers?
2. Was the quality level for a particular specification at the new plant acceptable to the customer?

In most cases the changes were acceptable, and the associated recommendations were implemented.

#### Additional Studies

Once the base study was completed, we were in a position to answer "What if" questions concerning additional equipment and new facilities. Not only were

| PRODUCT | DESTINATION | PRODUCING PLANT | | | TOTAL |
|---|---|---|---|---|---|
| | | SOUTH-EAST | NORTH-EAST | SOUTH-WEST | |
| 5 | Alabama | 10,081 | 0 | 0 | 10,081 |
| 5 | California | 0 | 5,524 | 0 | 5,524 |
| 5 | Colorado | 0 | 8,976 | 0 | 8,976 |
| 5 | Florida | 9,667 | 0 | 0 | 9,667 |
| 5 | Georgia | 12,084 | 0 | 0 | 12,084 |
| 5 | Illinois | 75,574 | 0 | 0 | 75,574 |
| 5 | Indiana | 0 | 30,381 | 0 | 30,381 |
| 5 | Kansas | 5,317 | 0 | 0 | 5,317 |
| 5 | Kentucky | 0 | 4,005 | 0 | 4,005 |
| 5 | Louisiana | 20,335 | 0 | 0 | 20,335 |
| 5 | Massachusetts | 0 | 13,602 | 0 | 13,602 |
| 5 | Minnesota | 0 | 5,455 | 0 | 5,455 |
| 5 | Missouri | 0 | 48,024 | 0 | 48,024 |
| 5 | New Jersey | 0 | 133,953 | 0 | 133,953 |
| 5 | New York | 0 | 5,869 | 0 | 5,869 |
| 5 | Ohio | 0 | 36,044 | 0 | 36,044 |
| 5 | Pennsylvania | 0 | 21,440 | 0 | 21,440 |
| 5 | Texas | 45,434 | 0 | 0 | 45,434 |
| 5 | Virginia | 0 | 10,147 | 0 | 10,147 |
| 5 | Washington | 0 | 1,588 | 0 | 1,588 |
| 5 | International—S.A. | 0 | 123,458 | 0 | 123,458 |
| 5 | International—Europe | 0 | 1,381 | 0 | 1,381 |
| Product Total | | 178,492 | 449,847 | 0 | 628,339 |
| 6 | Illinois | 0 | 138 | 0 | 138 |
| 6 | New Jersey | 0 | 552 | 0 | 552 |
| Product Total | | 0 | 690 | 0 | 690 |
| 10 | Georgia | 11,837 | 0 | 0 | 11,837 |
| 10 | Ohio | 0 | 137,267 | 0 | 137,267 |
| 10 | Wisconsin | 0 | 11,773 | 0 | 11,773 |
| Product Total | | 11,837 | 149,040 | 0 | 160,877 |
| 51 | Alabama | 6,905 | 0 | 0 | 6,905 |
| 51 | Florida | 22,786 | 0 | 0 | 22,786 |
| 51 | Georgia | 88,381 | 0 | 0 | 88,381 |
| 51 | Illinois | 2,071 | 0 | 0 | 2,071 |
| 51 | Kansas | 1,933 | 0 | 0 | 1,933 |
| 51 | Kentucky | 5,524 | 0 | 0 | 5,524 |
| 51 | Missouri | 5,110 | 2,762 | 0 | 7,872 |
| 51 | New Jersey | 0 | 88,520 | 0 | 88,520 |
| 51 | Ohio | 1,381 | 13,119 | 0 | 14,500 |
| 51 | Pennsylvania | 0 | 41,705 | 0 | 41,705 |
| 51 | Texas | 41,567 | 0 | 0 | 41,567 |
| 51 | Virginia | 7,747 | 0 | 0 | 7,747 |
| 51 | Wisconsin | 0 | 53,167 | 0 | 53,167 |
| 51 | International—S.A. | 587 | 0 | 0 | 587 |
| 51 | International—Europe | 0 | 1,381 | 0 | 1,381 |
| Product Total | | 183,992 | 200,654 | 0 | 384,646 |

**Figure 11.2**

Proposed Allocation of Certain Group B Products (in Thousands of Pounds) we interested in the impact on profitability of the current performance pattern, but also the impact on profitability in future years. For this study, the Product Sales Manager projected the demand by specification and geographical area for the fifth year in the future.

In all the following cases, studies were run for both the current year and the fifth year in the future. The profitability for intermediate years was interpolated. Working with just the two periods reduces the data gathering, manipulation, and computer time required. In similar studies on other product lines, more accuracy was required, so studies were run for each of the five years.

*Case I.* What if additional equipment is installed at the Southwest plant so that it can produce all Group A products?

Production costs for these specifications were estimated, and these variables, with the proper revenue and costs, were added to the linear programming matrix. Studies were run for both years and the resulting profits were compared with base studies described above. Of course, the increased profit could be ascribed to this additional equipment. Making an analysis using the required capital investment and any overhead costs, one could determine if this is a worthwhile investment.

*Case II.* What if additional equipment were installed at all three plants so that feedstock could be economically shipped from the Southwest plant to the other two plants? What if one or both of the primary facilities at the other two plants were shut down?

The analysis process is very much the same as in Case I. Additional production and transportation costs must be estimated, the matrix has to be reconstructed, and each of the studies has to be processed through the computer. Again, this investment opportunity was analyzed in terms of increased profit versus the additional capital investment.

*Case III.* What if one of the present facilities is shut down and some of the secondary equipment is moved to the other location?

*Case IV.* What if all facilities were moved to the Southwest location and production at the other two locations was terminated?

The analysis in Cases III and IV was very similar to that outlined in Cases I and II. All the cases were summarized, and a recommendation prepared and presented to top management.

These four additional studies indicated that there would be significant savings if the low cost single specification plant in the southwest were modified to produce and ship to customers all the Group A specifications. Since a relatively small investment was required to allow this plant to make all Group A products, the projected return on investment for this alternative was very favorable and the alternative was recommended to management.

Case II resulted in substantially greater profits than Case I. However, the investment required to ship the feedstock economically from the Southwest plant to the other two plants was substantial, so the return on the associated investment was marginal. The recommendation to management was that we should not close down the primary facilities at the other two plants at this time. However, since the operating savings are substantial, it was recommended that we continue to explore this alternative, especially ways of reducing the investment required to load and ship the feedstock.

The savings that would result from moving all production facilities to the Southwest location were not nearly as great as had been expected. This location was much farther from the market, so that increased outbound transportation costs offset a significant portion of the savings that resulted from lower overhead costs and lower inbound transportation costs. Also, the investment required to move this equipment was substantial. Thus, our recommendation to management was that we should not consolidate all our production facilities at the Southwest plant at this time.

Demand for the single product produced at the Southwest plant remains high, so management has postponed any modifications to the Southwest plant. However, current plans are to install additional equipment in the Southwest plant so that it can produce all Group A products.

#### Looking Back

The problem was tackled as a linear programming problem because linear programming was a feasible means of approaching the questions asked. In our formulation process it occurred to us that some potentially more efficient solution techniques may have been available (see problem 9 at the end of the chapter). However, computer programs for such methods were not available, and we did not attempt to develop one. (We essentially assumed that time and money costs for such development were prohibitive.)

At the outset of this study we were not fully aware of the nature of questions that management would ask during its course. Thus we did not build in certain capabilities for easy alterations in the matrix generator computer program. It is, of course, impossible to foresee all possible extensions of a given study. Nonetheless, in a future study of this nature we would attempt to utilize a matrix generator structure as flexible as is "reasonable" for possible alteration of the formulation. Similar comments may be made regarding the report writing program.

## 11.11 SUMMARY

In this chapter we have tried to create an awareness of some of the practical aspects and limitations of formulating and solving problems using linear programming. Although we have identified the various aspects of the problem-solving process and attempted to make apparent how much more difficult it is to formulate and solve a real problem, as opposed to a fictitious small problem, one is not fully aware of these differences until one actually applies linear programming to real problems.

### Selected Supplemental References

Section 11.2
[106], [107], [108], [145]
Section 11.5
[96], [312]
Section 11.9
[49], [50]
Applications
[49], [50], [88], [89], [91], [157], [176], [178], [180], [185], [189], [201], [205], [210], [211], [262], [265], [276], [291], [297]
General
[59]

## 11.12 PROBLEMS

**1.** For Example 11.2, prove that the omitted variables, such as cutting four 4 foot lengths from a 16 foot length, are unprofitable, i.e., dominated by more profitable combinations. Also show how the problem formulation is changed if there is a cost of $.05 for each cut made.

**2.** Scale the following problem using the method proposed in section 11.5:

$$\begin{aligned} \text{Maximize } z = 100x_1 + \quad 0.3x_2 & \\ \text{subject to:} \qquad x_1 + 0.001x_2 &\leq 4 \\ 20x_1 + \quad 0.01x_2 &\leq 60 \\ x_1, x_2 &\geq 0 \end{aligned}$$

**3.** A dog food manufacturer can purchase the following materials over the next two month period:

| | MONTH 1 | | MONTH 2 | |
|---|---|---|---|---|
| | QUANTITY | PRICE | QUANTITY | PRICE |
| Pure meat scraps | 150,000 lbs. | $.20/lb. | 100,000 lbs. | $.20/lb. |
| Meat by-products | 75,000 lbs. | $.12/lb. | 75,000 lbs. | $.15/lb. |
| Cereal | unlimited | $.08/lb. | unlimited | $.08/lb. |
| Water | unlimited | free | unlimited | free |

The manufacturer makes two brands of dog food: Albee and Bravo. Albee dog food must have at least 50 per cent meat scrap and at least 30 per cent meat by-products, no cereal, and not more than 5 per cent added water. Bravo dog food requires not more than 60 per cent cereal and not more than 10 per cent added water, the rest being meat or meat by-products. Both dog foods come in 2 pound packages only; Albee sells for $.50 per package, Bravo sells for $.40 per package. There are two operations in manufacturing the dog food:

1. Grinding and mixing the ingredients.
2. Packaging the foods.

The grinding facility can grind and mix up to 300,000 pounds of material (including water) per month. The packaging facility is common to both dog foods and requires 15 seconds per package of Albee and 10 seconds per package of Bravo. Monthly packaging time is 2 full 8 hour shifts, 5 days per week, 4 weeks per month. Changeovers and maintenance are accomplished in the remaining time. Variable grinding and packaging costs are about $.005 per package. Meat may be stored at a cost of $.005 per pound per month, and cereal and meat by-products may be stored at a cost of $.001 per pound per month. The packaged dog food may be stored at a cost of $.01 per package per month. Prior sales commitments are for the following amounts:

| | MONTH 1 | MONTH 2 |
|---|---|---|
| Albee | 40,000 packages | 50,000 packages |
| Bravo | 60,000 packages | 70,000 packages |

In addition, up to 50,000 additional packages of Albee can be sold each month, and up to 80,000 additional packages of Bravo can be sold each month. Finally, a supermarket chain has offered to buy up to 50,000 additional packages per month of Bravo at a price of $.35 per package.

a. What is the company's optimal production plan? (Formulate the problem as a linear programming problem.)
b. Identify the blending, resource allocation, and flow aspects of the problem.
c. Solve the problem via computer, examine the results, and formulate questions relevant to possible additional runs.
d. How does the problem solution change if unlimited additional meat by-products can be purchased each month at $.16 per pound?

**4.** Develop a scheme for including hiring and firing costs in a multiperiod production, inventory, and manpower scheduling linear programming problem.

**5.** Interstate Highway 17 has a toll booth whose 24 hour staffing demands are as follows:

| | TOLL COLLECTORS REQUIRED |
|---|---|
| 12 midnight—6 A.M. | 2 |
| 6 A.M.—10 A.M. | 6 |
| 10 A.M.—12 noon | 4 |
| 12 noon—2 P.M. | 5 |
| 2 P.M.—4 P.M. | 4 |
| 4 P.M.—6 P.M. | 6 |
| 6 P.M.—10 P.M. | 4 |
| 10 P.M.—12 midnight | 3 |

Each toll collector must work a complete 8 hour shift which may start at any hour.

a. If it is desired to have as few toll collectors as possible, what schedule should toll collectors work, and how many should work each shift?
b. How can we determine whether the solution to this problem is integer valued in general?
c. Devise a reasonable criterion for choosing among alternate optima. (An example might be to minimize the maximum number of toll collectors at any one time.)
d. Would the answer to part b change if all demands were in 4 hour intervals (e.g., 10-2, 2-6, 6-10, 10-2, 2-6, 6-10)?

**6.** An investor can borrow funds as follows:

a. \$2000 from an insurance policy at 4 per cent.
b. \$1000 from an insurance policy at 5 per cent.
c. up to 70 per cent of his total portfolio value from his broker at 6 per cent.
d. An unsecured bank loan at 8 per cent for up to \$10,000.

His current portfolio consists of \$10,000 in common stock, which he feels sure will earn him 8.5 per cent on what is invested. He would like to buy some more. The investor is very conservative and insists that he be liquid to the extent that he hold at least $\frac{1}{3}$ of total borrowed funds in bonds which pay 5.5 per cent.

a. Formulate the problem which will maximize his return as a linear programming problem.
b. Are the piecewise nonlinearities of the borrowing rates convex?
c. Solve the problem and pose questions for a postoptimal analysis.

**7.** A company has a monopoly for its two products. The products have the following demand curves during a period:

$$\text{Product 1} \quad p_1 = 12 - x_1$$

$$\text{Product 2} \quad p_2 = 18 - .5x_2$$

where $p_j$ is the price set on product $j$ and $x_j$ is the quantity of product $j$ demanded at that price. The total production of both products is limited to 12 units per time period, with the maximum production of product 1 further limited to 10 units per period.

a. Formulate the nonlinear programming problem of determining optimal prices and production. The constraints are linear, and the objective function is to maximize a separable concave function.

b. Approximate the objective function by piecewise linear approximations, and formulate accordingly. The resulting technique is a form of *separable* programming. (Hint: The term in the objective function for product 1 is $z_1 = (12 - x_1)x_1 = 12x_1 - x_1^2$. A graph of the piecewise linear approximation to this function is given below.)

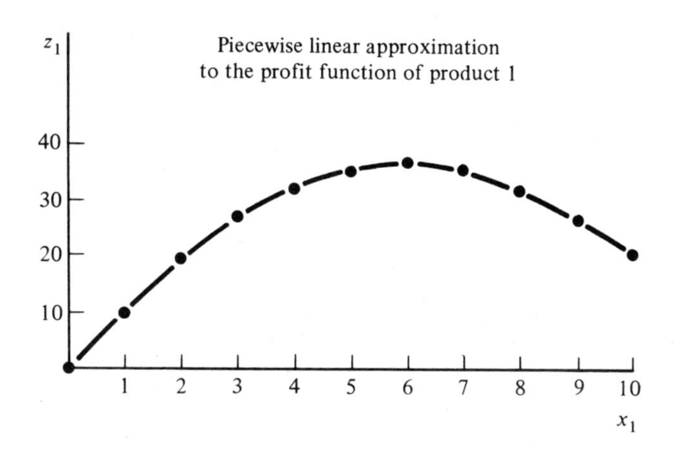


Verify that the above piecewise linear approximation using ten straight line segments makes the objective maximize $z_1$ approximately equivalent to

$$\text{Maximize } z_1 = 11x_{1a} + 9x_{1b} + 7x_{1c} + 5x_{1d} + 3x_{1e} + x_{1f} - x_{1g} - 3x_{1h} - 5x_{1k} - 7x_{1m}$$

$$\text{subject to: } 0 \leq x_{1j} \leq 1 \quad (j = a, b, c, d, e, f, g, h, k, m)$$

c. Formulate the problem using less than ten line segments for approximation for product 1. (For example, use five straight line segments in the graph above.)

d. Solve one or more of the approximations using a computer program, and compare with the continuous optimum of $x_1 = 2$, $x_2 = 10$, and total profits $= 150$.

**8.** A small producer of ready-mixed concrete in sacks purchases cement, sand, and gravel in large quantities. The sand and gravel cost the producer \$4.00 a ton, and the cement costs the producer \$10.00 a ton. The producer dry mixes the cement, sand, and gravel in the given proportions to make the following products, which are bagged by an automatic bagging machine for sale.

| | PRODUCT | | | |
|---|---|---|---|---|
| | SAND | CEMENT | HANDY CONCRETE | MORTAR MIX |
| Weight of package | 100 lbs. | 90 lbs. | 80 lbs. | 80 lbs. |
| Pounds of sand | 100 lbs. | — | 20 lbs. | 60 lbs. |
| Pounds of gravel | — | — | 50 lbs. | — |
| Pounds of cement | — | 90 lbs. | 10 lbs. | 20 lbs. |
| Bagging time | 25 seconds | 22 seconds | 15 seconds | 20 seconds |

The bags cost $.20 each; the direct cost of bagging is about $.01 per second. Sand sells for $1.00 a bag; cement sells for $1.20 a bag; handy concrete and mortar mix both sell for $1.10 a bag. A single bagging facility is used six days a week, eight hours per day. The producer has standing orders (which he must fill) for 3000 bags of sand per month, 6000 bags of cement per month, 10,000 bags of handy concrete, per month, and 5000 bags of mortar mix per month. (Assume $4\frac{1}{3}$ weeks per month). In addition, the sales department estimates that they can sell 2000 more bags of sand per month, 30,000 more bags of cement per month, 10,000 more bags of handy concrete per month, and 3000 more bags of mortar mix per month. All direct costs must be paid in the month accrued, whereas sales receipts occur in the following month. Because of recent heavy expenditures there is only $15,000 available for monthly expenses.

a. Formulate the optimal production plan problem for the producer. Solve the problem.
b. Suppose that the producer utilizes overtime operations at 1.5 times the day-time cost. (The machine can work at most six 24 hour days per week.) How does this change the solution?
c. Ignoring part b, suppose that the producer can borrow up to $20,000 for one month at .75 per cent per month. How is the optimal solution changed?
d. How do the changes in parts b and c together affect the optimal solution?
e. For parts a, b, c, and d, discuss the relative profitability of trying to generate additional sales by absorbing transportation costs, etc.
f. The producer is considering buying a ready-mixed concrete truck to supply contractors with ready-to-pour concrete on their jobs. The materials would be loaded dry into the truck (no processing is required) and water added. How can the solution to the linear programming problem assist the producer in determining whether or not to buy a truck? What additional data is needed?

**9.** Refer to the case history of section 11.10.

a. Can this problem be formulated using the method of generalized upper bounds described in Chapter 8? If so, formulate it accordingly.
b. Can it be partially or completely formulated using one of the network flow methods described in Chapter 9?

c. If a cost minimization objective function had been utilized, what change in the demand constraints would have been necessary? Why?
d. Discuss the use of shadow price information on future sales efforts for certain product-state combinations in the formulation, and for product-state combinations not in the formulation.

# 12

# Integer Programming —An Introduction

## 12.1 INTRODUCTION AND MOTIVATION

Thus far we have considered only linear programming problems and methods, as well as schemes directly based upon linear programming solution methods. With such minor exceptions as the transportation problem (with integral source and destination requirements), obtaining an optimal solution in which one or more variables are integer, as prespecified, is largely a question of chance (that is, if the problem happens to satisfy certain number theoretic conditions). More generally, for an *arbitrary* linear programming problem, having an optimal solution satisfy certain prespecified integrality conditions is very unlikely.

Provided that there is some valid reason why the class of linear programming problems with integer solutions is important, we can make a case for methods to solve such problems. It is useful to consider a few problems that require integer solutions. A manufacturer of automobiles, in planning production of an automobile, may find that a linear program solution calls for 589,635.23 automobiles of a particular model type to be produced in a year. By virtue of the large number, one would intuitively suggest that 589,635 or 589,636 or even 589,000 or 590,000 automobiles would be "close enough" to optimality. One would probably round such a solution because the input data which led to it (either a coefficient of the $\boldsymbol{A}$ matrix or of the $\boldsymbol{b}$ vector) might be ever so slightly altered to produce an integer answer, with virtually negligible effect on the optimality of the solution. Indeed, such an analysis would appear to be true for all linear programming problems with large integer solution values.

Practically speaking, the statement is correct, although pathological examples can be constructed in which the optimal integer solution is not at all "close" to the optimal (noninteger) solution to the linear programming problem.

On the other hand, the importance of an integer solution is usually much more evident for problems involving small integers than for problems involving large integers. Suppose that in choosing among capital investment projects the optimal solution to an associated linear programming problem is to build .527 of a new plant. Clearly, the choice must be to build either zero or one new plant. In such a case we may solve two linear programming problems: one assuming that the plant will be built, the other assuming that it will not be built.

If there are a relatively small number of such investment projects, all possible combinations may be examined and solved for, using one linear programming formulation for each. Solution of all possible linear programming problems is generally not a practical scheme because the number of possible problems that must be solved becomes enormous as the number of investment projects increases. For example, 20 different decisions each with 2 possible actions has $2^{20}$ (which is over one million) associated combinations of actions or strategies.

In this chapter and those that follow we shall consider linear programming problems in which some variables must be integer valued, and methods for solving integer programming problems. Before proceeding, it is useful to define a few terms.

**Definition 12.1**

*An integer variable is one which must take on an integer value in any feasible solution.*

**Definition 12.2**

*A continuous variable is one which may take on any real value in any feasible solution.*

Ordinarily, both integer and continuous variables are constrained to be nonnegative in value.

**Definition 12.3**

*An all-integer problem is one in which all variables are integer variables.*

**Definition 12.4**

*A mixed-integer problem is one in which some but not all variables are integer variables.*

**Definition 12.5**

*An optimal noninteger solution to an integer programming problem is an optimal solution obtained by ignoring any integrality requirements.*

**Definition 12.6**

*An optimal integer solution to an integer programming problem is a best integer solution to the problem.*

## 12.2 SIGNIFICANCE AND APPLICATIONS OF INTEGER PROGRAMMING METHODS

Reasons for solving integer programming problems were presented in the previous section. Even if the importance of solving integer programming problems were limited to those, there would still be sufficient reason for having integer programming methods! Interestingly enough, there are other important reasons for integer programming, many of which were originally put forth in an excellent paper by Dantzig [64]. It can be stated that *any deterministic problem* which can be precisely described in quantitative terms can be formulated approximately as accurately as desired as a mixed-integer programming problem. We use the word formulated rather than solved because there may be too many constraints and variables required for the approximation to be economically solved. Nonetheless, integer programming is valuable because virtually any deterministic problem may conceptually be formulated using it. The useful aspect of the conceptualization is that the problem once so formulated will often give rise to other methods, either exact or approximate, for solving the original problem.

An important reason for using integer variables is to represent constraint sets which are nonconvex. We shall now examine some specific cases of nonconvex constraint sets. First, consider a production process which has two cost components—a fixed cost which must be incurred if the process is utilized at all, and a variable or per unit cost which is the same for each unit produced. Production of zero units has a cost of zero, whereas both fixed and variable costs are incurred when production is positive. A graph of total cost versus number of units produced appears in Figure 12.1. Algebraically,

$$\text{total cost} = \begin{Bmatrix} 0, & x = 0 \\ c_f + c_v x, & x > 0 \end{Bmatrix}$$

It may appear at first that the fixed charge problem could be handled without integer variables. However, it can be shown that any noninteger scheme will either incur the full fixed cost $c_f$ or none of it for all levels of production including zero, or will charge a prorated amount of the fixed cost $c_f/u$ for each unit produced (or some combination of the two). All such noninteger formulations are incorrect. The following representation using an integer variable does accommodate the fixed cost problem:

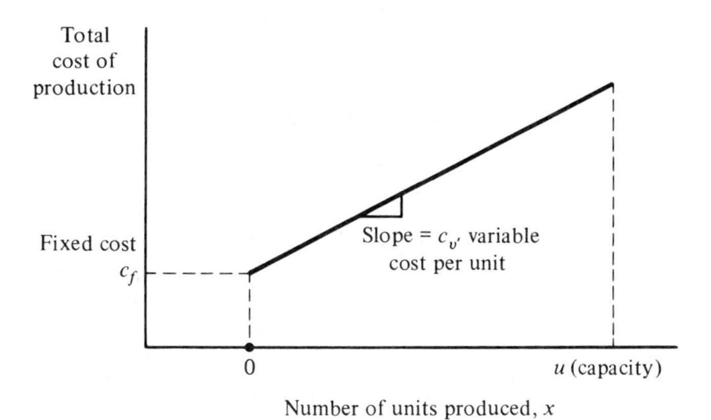


**Figure 12.1**

Total Costs as a Function of Units Produced for a Fixed Cost Process

$$
\begin{aligned}
&\text{Minimize } c_v x + c_f \delta \\
&\text{subject to: } x - u\delta \leq 0 \\
&\qquad\qquad \delta \leq 1 \\
&\qquad\qquad x, \delta \geq 0, \quad \delta \text{ integer}
\end{aligned}
\tag{12.1}
$$

(This formulation is part of a larger problem with other constraints and variables.) When $x$ is zero, $\delta$ is zero and total cost is zero. When $x$ is positive, $\delta$ is one and the fixed cost is incurred.

A generalization of the above may be made to an arbitrary nonconvex cost structure such as shown in Figure 12.2. To accommodate such a framework, $x_1$ (the amount that can be produced at cost $c_1$) must be $u_1$ before $x_2$ (the amount that can be produced at cost $c_2$) can be positive, and so on. Hence there is an upper limit constraint as before (i.e., $x_i \leq u_i\delta_{i-1}$), and in addition a lower limit constraint, i.e., $x_i \geq u_i\delta_i$ where the $\delta_i$'s are zero-one variables. Variable $\delta_i$ is one when production is in the appropriate range (e.g., $\delta_1 = 1$ when production is between $u_1$ and $u_1 + u_2$).

The following portion of a problem illustrates this:

$$
\begin{array}{lllllll}
\text{Minimize} & c_f\delta_0 & & & + c_1x_1 + c_2x_2 + c_3x_3 & \\
\text{subject to:} & -u_1\delta_0 & & & + x_1 & \leq 0 \\
& & - u_1\delta_1 & & + x_1 & \geq 0 \\
& & - u_2\delta_1 & & + x_2 & \leq 0 \\
& & & - u_2\delta_2 & + x_2 & \geq 0 \\
& & & - u_3\delta_2 & + x_3 & \leq 0 \\
& & & & 0 \leq \delta_0, \delta_1, \delta_2 \leq 1 \text{ and integer} & \\
& & & & x_1, x_2, x_3 \geq 0 &
\end{array}
\tag{12.2}
$$

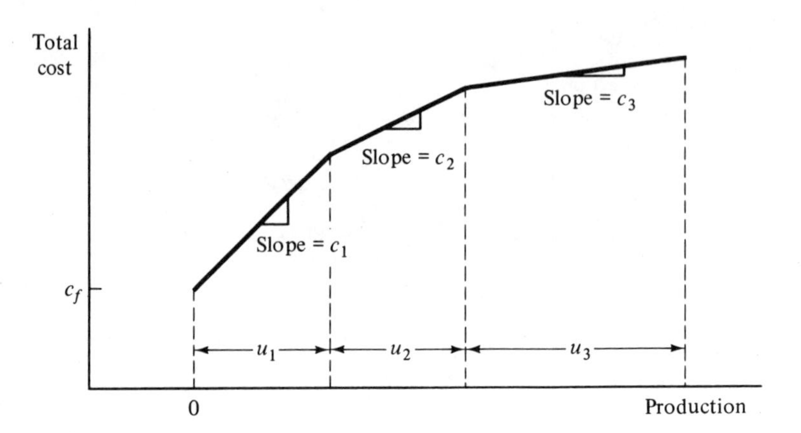


**Figure 12.2**
A More Complex Nonconvex Cost Structure

By observing that $x_1$ and $x_2$ are definitional in the second and fourth constraints respectively (as defined in Chap. 4), we may solve for them, using $s_2$ and $s_4$ as slack variables in the respective constraints, and eliminate $x_1$ and $x_2$ to achieve the following more economical representation:

$$
\begin{array}{llr}
\text{Minimize} & c_f\delta_0 + c_1u_1\delta_1 + c_2u_2\delta_2 + c_3x_3 + c_1s_2 + c_2s_4 & \\
\text{subject to:} & -u_1\delta_0 + u_1\delta_1 + s_2 & \leq 0 \\
& -u_2\delta_1 + u_2\delta_2 + s_4 & \leq 0 \\
(12.3) & -u_3\delta_2 + x_3 & \leq 0 \\
& & 0 \leq \delta_0, \delta_1, \delta_2 \leq 1 \\
& & x_3, s_2, s_4 \geq 0 \\
& & \delta_0, \delta_1, \delta_2 \text{ integer}
\end{array}
$$

which is an equivalent but more compact representation than (12.2). It appears as if further reduction can be achieved by noting that $\delta_0$, $\delta_1$, and $\delta_2$ are definitional. The variables are indeed definitional, but if eliminated there is no straightforward way to guarantee their integrality.

A graph of another type of nonconvex structure is shown in Figure 12.3. In this case the feasible region is nonconvex. Its representation is given by the portion of a problem in constraint set (12.4). The first three constraints in the formulation constitute the physical constraints when $x_1$ is in the appropriate range and the corresponding $\delta_j$ is 1. The last three constraints enforce the requirements on the $\delta_j$ and the relationship between $x_1$ and the $\delta_j$.

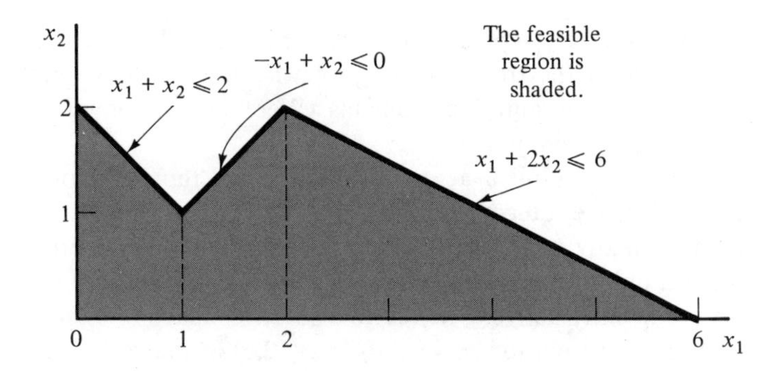


**Figure 12.3**

The Nonconvex Set Represented by the Constraint Set (12.4)

$$
\begin{array}{rrrrrrl}
8\delta_1 & & & + x_1 & + & x_2 & \leq 10 \\
& + 4\delta_2 & & - x_1 & + & x_2 & \leq 4 \\
& & & x_1 & + & 2x_2 & \leq 6 \\
\delta_1 & + \delta_2 & + \delta_3 & & & & = 1 \\
& \delta_2 & + 2\delta_3 & - x_1 & & & \leq 0 \\
-\delta_1 & - 2\delta_2 & - 6\delta_3 & + x_1 & & & \leq 0 \\
& & & & x_1, & x_2 & \geq 0 \\
& & & \delta_1, & \delta_2, & \delta_3 & \geq 0 \quad \text{and integer}
\end{array} \tag{12.4}
$$

For $x_1$ in the interval $0 \leq x_1 \leq 1$, $\delta_1 = 1$, and constraint 1 becomes $x_1 + x_2 \leq 2$. For $x_1$ in the interval $1 \leq x_1 \leq 2$, $\delta_2 = 1$, and constraint 2 becomes $-x_1 + x_2 \leq 0$. For $x_1$ in the interval $2 \leq x_1 \leq 6$, $\delta_3 = 1$ and only constraint $x_1 + 2x_2 \leq 6$ is in effect. Since the third constraint always holds, it does not contain $\delta_3$. The fourth constraint requires that exactly one $\delta_j$ be one, which corresponds to $x_1$ being in exactly one of the three ranges. The fifth constraint expresses a lower bound on $x_1$ ($x_1 \geq 0$ when $\delta_1 = 1$, $x_1 \geq 1$ when $\delta_2 = 1$, and $x_1 \geq 2$ when $\delta_3 = 1$). The last constraint expresses an upper bound on $x_1$ ($x_1 \leq 1$ when $\delta_1 = 1$, $x_1 \leq 2$ when $\delta_2 = 1$, and $x_1 \leq 6$ when $\delta_3 = 1$). As in the previous example, $x_1$ is definitional in the fifth constraint in (12.4). Denoting $s_5$ as the slack in that constraint and substituting for $x_1$ we obtain

$$
\begin{array}{rrrrrl}
8\delta_1 & + \delta_2 & + 2\delta_3 & + x_2 & + s_5 & \leq 10 \\
& 3\delta_2 & - 2\delta_3 & + x_2 & - s_5 & \leq 4 \\
& \delta_2 & + 2\delta_3 & + 2x_2 & + s_5 & \leq 6 \\
\delta_1 & + \delta_2 & + \delta_3 & & & = 1 \\
-\delta_1 & - \delta_2 & - 4\delta_3 & & + s_5 & \leq 0
\end{array} \tag{12.5}
$$

a more compact representation for the set of Figure 12.3. This approach is equivalent to the treatment of piecewise linear approximations for separable (nonlinear) programming problems. (Were the sets convex, integer programming would not be required.)

We can establish a general method for representation of nonconvex sets using zero-one integer variables, but the procedure is cumbersome and usually requires many constraints and variables. It involves representing a nonconvex set as a union of convex sets, using zero-one variables to indicate which convex set is currently active. The feasible solution space of Figure 12.3 and its representation by equation sets (12.4) and (12.5) is an example of this method.

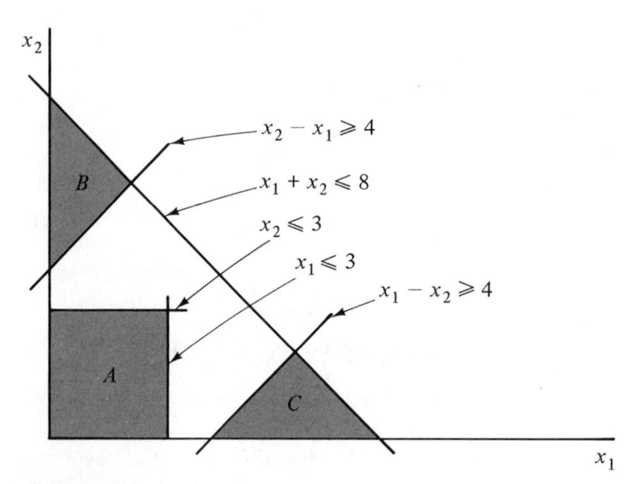


**Figure 12.4**
The Nonconvex Set Represented by Constraint Set (12.6)

A second example is shown in Figure 12.4. The three disjoint convex sets are shaded and their constraints are as in expression (12.6).

$$
\begin{aligned}
x_1 + x_2 &\leq 8 \\
x_1 + 7\delta_1 &\leq 10 \\
x_2 + 7\delta_1 &\leq 10 \\
-x_1 + x_2 - 19\delta_2 &\geq -15 \\
x_1 - x_2 - 19\delta_3 &\geq -15 \\
\delta_1 + \delta_2 + \delta_3 &= 1 \\
x_1, x_2, \delta_1, \delta_2, \delta_3 &\geq 0 \\
\delta_1, \delta_2, \delta_3 &\text{ integer}
\end{aligned}
\tag{12.6}
$$

When $\delta_1 = 1$, the solution is in region A; when $\delta_2 = 1$, the solution is in region B; when $\delta_3 = 1$, the solution is in region C.

Provided that the objective function is linear for the above type of problem (e.g., the problem given by expressions (12.4) and (12.6)), it is sufficient to represent only the convex hull of a nonconvex set as a convex set (which generally is not easy), and then solve the linear programming problem over the convex hull. In such instances, the integer variables are not required.

The above applications involve the use of zero-one variables, which are also called Boolean or Bivalent variables. The use of such variables permits, as has been shown, the enforcing of logical conditions. Such use may be extended to any operation from Boolean algebra. The following are a few examples:

| | |
|---|---|
| $\delta_1 + \delta_2 \leq 1$ | either $\delta_1 = 1$ or $\delta_2 = 1$ or neither |
| $\delta_1 + \delta_2 = 1$ | either $\delta_1 = 1$ or $\delta_2 = 1$ but one must equal one |
| $\delta_1 + \delta_2 \geq 1, \quad \delta_1 \leq 1, \quad \delta_2 \leq 1$ | either $\delta_1 = 1$ or $\delta_2 = 1$ or both |
| $\delta_1 + \delta_2 + \delta_3 = 2$ | any two of $\delta_1, \delta_2, \delta_3$ must be 1 |

($0 \leq \delta_1, \delta_2, \delta_3 \leq 1$ and integer in all cases.)

The first example constraint given above is equivalent to the nonlinear condition $\delta_1\delta_2 = 0$ where $\delta_1$ and $\delta_2$ are zero or one. Other constraints enforce similar logical conditions.

To treat nonlinear objectives and constraints, piecewise linear approximations may be utilized to approximate the functions as accurately as desired. Integer variables should be used to represent any nonconvexities. The approach would appear to be impractical for the most general class of problems, but for problems in which relatively few variables and constraints are added the method appears useful. (An example of a special class of nonlinear programming problems is given in problem 7 of Chapter 11.)

Two important classifications of problems amenable to integer programming formulation are covering problems and matching problems, both of which we shall consider now.

A weighted set covering problem has the following formulation:

$$\text{Minimize} \quad z = \sum_{j=1}^{n} c_j x_j$$

$$\text{subject to:} \sum_{j=1}^{n} a_{ij} x_j \geq a_{i0} \qquad i = 1, \ldots, m \tag{12.7}$$

$$x_j = 0 \text{ or } 1$$

$$\text{with all } a_{i0} = 1, \quad a_{ij}\ (j \neq 0) = 0 \text{ or } 1$$

This problem can often be simplified by using certain dominance and feasibility seeking considerations, but the problem form is not changed when such simplifications are employed. The set covering problem is the same as expression (12.7) with all $c_j = 1$, and the simple set covering problem is the set covering problem with the further requirement that exactly two entries $a_{ij}$ of each column $j$ be unity. A special algorithm for the simple set covering problem using the graph of a network associated with expression (12.7) is more efficient than integer programming methods. (See, for example, Normán and Rabin [219] or Balinski [15].)

The simple matching problem is closely related to the simple set covering problem, and an optimal solution to one may be found from an optimal solution to the other. The formulation of the simple matching problem is as follows:

$$\text{(12.8)} \qquad \begin{aligned} &\text{Maximize } \sum_{j=1}^{n} x_j \\ &\text{subject to: } \sum_{j=1}^{n} a_{ij} x_j \leq a_{i0} \\ &\qquad\qquad x_j = 0 \text{ or } 1 \end{aligned}$$

with all $a_{i0} = 1$, $a_{ij}(j \neq 0) = 1$ for precisely two entries of column $j$, and zero otherwise. There are special algorithms for solving simple covering and matching problems, and these are given by Balinski [15], among others.

An example of a simple matching problem is the assignment of men to jobs, assuming that every man can have at most one job and every job can have at most one man. The object is to maximize the number of men having (or matched to) jobs. Example 12.1 illustrates an application of the weighted set covering problem.

EXAMPLE 12.1:

A large film processor is selecting processing facility sites. Because of time constraints, the processor does not want to transport the film from user to processing facility more than 150 miles from a major city or more than 500 miles from any other location. Given the fixed costs for each facility, and assuming that any point will be served by the nearest processing facility, and that the *variable* cost of processing at every location is the same, what is the set of processing facility sites that minimizes the total fixed costs? In the associated formulation, a variable corresponds to a site, and a constraint corresponds to a user location.
The users (or set) are thus to be serviced (or covered) by a collection of processing facility sites, each having a different fixed cost (or weight).

Another important integer programming problem is the knapsack problem, which may be thought of as the problem facing a camper who is packing his knapsack in preparation for a hike. He has $n$ different types of items. Each unit of item $j$ weighs $a_j$ pounds. He can carry up to $p$ pounds. Each unit of item $j$ has value $c_j$ to him on the hike. The camper desires to know the maximum value knapsack he can carry satisfying the weight constraint. The corresponding formulation is as follows:

$$\text{Maximize } z = \sum_{j=1}^{n} c_j x_j$$

$$\text{(12.9)} \qquad \text{subject to: } \sum_{j=1}^{n} a_j x_j \leq p$$

$$x_j \geq 0 \text{ and integer}$$

$$j = 1, \ldots, n$$

A solution to this problem may be found efficiently via dynamic programming (see, for example, Gilmore and Gomory [106]) or implicit enumeration (see Chap. 16).

The last formulation which we shall consider briefly is the traveling salesman problem. This is the problem of a salesman who must visit $n$ towns and return to his starting point. He desires to minimize his total travel time. This problem has a number of variations which are real problems, and it has been extensively studied. Its formulation is asked in the problems at the end of the chapter. Many methods have been developed to find good but not necessarily optimal solutions, and a special branch-and-bound method is presented by Little *et al.* [199].

The application possibilities of integer programming methods to real problems are enormous. A survey of integer programming applications and uses (as well as methods) has been prepared by Balinski [15]. Additional applications are found in Dantzig [64]. Many problems are actually solved using integer programming, but the limiting factor is the large number of constraints and variables required for problem representation, and the limited capabilities of available methods. Some specific application types are given in the problems.

That a number of special forms of integer programming problems have special solution methods suggests that the perceived power and generality of integer programming may be a bit *too general* (and too expensive), and that special methods to solve certain categories of problems may be appropriate. This sentiment has been expressed by a number of researchers, including Graves and Whinston [131]. Of course, as better integer programming algorithms become available, the luxury of using a general purpose integer programming technique will become more affordable.

## 12.3 A SIMPLIFIED THEORETICAL FRAMEWORK—A UNIFIED APPROACH*

The methods for solving integer programming problems are seemingly very diverse. The purpose of this section is to introduce a simple theoretical frame-

* This section is based on Zionts [315].

work which is useful in explaining important aspects of many of the integer programming algorithms. This framework is intended as a unifying approach, and will be used to develop certain of the algorithms as we proceed.

To do this we shall present and explore aspects of the Extended Geometric Definition Method [313] as applied to integer linear programming problems. We shall show that a number of important integer programming results employ devices that can be related to the consequences of the Extended Geometric Definition Method. We shall first present some material on the Extended Geometric Definition Method, and then develop some results for integer variables.

Any linear programming problem can be written in the following canonical form *at any iteration*:

$$\text{(12.10)}\qquad \begin{aligned} &\text{Maximize } z = \sum_{j=1}^{n} c_j x_j \\ &\text{subject to:} \quad \sum_{j=1}^{n} a_{ij} x_j + x_{n+i} = b_i \qquad i = 1, \ldots, m \\ &\qquad\qquad\qquad\qquad x_j \geq 0, \qquad j = 1, \ldots, n + m \end{aligned}$$

Preserving this form (which is only for expository convenience) will require renumbering of the variables in general, after each iteration. $x_j$ $(j = 1, \ldots, n)$ are the nonbasic variables and $x_j$ $(j = n + 1, \ldots, n + m)$ are the basic variables. The $b_i$'s are the values of the basic variables, and the $c_j$'s are the negatives of the reduced costs. The matrix of $a_{ij}$'s are the matrix of coefficients. The tableau will be written in the usual manner.

### Outline of the Extended Geometric Definition Method

Every variable $x_j$ in a linear programming problem can be assumed to possess both an upper bound $u_j$ and a lower bound $h_j$, such that $h_j \leq x_j \leq u_j$. Note that $h_j = 0$, $u_j = M$, $M$ sufficiently large, corresponds to the most general case where only the (weakest) condition $x_j \geq 0$ is known to be fulfilled. (Tighter bounds may be found by solving ordinary linear programming problems, maximizing and minimizing the variable $x_j$ over the constraint set if desired.) The Extended Geometric Definition Method consists of setting bounds on every variable (both primal and dual) and then adjusting them where possible after each simplex method iteration by increasing lower bounds and reducing upper bounds, using results that will be presented. Denote the coefficients of the tableau as before and let

$$\text{(12.11)}\qquad a_{ij}^{+} = \begin{pmatrix} a_{ij} \text{ if } a_{ij} > 0 \\ 0 \text{ if } a_{ij} \leq 0 \end{pmatrix} \qquad a_{ij}^{-} = \begin{pmatrix} 0 \text{ if } a_{ij} > 0 \\ a_{ij} \text{ if } a_{ij} \leq 0 \end{pmatrix}$$

We may now state and prove the following theorems.

**THEOREM 12.1**

CONSIDER ANY LINEAR PROGRAMMING SOLUTION WHOSE TABLEAU ELEMENTS ARE $a_{ij}$. IF, FOR ANY CONSTRAINT $i$ THE FOLLOWING EXPRESSION IS TRUE:

$$h_{n+i} \equiv b_i - \sum_{j=1}^{n} a_{ij}^{+} u_j - \sum_{j=1}^{n} a_{ij}^{-} h_j > 0 \tag{12.12}$$

THEN $h_{n+i}$ IS A LOWER BOUND FOR $x_{n+i}$ (THE VARIABLES BASIC IN THE $i^{\text{th}}$ CONSTRAINT), AND THE CONSTRAINT MAY BE DROPPED FROM FURTHER CONSIDERATION BECAUSE $x_{n+i}$ WILL BE BASIC IN EVERY OPTIMAL SOLUTION.

PROOF:

Given a basic feasible solution arranged as in (12.10), we may write each basic variable as

$$x_{n+i} = b_i - \sum_{j=1}^{n} a_{ij} x_j \quad \text{or} \quad x_{n+i} = b_i - \sum_{j=1}^{n} a_{ij}^{+} x_j - \sum_{j=1}^{n} a_{ij}^{-} x_j$$

Since $h_j \leq x_j \leq u_j$ for all $j$, we may write

$$x_{n+i} = b_i - \sum_{j=1}^{n} a_{ij}^{+} x_j - \sum_{j=1}^{n} a_{ij}^{-} x_j \geq b_i - \sum_{j=1}^{n} a_{ij}^{+} u_j - \sum_{j=1}^{n} a_{ij}^{-} h_j = h_{n+i} > 0$$

Hence, $x_{n+i}$ must be basic in every optimal solution.

**THEOREM 12.2**

AT ANY LINEAR PROGRAMMING SOLUTION, AN UPPER BOUND FOR A BASIC VARIABLE $x_{n+i}$ IS GIVEN BY

$$u_{n+i} \equiv b_i - \sum_{j=1}^{n} a_{ij}^{+} h_j - \sum_{j=1}^{n} a_{ij}^{-} u_j$$

WHERE $i$ IS THE INDEX OF THE ROW IN WHICH $x_{n+i}$ IS BASIC AND $a_{ij}$ ARE THE TABLEAU COEFFICIENTS OF THE PARTICULAR SOLUTION.

PROOF:

As in the proof of Theorem 12.1

$$x_{n+i} = b_i - \sum_{j=1}^{n} a_{ij} x_j = b_i - \sum_{j=1}^{n} a_{ij}^{+} x_j - \sum_{j=1}^{n} a_{ij}^{-} x_j$$

Since $h_j \leq x_j \leq u_j$

$$x_{n+i} = b_i - \sum_{j=1}^{n} a_{ij}^{+} x_j - \sum_{j=1}^{n} a_{ij}^{-} x_j \leq b_i - \sum_{j=1}^{n} a_{ij}^{+} h_j - \sum_{j=1}^{n} a_{ij}^{-} u_j = u_{n+i}$$

or

$$x_{n+i} \leq u_{n+i}$$

Hence, $u_{n+i}$ is an upper bound for $x_{n+i}$.

We shall not pursue the Extended Geometric Definition Method further here, except to note that Theorems 12.1 and 12.2 hold for the dual problem, and that bounds for nonbasic variables can be computed in a manner similar to that used in these two theorems. The Extended Geometric Definition Method was proposed to be used in conjunction with linear progromming solution methods, with information on bounds to be used as developed. Although the computational aspects of the Extended Geometric Definition Method have not been fully explored, it does not appear promising as a means for solving linear programming problems efficiently. The interested reader should consult [313] for additional information on this topic.

### Problems Involving Integer Variables

Problems in which some or all variables must be integral can make use of sharpened versions of the above two theorems. Specifically, a lower bound for an integer variable $x_p$ is

$$h'_p = \langle h_p \rangle$$

where $h_p$ is the calculation given in Theorem 12.1 and $\langle x \rangle$ indicates the smallest integer $y$ not less than $x$, i.e.,

$$\langle x \rangle = \min y \geq x, y \text{ integer}$$

An upper bound for an integer variable $x_p$ is

$$u'_p = [u_p]$$

where $u$ is the calculation indicated in Theorem 12.2 and $[x]$ is the largest integer $y$ not greater than $x$, i.e.,

$$[x] = \text{Max } y \leq x, y \text{ integer}$$

Henceforth, we shall use $u_p$ and $h_p$ as the upper and lower bounds respectively, without distinguishing between integer and noninteger variables.

To make the results more concrete, suppose that we consider the $i^{\text{th}}$ constraint at any iteration as

$$\sum_{j=1}^{m+n} (a_{ij}^+ + a_{ij}^-)x_j = b_i$$

(Note that $a_{i,n+i}^+ = 1$, $a_{i,n+i}^- = 0$, and $a_{i,n+k}^+ = a_{i,n+k}^- = 0$, $k \neq i$.) Then we can state the following theorem.

**THEOREM 12.3**

UPPER AND LOWER BOUNDS FOR INTEGER VARIABLES $x_j$ ARE GIVEN AS BELOW.
1. FOR $a_{ij}^+ > 0$, A LOWER BOUND FOR INTEGRAL $x_j$ IS GIVEN BY

$$h_j = \max\left\{0, \left\langle \frac{b_i}{a_{ij}^+} - \frac{1}{a_{ij}^+}\sum_{k\neq j} a_{ik}^+ u_k - \frac{1}{a_{ij}^+}\sum_{k\neq j} a_{ik}^- h_k \right\rangle\right\}$$

AND AN UPPER BOUND FOR INTEGRAL $x_j$ IS GIVEN BY

$$u_j = \left[\frac{b_i}{a_{ij}^+} - \frac{1}{a_{ij}^+}\sum_{k\neq j} a_{ik}^+ h_k - \frac{1}{a_{ij}^+}\sum_{k\neq j} a_{ik}^- u_k\right]$$

2. FOR $a_{ij}^- < 0$, A LOWER BOUND FOR INTEGRAL $x_j$ IS GIVEN BY

$$h_j = \max\left\{0, \left\langle \frac{b_i}{a_{ij}^-} - \frac{1}{a_{ij}^-}\sum_{k\neq j} a_{ik}^+ h_k - \frac{1}{a_{ij}^-}\sum_{k\neq j} a_{ik}^- u_k \right\rangle\right\}$$

AND AN UPPER BOUND FOR INTEGRAL $x_j$ IS GIVEN BY

$$u_j = \left[\frac{b_i}{a_{ij}^-} - \frac{1}{a_{ij}^-}\sum_{k\neq j} a_{ik}^+ u_k - \frac{1}{a_{ij}^-}\sum_{k\neq j} a_{ik}^- h_k\right]$$

I.E., WE ROUND LOWER BOUNDS UP AND UPPER BOUNDS DOWN, IF THEY ARE NOT INTEGER.

The proof will not be given, since it follows in a straightforward manner from Theorems 12.1, 12.2, and earlier comments. Theorem 12.3 is applicable to both all-integer and mixed-integer problems. For problems of the latter class, the results of Theorems 12.1 and 12.2 are required to compute the bounds for noninteger variables. Logical inconsistencies (no feasible solution) are indicated by lower bounds greater than upper bounds.

In future chapters we shall use the results of Theorem 12.3 to explain a number of integer programming devices and methods. For now, we shall make a few remarks and then present an example. First note that the computations of Theorem 12.3 simplify greatly for variables which are zero-one. (All calculations reduce to additions and subtractions.) Second, a large number of published integer programming example problems (used to illustrate other methods) have been solved simply using the results of Theorem 12.3 in conjunction with the simplex method. It can easily be shown that problems cannot be solved in this way in general, but the fact that a number of published examples yield to the simplified method attests to the potential power of the framework. An example of the success of the simplified method follows.

EXAMPLE 12.2:

Solve the following problem of Gomory [123] using the simplex method plus implied integer bounds on variables. (The problem was used by Gomory to illustrate his fractional cut algorithm.)

$$\begin{aligned} \text{Maximize } z &= 4x_1 + 5x_2 + x_3 \\ \text{subject to:} \quad & 3x_1 + 2x_2 + x_4 = 10 \\ & x_1 + 4x_2 + x_5 = 11 \\ & 3x_1 + 3x_2 + x_3 + x_6 = 13 \\ & x_j \geq 0 \text{ and integer} \end{aligned}$$

We shall illustrate the development of the bounds. Initial bounds may be taken as zero and infinity. Then, by solving for $x_1$ in each constraint, we have

$$x_1 = \tfrac{10}{3} - \tfrac{2}{3}x_2 - \tfrac{1}{3}x_4 \text{ or } x_1 \leq 3$$
$$x_1 = 11 - 4x_2 - x_5 \text{ or } x_1 \leq 11$$
$$x_1 = \tfrac{13}{3} - x_2 - \tfrac{1}{3}x_3 - \tfrac{1}{3}x_6 \text{ or } x_1 \leq 4$$

Therefore, $x_1 \leq 3$ is the most restrictive constraint. Similarly, we have the following bounds:

$$x_2 \leq 2$$
$$x_3 \leq 13$$
$$x_4 \leq 10$$
$$x_5 \leq 11$$
$$x_6 \leq 13$$

No lower bounds can be deduced from the constraints. In order to avoid explicit representation of the upper bound constraints, $x_j \leq u_j$, we use the method of upper bounds described in Chapter 8. As was the convention there, $\bar{x}_j$ is the complement of $x_j$, with $\bar{x}_j = u_j - x_j$.
The following sequence of tableaus is obtained:

**Tableau 1**

| | | $x_1$ | $x_2$ ↓ | $x_3$ | $x_4$ | $x_5$ | $x_6$ |
|---|---|---|---|---|---|---|---|
| $z$ | 0 | −4 | −5 | −1 | 0 | 0 | 0 |
| $x_4$ | 10 | 3 | 2 | 0 | 1 | 0 | 0 |
| $x_5$ | 11 | 1 | 4 | 0 | 0 | 1 | 0 |
| $x_6$ | 13 | 3 | 3 | 1 | 0 | 0 | 1 |

The first iteration is an upper bound iteration, with $x_2$ going to its upper bound.

**Tableau 2**

| | | $x_1$ ↓ | $\bar{x}_2$ | $x_3$ | $x_4$ | $x_5$ | $x_6$ | |
|---|---|---|---|---|---|---|---|---|
| $z$ | 10 | −4 | 5 | −1 | 0 | 0 | 0 | |
| $x_4$ | 6 | [3] | −2 | 0 | 1 | 0 | 0 | ⟶ |
| $x_5$ | 3 | 1 | −4 | 0 | 0 | 1 | 0 | (Remember |
| $x_6$ | 7 | 3 | −3 | 1 | 0 | 0 | 1 | $\bar{x}_j = u_j - x_j$) |

**Tableau 3**

| | | $x_1$ | $\bar{x}_2$ | $x_3$ ↓ | $x_4$ | $x_5$ | $x_6$ | |
|---|---|---|---|---|---|---|---|---|
| $z$ | 18 | 0 | $\frac{7}{3}$ | −1 | $\frac{4}{3}$ | 0 | 0 | |
| $x_1$ | 2 | 1 | $-\frac{2}{3}$ | 0 | $\frac{1}{3}$ | 0 | 0 | |
| $x_5$ | 1 | 0 | $-\frac{10}{3}$ | 0 | $-\frac{1}{3}$ | 1 | 0 | |
| $x_6$ | 1 | 0 | −1 | [1] | −1 | 0 | 1 | ⟶ |

From Tableau 3 we can deduce that $x_5 \geq 1$ because of the second equation, although we don't make use of it here.

**Tableau 4**

| | | $x_1$ | $\bar{x}_2$ | $x_3$ | $x_4$ | $x_5$ | $x_6$ |
|---|---|---|---|---|---|---|---|
| $z$ | 19 | 0 | $\frac{4}{3}$ | 0 | $\frac{1}{3}$ | 0 | 1 |
| $x_1$ | 2 | 1 | $-\frac{2}{3}$ | 0 | $\frac{1}{3}$ | 0 | 0 |
| $x_5$ | 1 | 0 | $-\frac{10}{3}$ | 0 | $-\frac{1}{3}$ | 1 | 0 |
| $x_3$ | 1 | 0 | −1 | 1 | −1 | 0 | 1 |

Tableau 4 contains the optimal solution which is integer, which is $x_1 = 2$, $x_2 = 2$, $x_3 = 1$, $x_5 = 1$; all other variables are zero. If the bounds had not been employed, the linear programming solution would not have been integer.

## 12.4 A CATEGORIZATION OF INTEGER PROGRAMMING METHODS

We now present a framework in which the types of integer programming methods (by problem type) are indicated in terms of their essential characteristics. The typology is given in Table 12.1, showing the different classifications and their possible characteristics. We shall use this typology in categorizing the methods. For example, the Gomory all-integer algorithm that will be considered in Chapter 13 can be categorized as an all-integer cut method, starting from a dual feasible starting solution. Other methods can be similarly categorized.

**Table 12.1**

Classification of Integer Linear Programming Solution Methods

| | |
|---|---|
| *Type of Problem* | |
| All-integer | All variables are integer (not necessarily zero-one) |
| Zero-one integer | All variables are zero-one integer; slack variables of inequality constraints are not necessarily zero-one |
| Mixed-integer | Some, but not all variables, are integer |
| Mixed-zero-one integer | Mixed-integer, but all integer variables are zero-one |
| *Point at which Optimal Integer Solution Seeking Begins* | |
| Arbitrary | Any basic linear programming solution |
| Primal feasible | Any basic primal feasible linear programming solution |
| Dual feasible | Any basic dual feasible linear programming solution |
| Primal and dual feasible | An optimal (noninteger) solution to the linear programming problem |
| *Nature of Algorithm* | |
| Cut | Adds and enforces constraints which integer solutions must satisfy, but some noninteger solutions don't satisfy |
| Branch and bound or implicit enumeration | Implicitly enumerating all possible solutions by examining certain well-chosen solutions and using information generated from those solutions by dominance concepts and infeasibility arguments to eliminate other solutions and to establish tighter criteria for future solutions to be examined |
| Constructive | Constructs the optimal solution via an algorithm by systematically adjusting the values of variables |

The type of problem, as shown in Table 12.1, is self-explanatory. The point at which optimal integer solution seeking begins indicates the procedure necessary in preparing a problem for solution. For example, if a primal and dual feasible solution is required, we must first solve an ordinary linear programming problem to prepare the problem for solution. The nature of the algorithm refers to the technique employed to find integer solutions.

Cut algorithms add cut constraints which exclude noninteger solutions but do not exclude any integer solutions. Cut methods which begin with a dual feasible or a primal dual feasible starting solution are often referred to as dual methods, whereas cut methods which begin with a primal feasible starting solution are referred to as primal methods. Once a feasible integer solution is found using a dual cut method, it is optimal.

Implicit enumeration and branch and bound algorithms enumerate solutions in a highly efficient manner, generally eliminating (hopefully large) subsets of possible solutions without explicitly enumerating them. Once an integer solution is found using implicit enumeration or branch and bound methods, its optimality must subsequently be confirmed.

Constructive algorithms construct an optimal solution by systematically adjusting values of integer variables until a feasible integer solution is obtained. When such a solution is found, it is an optimal solution.

A typology such as this one is useful for categorizing the large number of integer programming approaches and algorithms. This is not the only such typology; for example, see Geoffrion and Marsten [103] for a different one. In addition, it is possible, by suitable manipulation, to categorize some methods into more than one type. However, it is believed that using this typology allows most methods to be naturally and readily categorized as being of a single type.

## 12.5 THE ROAD AHEAD

Unlike linear programming, given the present state of the art in integer programming, there is no assurance that an arbitrary integer programming problem of what seems like a modest size will be solvable in a reasonable amount of time, although certain large problems *having* particular structures are readily solvable. Therefore, in the chapters that follow, we present a collection of integer programming algorithms. These algorithms have been selected for a number of reasons:

1. They appear to be of practical value in solving problems.
2. They are of theoretical interest, and may be of interest in constructing more efficient algorithms.
3. They are currently being examined by researchers in order to develop efficient methods.
4. Though not of theoretical or practical interest, they were major breakthroughs at the time invented.

In Chapter 13 we explore cut methods and methods based on cuts; Chapter 14 examines the group theoretic approach to integer programming. The topic of Chapter 15 is branch and bound methods, and Chapter 16 concerns implicit enumeration methods and some related topics. Chapter 17 is a chapter on solving real integer programming problems, given the present state of the art.

Our approach is somewhat intuitive. Although some proofs are given, others are simply outlined, while still others are omitted completely, with appropriate references cited.

### Selected Supplemental References

Section 12.2
[63], [64], [73], [97], [109], [118], [170], [177], [190], [219], [269], [295]

Section 12.3
[315]

Applications
[18], [24], [42], [71], [74], [85], [94], [134], [163], [171], [198], [199], [213], [218], [224], [228], [234], [266], [268], [280]

General
[1], [7], [8], [15], [28], [39], [103], [135], [152], [158], [195], [209], [226], [233], [261]

## 12.6 PROBLEMS

**1.** Show how to use integer variables to represent a situation in which a number of production facilities may be used for production. Each one has a fixed cost that is incurred if the unit is in operation and a variable cost per unit to capacity. For technical reasons facility 1 must be producing at capacity before facility 2 is used, and so on. How does this formulation differ from that shown in Figure 12.1?

**2.** Develop the mixed integer constraint sets for the following feasible regions. (The feasible regions are shaded.)

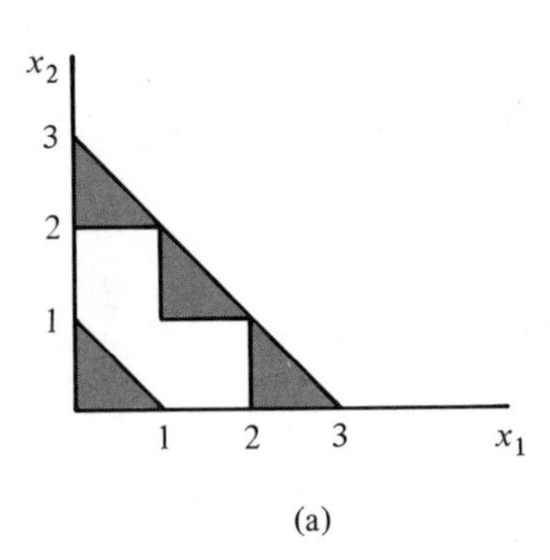


(a)

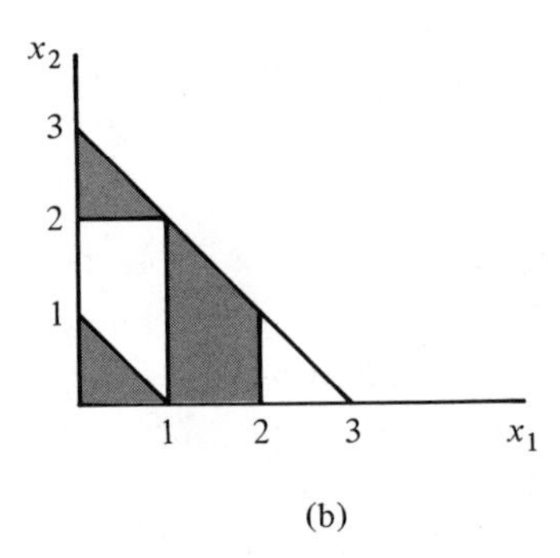


(b)

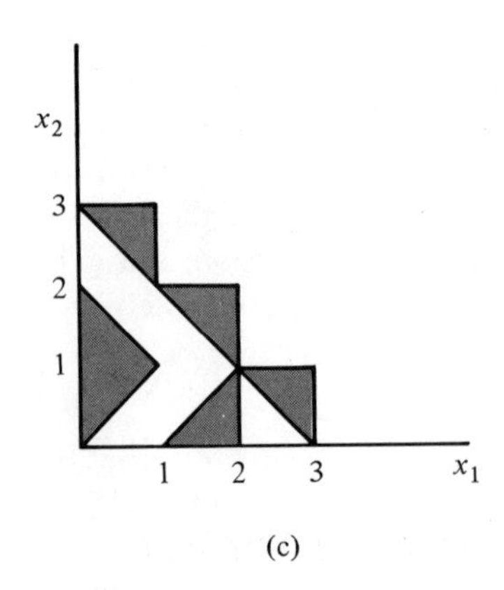


(c)

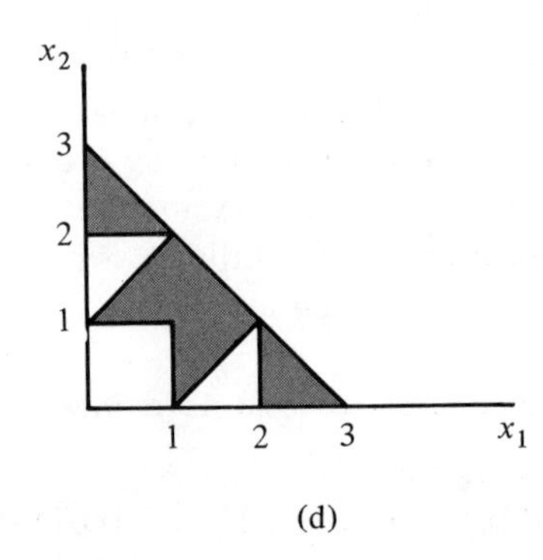


(d)

**3.** Prove the statement on page 327: "Provided that the objective function is linear (for nonconvex solution sets) . . . it is sufficient to represent only the convex hull of a nonconvex set as a convex set, and then solve the associated linear programming problem over the convex hull." If a simple scheme were known for generating the convex hull of the integer solutions, then integer programming methods would not be necessary. The convex hull is the integer programming polyhedron of Chapter 14.

**4.** Using the results of problem 3, develop the constraint sets for the feasible regions of problem 2, assuming that the objective function is linear in $x_1$ and $x_2$.

**5.** A company has $n$ different locations and wishes to connect them with a communication network so that every location has a direct connection (possibly through other locations) with every other location. Assuming that the volume of communication is sufficiently small so that the network, provided it exists, is adequate to handle all of the communication traffic, formulate the problem of determining the least cost communications network as an integer programming problem. Can you deduce a network method to solve this problem?

**6.** Formulate problem 5 for the case in which there is a demand for a specified capacity or number of lines between every pair of locations. Do this under two assumptions:

a. The cost of each line between two locations is independent of the number of lines installed between the two locations.
b. The cost of each line between two locations consists of a fixed component plus a variable component.

**7.** Formulate the knapsack problem assuming that there are nonincreasing (and possibly decreasing) returns to scale (e.g., the second toothbrush packed is not of greater value that the first toothbrush packed).

**8.** Show why the knapsack problem has a trivial solution when the integer requirements are relaxed. Why is this also true when each variable has an upper bound and the integer requirements are relaxed?

**9.** The traveling salesman problem is the problem of a salesman who must visit $n$ towns. To travel from town $i$ to town $j$ requires $c_{ij}$ minutes. The salesman desires to minimize his overall travel time, but he must visit every town. Formulate the problem as an integer programming problem. (Hint: First consider what constraint set types are required, then define the variables, then write the problem.)

**10.** Formulate the following quantity discount problem using integer variables: Less than $u_1$ units costs $c_1$ each. More than $u_1$ units costs $c_2 < c_1$ each.

**11.** Set up the problem of Example 12.1 as an integer programming problem.

**12.** An airline has flights into and out of 40 major cities. It has a schedule it must meet and it must assign flight crews to each of the flights. Each crew has a maximum number of flying hours per day, a maximum total duty time, a maximum number of flights per day, and minimum ground times between consecutive flights. Further, every crew has a home base and must return to its home base no less frequently than every three days. The airline pays for overnight expenses of the crews away from its home base. Formulate the problem of minimizing the cost to the airline of scheduling its crews.

**13.** A toothpaste manufacturer heard of the experience of the metalworking executive described in section 11.9. It occurred to him that he might use a similar approach to buying television spot commercials in a major metropolitan area. A spot commercial lasts for 60 seconds and has a cost and audience that depends on time of day, day of the week, and so on. The nature of audiences watching at different times of the day is available from the advertising agencies. The manufacturer would like to minimize his total costs of advertising on television, subject to requirements that the advertisements are seen by at least some specified number of housewives, teenagers, college students, and other audience segments. The manufacturer has specified minimum and maximum numbers of spot commercials on each station. Formulate the associated problem of choosing the spot commercial times that satisfy his coverage requirements at minimum cost.

**14.** A large corporation is considering the investment possibilities that it currently has available:

| INVESTMENT POSSIBILITY | EXPECTED NET PRESENT VALUE | EXPENDITURE REQUIRED |
|---|---|---|
| 1. Build a new warehouse | $p_1$ | $a_1$ |
| 2. Remodel the old warehouse | $p_2$ | $a_2$ |
| 3. Buy automation for the new warehouse | $p_3$ | $a_3$ |
| 4. Buy a company that supplies product *A* | $p_4$ | $a_4$ |
| 5. Build a plant to manufacture product *A* | $p_5$ | $a_5$ |
| 6. Refurbish corporate offices | $p_6$ | $a_6$ |

**a.** Assuming that the firm has funds to the extent of $a_0$, formulate the problem of choosing among the projects as an integer programming problem. (Projects 1 and 2 are mutually exclusive, as are projects 4 and 5; project 3 can only be undertaken if project 1 is undertaken—the automation is only feasible for a new warehouse.)

**b.** Extend the formulation of part a to a three period horizon, assuming that two additional investment possibilities become available in both periods 2 and 3, and that each investment project has both an inflow and an outflow in each period after the period in which it is undertaken. Further, assume that the funds available for investment in each period are limited to those made available by the corporation plus those unused in the previous period, as well as any cash inflows from investments. Assume there is another investment activity, such as a bank certificate of deposit paying a fixed rate of interest, which may be used as much or as little as desired.

**15.** An assembly line has eight jobs which are to be performed as indicated below:

| JOB | TIME REQUIRED | JOBS THAT MUST BE COMPLETED BEFORE STARTING THIS JOB |
|---|---|---|
| 1 | 7 min. | — |
| 2 | 6 min. | — |
| 3 | 8 min. | — |
| 4 | 8 min. | 1, 2 |
| 5 | 1 min. | 2, 3 |
| 6 | 6 min. | 4, 5 |
| 7 | 7 min. | 5 |
| 8 | 8 min. | 6, 7 |

One worker is positioned at each station and performs certain jobs at his station. Formulate the problem of determining how many stations should be set up and which jobs to assign to each station as an integer programming problem. The objective is to minimize total worker idle time. (Hint: Let $x_{ij}$ be a variable which is 1 if job $i$ is assigned to station $j$, and zero otherwise.)

**16.** A corporation is considering buying two warehouses in a region. There are 20 possible sites. Each site has a fixed cost of purchase and development. In addition, there is an associated capacity of each warehouse and costs of servicing each customer from each warehouse. Formulate the problem of determining the two best warehouses as an integer programming problem.

**17.** Show, using the following problem, that the use of the upper and lower integer bounds, in conjunction with the simplex method, is not sufficient to generate an integer optimum.

$$\begin{aligned} \text{Maximize } z = \; & 5x_1 + 4x_2 \\ \text{subject to:} \quad & 3x_1 + 3x_2 \leq 10 \\ & 12x_1 + 6x_2 \leq 24 \\ & x_1, x_2 \geq 0 \text{ and integer} \end{aligned}$$

# 13

# The Use of Cuts in Integer Programming

## 13.1 INTRODUCTION

In this chapter we shall explore the use of cuts in integer programming. A cut is a derived constraint which has the desirable property of "cutting off" part of the set of feasible solutions while not excluding any integer solutions. Most methods which employ cuts generate cut constraints and utilize them using linear programming methods until an optimal integer solution is obtained. There are some difficulties in doing this in general. For example, in some situations it may not be possible to obtain an integer solution using a finite number of cut constraints. In order for any method to be used for solving problems on a regular basis, such difficulties must of course be overcome.

Dantzig [63] observed that, given a noninteger optimal basic feasible solution to an integer program, a legitimate cut constraint is that the sum of the nonbasic variables must equal or exceed one. The justification is obvious: Under no circumstance could the sum of nonbasic variables be negative. Since all nonbasic variables equal to zero (which is the noninteger optimal) was not an integer solution, at least one nonbasic variable must equal or exceed one, which is exactly Dantzig's constraint. A procedure based on this cut was suggested:

1. Solve the linear programming problem, ignoring the integer constraints.
2. If the optimal solution is integer, stop. Otherwise go to step 3.

3. Add a constraint requiring that the sum of nonbasic variables be at least one.
4. Use the dual simplex method to solve the altered problem. Then go to step 2.

Although this procedure does converge to an optimum in a finite number of iterations in many instances, it has been proved (Gomory and Hoffman [128]) that it does not work in general. A variation of the Gomory all-integer algorithm (which will be considered in this chapter) may be viewed as an adaptation of Dantzig's method that does converge in every case.

In the remaining sections of the chapter we shall consider methods which do converge in a finite number of iterations to an integer optimum, provided one exists.

## 13.2 GOMORY'S CUT ALGORITHMS

Ralph Gomory has made many outstanding contributions to the field of integer programming. In this section we shall consider three of his algorithms which employ cut constraints.

1. The Gomory fractional algorithm.
2. The Gomory all-integer algorithm.
3. The Gomory mixed-integer algorithm.

Before we consider any of the methods, we shall explore the nature of the cuts employed, because the cuts derived here are the cuts employed by a number of the methods considered in this chapter.

Consider any constraint of the problem written in updated form corresponding to some basic solution of the linear programming problem. (Possibly some cut constraints have been incorporated in the problem; since the constraints are linear, this only alters the convex set by creating some new basic feasible solutions and making infeasible some previously basic feasible solutions.) A constraint at that stage can be written as follows:

$$a_{i1}x_1 + a_{i2}x_2 + \cdots + a_{in}x_n + x_{n+i} = b_i \tag{13.1}$$

where the nonbasic variables are written first as variables $x_1, \ldots, x_n$ and the basic variables are written last as $x_{n+1}, \ldots, x_{n+m}$. Then rewrite (13.1) as follows:

$$\sum_{j=1}^{n} [a_{ij}]x_j + \sum_{j=1}^{n} (a_{ij} - [a_{ij}])x_j + x_{n+i} = [b_i] + (b_i - [b_i]) \tag{13.2}$$

Rearranging (13.2), putting the integer expressions on the left side of the equality and the fractional expressions on the right side, yields the following equation:

$$\text{(13.3)} \qquad \sum_{j=1}^{n} [a_{ij}]x_j + x_{n+i} - [b_i] = (b_i - [b_i]) - \sum_{j=1}^{n} (a_{ij} - [a_{ij}])x_j$$

Now the left-hand side of (13.3) must be integral in an optimal integer solution because all the coefficients and variables must be integral. Because of the equality (13.3), the right-hand side must be integral also. Now a trivial application of Theorem 12.3 can be used to show that the upper bound of the right-hand side of (13.3) cannot be positive. (Obviously, the same is true for the left-hand side. That result is used by Gomory's all-integer algorithm and the primal algorithm.) That condition, namely expression (13.4), is the cut utilized by Gomory's fractional algorithm [123].

$$\text{(13.4)} \qquad (b_i - [b_i]) - \sum_{j=1}^{n} (a_{ij} - [a_{ij}])x_j \leq 0$$

or

$$\sum_{j=1}^{n} (a_{ij} - [a_{ij}])x_j \geq (b_i - [b_i])$$

#### Gomory's Fractional Algorithm [123]

Gomory's fractional algorithm can be classified as an all-integer, primal dual feasible starting solution, cut method. Proceeding from the previous development, the slack of constraint (13.4) is integer, by virtue of the fact that the left-hand side of (13.3) is integer.

EXAMPLE 13.1:

Derive a cut constraint of the form (13.4) from the following equality of an optimal noninteger solution:

$$2.75x_1 - 3.20x_2 - .6x_3 + 2x_4 + x_5 = 3.75$$

where $x_5$ is the basic variable. Using expression (13.4) we have

$$.75x_1 + .8x_2 + .4x_3 \geq .75$$

Gomory's fractional algorithm [123] may be described by the following sequence of steps:

1. Solve the corresponding linear programming problem using the simplex method or a variant.
2. If the solution is integer, halt; the optimal integer solution is at hand. Otherwise, go to step 3.

3. For any updated constraint whose $b_i$ value is fractional (including the objective function, since the objective function coefficients may be assumed to be integral),* generate a cut constraint of the form (13.4). (A common procedure here is to select the constraint in which $(b_i - [b_i])$ is maximized so as to utilize a "deep" cut.) Use the dual simplex method to obtain an optimal feasible solution. Then go to step 2.

This framework is rather simple and, except for the nature of the cut constraint in step 3, identical to the framework of Dantzig's method presented in the previous section. There are other aspects of the algorithm to be considered, but we shall first solve an example.

EXAMPLE 13.2:

Solve the following problem using Gomory's fractional algorithm.

$$\begin{aligned} \text{Maximize } z &= 5x_1 + 2x_2 \\ \text{subject to: } \quad & 2x_1 + 2x_2 + x_3 \phantom{+ x_4} = 9 \\ & 3x_1 + x_2 \phantom{+ x_3} + x_4 = 11 \\ & x_1, x_2, x_3, x_4 \geq 0 \text{ and integer} \end{aligned}$$

The following is the sequence of linear programming solutions.

**Tableau 1**

| | | $x_1$ ↓ | $x_2$ | $x_3$ | $x_4$ | |
|---|---|---|---|---|---|---|
| $z$ | 0 | −5 | −2 | 0 | 0 | |
| $x_3$ | 9 | 2 | 2 | 1 | 0 | |
| $x_4$ | 11 | [3] | 1 | 0 | 1 | → |

**Tableau 2**

| | | $x_1$ | $x_2$ ↓ | $x_3$ | $x_4$ | |
|---|---|---|---|---|---|---|
| $z$ | 18.33 | 0 | −0.33 | 0 | 1.67 | |
| $x_3$ | 1.67 | 0 | [1.33] | 1 | −0.67 | → |
| $x_1$ | 3.67 | 1 | 0.33 | 0 | 0.33 | |

* A constraint or objective function may be approximated as accurately as desired by an all-integer constraint, by multiplying the constraint by a large number and then rounding each coefficient to the nearest integer. Alternately, if the objective function coefficients are not integral, no approximation need be made; the objective function should in that case not be used to generate a cut.

**Tableau 3**

| | | $x_1$ | $x_2$ | $x_3$ | $x_4$ | |
|---|---|---|---|---|---|---|
| $z$ | 18.75 | 0 | 0 | 0.25 | 1.5 | |
| $x_2$ | 1.25 | 0 | 1 | 0.75 | −0.5 | optimal noninteger |
| $x_1$ | 3.25 | 1 | 0 | −0.25 | 0.5 | solution |

Any one of the constraints as well as the objective function can be used to generate a cut. Using the rule of thumb and choosing the greatest value of $(b_i - [b_i])$, we choose the objective function. Using the cut constraint (13.4) we have the following constraint:

$$18.75 - 18 \leq (0.25 - 0)x_3 + (1.5 - 1)x_4$$

or

$$0.75 \leq 0.25x_3 + 0.5x_4$$

Adjoining the constraint to the tableau and designating the slack as $s_1$, we have Tableau 4, which is no longer feasible.

**Tableau 4**

| | | $x_1$ | $x_2$ | $x_3$ ↓ | $x_4$ | $s_1$ | |
|---|---|---|---|---|---|---|---|
| $z$ | 18.75 | 0 | 0 | 0.25 | 1.5 | 0 | |
| $x_2$ | 1.25 | 0 | 1 | 0.75 | −0.5 | 0 | |
| $x_1$ | 3.25 | 1 | 0 | −0.25 | 0.5 | 0 | |
| $s_1$ | −0.75 | 0 | 0 | [−0.25] | −0.5 | 1 | → |

Using the dual simplex method, after one iteration we have Tableau 5, which is not feasible.

**Tableau 5**

| | | $x_1$ | $x_2$ | $x_3$ | $x_4$ ↓ | $s_1$ | |
|---|---|---|---|---|---|---|---|
| $z$ | 18 | 0 | 0 | 0 | 1 | 1 | |
| $x_2$ | −1 | 0 | 1 | 0 | [−2] | 3 | → |
| $x_1$ | 4 | 1 | 0 | 0 | 1 | −1 | |
| $x_3$ | 3 | 0 | 0 | 1 | 2 | −4 | |

Next $x_4$ enters the basis, replacing $x_2$ and yielding Tableau 6, which is optimal, but still not integer.

**Tableau 6**

| | | $x_1$ | $x_2$ ↓ | $x_3$ | $x_4$ | $s_1$ | $s_2$ | |
|---|---|---|---|---|---|---|---|---|
| $z$ | 17.5 | 0 | 0.5 | 0 | 0 | 2.5 | 0 | |
| $x_4$ | 0.5 | 0 | −0.5 | 0 | 1 | −1.5 | 0 | |
| $x_1$ | 3.5 | 1 | 0.5 | 0 | 0 | 0.5 | 0 | |
| $x_3$ | 2 | 0 | 1 | 1 | 0 | −1 | 0 | |
| $s_2$ | −0.5 | 0 | [−0.5] | 0 | 0 | −0.5 | 1 | → |

We again use the objective function to derive a constraint and get

$$(17.5 - 17) \leq (0.5 - 0)x_2 + (2.5 - 2)s_1$$

or

$$0.5 \leq 0.5x_2 + 0.5s_1$$

which is adjoined to Tableau 6. (There is a tie here; the first or second constraint could also have been used.) Designating the slack of this constraint as $s_2$, we utilize a dual simplex iteration in which $x_2$ replaces $s_2$. The result is Tableau 7, the optimal integer solution.

**Tableau 7**

| | | $x_1$ | $x_2$ | $x_3$ | $x_4$ | $s_1$ | $s_2$ |
|---|---|---|---|---|---|---|---|
| $z$ | 17 | 0 | 0 | 0 | 0 | 2 | 1 |
| $x_4$ | 1 | 0 | 0 | 0 | 1 | −1 | −1 |
| $x_1$ | 3 | 1 | 0 | 0 | 0 | 0 | 1 |
| $x_3$ | 1 | 0 | 0 | 1 | 0 | −2 | 2 |
| $x_2$ | 1 | 0 | 1 | 0 | 0 | 1 | −2 |

Consider the graph of the solutions to this problem, shown in Figure 13.1. The solutions correspond to the indicated tableaus. Each cut constraint is written in terms of the original structural variables.

There are a number of questions regarding the Gomory fractional algorithm that we have not yet considered. First, there is the question of how many cut constraints are to be added—the question of finiteness, which we shall consider

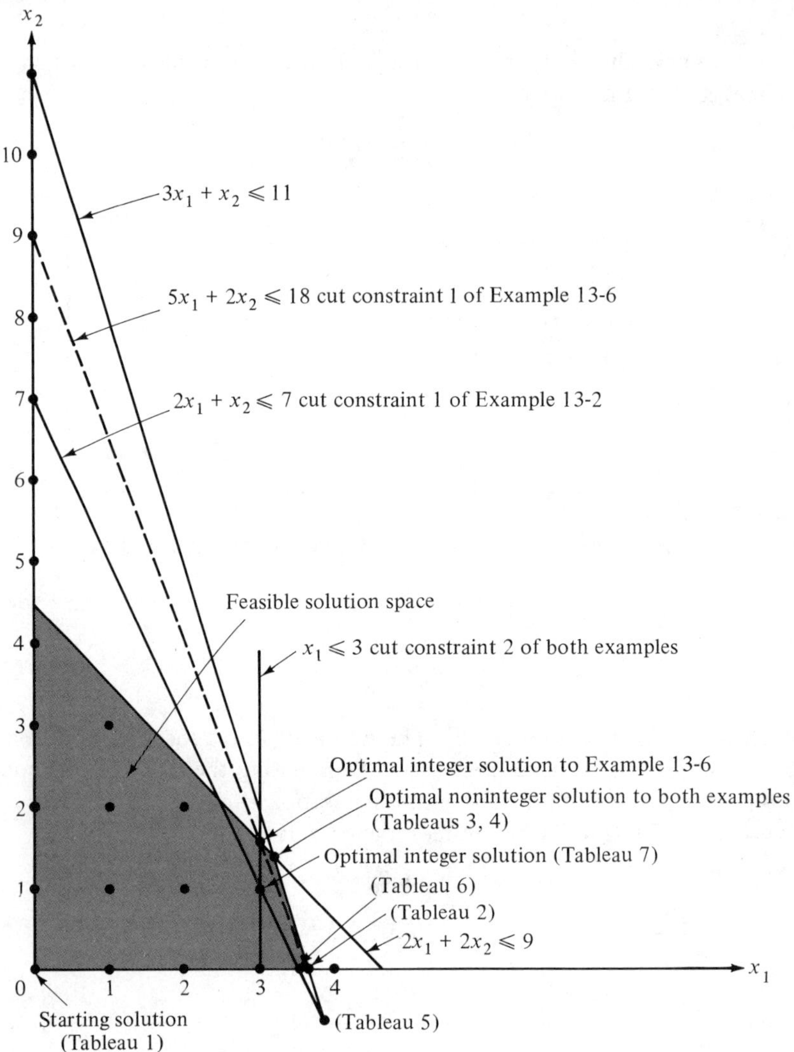


**Figure 13.1**

A Graph of Examples 13.2 and 13.6 Showing Solutions, Constraints, and Cut Constraints (tableau numbers are with reference to Example 13.2)

shortly. Second, it would appear as if the problem size grows without bound, since one constraint and one variable are added with the addition of each cut constraint. In the initial problem having $m$ equality constraints in $m + n$ variables, we can observe that there are $n$ nonbasic variables. When a cut constraint is adjoined, the number of equations and the number of variables will each be increased by one, but the number of nonbasic variables will always be $n$. Accordingly, *at most* $n$ slack variables of cut constraints can be nonbasic at any one time, and any additional cut constraints must be overly satisfied for a feasible solution. In other words, at most $n$ cut constraints can be binding at a time. If ever a cut constraint is overly satisfied (i.e., its slack is basic and positive), it must be implied by the remaining constraints. Hence, we drop any cut constraint once its slack becomes basic in a feasible solution.

Another tableau scheme may be used to avoid the problem of enlarging and decreasing tableau size with each additional cut constraint: Adjoin the nonnegativity constraints in the form $-x_j \leq 0$, omit the identity matrix, and drop the cut constraint after the pivoting operation. Then the adjoined constraint may be dropped after the iteration, and the nonnegativity constraint becomes the updated cut constraint. Only the designation of the nonbasic variable changes. Also, when an original variable becomes nonbasic, a nonnegativity constraint is restored in its row.

EXAMPLE 13.3:

Take the first iteration starting from the solution of Tableau 3 of Example 13.2, using the explicit representation of the nonnegativity constraints.

**Tableau 4′**

| | | $x_3$ ↓ | $x_4$ | |
|---|---|---|---|---|
| $z$ | 18.75 | .25 | 1.5 | |
| $x_2$ | 1.25 | .75 | −.5 | |
| $x_1$ | 3.25 | −.25 | .5 | |
| $x_3$ | 0 | −1 | 0 | |
| $x_4$ | 0 | 0 | −1 | |
| $s_1$ | −.75 | [−.25] | −.5 | → |

**Tableau 5′**

| | | $s_1$ | $x_4$ ↓ | |
|---|---|---|---|---|
| $z$ | 18 | 1 | 1 | |
| $x_2$ | −1 | 3 | [−2] | → |
| $x_1$ | 4 | −1 | 1 | |
| $x_3$ | 3 | −4 | 2 | |
| $x_4$ | 0 | 0 | −1 | |

Removing $s_1$ from the basis and introducing $x_3$ in Tableau 4′ leads to Tableau 5′, from which the adjoined constraint has been dropped. The rules for this iteration are otherwise the same as for the earlier iterations omitting the identity matrix, with the exception that when a problem variable leaves the basis, its nonnegativity constraint is explicitly restored.

### Convergence of Gomory's Fractional Algorithm

In order to prove convergence of Gomory's fractional algorithm, we must alter the algorithm slightly and make a few assumptions. We assume that every problem variable, including the objective function variable, has a lower bound.

(We could, if desired, use linear programming formulations to find such bounds.) Next we assume that the noninteger optimum is such that all nonbasic vectors are lexicographically positive.

**Definition 13.1** (Lexicographically positive)
*A vector is said to be lexicographically positive (negative) if its* first *nonzero component is positive (negative).*

If there is no dual degeneracy (i.e., no reduced costs of nonbasic variables are zero), then the vectors will all be lexicographically positive. If there is dual degeneracy, then any nonbasic variables whose columns are lexicographically negative must be introduced into the basis until all nonbasic variables become lexicographically positive. Alternatively, a lexicographic optimality condition can be enforced in the simplex method itself to insure that all nonbasic vectors are lexicographically positive. The dual simplex method using the dual of the perturbation methods (see Chap. 3) will preserve the lexicographical positivity of the nonbasic variables in the fractional algorithm.

We further use the tableau stating the nonnegativity constraints explicitly (as shown in Example 13.3), and use the first constraint having a noninteger basic variable to generate the cut. (This rule is not used in practice. To insure convergence, it is sufficient to generate every $k^{\text{th}}$ cut—where $k$ is arbitrary—according to this rule.) Consider the cut constraint and the variable selected to enter the basis. That variable will of necessity have a nonzero coefficient in the constraint used to generate the cut. There are two possibilities:

1. That coefficient is the first nonzero element in the incoming vector, in which case it is positive. When the first iteration is taken, the first noninteger variable will be reduced to the next lower integer or lower, preserving the integrality of all basic variables in preceding constraints.
2. That coefficient is not the first nonzero element in the incoming vector, in which case some preceding element is the first, and is, of course, positive. Accordingly, when the first iteration is taken, the corresponding basic variable is reduced from its integer value to some lower value.

Subsequent iterations for a given cut constraint, if necessary, will bring about a further lexicographic decrease in the solution vector. Hence, the solution vector is decreasing lexicographically with the addition and enforcement of each cut constraint. We can establish a finite upper bound on the number of cut constraints that must be added because, with the addition and enforcement of each cut constraint, either the first basic variable that is noninteger is reduced to at most the next lowest integer, or some earlier basic (integer) variable is reduced to a noninteger value. The number of iterations required for the enforcement of a cut constraint is thus finite because of the finiteness of the dual simplex method. Hence the procedure converges to an optimum, if one exists, with the addition of a finite number of cuts. As in the dual simplex method, if all of the entries in an outgoing row of a cut constraint are ever zero or positive (in a dual simplex iteration), there is no feasible solution to the integer programming problem.

### Gomory's All-Integer Algorithm [125]

Gomory's all-integer integer programming algorithm is an all-integer, dual feasible starting solution, cut method which appears analogous to the dual simplex method. All of the coefficients in the matrix are required to be integers, and the iterations preserve the integrality of the intermediate matrices. The method proceeds from a dual feasible starting solution, as in the dual simplex method. The most violated constraint is selected. If that constraint were to be used for an iteration, a fractional solution would result, unless the pivot element were $-1$. Gomory's approach derives a cut constraint from the original constraint so that a pivot element of $-1$ is assured. This is accomplished by first dividing the selected constraint by a constant $\lambda$ (which we shall demonstrate the procedure for determining). Then the cut constraint is derived using the nonpositivity of the left-hand side of (13.3), namely

$$\left[\frac{a_{ij}}{\lambda}\right]x_j + \left[\frac{1}{\lambda}\right]x_{n+i} - \left[\frac{b_i}{\lambda}\right] \leq 0 \tag{13.5}$$

The minimum value of $\lambda$ that assures a pivot element of $-1$ is chosen. To illustrate this we present an example.

EXAMPLE 13.4: (Gomory [125])

Maximize $z = -10x_1 - 14x_2 - 21x_3$ (or Minimize $10x_1 + 14x_2 + 21x_3$)

subject to:

$$\begin{aligned} 8x_1 + 11x_2 + 9x_3 &\geq 12 \\ 2x_1 + 2x_2 + 7x_3 &\geq 14 \\ 9x_1 + 6x_2 + 3x_3 &\geq 10 \\ x_1, x_2, x_3 &\geq 0 \text{ and integer} \end{aligned}$$

The initial tableau is given in Tableau 1, where the slack variables have been inserted. In the dual simplex method $x_5$ would normally be chosen to leave the basis; hence, that constraint is used to construct a cut using equation (13.5). For convenience, a selected number of cut constraints derived from the second constraint for the range of $\lambda$, together with the corresponding incoming variable and the pivot element, are shown in Table 13.1. To illustrate how one of the entries is determined, consider $\lambda = 10$. Dividing the second constraint by 10 yields

$$-.2x_1 - .2x_2 - .7x_3 \leq -1.4$$

**Tableau 1**

| | | $x_1$ | $x_2$ | $x_3$ | $x_4$ | $x_5$ | $x_6$ |
|---|---|---|---|---|---|---|---|
| $z$ | 0 | 10 | 14 | 21 | 0 | 0 | 0 |
| $x_4$ | −12 | −8 | −11 | −9 | 1 | 0 | 0 |
| $x_5$ | −14 | −2 | −2 | −7 | 0 | 1 | 0 |
| $x_6$ | −10 | −9 | −6 | −3 | 0 | 0 | 1 |

and applying equation (13.5) yields the following constraint:

$$-x_1 - x_2 - x_3 \leq -2$$

Any of the cut constraints in Table 13.1 with a pivot element of −1 will preserve the integrality conditions, but the strongest, in a sense, is the one having a pivot element of −1 with the smallest value of $\lambda$. Note also that the weakest cut constraint, with $\lambda \geq 14$, in this instance looks like Dantzig's cut, discussed in section 13.1.

**Table 13.1**

Various Cut Constraints Derived from the Second Constraint of Example 13.4

| RANGE OF $\lambda$ | CUT CONSTRAINT | INCOMING VARIABLE | PIVOT ELEMENT |
|---|---|---|---|
| $14 \leq \lambda < \infty$ | $-x_1 - x_2 - x_3 \leq -1$ | $x_1$ | −1 |
| $7 \leq \lambda < 14$ | $-x_1 - x_2 - x_3 \leq -2$ | $x_1$ | −1 |
| $4\frac{2}{3} \leq \lambda < 7$ | $-x_1 - x_2 - 2x_3 \leq -3$ | $x_1$ | −1 |
| $3\frac{1}{2} \leq \lambda < 4\frac{2}{3}$ | $-x_1 - x_2 - 2x_3 \leq -4$ | $x_1$ | −1 |
| $2\frac{4}{5} \leq \lambda < 3\frac{1}{2}$ | $-x_1 - x_2 - 3x_3 \leq -5$ | $x_3$ | −3 |
| $2\frac{1}{3} \leq \lambda < 2\frac{4}{5}$ | $-x_1 - x_2 - 3x_3 \leq -6$ | $x_3$ | −3 |
| $2 \leq \lambda < 2\frac{1}{3}$ | $-x_1 - x_2 - 4x_3 \leq -7$ | $x_3$ | −4 |

In general, the weakest Gomory all-integer cut (that is, with $\lambda$ sufficiently large) is that the sum of a partial set (corresponding to variables that are candidates for entry into the basis in the generator constraint in the dual simplex method) of nonbasic variables must equal or exceed one, and may be viewed as a variation of Dantzig's cut.

### A Dantzig Cut that Converges: A Digression

The observation that the weakest form of a Gomory all-integer cut looks very much like a Dantzig cut leads to a cut very similar to a Dantzig cut which will converge to a finite optimum in every case. The algorithm of section 13.1 is altered, first by requiring that the noninteger optimum and all subsequent solutions be lexicographically dual feasible (i.e., all nonbasic vectors must be lexicographically positive), and then by replacing step 3 of the algorithm with the following step:

3′. Choose the first constraint whose solution value is fractional. Denote the set of nonbasic variables with *noninteger* coefficients in that constraint as $J$. Then adjoin the constraint

$$\sum_{j \in J} x_j \geq 1$$

To illustrate the construction of such a constraint, the adjoined constraint, using step 3′ for the problem of Example 13.1, is

$$x_1 + x_2 + x_3 \geq 1$$

The derivation of the constraint of step 3′ is to first construct a Gomory fractional cut constraint from the selected constraint (with noninteger $b_i$). Then use a Gomory all-integer cut derived from the Gomory fractional cut with $\lambda = 1$. Convergence is assured by essentially the same argument as the Gomory fractional algorithm. However, the reduced $\boldsymbol{b}$ vector entry is not necessarily reduced to the next integer, but by at least $1/D$, where $D$ is the determinant of the optimal linear programming solution basis. Thus the procedure is finite, but generally converges no faster and usually slower than the Gomory fractional algorithm—that $D$ is bounded from above is true because all of the pivot entries for the cuts are unity. A justification for the above cut was independently derived by Bowman and Nemhauser [38].

### The All-Integer Method

In order to prove convergence of Gomory's all-integer method allowing dual degeneracy, we will have to assume as before that the nonbasic vectors are lexicographically positive, and utilize the dual of the perturbation methods (see Chap. 3), but we may view the procedure of constructing the cut constraint in general as being as explained, and in particular as follows:

1. Choose the most negative $b_i$. (To guarantee convergence, we should choose periodically the first constraint with negative $b_i$. See Gomory [125] for other rules that converge.) Designate the row as $r$. If the dual simplex method results in a pivot element of $-1$, use that pivot element for an iteration; then repeat step 1. Otherwise, go to step 2.
2. Choose the lexicographically smallest nonbasic column from those having $a_{rj} < 0$. Designate the column as $k$. Further, let the first nonzero element in column $k$ be $a_{pk}$ ($> 0$) and let the corresponding row be $p$.
3. For columns with $a_{rj} < 0$ compute $\mu_j = [a_{pj}/a_{pk}]$ if $a_{pj}$ is the first nonzero element in column $j$; otherwise $\mu_j = \infty$.

4. Compute $\lambda = \text{Max}\{|a_{rj}|/\mu_j\}$ such that $a_{rj} < 0$ and $\mu_j$ finite.
5. Derive the cut constraint $\sum [a_{rj}/\lambda]x_j - [b_r/\lambda] \leq 0$. Since $\lambda > 1$ except when the constraint itself is the cut constraint, $[1/\lambda]$ is zero and the second term in expression (13.5) vanishes.
6. Append the cut constraint and use the dual simplex method to take an iteration. If the resulting solution is feasible, stop. Otherwise, go to step 1.

Steps 2 through 5 give us an algorithm equivalent to the scheme used in Table 13.1 for deriving the cut constraint.

EXAMPLE 13.4: (continued)
We now proceed with the solution of Example 13.4. The sequence of iterations follows, as indicated in Tableaus 1 through 5 (with identity matrices omitted).

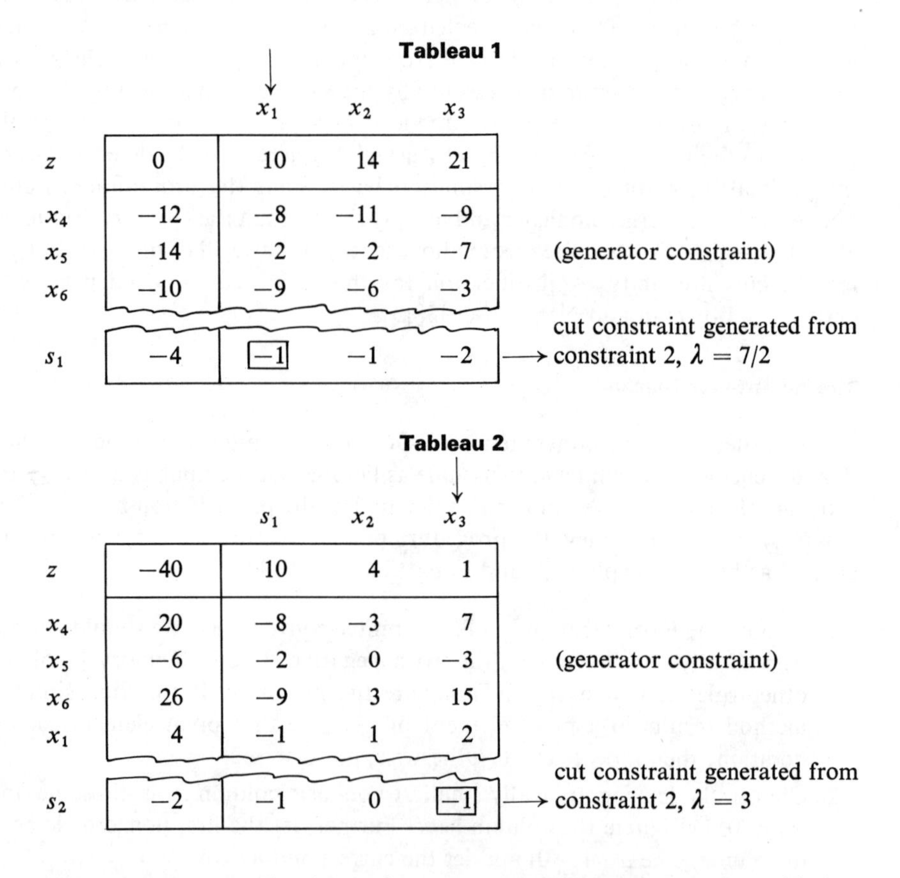

**Tableau 1**

| | | $x_1$ ↓ | $x_2$ | $x_3$ | |
|---|---|---|---|---|---|
| $z$ | 0 | 10 | 14 | 21 | |
| $x_4$ | −12 | −8 | −11 | −9 | |
| $x_5$ | −14 | −2 | −2 | −7 | (generator constraint) |
| $x_6$ | −10 | −9 | −6 | −3 | |
| $s_1$ | −4 | [−1] | −1 | −2 | → cut constraint generated from constraint 2, $\lambda = 7/2$ |

**Tableau 2**

| | | $s_1$ | $x_2$ | $x_3$ ↓ | |
|---|---|---|---|---|---|
| $z$ | −40 | 10 | 4 | 1 | |
| $x_4$ | 20 | −8 | −3 | 7 | |
| $x_5$ | −6 | −2 | 0 | −3 | (generator constraint) |
| $x_6$ | 26 | −9 | 3 | 15 | |
| $x_1$ | 4 | −1 | 1 | 2 | |
| $s_2$ | −2 | −1 | 0 | [−1] | → cut constraint generated from constraint 2, $\lambda = 3$ |

The second constraint is used to generate the cut constraint added to Tableau 2. The lexicographically minimum column $j$ for $a_{rj} < 0$ is column

3. Therefore $a_{pk} = a_{03} = 1$, $\mu_1 = 10$, $\mu_3 = 1$, and $\lambda = \text{Max}\{2/10, 3\} = 3$. Hence the cut constraint is $-s_1 - x_3 \leq -2$. Performing the indicated iteration leads to Tableau 3.

**Tableau 3**

| | | $s_1$ ↓ | $x_2$ | $s_2$ | |
|---|---|---|---|---|---|
| $z$ | −42 | 9 | 4 | 1 | |
| $x_4$ | 6 | −15 | −3 | 7 | |
| $x_5$ | 0 | 1 | 0 | −3 | |
| $x_6$ | −4 | −24 | 3 | 15 | (generator constraint) |
| $x_1$ | 0 | −3 | 1 | 2 | |
| $x_3$ | 2 | 1 | 0 | −1 | |
| $s_3$ | −1 | [−1] | 0 | 0 | → cut constraint generated from constraint 3, $\lambda = 24$ |

The third constraint is used to generate the cut constraint. The lexicographically minimum column for $a_{rj} < 0$ is the first column; hence $a_{pk} = a_{01} = 9$. Thus $\mu_1 = 1$; $\lambda = \text{Max}\{24/1\} = 24$. Hence the cut constraint is $-s_1 \leq -1$. Performing the iteration, $s_1$ enters the basis and the result is Tableau 4. Because $s_1$ has entered the basis, the constraint in which it is basic may be dropped (using exactly the same reasoning provided for Gomory's fractional algorithm).

**Tableau 4**

| | | $s_3$ | $x_2$ | $s_2$ ↓ | |
|---|---|---|---|---|---|
| $z$ | −51 | 9 | 4 | 1 | |
| $x_4$ | 21 | −15 | −3 | 7 | |
| $x_5$ | −1 | 1 | 0 | −3 | (generator constraint) |
| $x_6$ | 20 | −24 | 3 | 15 | |
| $x_1$ | 3 | −3 | 1 | 2 | |
| $x_3$ | 1 | 1 | 0 | −1 | |
| $s_4$ | −1 | 0 | 0 | [−1] | → cut constraint generated from constraint 2, $\lambda = 3$ |

The second constraint is used to generate the cut. The lexicographically minimum column $j$ for $a_{rj} < 0$ is column 3. Hence $\mu_3 = 1$, $\lambda = 3$, and the resulting cut constraint is $-s_2 \leq -1$. Performing the iteration leads to Tableau 5, which is feasible and hence the optimal integer solution. Because $s_2$ enters the basis we drop the constraint containing it, in the same manner as above.

**Tableau 5**

| | | $s_3$ | $x_2$ | $s_4$ |
|---|---|---|---|---|
| $z$ | −52 | 9 | 4 | 1 |
| $x_4$ | 14 | −15 | −3 | 7 |
| $x_5$ | 2 | 1 | 0 | −3 |
| $x_6$ | 5 | −24 | 3 | 15 |
| $x_1$ | 1 | −3 | 1 | 2 |
| $x_3$ | 2 | 1 | 0 | −1 |

The solution is optimal

As we did with Gomory's fractional algorithm, we can avoid the problem of added constraints by adjoining the nonnegativity constraints explicitly. A second example illustrating this is now presented, and the corresponding solution is plotted in Figure 13.2.

EXAMPLE 13.5:

$$\begin{aligned} \text{Maximize } z &= -5x_1 - 2x_2 \\ \text{subject to: } & -2x_1 - 2x_2 \leq -9 \\ & -3x_1 - x_2 \leq -11 \\ & x_1, x_2 \geq 0 \text{ and integer} \end{aligned}$$

The sequence of iterations follows in Tableaus 1 through 3, with the slacks of the first and second constraints being $x_3$ and $x_4$, respectively.

**Tableau 1**

| | | $x_1$ | $x_2$ ↓ | |
|---|---|---|---|---|
| $z$ | 0 | 5 | 2 | |
| $x_3$ | −9 | −2 | −2 | |
| $x_4$ | −11 | −3 | −1 | (generator constraint) |
| $x_1$ | 0 | −1 | 0 | |
| $x_2$ | 0 | 0 | −1 | |
| $s_1$ | −8 | −2 | [−1] | ⟶ cut constraint generated from constraint 2 with $\lambda = 1.5$ |

In this case, the second constraint can be used directly for the iteration (since $\lambda = 1$).

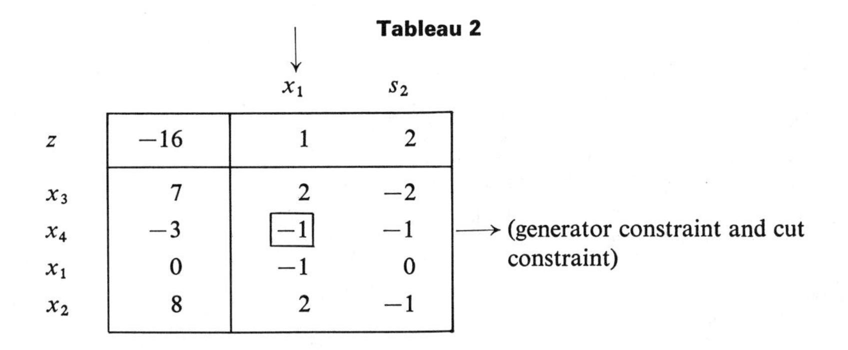

**Tableau 2**

| | | ↓ $x_1$ | $s_2$ | |
|---|---|---|---|---|
| $z$ | $-16$ | 1 | 2 | |
| $x_3$ | 7 | 2 | $-2$ | |
| $x_4$ | $-3$ | $\boxed{-1}$ | $-1$ | → (generator constraint and cut constraint) |
| $x_1$ | 0 | $-1$ | 0 | |
| $x_2$ | 8 | 2 | $-1$ | |

**Tableau 3**

| | | $x_4$ | $s_1$ | |
|---|---|---|---|---|
| $z$ | $-19$ | 1 | 1 | |
| $x_3$ | 1 | 2 | $-4$ | |
| $x_4$ | 0 | $-1$ | 0 | The solution is optimal |
| $x_1$ | 3 | $-1$ | 1 | |
| $x_2$ | 2 | 2 | $-3$ | |

Notice that at times (as in Tableaus 1 and 2) the same constraint generates successive cuts.

### The Convergence of Gomory's All-Integer Algorithm

The proof for the convergence of Gomory's all-integer algorithm is analogous to the proof for the fractional algorithm. First assume that the problem is not dual degenerate, and that a finite lower-limit on the objective function is known a priori. Then, if an optimal feasible solution exists, it will be found in a finite number of iterations because the objective function decreases by at least one for each added constraint. Since the objective function has a lower bound, convergence has been proved.

If dual nondegeneracy cannot be assumed, then the nonnegativity constraints of the original nonbasic variables must be written explicitly. We require, without loss of generality, that the columns be lexicographically positive. Consequently, using the dual of the perturbation methods (see Chap. 3), the first column of the matrix (the updated $\boldsymbol{b}$ vector which we shall refer to as $\boldsymbol{a}_0$) must decrease lexicographically with each iteration. We further stipulate that the

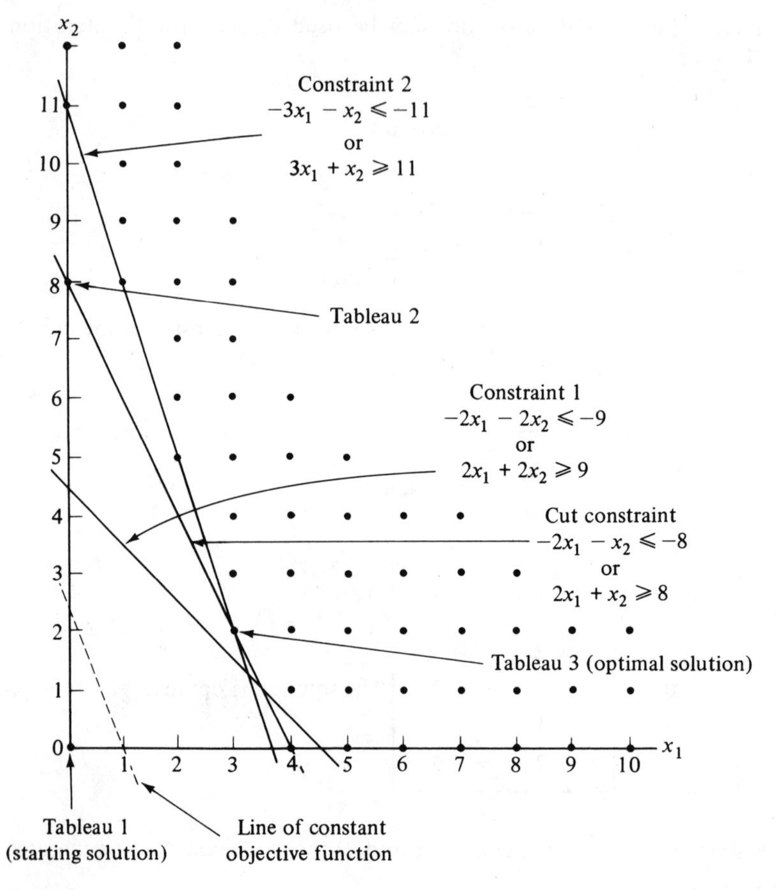


**Figure 13.2**

A Plot of the Solutions to Example 13.5

*first negative entry* in $\boldsymbol{a}_0$ (excluding the evaluator row) be used to generate the cut constraint. The first component of $\boldsymbol{a}_0$ cannot decrease indefinitely because it is bounded by the lower bound of the objective function. In the case of dual degeneracy, the first component of $\boldsymbol{a}_0$ which is altered is reduced. If some earlier component of $\boldsymbol{a}_0$ becomes negative, then the row corresponding to the first such entry is used to generate a cut constraint, thereby reducing some preceding entry. The analysis may be pursued to show that the vector $\boldsymbol{a}_0$ can decrease lexicographically only a finite number of times to an integer optimum, if one exists. The occurrence of a dual unbounded condition indicates that no optimal feasible integer solution exists. It should be noted for this method of proof that the usual criterion for the dual simplex method—selecting the most negative $b_i$—will not necessarily converge, although it seems satisfactory in practice.

As before, to ensure convergence computationally, it is sufficient to generate every $k^{\text{th}}$ cut according to the procedure necessary for convergence ($k$ arbitrary).

**Gomory's Mixed-Integer Algorithm**

A third algorithm developed by Ralph Gomory is his mixed-integer programming algorithm [124]. That algorithm may be considered as a mixed-integer, primal dual feasible starting solution, cut method. It is similar to the fractional algorithm but, in this case, not all of the variables are required to be integral. The first step in this algorithm is to solve the corresponding linear programming problem. Consider an *updated* constraint equation at that point of the following form:

$$\sum (a_{ij}^{+} + a_{ij}^{-})x_j + x_{n+i} = (b_i - [b_i]) + [b_i]$$

where

$$a_{ij}^{+} = \begin{Bmatrix} a_{ij}, \text{ if } a_{ij} \geq 0 \\ 0 \text{ otherwise} \end{Bmatrix}, \quad a_{ij}^{-} = \begin{Bmatrix} 0, \text{ if } a_{ij} \geq 0 \\ a_{ij} \text{ otherwise} \end{Bmatrix}$$

Assume that $x_{n+i}$ is an integer variable, but $b_i$ is not integer. Then, designating $b_i - [b_i]$ as $f$, we have

$$\sum (a_{ij}^{+} + a_{ij}^{-})x_j + x_{n+i} = f + \delta \tag{13.6}$$

where $\delta$ is the integer $[b_i]$. If the left-hand side of (13.6) were positive, we would have it equal to $f$, $1 + f, \ldots$ and the following expression would be satisfied:

$$\sum (a_{ij}^{+} + a_{ij}^{-})x_j \geq f$$

Now, since $\sum (a_{ij}^{+})x_j \geq \sum (a_{ij}^{+} + a_{ij}^{-})x_j$, we have

$$\sum a_{ij}^{+}x_j \geq f \tag{13.7}$$

If the left-hand side of (13.6) were negative, we would have it equal to $-1 + f$, $-2 + f, \ldots$ and the following expression would be satisfied:

$$\sum (a_{ij}^{+} + a_{ij}^{-})x_j \leq -1 + f$$

Now, since $\sum (a_{ij}^{-})x_j \leq \sum (a_{ij}^{+} + a_{ij}^{-})x_j$, we have

$$\sum (a_{ij}^{-})x_j \leq f - 1$$

or, by dividing by $f - 1$

$$\frac{\sum (a_{ij}^{-})x_j}{(f - 1)} \geq 1 \quad \text{(remember that } f - 1 \text{ is negative)}$$

Further, multiplying the above constraint by $f$, we have the following:

$$\frac{\sum f(-a_{ij}^{-})x_j}{1-f} \geq f \tag{13.8}$$

Now either (13.7) or (13.8) holds; hence the following must hold:

$$\sum a_{ij}^{+}x_j + \sum \frac{f}{1-f}(-a_{ij}^{-})x_j \geq f \tag{13.9}$$

Thus far we have not made use of the fact that some of the variables in the constraint other than the basic variables $x_{n+i}$ might be integer variables. We may add or subtract any integer value to the coefficient of $a_{ij}$ $(= a_{ij}^{+} + a_{ij}^{-})$ of an *integer* variable in (13.6) and subtract or add, respectively, the same value to $x_{n+i}$ in that expression. Grouping the terms accordingly, the logic is the same and expression (13.9) remains intact except that the coefficients of integer variables may be altered by integer amounts. In an effort to make the cut constraint (13.9) deep, we want the coefficient of an integer variable in that expression to be as small as possible. The smallest positive coefficient we may obtain in (13.6) is $(a_{ij} - [a_{ij}])$, and the largest negative coefficient we may obtain in (13.6) is $(1 - a_{ij} + [a_{ij}])$. Thus for integer variables in (13.9), we have a choice of two expressions:

$$(a_{ij} - [a_{ij}]), \quad \frac{f}{(1-f)}(1 - a_{ij} + [a_{ij}])$$

We choose the smaller of the two to get the deeper cut constraint that can be generated from constraint (13.9). The first is smaller when

$$(a_{ij} - [a_{ij}]) < \frac{f(1 - a_{ij} + [a_{ij}])}{1-f}$$

or

$$(a_{ij} - [a_{ij}]) < f$$

The cut constraint thus becomes

(13.10)

$$\underbrace{\sum a_{ij}^{+}x_j + \sum \frac{f}{1-f}(-a_{ij}^{-})x_j}_{\text{for noninteger variables } x_j} + \underbrace{\underbrace{\sum (a_{ij} - [a_{ij}])x_j}_{\text{for } (a_{ij} - [a_{ij}]) \leq f} + \underbrace{\sum \frac{f}{1-f}(1 - a_{ij} + [a_{ij}])x_j}_{\text{for } (a_{ij} - [a_{ij}]) > f}}_{\text{for integer variables } x_j} \geq f$$

The slack of the cut constraint is not required to be integer. Otherwise, the mixed-integer algorithm proceeds exactly as does the fractional algorithm.

First the optimal noninteger solution is found using the simplex method or a variant. Then cut constraints are generated, appended, and a new optimal solution found, until an optimal solution is generated which satisfies the integrality requirements, or evidence is generated which indicates that there is no feasible integer solution (a dual unbounded condition).

To illustrate the method, we consider an example.

EXAMPLE 13.6:

Solve the following problem:

$$\begin{aligned} \text{Maximize } z &= 5x_1 + 2x_2 \\ \text{subject to:} \quad & 2x_1 + 2x_2 + x_3 = 9 \\ & 3x_1 + x_2 + x_4 = 11 \\ & x_1, x_2, x_3, x_4 \geq 0 \\ & x_1, x_3 \text{ integer} \end{aligned}$$

For this problem we shall omit the slack variables and display the nonnegativity constraints. (We list the integer variables first.)

**Tableau 1**

| | | ↓ $x_1$ | $x_2$ |
|---|---|---|---|
| $z$ | 0 | −5 | −2 |
| $x_1$ | 0 | −1 | 0 |
| $x_3$ | 9 | 2 | 2 |
| $x_2$ | 0 | 0 | −1 |
| $x_4$ → | 11 | [3] | 1 |

**Tableau 2**

| | | $x_4$ | ↓ $x_2$ |
|---|---|---|---|
| $z$ | $18\frac{1}{3}$ | $\frac{5}{3}$ | $-\frac{1}{3}$ |
| $x_1$ | $\frac{11}{3}$ | $\frac{1}{3}$ | $\frac{1}{3}$ |
| $x_3$ → | $\frac{5}{3}$ | $-\frac{2}{3}$ | [$\frac{4}{3}$] |
| $x_2$ | 0 | 0 | −1 |
| $x_4$ | 0 | −1 | 0 |

Ordinary simplex method iterations are utilized until the noninteger optimum is reached in Tableau 3. Using the indicated generating equation, the cut constraint is generated from the first constraint as follows, using expression (13.10).

$$\frac{1}{2}x_4 + \frac{\frac{1}{4}}{\frac{3}{4}}\left(1 - \frac{3}{4}\right)x_3 \geq \frac{1}{4}$$

or

$$\frac{1}{2}x_4 + \frac{1}{12}x_3 \geq \frac{1}{4}$$

which is appended to Tableau 3. After two lexicographic dual simplex iterations (i.e., using the pertubation methods on the dual when determining which variable is to enter the basis using the dual simplex method), an optimal solution is achieved in Tableau 5. Since it does not as yet satisfy the integrality constraints, an additional cut constraint is generated using the indicated constraint using (13.10), namely

$$\frac{2}{5}x_2 + \frac{3}{5}s_1 \geq \frac{3}{5}$$

It is appended to Tableau 5, and after one dual simplex iteration the optimal integer solution is achieved. Note that this is the same problem as that in Example 13.2, except that only $x_1$ and $x_3$ are required to be integer. It should not be surprising that the solutions are not identical. It can be seen that the first cut constraint

$$\frac{1}{12}x_3 + \frac{1}{2}x_4 \geq \frac{1}{4}$$

is

$$\frac{1}{12}(9 - 2x_1 - 2x_2) + \frac{1}{2}(11 - 3x_1 - x_2) \geq \frac{1}{4}$$

or

$$5x_1 + 2x_2 \leq 18$$

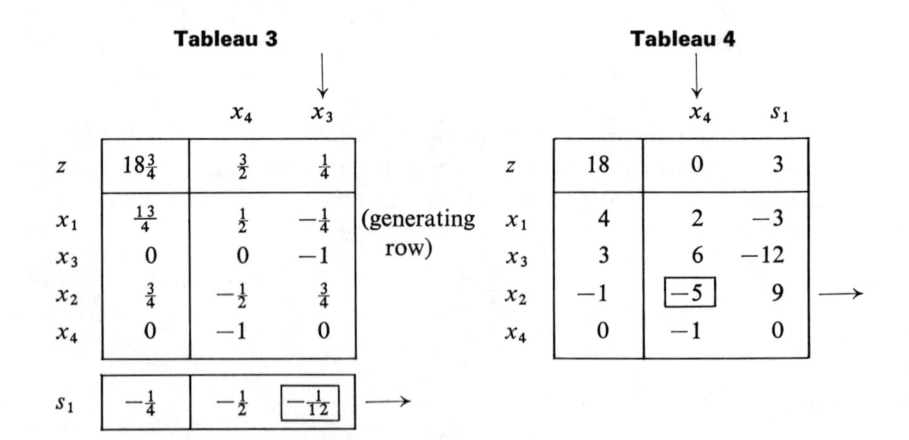

**Tableau 3**

| | | $x_4$ | $x_3$ ↓ | |
|---|---|---|---|---|
| $z$ | $18\frac{3}{4}$ | $\frac{3}{2}$ | $\frac{1}{4}$ | |
| $x_1$ | $\frac{13}{4}$ | $\frac{1}{2}$ | $-\frac{1}{4}$ | (generating row) |
| $x_3$ | 0 | 0 | $-1$ | |
| $x_2$ | $\frac{3}{4}$ | $-\frac{1}{2}$ | $\frac{3}{4}$ | |
| $x_4$ | 0 | $-1$ | 0 | |
| $s_1$ | $-\frac{1}{4}$ | $-\frac{1}{2}$ | [$-\frac{1}{12}$] | ⟶ |

**Tableau 4**

| | | $x_4$ ↓ | $s_1$ | |
|---|---|---|---|---|
| $z$ | 18 | 0 | 3 | |
| $x_1$ | 4 | 2 | $-3$ | |
| $x_3$ | 3 | 6 | $-12$ | |
| $x_2$ | $-1$ | [$-5$] | 9 | ⟶ |
| $x_4$ | 0 | $-1$ | 0 | |

**Tableau 5**

|  |  | ↓ $x_2$ | $s_1$ |  |
|---|---|---|---|---|
| $z$ | 18 | 0 | 3 |  |
| $x_1$ | $\frac{18}{5}$ | $\frac{2}{5}$ | $\frac{3}{5}$ | (generating row) |
| $x_3$ | $\frac{9}{5}$ | $\frac{6}{5}$ | $-\frac{6}{5}$ |  |
| $x_2$ | 0 | $-1$ | 0 |  |
| $x_4$ | $\frac{1}{5}$ | $-\frac{1}{5}$ | $-\frac{9}{5}$ |  |
| $s_2$ | $-\frac{3}{5}$ | $\boxed{-\frac{2}{5}}$ | $-\frac{3}{5}$ | → |

**Tableau 6**

|  |  | $s_2$ | $s_1$ |
|---|---|---|---|
| $z$ | 18 | 0 | 3 |
| $x_1$ | 3 | 1 | 0 |
| $x_3$ | 0 | 3 | $-3$ |
| $x_2$ | $\frac{3}{2}$ | $-\frac{5}{2}$ | $\frac{3}{2}$ |
| $x_4$ | $\frac{1}{2}$ | $-\frac{1}{2}$ | $-\frac{3}{2}$ |

and that the second cut constraint

$$\frac{2}{5}x_2 + \frac{3}{5}s_1 \geq \frac{3}{5}$$

is

$$\frac{2}{5}x_2 + \frac{3}{5}\left(6 - \frac{5}{3}x_1 - \frac{2}{3}x_2\right) \geq \frac{3}{5}$$

or

$$x_1 \leq 3$$

The solution path for this problem can be seen by reference to Figure 13.1. Tableaus 1, 2, and 3 in Example 13.6 correspond exactly to Tableaus 1, 2, and 3 in Example 13.2. The first cut constraint of Example 13.6 is $5x_1 + 2x_2 \leq 18$, and is shown using a broken line.
Tableau 4 of Example 13.6 corresponds to Tableau 5 of Example 13.2. Tableau 5 of Example 13.6 is a point between Tableaus 2 and 6 of Example 13.2. The second cut constraints are the same for both problems, and the optimal solution is as indicated.

The proof of this algorithm follows closely that of the Gomory fractional algorithm (having the integer variables first in the tableau with nonnegativity constraints stated explicitly) and will not be given here. See Gomory [124] for more information.

## 13.3 THE INTERSECTION CUT

Egon Balas [11] recently proposed the use of an intersection cut for solving integer programming problems. His method, which may be classified as an all-integer, primal dual feasible, cut method, is the same as the Gomory fractional algorithm except for the cut. (The method is extendable to the mixed-integer case as well.)

The logic of a cut may be seen by a geometric construct (which is not necessary for proof of convergence). Assume for now that the linear programming solution has no basic variables integer. Then consider the unique integer hypercube—in terms of basic variables—which contains the optimal linear programming solution with integer solutions only at its extreme points. (A two-dimensional example is shown in Fig. 13.3.) Circumscribe the hypercube by

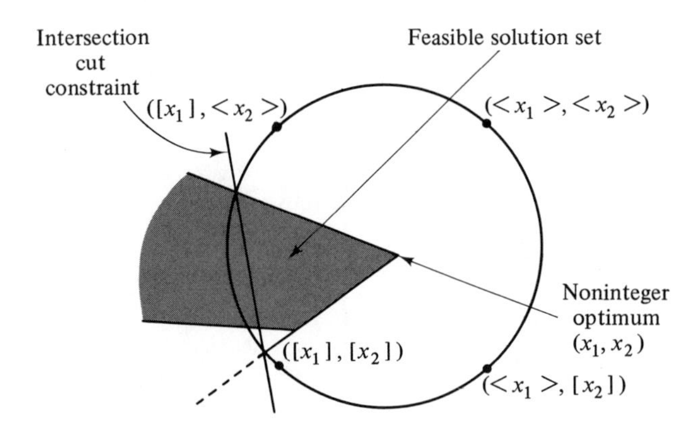


**Figure 13.3**

A Graph of a Two-Dimensional Intersection Cut

a hypersphere. Now consider halflines leading from the noninteger optimum to adjacent extreme points (but ignoring any constraints not binding at the noninteger optimum). Construct a hyperplane through the intersection of the halflines with the hypersphere. That hyperplane defines a legitimate (intersection) cut constraint.

If one or more (but not all) basic integer variables happen to be integer, there are a number of possible hypercubes that may be selected, each hypercube having its associated cut. Criteria are developed for choosing the best such cut. The intersection cut may be strengthened by integerization or by developing a Gomory all-integer cut from it. The resulting pivot element need not be $-1$, and the strongest integerized cut is the one which brings about the greatest (lexicographic) change in the objective function. Balas recommends selecting the strongest cut, but does not indicate a way of finding it (other than complete enumeration). He does propose a heuristic for finding a strong cut.

We now briefly develop the intersection cut. Let the noninteger optimum be $\boldsymbol{x}^*$. Let $x_j$ be a nonbasic variable at that optimum, with the associated vector $\boldsymbol{a}_j$ expressed in terms of the noninteger optimal basis. The halfline described above corresponding to increasing $x_j$ from zero is as follows (expressed in terms of basic variables):

$$\boldsymbol{\alpha}_j = \boldsymbol{x}^* - \boldsymbol{a}_j \lambda_j \text{ with } \lambda_j \geq 0$$

For convenience, let

$$\boldsymbol{\beta} = \boldsymbol{x}^* - [\boldsymbol{x}^*] - \frac{1}{2}(\mathbf{1})$$

$$\gamma = (\boldsymbol{x}^* - [\boldsymbol{x}^*])'(\mathbf{1} - \boldsymbol{x}^* + [\boldsymbol{x}^*])$$

Then it can be shown (see Balas [11]) that the values of $\lambda_j$ corresponding to the intersection of the halflines with the hypersphere are

$$\lambda_j = \frac{1}{(\boldsymbol{a}_j'\boldsymbol{a}_j)}(\boldsymbol{\beta}'\boldsymbol{a}_j + \sqrt{(\boldsymbol{\beta}'\boldsymbol{a}_j)^2 + \gamma \boldsymbol{a}_j'\boldsymbol{a}_j}) \tag{13.11}$$

The derived cut constraint must be satisfied:

$$\sum_{j \in N} \frac{x_j}{\lambda_j} \geq 1 \tag{13.12}$$

where $N$ is the set of nonbasic variables. An example follows.

EXAMPLE 13.7:

Using intersection cuts, derive a cut constraint and perform the first integer iteration for the problem of Example 13.2. We begin with the noninteger optimal solution given in Tableau 1 (ignoring the adjoined constraint).

**Tableau 1**

| | | $x_3$ ↓ | $x_4$ |
|---|---|---|---|
| $z$ | $18\frac{3}{4}$ | $\frac{1}{4}$ | $\frac{3}{2}$ |
| $x_2$ | $\frac{5}{4}$ | $\frac{3}{4}$ | $-\frac{1}{2}$ |
| $x_1$ | $\frac{13}{4}$ | $-\frac{1}{4}$ | $\frac{1}{2}$ |
| $s_1$ | $-1$ | $\boxed{-1}$ | $-1$ → |

The computations are as follows, using expressions (13.11) and (13.12).

$$x^* = \begin{pmatrix} \frac{5}{4} \\ \frac{13}{4} \end{pmatrix}, \quad \boldsymbol{\beta} = \begin{pmatrix} \frac{1}{4} \\ \frac{1}{4} \end{pmatrix} - \begin{pmatrix} \frac{1}{2} \\ \frac{1}{2} \end{pmatrix} = \begin{pmatrix} -\frac{1}{4} \\ -\frac{1}{4} \end{pmatrix}$$

$$\gamma = \begin{pmatrix} \frac{1}{4} & \frac{1}{4} \end{pmatrix} \begin{pmatrix} \frac{3}{4} \\ \frac{3}{4} \end{pmatrix} = \frac{3}{8}$$

$$\lambda_3 = \frac{3}{5}, \quad \lambda_4 = \frac{\sqrt{3}}{2}$$

and the cut constraint is

$$-\frac{5}{3}x_3 - \left(\frac{2}{3}\right)\sqrt{3}\,x_4 \leq -1$$

or approximately

$$-5x_3 - 3.47x_4 \leq -3$$

(Balas shows that any irrational $\lambda_j$ may be approximated by a rational number not larger than $\lambda_j$ without excluding any integer solutions.) We integerize the solution by using a divisor of 10, which gives

$$[-.5]x_3 + [-.347]x_4 \leq [-.3]$$

or $-x_3 - x_4 \leq -1$, which happens to be the strongest cut obtainable by integerization of this constraint. (It happens to be the same as the Dantzig cut for this example.) The constraint is adjoined to Tableau 1, and the indicated iteration leads to Tableau 2, which is still not integer.

**Tableau 2**

| | | $s_1$ | $x_4$ |
|---|---|---|---|
| $z$ | $18\frac{1}{2}$ | $\frac{1}{4}$ | $\frac{5}{4}$ |
| $x_2$ | $\frac{1}{2}$ | $\frac{3}{4}$ | $-\frac{5}{4}$ |
| $x_1$ | $\frac{7}{2}$ | $-\frac{1}{4}$ | $\frac{3}{4}$ |
| $x_3$ | 1 | $-1$ | 1 |

We leave completion of the solution to the reader, noting only that there are two hypercubes which may be used to construct the next iteration's intersection cut. A way of extending the method to the mixed-integer case is given by Balas, who also shows how Gomory's fractional and mixed-integer cuts are special cases of intersection cuts, using hypercylinders instead of hyperspheres. He also proves convergence. In a sequel paper, Balas, Bowman, Glover, and Sommer [13] find that even stronger cuts may be found by using a hypercube dual to the original instead of a hypersphere. Still stronger cuts are reported in a subsequent paper by Balas [10].

No computational experience is reported for these cuts.

## 13.4 PRIMAL INTEGER PROGRAMMING ALGORITHMS

Much of the research on integer programming algorithms has concentrated on dual algorithms. The importance of primal algorithms has been stressed, but until recently only very rudimentary primal algorithms were available. (See, for example, Ben-Israel and Charnes [28].) Richard D. Young [307] produced a breakthrough in this area in the form of a very complex algorithm, and more recently Young [308] and Fred Glover [117] developed simplified primal algorithms. In this section we shall consider primal algorithms in general, and focus on the algorithms of Young and Glover. These algorithms may be considered as all-integer, primal feasible starting solution, cut methods.

The basic idea of the primal algorithms is completely analagous to the primal simplex method, and is relatively straightforward. The problem is required to have all coefficients integer (which is not a limitation in general). The algorithm is indicated by the following steps:

1. Set up a basic feasible solution (possibly using an artificial basis).
2. Check for optimality in the usual way. Determine an ordinary simplex method iteration to be taken. Designate the incoming column as $k$. If all coefficients $a_{ik}$ are nonpositive, stop; there is no finite optimal solution. Otherwise, choose a trial outgoing row $r$ as in the simplex method.
3. a) If $a_{rk}$ is $+1$, perform the associated ordinary simplex iteration, and go to step 2.
   b) If $a_{rk}$ is greater than $+1$ (remember that all coefficients are integer, and it can be shown that with a pivot element of $+1$, the integrality from iteration to iteration will be preserved), set $\lambda = a_{rk}$ and derive a cut constraint of the form (13.5) for constraint $r$. Append and introduce $x_k$ into the cut constraint. Go to step 2.

An example using this algorithm will be given.

EXAMPLE 13.8:

Solve the problem of Example 13.2 using the primal algorithm presented above. This example is solved using the tableau omitting the identity matrix, but stating the nonnegativity constraints explicitly.

**Tableau 1**

| | | ↓ $x_1$ | $x_2$ | |
|---|---|---|---|---|
| $z$ | 0 | $-5$ | $-2$ | |
| $x_3$ | 9 | 2 | 2 | |
| $x_4$ | 11 | [3] (dashed box) | 1 | (generating constraint) |
| $x_1$ | 0 | $-1$ | 0 | |
| $x_2$ | 0 | 0 | $-1$ | |
| $s_1$ | 3 | [1] (boxed) | 0 | ⟶ |

For the ordinary simplex method the potential pivot element in the dashed box would be selected. Since it is greater than $+1$, the cut constraint indicated is generated (with $\lambda = 3$) and the pivot is performed, yielding Tableau 2.

**Tableau 2**

| | | $s_1$ | ↓ $x_2$ | |
|---|---|---|---|---|
| $z$ | 15 | 5 | $-2$ | |
| $x_3$ | 3 | $-2$ | [2] (dashed box) | (generating constraint) |
| $x_4$ | 2 | $-3$ | 1 | |
| $x_1$ | 3 | 1 | 0 | |
| $x_2$ | 0 | 0 | $-1$ | |
| $s_2$ | 1 | $-1$ | [1] (boxed) | ⟶ |

**Tableau 3**

| | | $s_1$ | $s_2$ |
|---|---|---|---|
| $z$ | 17 | 3 | 2 |
| $x_3$ | 1 | 0 | $-2$ |
| $x_4$ | 1 | $-2$ | $-1$ |
| $x_1$ | 3 | 1 | 0 |
| $x_2$ | 1 | $-1$ | 1 |

Similarly, Tableau 3 follows from Tableau 2. The solution in Tableau 3 is the same as that in Tableau 7 of Example 13.2, but the cut constraints are different. The first in this example is $x_1 \leq 3$, and the second is $x_1 + x_2 \leq 4$.

From the above example we can observe some advantages of a primal algorithm.

1. All solutions are feasible and integer, assuming that the starting solution is feasible (i.e., not artificial).
2. Because all solutions are feasible and integer, the solution may be discontinued after a number of iterations, and a good or near optimal solution will be available. Given the optimal noninteger solution, we have a bound on how far from optimal the best integer solution may be.

Unfortunately, the algorithm presented above does not necessarily converge to an optimal integer solution because the cut constraints become degenerate, and cycling cannot be prevented in general.

### Young's Primal Algorithm [308]

To remedy the defect of a cycling possibility, Young differentiates between two types of iterations in the above framework.

1. A transition cycle or iteration in which the objective function increases (by an integer amount).
2. A stationary cycle or iteration in which the objective function remains the same.

He then observes that a finite number of transition cycles are sufficient to assure convergence, or, equivalently, that convergence is assured if every sequence of stationary cycles is finite, i.e., each results in a transition cycle. He then shows how to assure that every sequence of stationary cycles is finite.

The transition cycles are taken exactly as in the algorithm above. In fact, Young proposes a rule which utilizes a transition cycle if one is available. The rule involves seeing whether any ordinary simplex method iteration would have a variable enter the basis at some value of one or greater. If so, that row is used to generate a cut for a transition cycle. If a transition cycle is not available, then a regularization constraint limiting the sum of the nonbasic variables to a specified upper limit is added. This upper limit can be determined by inspection or by solving a linear programming problem. Such a constraint is called an $L$-row, and is appended anew (replacing the old $L$-row) at the end of every transition cycle.

To determine the incoming variable for a stationary cycle, all columns with *positive* coefficients in the updated $L$-rows are considered. Each such column $j$ is divided by its positive $L$-row coefficient to yield a normalized column $\boldsymbol{\alpha}_j$ (only for computational purposes; the tableau is not altered at this point). The variable $x_j$ corresponding to the lexicographically minimum $\boldsymbol{\alpha}_j$ is selected to enter the basis. Denote the incoming column as $k$.

To determine the source constraint for the cut constraint in a stationary iteration (remember that a stationary iteration is utilized *only* when a transition iteration is not available), choose

1. The row $r$ used for the cut for the previous iteration, if the previous iteration was a stationary iteration and if
   a) $a_{rk} > 0$
   b) $a_{rk} > a_{r0}$
2. Otherwise any row $r$ for which
   a) $a_{rk} > 0$
   b) $a_{rk} > a_{r0}$

The proof for Young's method was outlined briefly above, and will not be given here. The interested reader may refer to the proof in Young's paper [308].

### Glover's Primal Algorithm [117]

Fred Glover also uses an auxiliary constraint in his algorithm. After observing that Young's $L$-row is a legitimate auxiliary constraint, he proves that *any* feasible solution to the dual of the original linear programming problem can be used as multipliers of the constraints to form an auxiliary constraint. Thus Glover's auxiliary constraint is a *surrogate* constraint, as defined in Chapter 16. He proposes using the optimal solution to the dual as multipliers of the constraints, but does not imply that this is necessarily superior to using any feasible solution. He does not propose changing the constraint periodically, as does Young, but indicates the possibility of doing so.

The selection of an incoming variable in Glover's algorithm is exactly the same as in Young's, with the auxiliary constraint in place of the $L$-row, except that Glover does *not* seek out transition cycles.

Selection of the constraint to be used for generating a cut constraint is the same as that of Young except that the previous iteration is not considered. Glover breaks any ties by choosing the iteration which results in the maximum sum of the negative $a_{oj}$ in the next tableau. In any resulting ties the equation with the smallest index is chosen. Although the latter procedure is heuristic in nature, it does ensure convergence, proof of which is given in Glover's paper [117].

EXAMPLE 13.9:

Solve the following problem using Young's and Glover's algorithms:

$$
\begin{aligned}
\text{Maximize } z &= 5x_1 + 4x_2 \\
\text{subject to:} \quad 3x_1 + 3x_2 &\leq 10 \\
12x_1 + 6x_2 &\leq 24 \\
x_1, x_2 &\geq 0
\end{aligned}
$$

We shall first form Glover's auxiliary constraint by noting that a feasible (nonoptimal) solution to the dual problem is $y_1 = 1$, $y_2 = 1$. Hence we

**Young's Method**

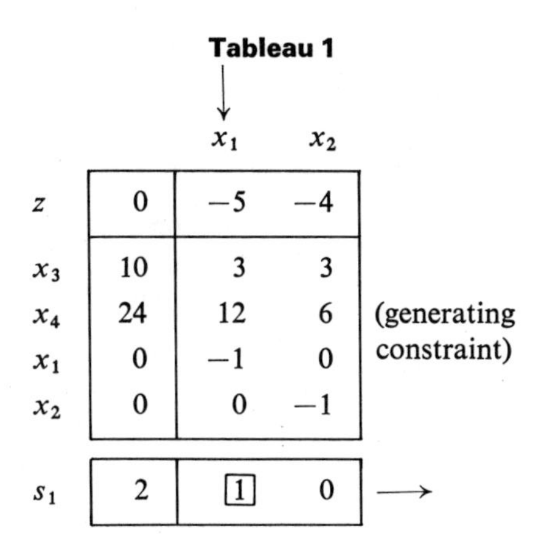

**Tableau 1** (↓ pivot column $x_1$)

| | | $x_1$ ↓ | $x_2$ | |
|---|---|---|---|---|
| $z$ | 0 | −5 | −4 | |
| $x_3$ | 10 | 3 | 3 | |
| $x_4$ | 24 | 12 | 6 | (generating constraint) |
| $x_1$ | 0 | −1 | 0 | |
| $x_2$ | 0 | 0 | −1 | |
| $s_1$ | 2 | [1] | 0 | ⟶ |

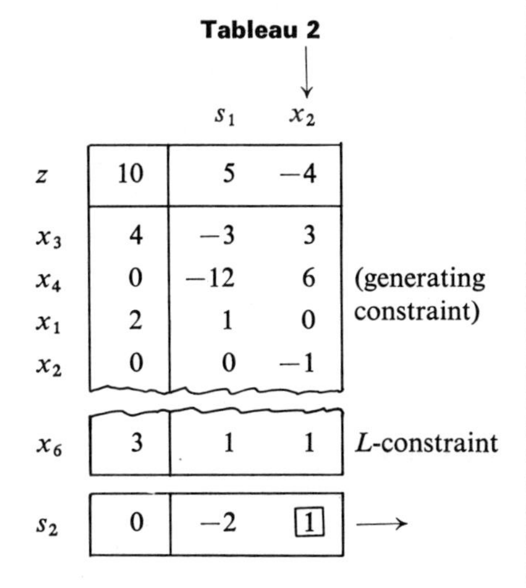

**Tableau 2** (↓ pivot column $x_2$)

| | | $s_1$ | $x_2$ ↓ | |
|---|---|---|---|---|
| $z$ | 10 | 5 | −4 | |
| $x_3$ | 4 | −3 | 3 | |
| $x_4$ | 0 | −12 | 6 | (generating constraint) |
| $x_1$ | 2 | 1 | 0 | |
| $x_2$ | 0 | 0 | −1 | |
| $x_6$ | 3 | 1 | 1 | $L$-constraint |
| $s_2$ | 0 | −2 | [1] | ⟶ |

The sum of $s_1$ and $x_2$ cannot exceed 3. That constraint is used as the $L$-constraint. (An easier to see upper bound is 5 because $s_1 \leq 2$ and $x_2 \leq 3$.)

**Glover's Method**

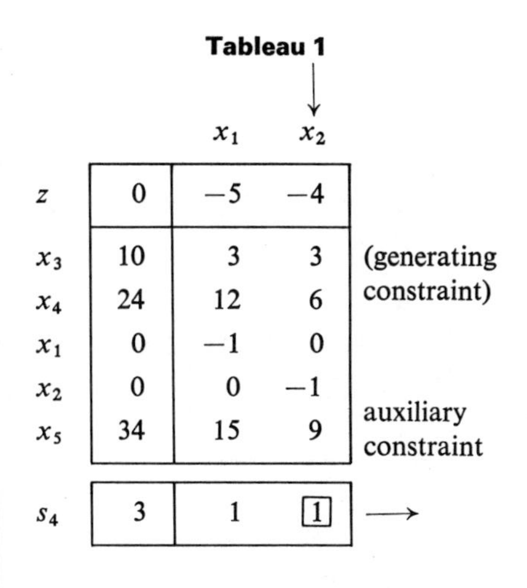

**Tableau 1** (↓ pivot column $x_2$)

| | | $x_1$ | $x_2$ ↓ | |
|---|---|---|---|---|
| $z$ | 0 | −5 | −4 | |
| $x_3$ | 10 | 3 | 3 | (generating constraint) |
| $x_4$ | 24 | 12 | 6 | |
| $x_1$ | 0 | −1 | 0 | |
| $x_2$ | 0 | 0 | −1 | |
| $x_5$ | 34 | 15 | 9 | auxiliary constraint |
| $s_4$ | 3 | 1 | [1] | ⟶ |

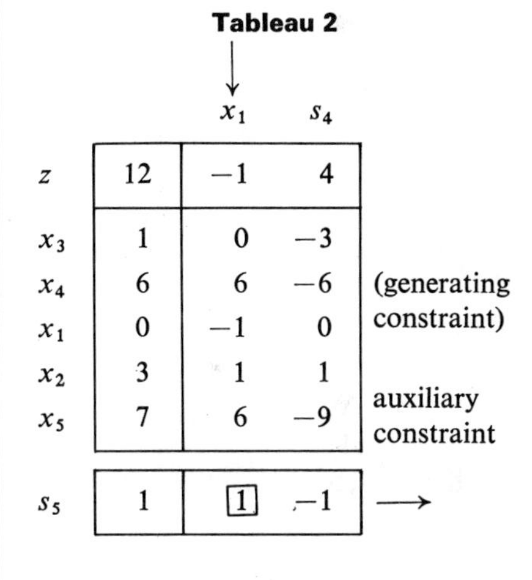

**Tableau 2** (↓ pivot column $x_1$)

| | | $x_1$ ↓ | $s_4$ | |
|---|---|---|---|---|
| $z$ | 12 | −1 | 4 | |
| $x_3$ | 1 | 0 | −3 | |
| $x_4$ | 6 | 6 | −6 | (generating constraint) |
| $x_1$ | 0 | −1 | 0 | |
| $x_2$ | 3 | 1 | 1 | |
| $x_5$ | 7 | 6 | −9 | auxiliary constraint |
| $s_5$ | 1 | [1] | −1 | ⟶ |

**Figure 13.4**

The Solution to the Problem of Example 13.9 Using Primal Algorithms

**Young's Method**

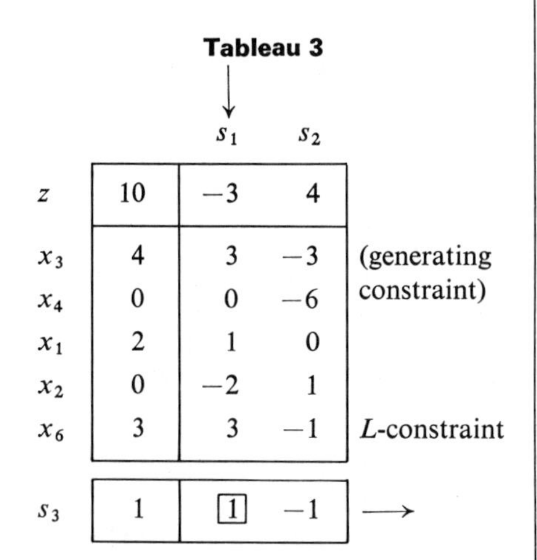

**Tableau 3**

| | | $s_1$ ↓ | $s_2$ | |
|---|---|---|---|---|
| $z$ | 10 | −3 | 4 | |
| $x_3$ | 4 | 3 | −3 | (generating constraint) |
| $x_4$ | 0 | 0 | −6 | |
| $x_1$ | 2 | 1 | 0 | |
| $x_2$ | 0 | −2 | 1 | |
| $x_6$ | 3 | 3 | −1 | $L$-constraint |
| $s_3$ | 1 | [1] | −1 | ⟶ |

**Tableau 4**

| | | $s_3$ | $s_2$ | |
|---|---|---|---|---|
| $z$ | 13 | 3 | 1 | |
| $x_3$ | 1 | −3 | 0 | |
| $x_4$ | 0 | 0 | −6 | optimal |
| $x_1$ | 1 | −1 | 1 | integer |
| $x_2$ | 2 | 2 | −1 | solution |
| $x_6$ | 0 | −3 | 2 | |

**Glover's Method**

**Tableau 3**

| | | $s_5$ | $s_4$ | |
|---|---|---|---|---|
| $z$ | 13 | 1 | 3 | |
| $x_3$ | 1 | 0 | −3 | |
| $x_4$ | 0 | −6 | 0 | optimal |
| $x_1$ | 1 | 1 | −1 | integer |
| $x_2$ | 2 | −1 | 2 | solution |
| $x_5$ | 1 | −6 | −3 | |

**Figure 13.4 (concluded)**

have as an auxiliary constraint

$$3x_1 + 3x_2 + 12x_1 + 6x_2 \leq 10 + 24$$

or

$$15x_1 + 9x_2 \leq 34$$

We denote the slack of the auxiliary constraint by $x_5$ and the slacks of the constraints as $x_3$ and $x_4$, respectively. Initially, we need no $L$-constraint for Young's method. The iterations for both solutions are given in Figure 13.4.

By way of explanation, Young's method initially considers either $x_1$ or $x_2$ to enter the basis, since either results in a transition cycle. We choose $x_1$ because it has the most negative reduced cost. Then, in Tableau 2 no transition cycle can be achieved, so an $L$-constraint is developed and a stationary cut constraint utilized. In Tableau 3 a transition cycle may be taken, leading to Tableau 4, which is optimal.

For Glover's method, we choose the lexicographical minimum normalized nonbasic vector (having divided the elements of the vector by the positive element in the auxiliary constraint). The cut constraints are then derived, and the iterations taken as indicated.

### A Simplified Iteration

Glover also observed a shorthand way of performing an iteration. Instead of adjoining the cut constraint and pivoting (provided the nonnegativity constraints are explicitly stated), it is sufficient to treat the coefficient $a_{rk}$ as a pseudo-pivot. The pivot is then performed by adding column $\boldsymbol{a}_k$ to any column $\boldsymbol{a}_j$ with $a_{rj} < 0$ as many times as necessary until $\bar{a}_{rj}$ (the updated value of $a_{rj}$) first becomes nonnegative, and by subtracting column $\boldsymbol{a}_k$ from any column $\boldsymbol{a}_j$ with $a_{rj} \geq a_{rk}$ until $\bar{a}_{rj} \geq 0$, but one more subtraction would make $\bar{a}_{rj}$ negative. Finally, $\bar{\boldsymbol{a}}_k$, the new column $k$, is equal to the negative of the old. This result can be easily proved.

EXAMPLE 13.10:

We illustrate the simplified iteration on the first iteration of Glover's example [117]. Since $\boldsymbol{a}_3/51$ is the lexicographically smallest vector ($x_8$ is the slack of the auxiliary constraint), $x_3$ is selected to enter the solution. The third constraint is selected uniquely to generate the cut with $\lambda = 2$. Performing the iteration indicated yields the following tableaus. Observe that because no element $a_{3j}$ $(j \neq 3) \geq 2$, no subtractions can be done, but twice the third column is added to the first, and twice the third column added to the second. To convince yourself, do the next iteration.

**Tableau 0 (with cut)**

| | | $x_1$ | $x_2$ | $x_3$ |
|---|---|---|---|---|
| $z$ | 0 | −4 | −6 | −3 |
| $x_4$ | 5 | 1 | 2 | 0 |
| $x_5$ | 8 | 9 | 2 | −4 |
| $x_6$ | 1 | −3 | −2 | 2 |
| $x_7$ | 16 | −5 | 4 | 6 |
| $x_8$ | 500 | 68 | 125 | 51 |
| $x_1$ | 0 | −1 | 0 | 0 |
| $x_2$ | 0 | 0 | −1 | 0 |
| $x_3$ | 0 | 0 | 0 | −1 |
| $s_1$ | 0 | −2 | −1 | [1] |

**Tableau 0 (without cut)**

| | | $x_1$ | $x_2$ | $x_3$ |
|---|---|---|---|---|
| $z$ | 0 | −4 | −6 | −3 |
| $x_4$ | 5 | 1 | 2 | 0 |
| $x_5$ | 8 | 9 | 2 | −4 |
| $x_6$ | 1 | −3 | −2 | [2] |
| $x_7$ | 16 | −5 | 4 | 6 |
| $x_8$ | 500 | 68 | 125 | 51 |
| $x_1$ | 0 | −1 | 0 | 0 |
| $x_2$ | 0 | 0 | −1 | 0 |
| $x_3$ | 0 | 0 | 0 | −1 |

**Tableau 1 (in both cases)**

| | | $x_1$ | $x_2$ | $s_1$ |
|---|---|---|---|---|
| $z$ | 0 | −10 | −9 | 3 |
| $x_4$ | 5 | 1 | 2 | 0 |
| $x_5$ | 8 | 1 | −2 | 4 |
| $x_6$ | 1 | [1] | 0 | −2 |
| $x_7$ | 16 | 7 | 10 | −6 |
| $x_8$ | 500 | 170 | 176 | −51 |
| $x_1$ | 0 | −1 | 0 | 0 |
| $x_2$ | 0 | 0 | −1 | 0 |
| $x_3$ | 0 | −2 | −1 | 1 |
| $s_2$ | 1 | [1] | 0 | −2 |

(In this case, since the potential pivot element is 1, there is no need to generate the cut)

Glover further observed that it is not necessary to retain the identity of the nonbasic variables because all original variables are retained in the expanded representation. Glover's simplified iteration is suggestive of Glover's bound escalation method [111], which we consider in the next section.

On the question of efficiency, no empirical results have been given, and Glover and Young do not consider this question. From what they do say, however, drawn primarily from a few example problems, it would appear that Glover's approach has more promise that Young's, but further testing is required to substantiate or disprove this.

## 13.5 GLOVER'S BOUND ESCALATION METHOD [111]

The bound escalation method of Glover [111] requires a dual feasible starting solution. It may be classified as an all-integer, dual feasible starting solution, constructive method. (It is presented along with the cut methods because, as we shall see, it may be shown to be equivalent to a cut method.) The problem is written at *every* iteration in the following form, which is equivalent to the tableau format we have been using all along:

$$\text{Minimize } z = \sum_{j=1}^{n} c_j x_j$$

$$\text{subject to:} \quad \sum_{j=1}^{n} a_{ij} x_j + x_{n+i} = b_i, \quad i = 1, \ldots, m$$

$$x_j \geq 0 \quad j = 1, \ldots, n$$

All $a_{ij}$, $c_j$, and $b_i$ are integers, which is no limitation in practice. Further, we assume $c_j \geq 0, j = 1, \ldots, n$. If, for any constraint $i$, all coefficients $a_{ij}$ are nonnegative except one, and coefficient $a_{ik}$ is negative, and $b_i < 0$, then by Theorem 12.3:

$$x_k \geq \left\langle \frac{b_i}{a_{ik}} \right\rangle$$

The bound escalation method alternatively constructs and enforces such bounds. The method is presented very generally, and there is considerable flexibility in the choice of specific rules. We shall present one of the simpler sets of rules. Initially, at least one $b_i$ must be negative, for otherwise an optimal integer solution is available.

1. For rows with $b_i < 0$, choose the row with the fewest negative $a_{ij}$. (If there is no negative $a_{ij}$, then it can be shown that the problem has no feasible integer solution.) In case of a tie, choose the minimum $b_i$. Denote the row as $r$.
2. For $a_{rj} < 0$, choose the lexicographically smallest $m + 1$ vector $\begin{pmatrix} c_j \\ \boldsymbol{a}_j \end{pmatrix}$. Denote the corresponding column as column $k$.
3. Subtract column $k$ of the matrix from every column $j$ that has $a_{rj} < 0$ some number of times until *either*
   a) $\bar{a}_{rj} = a_{rj} - pa_{rk}$ (where $p$ is the number of subtractions) becomes zero or positive, *or*
   b) $\begin{pmatrix} \bar{c}_j \\ \bar{\boldsymbol{a}}_j \end{pmatrix} = \begin{pmatrix} c_j \\ \boldsymbol{a}_j \end{pmatrix} - p\begin{pmatrix} c_k \\ \boldsymbol{a}_k \end{pmatrix}$ (where $p$ is as above) would become lexicographically negative with the next subtraction

   *whichever occurs first.*
4. Steps 1 through 3 are repeated until only one coefficient in a constraint having a negative $b_i$ is negative. In general, one or more rows will exhibit that form at the completion of steps 1 through 3.
5. Consider rows in which all elements but one other than the $b_i$ are nonnegative. (Glover calls such a set of constraints a bounding form.) Omit any nonnegativity conditions.
6. For those rows omit any columns that have no negative entries.
7. Omit some rows so that each column remaining has exactly one negative coefficient. The remaining structure is called a prime bounding form (providing that at *least* one row $i$ has a negative $b_i$).
8. The bounds from each constraint of the prime bounding form are enforced sequentially upon the entire problem, by subtracting the appropriate column from the $\boldsymbol{b}$ vector an appropriate number of times, until all constraints of the prime bounding form are satisfied. Then go to step 1.

Step 8 may be replaced, with some gain in efficiency, by application of the inverse of the prime bounding form. More explicitly, by using the inverse of the prime bounding form, a set of fractional lower bounds may be found and rounded up. There are also questions of strategy regarding, for example, the possibility of increasing the size of the prime bounding form. We shall not go into detail on such topics. The interested reader is referred to Glover [111].

Some examples will precede our consideration of certain theoretical aspects of the bound escalation method. As in the earlier methods, this tableau form assures us that the values of the original variables will always be available.

EXAMPLE 13.11:
(This is Example 13.4 solved by Glover's bound escalation method.)

**Tableau 1**

| | | $x_1$ | $x_2$ | $x_3$ | |
|---|---|---|---|---|---|
| | 0 | 10 | 14 | 21 | |
| $x_4$ | −12 | −8 | −11 | −9 | |
| $x_5$ | −14 | −2 | −2 | −7 | (generating constraint) |
| $x_6$ | −10 | −9 | −6 | −3 | |
| $x_1$ | 0 | −1 | 0 | 0 | |
| $x_2$ | 0 | 0 | −1 | 0 | |
| $x_3$ | 0 | 0 | 0 | −1 | |

The second row is selected for generation of the bounding form since it is tied for having the least number of negative coefficients, but has the most negative $b_i$. $x_1$ has the lexicographically smallest column, and hence the first column is subtracted once from the second column and twice from the third column, yielding Tableau 2. As we shall show shortly, the columns subtracted in order to establish bounding forms, as well as columns subtracted from the $\boldsymbol{b}$ vector when enforcing bounds, no longer represent the variables they did prior to the subtraction. Because all of the variables are represented explicitly (as basic), the loss of identity of nonbasic variables is of no consequence; all values of variables of concern are readily available. Thus $x_1'$, rather than $x_1$, appears in the appropriate column of Tableau 2. A bounding form is then available from the third row in Tableau 2, and it is applied to yield $x_1' \geq 2$. (The constraint $-9x_1' + 3x_2 + 15x_3 \leq -10$ implies $x_1' \geq 2$.) Enforcing this bound yields the adjusted $\boldsymbol{b}$ vector, which is shown alongside Tableau 2 and replaces the $\boldsymbol{b}$ vector. Above the adjusted $\boldsymbol{b}$ vector is shown the vector that has been subtracted from the $\boldsymbol{b}$ vector (enforcing the bound) and the number of times subtracted. In the adjusted Tableau 2, the second constraint is again selected; this time $x_3$ is the lexicographically smallest column, and hence is subtracted from the first column, yielding Tableau 3. At this point we can observe that there are two possible prime bounding forms:

**Tableau 2**

| | $\binom{2}{x_1'}$ | | $x_1'$ | $x_2$ | $x_3$ | |
|---|---|---|---|---|---|---|
| | −20 | 0 | 10 | 4 | 1 | |
| $x_4$ | 4 | −12 | −8 | −3 | 7 | |
| $x_5$ | −10 | −14 | −2 | 0 | −3 | (generating constraint) |
| $x_6$ | 8 | −10 | −9 | 3 | 15 | |
| $x_1$ | 2 | 0 | −1 | 1 | 2 | |
| $x_2$ | 0 | 0 | 0 | −1 | 0 | |
| $x_3$ | 0 | 0 | 0 | 0 | −1 | (the primes indicate that these variables are no longer the original variables) |

**Tableau 3**

| | $\binom{1}{x_3'}$ | $\binom{3}{x_1''}$ | $\binom{4}{x_3'}$ | | $x_1''$ | $x_2$ | $x_3'$ |
|---|---|---|---|---|---|---|---|
| | −52 | −51 | −24 | −20 | 9 | 4 | 1 |
| $x_4$ | 14 | 21 | −24 | 4 | −15 | −3 | 7 |
| $x_5$ | 2 | −1 | 2 | −10 | 1 | 0 | −3 |
| $x_6$ | 5 | 20 | −52 | 8 | −24 | 3 | 15 |
| $x_1$ | 1 | 3 | −6 | 2 | −3 | 1 | 2 |
| $x_2$ | 0 | 0 | 0 | 0 | 0 | −1 | 0 |
| $x_3$ | 2 | 1 | 4 | 0 | 1 | 0 | −1 |

1) the one formed by $x_1''$ and $x_3'$ and the second and third constraints; and
2) the one formed by $x_1''$ and $x_3'$ and the second and fourth constraints.

Using the first of these two yields the sequence of adjusted $\boldsymbol{b}$ vectors reading from right to left adjacent to Tableau 3.

First, we deduce from the second constraint that $x_3' \geq 4$. We enforce this constraint by subtracting four times the $x_3'$ column from the $\boldsymbol{b}$ vector. Next, using the adjusted $\boldsymbol{b}$ vector to the left of Tableau 3, we deduce from the third constraint that $x_1'' \geq 3$. We enforce this constraint by subtracting three times the $x_1''$ column from the adjusted $\boldsymbol{b}$ vector. Finally, using the second adjusted $\boldsymbol{b}$ vector to the left of Tableau 3, we deduce from the second constraint that the lower bound on $x'_3$ may be increased by one. We then subtract the $x'_3$ column from the second adjusted $\boldsymbol{b}$ vector to the left of Tableau 3 to obtain the left-most $\boldsymbol{b}$ vector which is feasible and hence optimal. (Above the additional $\boldsymbol{b}$ vectors are indicated the bounds which are being escalated.) The solution values are read corresponding to the original variables along the left side of the tableau ($x_4 = 14, x_5 = 2, x_6 = 5, x_1 = 1, x_2 = 0, x_3 = 2$), and the optimal objective function value is 52. To see the alternative to step 8 in the method, consider the prime bounding form utilized in the above problem in Tableau 3 (prior to the escalation), namely

$$-x''_1 + 3x'_3 \geq 10$$

$$24x''_1 - 15x'_3 \geq -8$$

The inverse to the matrix

$$\begin{pmatrix} -1 & 3 \\ 24 & -15 \end{pmatrix} \quad \text{is} \quad \begin{pmatrix} \frac{15}{57} & \frac{3}{57} \\ \frac{24}{57} & \frac{1}{57} \end{pmatrix}$$

and hence

$$\begin{pmatrix} x''_1 \\ x'_3 \end{pmatrix} \geq \begin{pmatrix} \frac{15}{57} & \frac{3}{57} \\ \frac{24}{57} & \frac{1}{57} \end{pmatrix} \begin{pmatrix} 10 \\ -8 \end{pmatrix} = \begin{pmatrix} \frac{126}{57} \\ \frac{232}{57} \end{pmatrix}$$

or

$$\begin{pmatrix} x''_1 \\ x'_3 \end{pmatrix} \geq \begin{pmatrix} 3 \\ 5 \end{pmatrix} \geq \begin{pmatrix} \frac{126}{57} \\ \frac{232}{57} \end{pmatrix}$$

In this case, all the bounds to be achieved from the form are found at once. In general, additional bounds may have to be enforced after the initial escalation using the inverse is enforced.

### An Exploration of the Nature of the Linear Transformations Used by the Bound Escalation Method

As seen above, the bound escalation method uses a linear transformation of variables which achieves a bounding form. The nature of this transformation is to subtract one column from another. More formally, the original variables are defined as a linear function of new variables, whose nonnegativity is implied. (The nonnegativity of the original variables is assured by virtue of their explicit representation.) For example, in Example 13.10, $x_1$ is defined as

$$x_1 = x'_1 - x_2 - 2x_3 \tag{13.13}$$

In every such case, the column being subtracted is the one whose variable is being redefined. (Make the indicated substitution and see.) That $x'_1$ is required to be nonnegative may be seen by virtue of $x'_1$ being definitional in constraint (13.13). In general terms

$$x'_s = x_s + \sum_{j \neq s} p_j x_j, \quad p_j \geq 0 \text{ and integer} \tag{13.14}$$

where $s$ is the column being subtracted and $p_j$ is the integer multiple of column $s$ subtracted from column $j$. Glover raised the possibility of adding columns. This would mean that in the above framework $p_j$ would be nonpositive, in which case (13.14) would not imply the nonnegativity of $x'_s$. Glover observed that if a constraint of the problem at some stage were in a bounding form, (i.e., it were definitional), and if multiples of the column corresponding to the negative element in that constraint were added to columns corresponding to the positive elements in that constraint *without destroying the bounding form* (i.e., preserving the definitional character), then the nonnegativity of the new variable would be assured by the form of the constraint. Such additions of columns were found to be very powerful in general. An example of this is taken from Glover [111] (p. 148), and is given in Tableaus 1 and 2.

**Tableau 1**

| | | $x_1$ | $x_2$ | $x_3$ |
|---|---|---|---|---|
| | 0 | 1 | 4 | 8 |
| $x_4$ | −7 | −25 | 8 | −7 |
| $x_5$ | −6 | 20 | −7 | 9 |
| $x_6$ | −9 | −1 | 1 | −6 |
| $x_1$ | 0 | −1 | 0 | 0 |
| $x_2$ | 0 | 0 | −1 | 0 |
| $x_3$ | 0 | 0 | 0 | −1 |

**Tableau 2**

| | | $x_1$ | $x'_2$ | $x_3$ |
|---|---|---|---|---|
| | 0 | 9 | 4 | 12 |
| $x_4$ | −7 | −9 | 8 | 1 |
| $x_5$ | −6 | 6 | −7 | 2 |
| $x_6$ | −9 | 1 | 1 | −5 |
| $x_1$ | 0 | −1 | 0 | 0 |
| $x_2$ | 0 | −2 | −1 | −1 |
| $x_3$ | 0 | 0 | 0 | −1 |

We observe that the second constraint exhibits a bounding form. By adding twice the second column to the first column and adding the second column to the third column, Tableau 2 is obtained. The result is a bounding form in all three columns; in Glover's terminology, it is a maximal bounding form. Further, as required, in Tableau 2 the second constraint retains its bounding form on $x'_2$. Glover also discusses other computational devices and considerations, including an improved strategy for generating a bounding form which is not unlike the dual simplex method. For more information see Glover [111].

#### Convergence of the Bound Escalation Method

To prove convergence of the bound escalation method we require that the vectors of the problem be lexicographically positive. As before, this may be achieved by using a regularization constraint, if necessary. Assume further that a maximum possible objective function value of a solution is known, and that upper and lower bounds are known for all variables. We shall choose as the (only) constraint to generate the bounding form the row closest to the top of the tableau with a negative $b_i$ entry. On execution of the bounding form of that constraint, the $\boldsymbol{b}$ vector entry corresponding to the first nonzero element of the variable whose bound is being escalated will strictly decrease by an integer amount. Thus the $\boldsymbol{b}$ vector column will decrease lexicographically an integer amount with each bound escalation. Since the vector can only decrease lexicographically a finite number of times, convergence is proven.

#### Gomory's All-Integer Algorithm as a Special Case of the Bound Escalation Method

Glover discusses a way in which Gomory's all-integer method can be cast into the framework of the bound escalation method, in that it yields the same sequence of iterations. (This implies that the bound escalation method may be viewed as a cut method, which is true, although the cuts are implicit rather than explicit.) The algorithm proceeds as follows:

1. Select a negative $b_i$. Denote the corresponding row as $r$. If none exists, the solution is feasible and hence optimal.
2. Apply steps 2 and 3 of the bound escalation method to the tableau, and also to the (added) constraint $x_k \geq 0$. (Recall that column $k$ is the lexicographically minimum column and that $x_k$ is the variable represented by that column.)
3. Add to constraint $r$ the smallest nonnegative integer multiple of the updated constraint $x_k \geq 0$ so that the resulting constraint achieves a bounding form.
4. Make all *permissible* column additions with respect to the new constraint (i.e., preserving both the bounding form and the dual feasibility), escalate the lower bound, and go to step 1.

EXAMPLE 13.12:

Use the bound escalation method altered to yield the same sequence of iterations as Gomory's all-integer algorithm for the problem of Example 13.4. In Tableau 1 we designate the second constraint to be used for generating the bound. Because the $x_1$ column is the lexicographical minimum, we adjoin the constraint $x_1 \geq 0$ to Tableau 1. (We cannot use the constraint $x_1 \geq 0$ in the tableau for two reasons:

1. It may have been updated.
2. We destroy the adjoined constraint in the computational process.)

Performing legitimate column subtractions leads to Tableau 2. We then add twice the updated adjoined constraint to the updated original constraint (to get a bounding form) and replace the adjoined constraint in Tableau 3 by the result.
Since no column additions are possible in Tableau 3, we escalate the bound on the adjoined constraint ($x'_1 \geq 4$). Using the new $\boldsymbol{b}$ vector in Tableau 3,

**Tableau 1**

| | | $x_1$ | $x_2$ | $x_3$ | |
|---|---|---|---|---|---|
| $z$ | 0 | 10 | 14 | 21 | |
| $x_4$ | −12 | −8 | −11 | −9 | |
| $x_5$ | −14 | −2 | −2 | −7 | (generating constraint) |
| $x_6$ | −10 | −9 | −6 | −3 | |
| $x_1$ | 0 | −1 | 0 | 0 | |
| $x_2$ | 0 | 0 | −1 | 0 | |
| $x_3$ | 0 | 0 | 0 | −1 | |
| | 0 | −1 | 0 | 0 | adjoined constraint $x_1 \geq 0$ |

**Tableau 2**

| | | $x'_1$ | $x_2$ | $x_3$ |
|---|---|---|---|---|
| $z$ | 0 | 10 | 4 | 1 |
| $x_4$ | −12 | −8 | −3 | 7 |
| $x_5$ | −14 | −2 | 0 | −3 |
| $x_6$ | −10 | −9 | 3 | 15 |
| $x_1$ | 0 | −1 | 1 | 2 |
| $x_2$ | 0 | 0 | −1 | 0 |
| $x_3$ | 0 | 0 | 0 | −1 |
| | 0 | −1 | 1 | 2 |

add twice this constraint to $x_5$ row and replace the adjoined constraint by it

**Tableau 3**

| | $\binom{4}{x'_1}$ | | $x'_1$ | $x_2$ | $x_3$ | |
|---|---|---|---|---|---|---|
| $z$ | −40 | 0 | 10 | 4 | 1 | |
| $x_4$ | 20 | −12 | −8 | −3 | 7 | |
| $x_5$ | −6 | −14 | −2 | 0 | −3 | (generating constraint) |
| $x_6$ | 26 | −10 | −9 | 3 | 15 | |
| $x_1$ | 4 | 0 | −1 | 1 | 2 | |
| $x_2$ | 0 | 0 | 0 | −1 | 0 | |
| $x_3$ | 0 | 0 | 0 | 0 | −1 | |
| | 2 | −14 | −4 | 2 | 1 | |
| | 0 | | 0 | 0 | −1 | adjoined constraint $x_3 \geq 0$ |

we discard the adjoined constraint and again use the second constraint to develop a bounding form. Because the $x_3$ column is the lexicographical minimum we adjoin the constraint $x_3 \geq 0$. After subtracting the $x_3$ column from the $x'_1$ column we have Tableau 4. No column additions are possible, nor are any additions of the updated $x_3 \geq 0$ constraint necessary. We therefore escalate the second adjoined constraint which is the updated generating constraint (plus *zero* times the updated $x_3 \geq 0$ constraint). Using the first added $\boldsymbol{b}$ vector in Tableau 4 we discard the adjoined constraints and use the third constraint to develop a bounding form. It is already in a bounding form, hence no additions of the $x'_1 \geq 0$ constraint are required. Further, no new additions are possible, and we can directly escalate the bounds on $x'_1$, yielding the second added $\boldsymbol{b}$ vector to the left of Tableau 4. We are in a similar situation with respect to the second constraint, and escalate the bound on $x'_3$, which gives the third added $\boldsymbol{b}$ vector to the left of Tableau 4. That solution is feasible and hence optimal. It is useful to compare the sequence of iterations here with that of Example 13.4. It is also useful to compare the results with those of Example 13.11.

**Tableau 4**

| | $\binom{1}{x'_3}$ | $\binom{1}{x''_1}$ | $\binom{2}{x'_3}$ | | $x''_1$ | $x_2$ | $x'_3$ |
|---|---|---|---|---|---|---|---|
| $z$ | −52 | −51 | −42 | −40 | 9 | 4 | 1 |
| $x_4$ | 14 | 21 | 6 | 20 | −15 | −3 | 7 |
| $x_5$ | 2 | −1 | 0 | −6 | 1 | 0 | −3 |
| $x_6$ | 5 | 20 | −4 | 26 | −24 | 3 | 15 |
| $x_1$ | 1 | 3 | 0 | 4 | −3 | 1 | 2 |
| $x_2$ | 0 | 0 | 0 | 0 | 0 | −1 | 0 |
| $x_3$ | 2 | 1 | 2 | 0 | 1 | 0 | −1 |
| | | | | 0 | 1 | 0 | −1 |
| | 2 | −1 | 0 | −6 | 1 | 0 | −3 |

add zero times this constraint to $x_5$ row, and replace the adjoined constraint by it, which yields the second adjoined constraint here

Glover discusses the pros and cons of the two methods, and suggests that, since Gomory's all-integer method may be cast as a special case of the bound escalation method, there is reason for optimism about the latter. To quote Professor Glover [111]: "The only real answer to these issues (of computational efficiency) must, of course, come from empirical studies of a wide range of problems of varying size and structure." To my knowledge, no such studies have yet been carried out.

**Selected Supplemental References**

Section 13.2
[19], [38], [61], [123], [124], [125], [128], [207]
Section 13.3
[11], [13]
Section 13.4
[28], [117], [307], [308]
Section 13.5
[111]
General
[14], [110]

## 13.6 PROBLEMS

**1.** Complete the computations of Example 13.3.

**2.** Solve the following problem using Gomory's fractional algorithm in two forms:

a. In the manner originally presented.
b. In the tableau form omitting basic variables and writing the non-negativity constraints explicitly.

$$\begin{aligned} \text{Maximize } z &= 5x_1 + 4x_2 \\ \text{subject to:} \quad 3x_1 + 3x_2 &\leq 10 \\ 12x_1 + 6x_2 &\leq 24 \\ x_1, x_2 &\geq 0 \text{ and integer} \end{aligned}$$

**3.** Solve Example 13.4, having written the nonnegativity constraints explicitly.

**4.** Solve Example 13.5, not having written the nonnegativity constraints explicitly.

**5.** Solve the problem of Example 13.8 using Glover's suggested simplification illustrated in Example 13.9.

**6.** Solve the problem of Example 13.5 using the bound escalation method.

**7.** Solve the following problem:

$$\begin{aligned} \text{Maximize } z &= 2x_1 + 3x_2 + x_3 \\ \text{subject to:} \quad x_1 + 2x_2 + 3x_3 &\leq 7 \\ 3x_1 + 2x_2 + 5x_3 &\leq 11 \\ x_1, x_2, x_3 &\geq 0 \text{ and integer} \end{aligned}$$

a. using Gomory's fractional algorithm.
b. using Young's primal algorithm.
c. using Glover's primal algorithm.
d. using integer upper bounds on variables only.

**8.** Solve the following problem:

$$\begin{aligned} \text{Maximize } z = {} & 3x_1 + 4x_2 + 3x_3 \\ \text{subject to:} \quad & 3x_1 + 2x_2 + 2x_3 \leq 13 \\ & 2x_1 + 5x_2 + 3x_3 \leq 15 \\ & 2x_1 + x_2 + 2x_3 \leq 9 \\ & x_1, x_2, x_3 \geq 0 \quad \text{and integer} \end{aligned}$$

a. using Gomory's fractional algorithm.
b. using Young's primal algorithm.
c. using Glover's primal algorithm.

**9.** Solve the following problem:

$$\begin{aligned} \text{Minimize } z = {} & 3x_1 + 4x_2 + 3x_3 \\ \text{subject to:} \quad & 3x_1 + 2x_2 + 2x_3 \geq 13 \\ & 2x_1 + 5x_2 + 3x_3 \geq 15 \\ & 2x_1 + x_2 + 2x_3 \geq 9 \\ & x_1, x_2, x_3 \geq 0 \text{ and integer} \end{aligned}$$

a. using Gomory's all-integer method.
b. using Glover's bound escalation method.

**10.** Solve the following problem:

$$\begin{aligned} \text{Minimize } z = {} & 2x_1 + 3x_2 + 5x_3 \\ \text{subject to:} \quad & x_1 + 2x_2 + 3x_3 \geq 7 \\ & 3x_1 + 2x_2 + 3x_3 \geq 11 \\ & x_1, x_2, x_3 \geq 0 \text{ and integer} \end{aligned}$$

a. using Gomory's all-integer method.
b. using Glover's bound escalation method.

**11.** Solve problem 9 using Gomory's mixed-integer algorithm:
a. with $x_1$ and $x_3$ only constrained to be integer.
b. with $x_1$ and $x_2$ only constrained to be integer.
c. with $x_2$ and $x_3$ only constrained to be integer.

**12.** Solve problem 10 using Gomory's mixed-integer algorithm:
a. with $x_2$ only constrained to be integer.

b. with $x_1$ and $x_2$ only constrained to be integer.
c. with $x_2$ and $x_3$ only constrained to be integer.

**13.** Solve Example 13.6 requiring that $x_2$ and $x_4$ only are integer.

**14.** Complete Example 13.7.

**15.** Complete Example 13.10.

**16.** Consider the following problem (Gomory and Hoffman [128]):

$$\begin{aligned} \text{Maximize } z = x_1 + x_2 & \\ \text{subject to:} \quad 2x_1 \qquad & \leq 3 \\ 2x_2 & \leq 3 \\ x_1, x_2 & \geq 0 \text{ and integer} \end{aligned}$$

This problem cannot be solved using the original cut proposed by Dantzig. Show that it is solved by the modified procedure of section 13.2.

**17.** Show how the cut constraints (e.g., in Example 13.2) are expressed in terms of the original structural variables.

**18.** Prove the statement on p. 349 that if ". . . ever a cut constraint is overly satisfied . . . it must be implied by the remaining constraints."

**19.** Develop a lexicographic dual simplex method, which uses the dual of the perturbation methods (the perturbation methods are described in Chap. 3), to use in conjunction with the various integer programming dual algorithms (e.g., Gomory's fractional algorithm).

**20.** Show that the same device used to generate a dual feasible solution for the primal dual algorithm of Chapter 6 may be used to generate a dual feasible solution for Gomory's all-integer method. (There need not be an artificial variable in the regularization constraint.)

**21.** Prove that expression (13.10) follows from (13.9).

**22.** Prove expression (13.11).

**23.** Show that any $\lambda$ in the range $a_{rk}/2 < \lambda \leq a_{rk}$ will give a legitimate cut for the primal all-integer method. Why is the one using $\lambda$ as *large* as possible chosen for the primal method, in contrast to the Gomory all-integer method, in which $\lambda$ is chosen as small as possible?

**24.** Show why a negative reduced cost implies a positive $L$-constraint or auxiliary constraint element for either Young's or Glover's primal algorithm, respectively.

**25.** G.T. Martin [207] has developed an accelerated Euclidean algorithm which accelerates Gomory's fractional algorithm. Assuming a noninteger optimal solution is available, the steps are as follows:

a. Select a generator row $r$ from the problem.
b. Abstract a Gomory cut constraint $s$ from constraint $r$. Determine the dual simplex pivot column, $k$.

c. Enforce the cut constraint on only the row $r$.
d. If the updated row $r$ is all integer, go to step e; otherwise, abstract a new Gomory cut constraint from the updated row $r$, choose $k$ as the pivot column, and go to step c.
e. Determine a single composite constraint that produces the integer row $r$ in one iteration by using all of the cut constraints just derived from row $r$.
f. Append the composite constraint to the entire problem, and reoptimize.
g. If the result is integer, stop. Otherwise, go to step a.

Martin solved the following problem:

$$\text{Maximize } z = 2x_1 + 3x_2$$
$$\text{subject to:} \quad 2x_1 + 5x_2 \leq 8$$
$$3x_1 + 2x_2 \leq 9$$
$$x_1, x_2 \geq 0 \text{ and integer}$$

The optimal noninteger solution omitting the identity matrix is given in Tableau 1.

**Tableau 1**

| | | $x_4$ | $x_3$ |
|---|---|---|---|
| $z$ | $\frac{76}{11}$ | $\frac{4}{11}$ | $\frac{5}{11}$ |
| $x_2$ | $\frac{6}{11}$ | $-\frac{2}{11}$ | $\frac{3}{11}$ |
| $x_1$ | $\frac{29}{11}$ | $\frac{5}{11}$ | $-\frac{2}{11}$ |

Choosing $x_1$ as the source row we have the cut constraint

$$\frac{7}{11} \leq \frac{5}{11}x_4 + \frac{9}{11}x_3 \; (s_1 \text{ is the slack variable})$$

and the partial iteration proceeding from Tableau 1′ (containing only the generating and cut constraints) to Tableau 2, updating only the $x_1$ row.

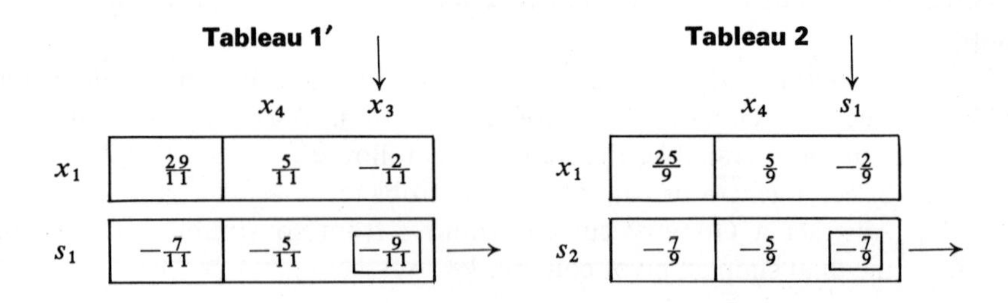

**Tableau 1′**

| | | $x_4$ | $x_3$ ↓ |
|---|---|---|---|
| $x_1$ | $\frac{29}{11}$ | $\frac{5}{11}$ | $-\frac{2}{11}$ |
| $s_1$ | $-\frac{7}{11}$ | $-\frac{5}{11}$ | $\boxed{-\frac{9}{11}}$ → |

**Tableau 2**

| | | $x_4$ | $s_1$ ↓ |
|---|---|---|---|
| $x_1$ | $\frac{25}{9}$ | $\frac{5}{9}$ | $-\frac{2}{9}$ |
| $s_2$ | $-\frac{7}{9}$ | $-\frac{5}{9}$ | $\boxed{-\frac{7}{9}}$ → |

We abstract a constraint from the $x_1$ row of Tableau 2, retain the same pivot column as before, and perform the partial iteration from Tableau 2 to Tableau 3, updating only the $x_1$ row. We similarly proceed from Tableau 3 to Tableaus 4, 5, and 6. We next derive the single constraint which transforms the $x_1$ row

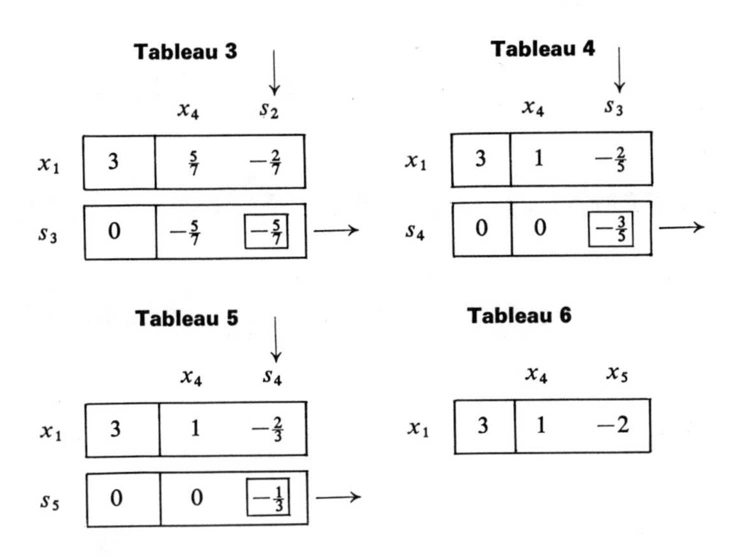

**Tableau 3** (pivot column $s_2$)

| | | $x_4$ | $s_2$ |
|---|---|---|---|
| $x_1$ | 3 | $\frac{5}{7}$ | $-\frac{2}{7}$ |
| $s_3$ | 0 | $-\frac{5}{7}$ | $\boxed{-\frac{5}{7}}$ → |

**Tableau 4** (pivot column $s_3$)

| | | $x_4$ | $s_3$ |
|---|---|---|---|
| $x_1$ | 3 | 1 | $-\frac{2}{5}$ |
| $s_4$ | 0 | 0 | $\boxed{-\frac{3}{5}}$ → |

**Tableau 5** (pivot column $s_4$)

| | | $x_4$ | $s_4$ |
|---|---|---|---|
| $x_1$ | 3 | 1 | $-\frac{2}{3}$ |
| $s_5$ | 0 | 0 | $\boxed{-\frac{1}{3}}$ → |

**Tableau 6**

| | | $x_4$ | $x_5$ |
|---|---|---|---|
| $x_1$ | 3 | 1 | $-2$ |

of Tableau 1 to that of Tableau 6 in one iteration. We obtain this by back substitution. The fifth cut constraint from Tableau 5 is $-\frac{1}{3}s_4 \leq 0$. We substitute for $s_4$ from Tableau 4: $s_4 - \frac{3}{5}s_3 = 0$, which gives $-\frac{1}{5}s_3 \leq 0$. Similarly, we substitute for $s_3$ from Tableau 3: $s_3 - \frac{5}{7}x_4 - \frac{5}{7}s_2 = 0$, to get $-\frac{1}{7}x_4 - \frac{1}{7}s_2 \leq 0$. We then substitute for $s_2$ from Tableau 2: $s_2 - \frac{5}{9}x_4 - \frac{7}{9}s_1 = \frac{7}{9}$, to get $-\frac{2}{9}x_4 - \frac{1}{9}s_1 \leq -\frac{1}{9}$. We then substitute for $s_1$ from Tableau 1′: $s_1 - \frac{5}{11}x_4 - \frac{9}{11}x_3 = -\frac{7}{11}$, to get $-\frac{3}{11}x_4 - \frac{1}{11}x_3 \leq -\frac{2}{11}$, which is the desired constraint. This constraint is then appended to Tableau 1 and the problem resolved. If the solution is not integer (it is not for this problem), the process is repeated, starting at step 1.

a. Complete the solution to this problem using the accelerated Euclidean algorithm.

b. Solve the problem using Gomory's fractional algorithm and compare the solution with the solution in part a.

# 14

# The Group Theoretic Approach to Solving Integer Programming Problems

## 14.1 INTRODUCTION

This chapter* introduces the group theoretic approach to integer programming, based on the pioneering work of Gomory [126]. The approach may be viewed as an all-integer, primal dual feasible starting solution, constructive method. Our procedure in pursuing the method will be to begin with Gomory's fractional algorithm and proceed to the development of a group of cut constraints. Using an example, we shall then develop a geometric interpretation which will be used as an aid to introduce the group problem. A method for solving the group problem will be posed, and then sufficient conditions will be developed indicating when the optimal solution to the group problem is an optimal integer solution to the original problem. Strategies for solving the problem in general are then discussed, and a number of examples presented.

Let the problem be written as:

$$\text{Minimize } z = \boldsymbol{c}'\boldsymbol{x}$$
$$(14.1) \qquad \text{subject to: } \boldsymbol{A}\boldsymbol{x} = \boldsymbol{b}$$
$$\boldsymbol{x} \geq \boldsymbol{0} \text{ and integer}$$

where $\boldsymbol{A}$ is a matrix of rank $m$ of order $m$ by $n$ $(m < n)$, $\boldsymbol{x}$ and $\boldsymbol{c}$ are $n$ vectors

* This chapter is taken from a paper [56] coauthored with Prof. Der-San Chen, Department of Industrial Engineering, University of Alabama.

and $\boldsymbol{b}$ is an $m$ vector. Further, let $\boldsymbol{A}$ be partitioned as $\boldsymbol{B}$ and $\boldsymbol{N}$, $\boldsymbol{B}$ being the optimal linear programming basis. Vectors $\boldsymbol{x}$ and $\boldsymbol{c}$ are similarly partitioned into $\boldsymbol{x}_B$, $\boldsymbol{x}_N$, $\boldsymbol{c}_B$, and $\boldsymbol{c}_N$, respectively. Without loss of generality, assume that all the coefficients of $\boldsymbol{A}$ and $\boldsymbol{b}$ are integer. (This is equivalent to assuming that $\boldsymbol{A}$ and $\boldsymbol{b}$ consist of rational numbers.)

Expression (14.1) may be written as follows:

$$\begin{aligned} &\text{Minimize } z = \boldsymbol{c}'_B\boldsymbol{x}_B + \boldsymbol{c}'_N\boldsymbol{x}_N \\ &\text{subject to:} \quad \boldsymbol{B}\boldsymbol{x}_B + \boldsymbol{N}\boldsymbol{x}_N = \boldsymbol{b} \\ &\qquad\qquad \boldsymbol{x}_B, \boldsymbol{x}_N \geq \boldsymbol{0} \text{ and integer} \end{aligned} \tag{14.2}$$

where $\boldsymbol{B}$ is of order $m$ by $m$ and nonsingular, $\boldsymbol{N}$ is of order $m$ by $(n - m)$, $\boldsymbol{c}_B$ and $\boldsymbol{x}_B$ are of order $m$ by 1, and $\boldsymbol{c}_N$ and $\boldsymbol{x}_N$ are of order $(n - m)$ by 1.

Consider the linear programming problem (14.2) in the updated form (14.3).

$$\begin{aligned} &\text{Minimize } z = (\boldsymbol{c}_N - \boldsymbol{c}_B\boldsymbol{B}^{-1}\boldsymbol{N})\boldsymbol{x}_N + \boldsymbol{c}_B\boldsymbol{B}^{-1}\boldsymbol{b} \\ &\text{subject to:} \quad \boldsymbol{x}_B + \boldsymbol{B}^{-1}\boldsymbol{N}\boldsymbol{x}_N = \boldsymbol{B}^{-1}\boldsymbol{b} \\ &\qquad\qquad \boldsymbol{x}_B, \boldsymbol{x}_N \geq \boldsymbol{0} \end{aligned} \tag{14.3}$$

The optimal conditions of a linear programming problem solution, $\boldsymbol{c}_N - \boldsymbol{c}_B\boldsymbol{B}^{-1}\boldsymbol{N} \geq \boldsymbol{0}$, must be satisfied, and the noninteger optimum is $\boldsymbol{x}_B = \boldsymbol{B}^{-1}\boldsymbol{b}$ and $\boldsymbol{x}_N = \boldsymbol{0}$. In all but trivial cases, $\boldsymbol{B}^{-1}\boldsymbol{b}$ will not be all-integer.

## 14.2 GOMORY'S FRACTIONAL ALGORITHM AND THE GROUP OF CUT CONSTRAINTS

Recall from Chapter 13 that Gomory's fractional algorithm required first solving the linear programming problem. Then a cut constraint was generated and added to the problem, and the problem was re-solved. By means of an example, we shall show that, corresponding to a noninteger optimum, there exists a finite algebraic group of cuts which can be generated. Under certain circumstances, the solution appending only those cut constraints to the original problem will lead to an optimal integer solution.

Consider the following example problem (Example 13.2):

$$\begin{aligned} &\text{Minimize } z = -5x_1 - 2x_2 \\ &\text{subject to:} \quad 2x_1 + 2x_2 + x_3 \phantom{+ x_4} = 9 \\ &\qquad\qquad 3x_1 + x_2 \phantom{+ x_3} + x_4 = 11 \\ &\qquad\qquad x_1, x_2, x_3, x_4 \geq 0 \text{ and integer} \end{aligned} \tag{14.4}$$

Neglecting the integer restrictions, we solve it as a linear programming problem.

Then we obtain the optimal noninteger solution, written in equation form, as follows:

$$
\begin{array}{llll}
 & \text{Minimize } z + 18.75 = & & \tfrac{1}{4}x_3 + \tfrac{3}{2}x_4 \\
(14.5) & \text{subject to:} \quad 1.25 = & & x_2 + \tfrac{3}{4}x_3 - \tfrac{1}{2}x_4 \\
 & \qquad\qquad\quad 3.25 = x_1 & & \quad - \tfrac{1}{4}x_3 + \tfrac{1}{2}x_4
\end{array}
$$

By taking the positive fractional portions of the coefficients of the objective function, we derive a Gomory cut constraint:

$$\tfrac{3}{4} \leq \tfrac{1}{4}x_3 + \tfrac{1}{2}x_4$$

If, instead of using the objective function, we had first doubled the objective function and then derived a cut, we would have had $2z + 37.5 = .5x_3 + 3x_4$, with the cut as follows:

$$.5 \leq .5x_3$$

Similarly, using three times the objective function, the cut constraint is

$$.25 \leq .75x_3 + .5x_4$$

and using four times the objective function the cut constraint is $0 \leq 0$, a vacuous constraint. Using five times the objective function yields the same cut constraint as using one times the objective function. For this problem, using any integer multiple of the objective function yields one of the four cut constraints given above.

We can generate cuts from any of the constraints of the problem in addition to the objective function. Multiples of constraints can also be added to multiples of the objective function or constraints, and cuts derived from the result. Interestingly, for this example, the four cut constraints derived from the objective function are the *only* ones that are obtained. (The reader may wish to convince himself of this before proceeding.)

In general, the complete set of cut constraints derived in the above manner forms a *finite additive Abelian group** of order $D$ (or less) where $D$ is the absolute value of the determinant of the optimal linear program basis. (For the group to be finite, $D$ must be finite, or, equivalently, all coefficients must be rational.) For example problem (14.5) just given, the group of cut constraints is a cyclic group of order 4. In general, the group of cuts need not be cyclic.

It can be shown [126] that the order of the group equals or exactly divides $D$, the absolute value of the determinant of the optimal noninteger basis. A sufficient though not necessary condition for the order of the group to equal $D$ is that $A$ contains an identity matrix. Gomory's form of the problem assumes this.

* See the appendix to this chapter for definitions of this and other group theoretic terms that we use.

## 14.3 AN ALGEBRAIC—GEOMETRIC STRUCTURE RELATING NONINTEGER AND INTEGER OPTIMA

From (14.3) we know that for a specified solution value of $\boldsymbol{x}_N$, $\boldsymbol{x}_B$ is uniquely determined; therefore the strategy for finding an integer optimum will be to examine certain solutions of the set $\boldsymbol{x}_N \geq \boldsymbol{0}$ and integer. There are three problems in examining such solutions in general:

1. $\boldsymbol{x}_N \geq \boldsymbol{0}$ and integer are not sufficient to assure that $\boldsymbol{x}_B$ will be integer.
2. $\boldsymbol{x}_N \geq \boldsymbol{0}$ and integer are not sufficient to assure that the inequalities $\boldsymbol{x}_B \geq \boldsymbol{0}$ will be satisfied.
3. $\boldsymbol{x}_N \geq \boldsymbol{0}$ and integer are not sufficient to assure the optimality of an integral solution to (14.3), *even ignoring* the constraints $\boldsymbol{x}_B \geq \boldsymbol{0}$ and integer.

When a solution $\boldsymbol{x}_N \geq \boldsymbol{0}$ and integer overcomes these three problems simultaneously, such an $\boldsymbol{x}_N$ determines an optimal integer solution. Before attempting to resolve these problems, consider a geometric interpretation. We superimpose two sets of integer solution points on the solution space (obviously a visual geometrical representation can be made only in two or three dimensions). One set is $\boldsymbol{x}_B \geq \boldsymbol{0}$ and integer; the other set is $\boldsymbol{x}_N \geq \boldsymbol{0}$ and integer. Obviously, both point sets must be represented in the same space. Of particular interest are points of intersection of both sets. One such point which is particularly desirable is the one "closest" in the "direction" of the objective function to the objective function value of the noninteger optimum. That point is the integer optimum.

In expression (14.5), $x_3, x_4 \geq 0$ corresponds to $\boldsymbol{x}_N \geq \boldsymbol{0}$ in expression (14.3). Similarly, $x_1, x_2 \geq 0$ corresponds to $\boldsymbol{x}_B \geq \boldsymbol{0}$. (It is a coincidence that the nonbasic variables are the original slack variables, and that the basic variables are the original structural variables.) The two sets of points are shown in the space of basic variables ($x_1$ and $x_2$) in Figure 14.1. The squares correspond to the integer points $\boldsymbol{x}_B \geq \boldsymbol{0}$; the dots correspond to the integer points $\boldsymbol{x}_N \geq \boldsymbol{0}$. Thus the superimposed integer points (dots within squares) of these two types are feasible solutions to the integer programming problem. The superimposed point closest to the objective hyperplane at the linear programming optimum is the integer optimum (i.e., $x_3 = 1, x_4 = 1, x_1 = 3, x_2 = 1$). For this example (though not in general) all integer solutions $\boldsymbol{x}_B \geq \boldsymbol{0}$ also have integer values of $\boldsymbol{x}_N$, since all the nonbasic variables happen to be slack variables.

We pause for some useful geometric interpretations. By reference to Figure 14.1, the set of feasible linear programming solutions is called the linear programming polyhedron. Inside the linear programming polyhedron we have the integer programming polyhedron, a convex set formed by convex combinations of all feasible integer solutions to the original problem. The integer programming polyhedron for the example is shown in Figure 14.1 as bounded

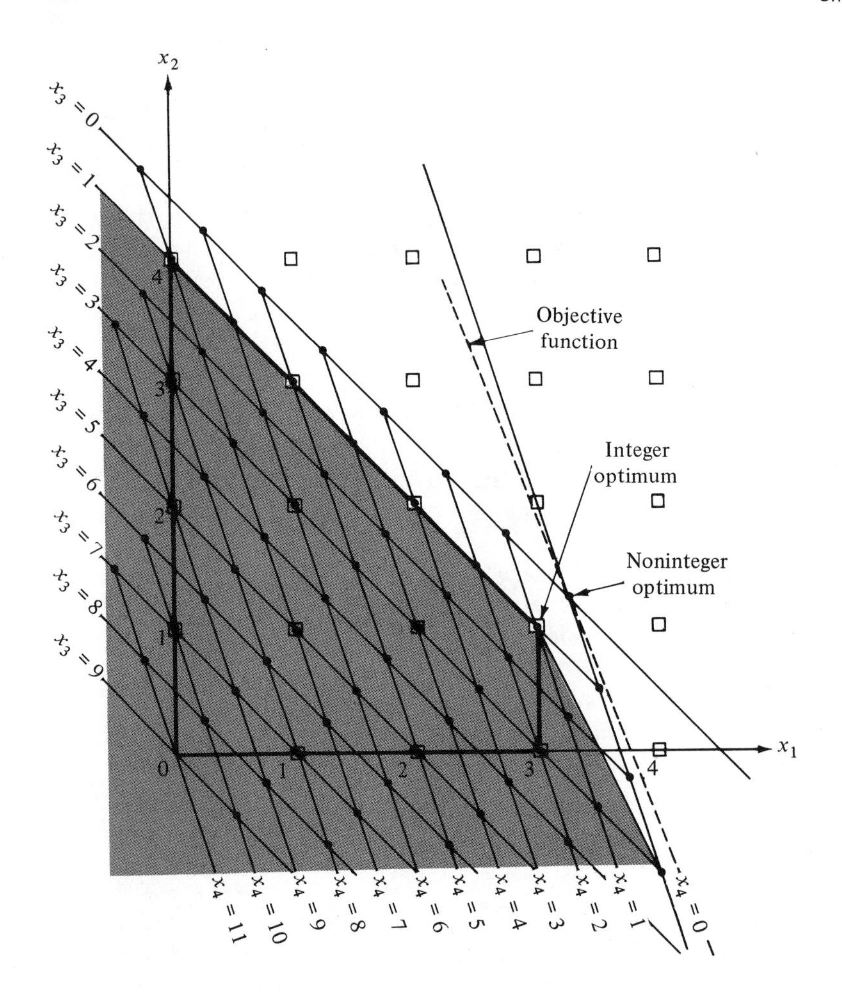


**Figure 14.1**
Graph of Example Problem (14.4).

by the bold line. Minimizing the objective function over the integer programming polyhedron is an ordinary linear programming problem whose optimal solution is the optimal solution to the integer programming problem. Accordingly, every feasible basis of the integer programming polyhedron is integer. However, the relationship between the linear programming polyhedron and the integer programming polyhedron is not simple, and there is no known computationally inexpensive method for generating the constraints for the integer programming polyhedron.

By reference to the optimal *linear* programming solution whose basis is $\boldsymbol{B}$, we now ignore the constraints $\boldsymbol{x}_B \geq \boldsymbol{0}$. The convex set that remains is the set of solutions to $\boldsymbol{x}_N \geq \boldsymbol{0}$ and is an unbounded polyhedral cone. For example problem (14.4), the unbounded polyhedral cone is the set of solutions to $x_3 \geq 0$ (alternatively, $2x_1 + 2x_2 \leq 9$) and $x_4 \geq 0$ (alternatively, $3x_1 + 2x_2 \leq 11$), with $x_1$ and $x_2$ unrestricted in sign. Note that minimizing the objective function over the unbounded polyhedral cone is equivalent to finding the optimal linear programming solution, because only constraints nonbinding at the optimum have been eliminated. Analogous to the unbounded polyhedral cone, we have the convex set formed by convex combinations of integer solutions (solutions which have both $\boldsymbol{x}_B$ and $\boldsymbol{x}_N$ integer) in the unbounded polyhedral cone. That set is called the *corner polyhedron.* A portion of the corner polyhedron for the example problem is the shaded area in Figure 14.1. There are two vertices of this corner polyhedron: $x_3 = 1$, $x_4 = 1$; and $x_3 = 3$, $x_4 = 0$. Minimizing the objective function over the corner polyhedron gives the optimal solution to the integer programming problem, *provided* that the resulting optimum happens to satisfy the constraints $\boldsymbol{x}_B \geq \boldsymbol{0}$. (As we shall see, the latter condition is not fulfilled in our example.)

We have seen two possible ways of solving the integer programming problem:

1. Determine the set of constraints corresponding to the integer programming polyhedron. Minimize the objective function subject to that set of constraints. The resulting solution is the optimal integer solution to the original problem.
2. Determine the set of constraints corresponding to the corner polyhedron. Minimize the objective function subject to that set of constraints requiring that $\boldsymbol{x}_B$ and $\boldsymbol{x}_N$ be integer. If the resulting solution satisfies $\boldsymbol{x}_B \geq \boldsymbol{0}$, it is the optimal integer solution to the original problem.

Unfortunately, neither of these procedures appears feasible at this point. We shall show later that the second procedure is indeed feasible, and is the basis of the group theoretic approach.

## 14.4 THE GROUP THEORETIC ALGORITHM

Returning now to the constraints (14.3), it can be shown, using an argument analogous to the derivation of the Gomory cut constraint, that the first problem can be resolved by adding the constraints

$$\sum_{j \in N} (a_{ij} - [a_{ij}])x_j = (b_i - [b_i]) \pmod 1, \; i = 1, \ldots, m \tag{14.6}$$

where $a = b \bmod c$ means that $a$ and $b$ are congruent modulo $c$, or that $a$ and $b$ differ by an integer multiple of $c$ (i.e., $a - b = rc$, $r$ integer). In addition, $a_{ij}$ are the updated matrix coefficients of (14.2) and $[a_{ij}]$ is the largest integer not larger than $a_{ij}$. (Note that no component of $\boldsymbol{x}_B$ appears in (14.6).)

By way of proof, the $i^{\text{th}}$ original constraint in terms of the optimal noninteger basis is

$$\sum_{j \in N} (a_{ij} - [a_{ij}])x_j + \sum_{j \in N} [a_{ij}]x_j + x_i = (b_i - [b_i]) + [b_i]$$

where the nonbasic variables $x_j$ are indexed $m + 1$ through $n$. Writing all of the integer terms on one side of the equation and all of the noninteger terms on the other side yields

$$\sum_{j \in N} (a_{ij} - [a_{ij}])x_j - (b_i - [b_i]) = -\sum_{j \in N} [a_{ij}]x_j + [b_i] - x_i$$

The right-hand side and hence the left-hand side must be integer or equivalently

$$\sum_{j \in N} (a_{ij} - [a_{ij}])x_j - (b_i - [b_i]) = 0 \pmod 1$$

or, equivalently, expression (14.6) must be satisfied. Returning to the original problem, the objective function is still appropriate for optimization.

Let us consider what we have accomplished. Satisfaction of constraints (14.6) assures us that the first problem above is solved. In other words, for integer $\boldsymbol{x}_N$ only integer values of $\boldsymbol{x}_B$ will be considered, and the objective function categorizes the optimal solution, therefore solving problem 3. Thus, if we could solve the following problem, we would overcome problems 1 and 3.

$$\begin{aligned}
&\text{Minimize } z = \sum_{j \in N} (c_j - z_j)x_j + \boldsymbol{c}_B\boldsymbol{B}^{-1}\boldsymbol{b} \\
(14.7)\quad &\text{subject to:} \quad \sum_{j \in N} (a_{ij} - [a_{ij}])x_j = b_i - [b_i] \quad \pmod 1 \\
&\qquad\qquad\qquad\qquad\qquad\qquad\qquad i = 1, \ldots, m \\
&\qquad\qquad\qquad x_j \geq 0 \text{ and integer}
\end{aligned}$$

where $c_j - z_j$ are elements of $(\boldsymbol{c}_N - \boldsymbol{c}_B\boldsymbol{B}^{-1}\boldsymbol{N})$.

It is usually possible to eliminate some of the constraints of (14.7). Any constraints which can be shown to be congruent modulo one to other equations or congruent modulo one to linear combinations are redundant and may be dropped.

Some other ways of reducing the size of the group problem are given in Hu [158]. We illustrate our method on the current example problem. Applying formulation (14.7) to problem (14.5) we get

$$\begin{aligned}
&\text{Minimize } z' = \tfrac{1}{4}x_3 + \tfrac{3}{2}x_4 \\
&\text{subject to:} \quad \tfrac{1}{4}x_3 + \tfrac{1}{2}x_4 = \tfrac{3}{4} \pmod 1 \quad \text{(from objective function)} \\
&\qquad\qquad\quad \tfrac{3}{4}x_3 + \tfrac{1}{2}x_4 = \tfrac{1}{4} \pmod 1 \quad \text{(from first constraint)} \\
&\qquad\qquad\quad \tfrac{3}{4}x_3 + \tfrac{1}{2}x_4 = \tfrac{1}{4} \pmod 1 \quad \text{(from second constraint)} \\
&\qquad\qquad\qquad x_3, x_4 \geq 0 \text{ and integer}
\end{aligned}$$

Notice that the second and third constraints are identical; hence, one can be dropped. In addition, the first or second constraint is congruent modulo one to three times the other (that is, the second or first constraint, respectively), so that in this case only one constraint need be written. In general, constraints which can be constructed by multiplication and addition modulo one from other constraints may be deleted. The constraints that cannot be deleted are generating constraints for the group, and are sufficient to admit only valid solutions to the group of constraints. Thus, when the group is cyclic, there is only one constraint necessary to solve the group problem. For the example problem the group problem is

$$\begin{aligned} &\text{Minimize } z = \tfrac{1}{4}x_3 + \tfrac{3}{2}x_4 - 18\tfrac{3}{4} \\ &\text{subject to:} \quad \tfrac{1}{4}x_3 + \tfrac{1}{2}x_4 = \tfrac{3}{4} \pmod 1 \\ &\qquad\qquad x_3, x_4 \geq 0 \text{ and integer} \end{aligned}$$

For convenience we multiply the constraint through by the *least common denominator*:

$$\begin{aligned} &\text{Minimize } z = \tfrac{1}{4}x_3 + \tfrac{3}{2}x_4 - 18\tfrac{3}{4} \\ &\text{subject to:} \quad x_3 + 2x_4 = 3 \pmod 4 \\ &\qquad\qquad x_3, x_4 \geq 0 \text{ and integer} \end{aligned} \tag{14.8}$$

For any pair of integer values $x_3, x_4 \geq 0$ in (14.8) we have 4 possible right-hand side values of $x_3 + 2x_4$ (mod 4)—the group elements, 0, 1, 2, and 3. Further, a unit of $x_3$ corresponds to group element "1," and a unit of $x_4$ corresponds to group element "2." Figure 14.2 illustrates these values distinctively with the following symbols:

| SYMBOLS | CONGRUENCE VALUES |
|---|---|
| ⊙ | $x_3 + 2x_4 = 0 \pmod 4$ |
| △ | $x_3 + 2x_4 = 1 \pmod 4$ |
| • | $x_3 + 2x_4 = 2 \pmod 4$ |
| ⊡ | $x_3 + 2x_4 = 3 \pmod 4$ |

Figure 14.2 is identical to Figure 14.1 except that the congruence of some solutions $\mathbf{x}_N \geq \mathbf{0}$ and integer are indicated. We may treat Figure 14.2 as an infinite directed network with the noninteger optimum as the source by interpreting dots as nodes and line segments between all pairs of nodes as directed arcs in the direction indicated.

In the event that more than one variable corresponds to the same group element, we need consider only the one whose cost is minimum. Suppose, for

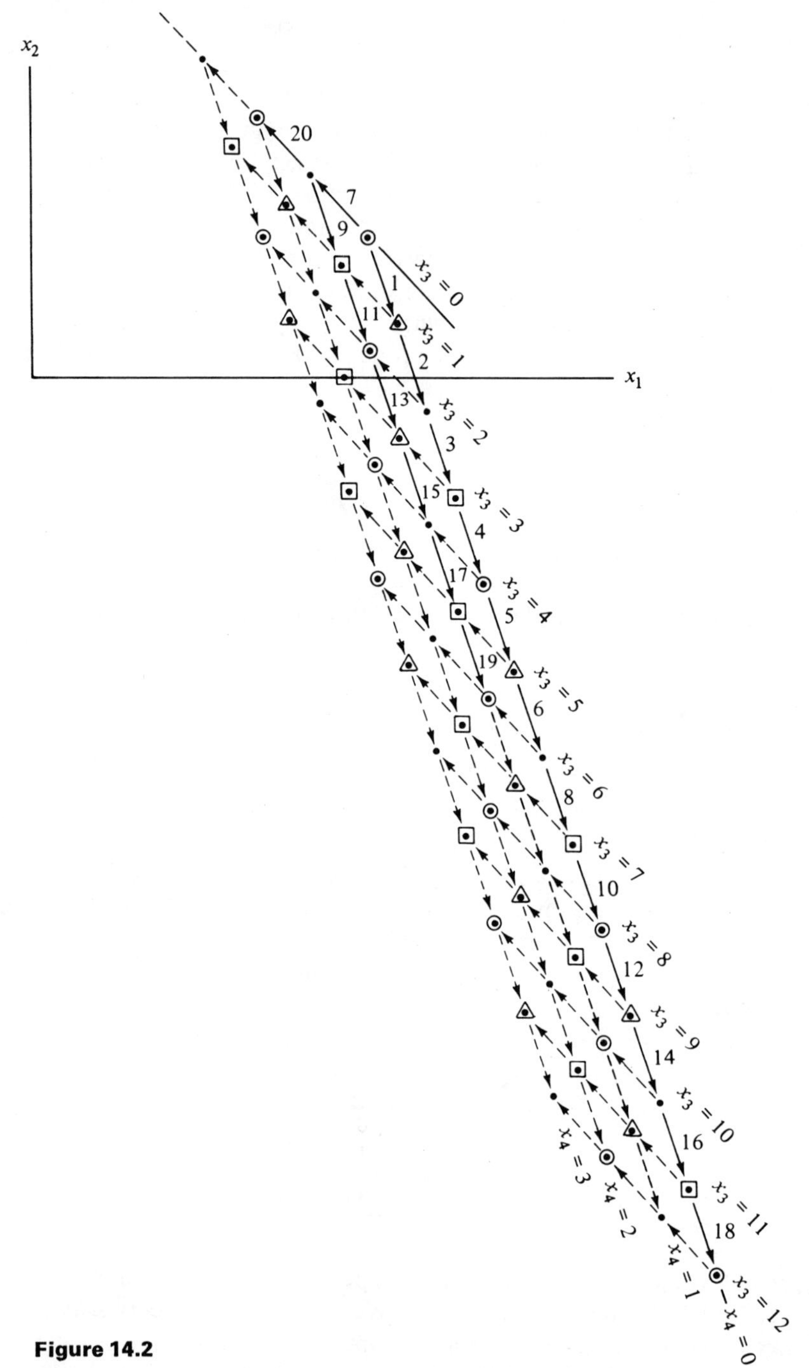


**Figure 14.2**

A Directed Network Imposed on the $\mathbf{x}_N$ Coordinates in $\mathbf{x}_B$ Space. Graph of the example problem in $\mathbf{x}_B$ space with the congruence of some solutions $\mathbf{x}_N \geq \mathbf{0}$ and integer indicated. (Note: All arrows [including solid arrows] indicate the directed arcs in the network. Numbers along the arcs indicate the sequence of solutions obtained.)

example, that formulation (14.8) were instead

$$\begin{aligned} &\text{Minimize } z' = \tfrac{1}{4}x_3 + \tfrac{3}{2}x_4 + \tfrac{5}{2}x_5 \\ &\text{subject to:} \qquad x_3 + 2x_4 + 2x_5 = 3 \pmod 4 \\ &\qquad\qquad\qquad x_3, x_4, x_5 \geq 0 \text{ and integer} \end{aligned}$$

in which case both $x_4$ and $x_5$ correspond to group element "2." Since $x_4$ has a lower cost, we may therefore drop $x_5$ in finding the optimal group problem solution. (If the optimal solution to the group problem does not have a corresponding $\boldsymbol{x}_B \geq \mathbf{0}$, then any variables which correspond to the same group elements cannot be deleted, because a solution other than the minimum solution to the group problem will be optimal.) Even though the network consists of an infinite number of nodes and arcs, there are only four types of nodes (corresponding to the order of the group), and two types of arcs (corresponding to at most the number of nonbasic variables). The "distances" or costs associated with arcs correspond to the respective costs of the objective function of (14.8). In Figure 14.2, the solution to (14.7) in general and (14.8) in particular is the node of the appropriate congruence—in the example, a square—whose objective function value is closest to that of the noninteger optimum. Considering $\boldsymbol{x}_N = \mathbf{0}$ as a source, as shown in Figure 14.2, an alternate interpretation is that the solution to (14.7) is a "shortest route" in terms of cost from the source to a node of desired congruence.

As mentioned before, we can overcome problems 1 and 3 by finding a shortest route from the origin to a node (of the infinite network) having the proper congruence. To find the shortest route in the network, any standard algorithm can be used. However, in consideration of the special properties of the network, more efficient algorithms have been developed.

### Hu's Group Minimization Algorithm

We shall now introduce a special shortest route method developed by T.C. Hu [158]. The method takes advantage of the symmetry of every node having the same number and composition of arcs entering and leaving.

Initially, every node but the origin is unlabeled, and the origin is labeled zero. Each node (numbered $k$) has three attributes:

1. A congruence $m(k)$. It is usually convenient to assign node numbers $k$ to correspond to congruences in some systematic manner. Because our example group is cyclic, we shall let the node number be the congruence.
2. The cost from the origin $\pi(k)$.
3. An index $i(k)$ to indicate an intermediate node along the shortest route from the origin.

Initially, all indices are initialized to the node numbers. A node on which an arc from the origin terminates has its cost $\pi(k)$ initialized to the cost of that arc. Other nodes have the costs $\pi(k)$ initialized as infinity. The index $i(k)$ is initialized as $k$. We shall label a node using an asterisk (*). Once a node is labeled its entries remain unchanged. Let $J$ be the set of labeled nodes. Select the node with minimum cost $\pi(k)$ and label it. Then, using the convention that $a \oplus b = a + b \bmod D$ (this convention must be extended for noncyclic groups, since noncyclic groups require vector representation), perform the following steps, letting $h$ be the index of the newly labeled node.

1. For all labeled nodes $j \in J$, set $\pi(j \oplus h) = \text{Min}\,\{\pi(j \oplus h), \pi(j) + \pi(h)\}$ and set $i(j \oplus h) = h$ only if $\pi(j) + \pi(h) < \pi(j \oplus h)$.
2. Label the unlabeled node whose cost is minimum.
3. If the desired node (the node of appropriate congruence) is labeled, halt and backtrack to find the minimum cost route from the origin. Otherwise, designate the newly labeled node as node $h$ and go to step 1.

EXAMPLE 14.1:

Solve the following problem using the method just presented:

$$\begin{aligned} \text{Minimize } z &= 2y_1 + 4y_3 + y_5 \\ \text{subject to:} \quad & y_1 + 3y_3 + 5y_5 = 6 \bmod 7 \end{aligned}$$

Notice that the indices have been arranged so that $y_1$ corresponds to element "1," $y_3$ to element "3," and so on. A graph of the network of this problem is shown in Figure 14.3. We want the least cost route from node 0

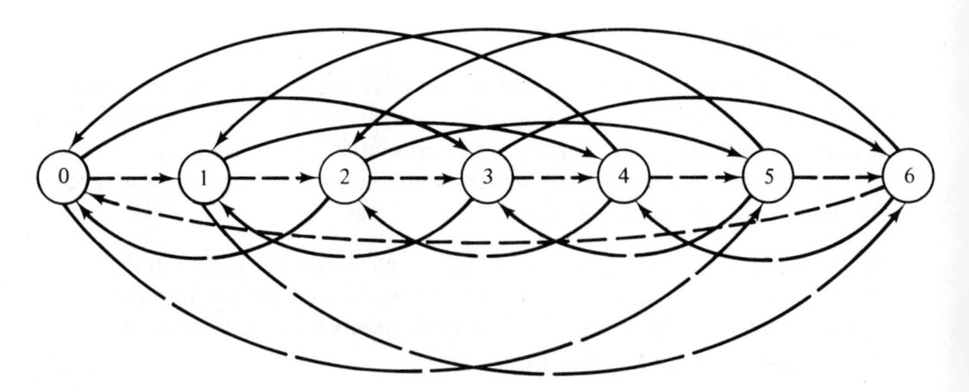


**Figure 14.3**

Graph of Example 14.1. A solid arc corresponds to an increase in $y_3$ by 1 at a cost of 4; a short dashed arc corresponds to an increase in $y_1$ by 1 at a cost of 2; a long dashed arc corresponds to an increase in $y_5$ by 1 at a cost of 1.

to node 6, and the associated cost. We have the following initial values:

| Node number $k$ | 1 | 2 | 3 | 4 | 5 | 6 |
|---|---|---|---|---|---|---|
| Congruence $m(k)$ | 1 | 2 | 3 | 4 | 5 | 6 |
| Cost $\pi(k)$ | 2 | $\infty$ | 4 | $\infty$ | 1 | $\infty$ |
| Index $i(k)$ | 1 | 2 | 3 | 4 | 5 | 6 |

Since the congruence is the same as the node number in any cyclic group, and our group is cyclic, we combine the first two rows in the table in subsequent iterations. The minimum $\pi(k)$ is $\pi(5) = 1$ for $k = 5$, so we label $k = 5$.

$$J = \{5\}, \quad h = 5$$

We then compute the following:

$$\pi(5 \oplus 5) = \pi(3) = \text{Min}\,\{\pi(3) = 4, \quad \pi(5) + \pi(5) = 2\} \text{ and set } i(3) = 5$$

There is a tie for the node to be labeled; arbitrarily we break the tie and label node 3, which gives the following table:

| $k, m(k)$ | 1 | 2 | 3* | 4 | 5* | 6 |
|---|---|---|---|---|---|---|
| $\pi(k)$ | 2 | $\infty$ | 2 | $\infty$ | 1 | $\infty$ |
| $i(k)$ | 1 | 2 | 5 | 4 | 5 | 6 |

$$J = \{3, 5\}, \quad h = 3$$

We then compute the following:

$$\pi(3 \oplus 3) = \pi(6) = \text{Min}\,\{\pi(6) = \infty, \quad \pi(3) + \pi(3) = 4\} \text{ and set } i(6) = 3$$

$$\pi(5 \oplus 3) = \pi(1) = \text{Min}\,\{\pi(1) = 2, \quad \pi(5) + \pi(3) = 3\} \text{ and leave } i(1) \text{ unchanged}$$

We now label node 1, which yields the following table:

| $k, m(k)$ | 1* | 2 | 3* | 4 | 5* | 6 |
|---|---|---|---|---|---|---|
| $\pi(k)$ | 2 | $\infty$ | 2 | $\infty$ | 1 | 4 |
| $i(k)$ | 1 | 2 | 5 | 4 | 5 | 3 |

$$J = \{1, 3, 5\}, \quad h = 1$$

Next, we compute

$$\pi(1 \oplus 1) = \pi(2) = \text{Min}\,\{\pi(2) = \infty, \quad \pi(1) + \pi(1) = 4\} \text{ and set } i(2) = 1$$
$$\pi(3 \oplus 1) = \pi(4) = \text{Min}\,\{\pi(4) = \infty, \quad \pi(3) + \pi(1) = 4\} \text{ and set } i(4) = 1$$
$$\pi(5 \oplus 1) = \pi(6) = \text{Min}\,\{\pi(6) = 4, \quad \pi(5) + \pi(1) = 3\} \text{ and set } i(6) = 1$$

Node 6 is labeled next, and we have the following table:

| $k, m(k)$ | 1* | 2 | 3* | 4 | 5* | 6* |
|---|---|---|---|---|---|---|
| $\pi(k)$ | 2 | 4 | 2 | 4 | 1 | 3 |
| $i(k)$ | 1 | 1 | 5 | 1 | 5 | 1 |

At this point we may stop because node 6 is labeled. Were we to proceed, however, we would make nine more comparisons and label nodes 2 and 4, but not change any $\pi$'s or $i$'s. Hence the above table is correct for all nodes. Now we shall determine the composition of the shortest path. The index of a node gives either its own number or some other. If the index is equal to the node number, there is no intermediate node on the shortest path from the origin. Otherwise, the number given is the node number of an intermediate node. We may proceed from the intermediate node back to the origin in a manner similar to the present analysis. But how do we find the route to the terminal node from the intermediate node? That route is equivalent to a route from the origin to some other node. Specifically, if the intermediate node is $u$ and the terminal node is $v$, the route from $u$ to $v$ is equivalent to the route from node zero to node $v \ominus u$ or $v - u \bmod D$. We shall use this scheme to find an optimal solution to our example. We want the optimal path from node 0 to node 6. It is helpful to refer to Figure 14.3 in tracing out the path. We see that $i(6) = 1$, which means that node 1 is intermediate. The route from node 0 to 1 may be found using $i(1) = 1$, which means that the shortest route from node 0 to node 1 has one arc corresponding to $y_1$. Thus $y_1$ must be at least 1. The route from node 1 to node 6 is equivalent to the route from node 0 to node $6 \ominus 1$, or node 5. Hence we now find the optimal route from node 0 to node 5. Since $i(5) = 5$, that route has only one arc, corresponding to $y_5$. Hence the optimal solution to the example problem is $y_1 = 1$, $y_5 = 1$, with the associated cost $\pi(6) = 3$.

For illustration purposes only, we consider a slightly more complicated route. Consider the optimal route from node 0 to node 4. Since $i(4) = 1$, node 1 is on the optimal path from 0 to 4. Since $i(1) = 1$, there is no node between nodes 0 and 1 on the path. Hence $y_1$ is at least 1. The optimal path from node 1 to node 4 corresponds to the optimal path from node 0 to node $4 \ominus 1$, or node 3. Since $i(3) = 5$, node 5 is on the optimal path from node 0 to node 3. Since $i(5) = 5$, there is no node between node 0 and node 5; thus $y_5$ is at least 1. The optimal path from node 5 to node 3 corresponds to the optimal path from node 0 to node $3 \ominus 5$, or node 5. Since $i(5) = 5$, $y_5$ must now be one plus what we found before, or at least 2. Further, there is no intermediate node along the arc from node 0 to node 5. Hence, $y_1 = 1$, $y_5 = 2$ is the minimal cost solution for a path from node 0 to node 4.

We now consider the group problem (14.8). Solving it using Hu's method, we have the following optimal solution:

| $k, m(k)$ | 1* | 2* | 3* |
|---|---|---|---|
| $\pi(k)$ | $\frac{1}{4}$ | $\frac{1}{2}$ | $\frac{3}{4}$ |
| $i(k)$ | 1 | 1 | 2 |

The optimal solution to group problem (14.8) can be shown to be $x_3 = 3$, $x_4 = 0$.

T.C. Hu's method has been further refined by D. Chen [55], who reduces the comparisons by about half by considering only unlabeled nodes for possible alteration.

Notice that the solution to (14.7) does not necessarily overcome problem 2, that is, we cannot be sure that the solution satisfies the constraints $\boldsymbol{x}_B \geq \boldsymbol{0}$. If we somehow incorporate the restrictions $\boldsymbol{x}_B \geq \boldsymbol{0}$ into the infinite network, we reduce the network to a subgraph (possibly a disjoint subgraph) of the original network. Thus the shortest path from the source to a node of the appropriate congruence in the finite subgraph will be the integer optimum of the original integer programming problem. Before we consider how to find a solution overcoming problem 2, we first consider under what conditions problem 2 will be automatically solved.

## 14.5 SUFFICIENT CONDITIONS FOR AN OPTIMUM

For the example problem (14.8), the shortest route from the origin to the closest point of desired congruence (as may be seen in Fig. 14.2) is three units along the $x_3$ direction, or $x_3 = 3$, $x_4 = 0$, with an associated cost of $\frac{3}{4}$. Relating the solution back to the basic variables, either by checking the conditions $\boldsymbol{x}_B = \boldsymbol{B}^{-1}\boldsymbol{b} - \boldsymbol{B}^{-1}\boldsymbol{N}\boldsymbol{x}_N \geq \boldsymbol{0}$, or by using Figure 14.2, it is found that the solution is $x_1 = 4$ and $x_2 = -1$, which is not a feasible solution to $\boldsymbol{x}_B \geq \boldsymbol{0}$.

Under what conditions will the above procedure yield the optimal solution? Note that finding a shortest route from an origin in a network is equivalent to finding a tree of shortest routes. Assuming that the group is of order $D$, the tree may therefore connect at most $D$ nodes. Since a tree connecting $D$ nodes has $D-1$ arcs, a sufficient but not necessary condition can be derived. Each arc corresponds to a nonbasic variable incremented by one. Hence the total number of increments to nonbasic variables cannot exceed $D-1$ or, equivalently, the sum of nonbasic variables cannot exceed $D-1$. We may therefore conclude that if all solutions which satisfy the sum of nonbasic variables not exceeding $D-1$ also satisfy $\boldsymbol{x}_B \geq \boldsymbol{0}$, then the solution to (14.7) solves the original integer programming problem.

To test the sufficient condition is relatively simple. Once the noninteger optimum has been achieved, the determinant $D$ of the optimal basis $\boldsymbol{B}$ is computed. Since $\sum_{j \in N} x_j \leq D-1$ has as its extreme points with $x_j \geq 0$, $j \in N$, the points $x_h = D-1$, $x_j = 0$ $(j \neq h;\ j, h \in N)$, all that is to be done is to substitute $D-1$ as the value of each nonbasic variable—one at a time—holding the others at zero into the constraints of (14.3) to see if any of the solutions have any basic variables negative. *If any* of the *basic variables are negative* for such solutions, *the sufficient conditions do not hold.* Only if all the basic variables are nonnegative for such solutions do the sufficient conditions hold, for then no convex combination of the extreme points of the set $\sum_{j \in N} x_j \leq D-1$, $x_j \geq 0$, $j \in N$ can have a negative variable.

The solution to problem (14.7) solves general integer linear programming problems if the solution set $\boldsymbol{x}_B \geq \boldsymbol{0}$ contains the intersection of two sets, $\sum_{j \in N} x_j \leq D-1$ and $\boldsymbol{x}_N \geq \boldsymbol{0}$. Intuitively, the method works if all the constraints, $\boldsymbol{x}_B = \boldsymbol{B}^{-1}\boldsymbol{b} - \boldsymbol{B}^{-1}\boldsymbol{N}\boldsymbol{x}_N \geq \boldsymbol{0}$, are "far away" from the noninteger optimum. More precisely, $\boldsymbol{x}_B \geq \boldsymbol{0}$ corresponds to $\boldsymbol{B}^{-1}\boldsymbol{N}\boldsymbol{x}_N \leq \boldsymbol{B}^{-1}\boldsymbol{b}$, which is satisfied for $\boldsymbol{x}_N = \boldsymbol{0}$. For $\boldsymbol{x}_N \geq \boldsymbol{0}$, then for $\boldsymbol{b}$ sufficiently large the sufficient condition, $\boldsymbol{B}^{-1}\boldsymbol{N}\boldsymbol{x}_N \leq \boldsymbol{B}^{-1}\boldsymbol{b}$ or, equivalently, $\boldsymbol{x}_B \geq \boldsymbol{0}$, will be satisfied. For this reason the associated algorithm has been called the asymptotic algorithm.

EXAMPLE 14.2:

That the above condition is not necessary can be seen by replacing the objective function of example problem (14.4) with the following function:

$$\text{Minimize } z = -15x_1 - 13x_2$$

The optimal noninteger solution to that problem is solution (14.5) with the objective function replaced by the following function:

$$\text{Minimize } z + 61 = 6x_3 + x_4$$

The appropriate group thus has $D = 4$ and the associated group problem is as follows:

$$\begin{aligned} &\text{Minimize } z = -61 + 6x_3 + x_4 \\ (14.9) \qquad &\text{subject to:} \qquad x_3 + 2x_4 = 3 \pmod 4 \\ &\qquad\qquad x_3, x_4 \geq 0 \text{ and integer} \end{aligned}$$

For this problem the sufficient condition is not satisfied. The two extreme points (other than $x_3 = x_4 = 0$) of the set $x_3 + x_4 \leq D - 1 = 3$, $x_3 \geq 0$, $x_4 \geq 0$, are $x_3 = 0$, $x_4 = 3$, and $x_3 = 3$, $x_4 = 0$. The solution $x_3 = 0$, $x_4 = 3$ corresponds to $x_1 = 1.75$, $x_2 = 2.75$ (and hence $\boldsymbol{x}_B \geq \mathbf{0}$). The solution $x_3 = 3$, $x_4 = 0$ corresponds to $x_1 = 4$, $x_2 = -1$, which is not feasible ($\boldsymbol{x}_B \ngeq \mathbf{0}$). Thus, the sufficient condition is not satisfied. However, the optimal solution to (14.9)—($x_3 = 1$, $x_4 = 1$, $x_1 = 3$, $x_2 = 1$)—happens to satisfy the condition $\boldsymbol{x}_B \geq \mathbf{0}$ and is the optimal integer solution to this problem.

For completeness we include a sufficient condition developed by Gomory. As a result of the group problem we find a solution $\boldsymbol{x}_N^*$ which may be viewed as a correction to the noninteger optimum. From (14.3) we have that the optimal integer solution is

$$\boldsymbol{x}_B = \boldsymbol{B}^{-1}(\boldsymbol{b} - \boldsymbol{N}\boldsymbol{x}_N^*)$$

If $\boldsymbol{x}_B$ is to be feasible, then $\boldsymbol{B}^{-1}(\boldsymbol{b} - \boldsymbol{N}\boldsymbol{x}_N) \geq \mathbf{0}$, and the vector $\boldsymbol{b} - \boldsymbol{N}\boldsymbol{x}_N$ must be contained in the cone corresponding to the optimal basis of the linear programming problem, i.e., the solution set $\boldsymbol{q}$ to $\boldsymbol{B}^{-1}\boldsymbol{q} \geq \mathbf{0}$. Denote that cone as $K_B(0)$. We do not know a priori the value of $\boldsymbol{x}_N^*$ and $\boldsymbol{N}\boldsymbol{x}_N^*$. We can, however, establish a bound on the Euclidian norm (length) of $\boldsymbol{N}\boldsymbol{x}_N^*$.

Of all the columns of $\boldsymbol{N}$, consider the one whose vector has the maximum Euclidean norm (length), that is, column $j$ for which

$$\left(\sum_{i=1}^{m} a_{ij}^2\right)^{1/2}$$

is maximum. Denote the maximum as $s$. Then an upper limit to the norm of $\boldsymbol{N}\boldsymbol{x}_N^*$ is $(D - 1)s$. (Recall that the maximum sum $\mathbf{1}\boldsymbol{x}_N$ is $D - 1$.)

We would like to show that if $b$ less $(D - 1)s$ in *any* direction is still contained in the convex cone, then the asymptotic solution will be optimal. The easiest way to do that is to check whether the vector $\boldsymbol{b}$ lies within a distance $(D - 1)s$ from the boundaries of the cone $K_B(0)$. Denoting the restricted cone as $K_B((D - 1)s)$, we have informally proved Gomory's sufficient condition:

If $\boldsymbol{b} \in K_B((D - 1)s)$, where $D$ is the determinant of the basis $\boldsymbol{B}$, then the optimal solution to the group problem is the optimal solution to the integer programming problem.

Unfortunately, this sufficient condition is not as easily checked as the earlier one. Still another sufficient condition, given by Hu [158], is included in the problems at the end of the chapter.

## 14.6 THE CORNER POLYHEDRON, ITS FACES, AND AN ALTERNATE APPROACH

We stated earlier that if the solution to a linear programming problem over its corner polyhedron satisfies the constraints $\boldsymbol{x}_B \geq \boldsymbol{0}$, then that solution is an optimal integer solution to the original problem. Such a procedure which is an alternative to using a shortest route procedure, is particularly attractive if there is some inexpensive way of generating the constraints that define the corner polyhedron. Accordingly, tables have been compiled which give these face constraints and corresponding extreme point solutions for groups of different orders. (There is more than one group for certain nonprime orders, and hence the appropriate group must be used.) Such tables (available for groups up to and including order 11) in Gomory [127] and Hu [158] can be employed to find a solution by using the constraints and eliminating any variables of the set which do not appear. For example, the general constraints from Gomory [127] for a cyclic group of order 4 with congruence 3 are

$$\begin{aligned} t_1 \qquad + t_3 &\geq 1 \\ t_1 + 2t_2 + 3t_3 &\geq 3 \end{aligned}$$

where $t_1$, $t_2$, and $t_3$ are the increments in different dimensions or for corresponding group elements. Methods for generating constraint faces are found in Glover [112], Gomory [127], and Hu [158]. Our group elements for the example correspond as $x_3$ to $t_1$ and $x_4$ to $t_2$. No element corresponds to $t_3$. Hence the constraints for the example of expression (14.8) are

$$\begin{aligned} x_3 \qquad &\geq 1 \\ x_3 + 2x_4 &\geq 3 \\ x_3, x_4 &\geq 0. \end{aligned} \tag{14.10}$$

We can then solve for the optimal solution to the group problem via linear programming by using the face constraints (14.10) and the objective function of (14.8).

The group of cut constraints derived in section 14.2 are somewhat weaker than the face constraints, though stronger than the ordinary set of Gomory cut constraints. The two sets may be shown to be equivalent, *provided that* the slacks of the cut constraints of the group are integer valued. There is no simple way of insuring this in general.

## 14.7 MODIFICATIONS TO THE ALGORITHM WHEN THE ORIGINAL GROUP THEORETIC SOLUTION IS NOT FEASIBLE

In our example problem (14.4) the optimal solution to the group theoretic problem (14.6) violated one of the constraints $\boldsymbol{x}_B \geq \mathbf{0}$. In such a case it appears that the group theoretic method does not help in finding an integer optimum. However, as we shall now show, it can help.

Instead of limiting the nodes to the $D$ distinct elements which have the lowest cost, we shall expand the set to the infinite set whose nodes are the integer solutions of $\boldsymbol{x}_N \geq \mathbf{0}$, but we shall enumerate them in the same way as before. That is, we add a node to the tree if its cost from the origin is minimum. There is a problem, in that the same node may be reached in a number of different but equivalent ways, and an efficient method must not duplicate a node. To illustrate the difficulty, observe that in Figure 14.2 the node $x_3 = 3$, $x_4 = 2$ may be reached in 10 distinct ways, all of which are equivalent: (1) $x_4 = 1$, $x_4 = 1$, $x_3 = 1$, $x_3 = 1$, $x_3 = 1$; (2) $x_4 = 1$, $x_3 = 1$, $x_4 = 1$, $x_3 = 1$, $x_3 = 1$; (3) $x_4 = 1$, $x_3 = 1$, $x_3 = 1$, $x_4 = 1$, $x_3 = 1$; (4) $x_4 = 1$, $x_3 = 1$, $x_3 = 1$, $x_3 = 1$, $x_4 = 1$; (5) $x_3 = 1$, $x_4 = 1$, $x_4 = 1$, $x_3 = 1$, $x_3 = 1$; (6) $x_3 = 1$, $x_4 = 1$, $x_3 = 1$, $x_4 = 1$, $x_3 = 1$; (7) $x_3 = 1$, $x_4 = 1$, $x_3 = 1$, $x_3 = 1$, $x_4 = 1$; (8) $x_3 = 1$, $x_3 = 1$, $x_4 = 1$, $x_4 = 1$, $x_3 = 1$; (9) $x_3 = 1$, $x_3 = 1$, $x_4 = 1$, $x_3 = 1$, $x_4 = 1$; (10) $x_3 = 1$, $x_3 = 1$, $x_3 = 1$, $x_4 = 1$, $x_4 = 1$. In general, the number of such equivalent solutions $x_1, x_2, \ldots, x_n$ can be shown to be

$$\frac{\left(\sum_{i=1}^{n} x_j\right)!}{x_1!\, x_2! \ldots x_n!}$$

A number of authors [115, 257] have proposed an enumeration scheme that avoids the duplication. First, the nonbasic variables are ordered in an arbitrary manner from 1 to $p$. For ease of exposition, we order them in ascending order of the indices associated with the nonbasic variables. For example problem (14.5) the order is: (1) $x_3$, (2) $x_4$, where $p = 2$.

If a node has been reached by incrementing a nonbasic variable with index $k^*$, then only nonbasic variables with index $k = 1, \ldots, k^*$, can be increased. The tree thus generalized for the nodes up to and including an objective value of 3 is given by the solid arrows, as shown in Figure 14.2. The node $x_3 = 3$, $x_4 = 2$ will be reached in the order (1) above.

Ties may be broken arbitrarily, as we have done in Figure 14.2. Using Figure 14.2, we see that the lowest cost solution, the third labeled node, is $x_3 = 3$, $x_4 = 0$, and has a cost above the noninteger optimal solution of $\frac{3}{4}$. That solution is not feasible because the corresponding values of $\boldsymbol{x}_B$ are negative

($x_1 = 4$, $x_2 = -1$), as was pointed out earlier. Solutions which satisfy the congruence relationship in that tree are indicated in Table 14.1. The first *feasible* solution (with respect to $\mathbf{x}_B \geq \mathbf{0}$) to the group problem is the third shortest route (it is tied with the second), namely $x_3 = 1$, $x_4 = 1$, which corresponds to $x_1 = 3$, $x_2 = 1$.

**Table 14.1**

The Eight Best Integer Solutions to the Group Theoretic Problem for the Example Problem

| NUMBER OF SOLUTION | NODE POSITION IN LABELING SEQUENCE | OBJECTIVE FUNCTION VALUE | VALUES OF $x_N$ | | VALUES OF $x_B$ | | FEASIBLE ? |
|---|---|---|---|---|---|---|---|
| | | | $x_3$ | $x_4$ | $x_1$ | $x_2$ | ($x_B \geq 0$ ?) |
| 1 | 3 | $\frac{3}{4}$ | 3 | 0 | 4 | $-1$ | No |
| 2 | 8 | $\frac{7}{4}$ | 7 | 0 | 5 | $-4$ | No |
| 3 | 9 | $\frac{7}{4}$ | 1 | 1 | 3 | 1 | Yes |
| 4 | 16 | $\frac{11}{4}$ | 11 | 0 | 6 | $-7$ | No |
| 5 | 17 | $\frac{11}{4}$ | 5 | 1 | 4 | $-2$ | No |
| 6* | 27 | $\frac{15}{4}$ | 15 | 0 | 7 | $-10$ | No |
| 7* | 28 | $\frac{15}{4}$ | 9 | 1 | 5 | $-5$ | No |
| 8* | 29 | $\frac{15}{4}$ | 3 | 2 | 3 | 0 | Yes |

* Not indicated in Figure 14.2.

Suppose, however, the original problem is augmented by the additional constraint $2x_1 + 6x_2 \geq 13$. Although the noninteger optimum and the group theoretic problem are unchanged, none of the first 8 optimal solutions to the group problem (as shown in Table 14.1) satisfy the additional constraint. The optimal solution, given this added constraint, corresponds to the 12$^{\text{th}}$ optimal solution, which is $x_3 = 1$, $x_4 = 3$, $x_1 = 2$, $x_2 = 2$. The method may appear unwieldy in light of an example such as this, although other devices may be employed to accelerate the solution process. There are statements by some authors (see, e.g., Hu [158]) to the effect that such problems are rare in practice. In addition, rules have been developed that reduce search by excluding certain infeasible regions. See, for example, Chen [55].

### 14.8 AN EXAMPLE OF A NONCYCLIC GROUP

For completeness we present an example of a problem for which the group is noncyclic.

EXAMPLE 14.3:

Consider the following problem:

$$\begin{aligned}
\text{Minimize } z &= -5x_1 - 4x_2 \\
\text{subject to:} \quad & 3x_1 + 3x_2 + x_3 \qquad = 10 \\
& 12x_1 + 6x_2 \qquad + x_4 = 24 \\
& x_1, x_2, x_3, x_4 \geq 0 \text{ and integer}
\end{aligned}$$

The noninteger optimum is given below in equation form:

$$\begin{aligned}
\text{Minimize } z &= \qquad x_3 + \tfrac{1}{6}x_4 - 14 \\
\text{subject to:} \quad & x_2 + \tfrac{2}{3}x_3 - \tfrac{1}{6}x_4 = \tfrac{8}{3} \\
x_1 \quad & \quad - \tfrac{1}{3}x_3 + \tfrac{1}{6}x_4 = \tfrac{2}{3}
\end{aligned}$$

for which $D$, the determinant of the optimal basis, is 18. The following group problem is then derived:

$$\begin{aligned}
\text{Minimize } z &= x_3 + \tfrac{1}{6}x_4 \\
\text{subject to:} \quad & \tfrac{1}{6}x_4 = 0 \bmod 1 \text{ (from objective function)} \\
& \tfrac{2}{3}x_3 + \tfrac{5}{6}x_4 = \tfrac{2}{3} \bmod 1 \text{ (from first constraint)} \\
& \tfrac{2}{3}x_3 + \tfrac{1}{6}x_4 = \tfrac{2}{3} \bmod 1 \text{ (from second constraint)}
\end{aligned}$$

Adding four times the first constraint to the third yields the second; hence, the second constraint may be dropped. We then multiply the constraints by 6, thereby clearing fractions. The result is:

$$\begin{aligned}
\text{Minimize } z &= x_3 + \tfrac{1}{6}x_4 \\
\text{subject to:} \quad & x_4 = 0 \bmod 6 \\
& 4x_3 + x_4 = 4 \bmod 6
\end{aligned}$$

The group corresponding to this problem is of order 18 and not cyclic. The shortest route tree is shown in Figure 14.4. The notation used in the figure is as follows:

1. The vector at the node $\binom{a}{b}$ is the value of $\binom{x_4}{4x_3 + x_4}$ mod 6, and the circled entry at the node is the objective function value.
2. The arrow leads from the predecessor node to the successor node.
3. Along the arcs is indicated the variable being incremented.

In addition, a number is shown which indicates the order in which the arcs were added, so that the steps can be easily followed by the reader. Note that since the optimal solution is found at step six, none of the subsequent steps (indicated by dotted arrows) need have been computed.

By traversing the tree in the reverse direction from the destination vector

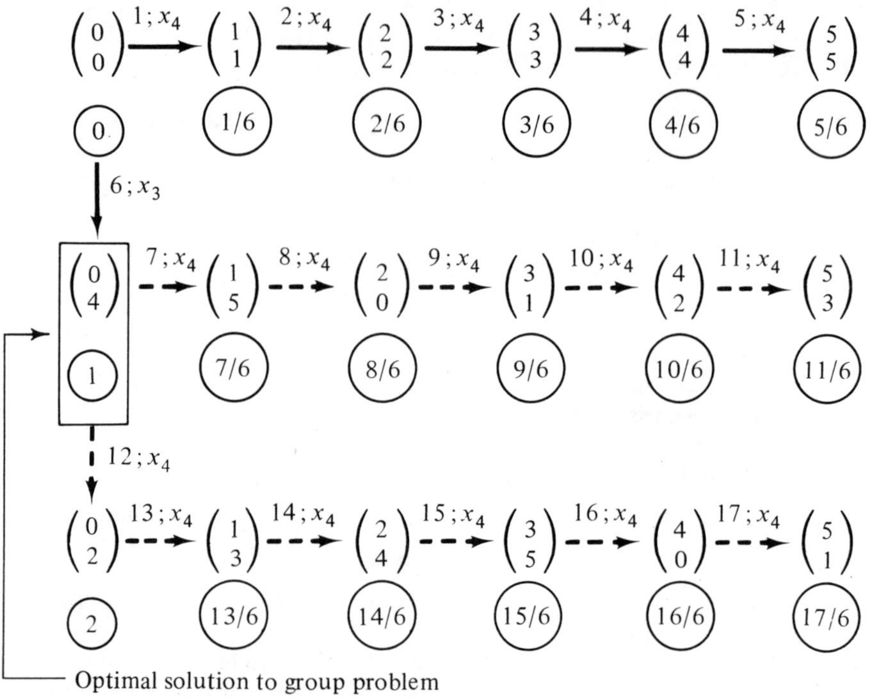

Optimal solution to group problem

**Figure 14.4**
The Development of the Integer Optimum for Example 14.3

$\binom{0}{4}$ to the origin, the solution may be found to be $x_3 = 1$, $x_4 = 0$. Upon checking the original constraints, as in the noninteger optimum, it can be seen that the solution is as follows:

$$x_3 = 1, \quad x_4 = 0, \quad x_2 = 2, \quad x_1 = 1$$

Since it satisfies the nonnegativity constraints $\boldsymbol{x}_B \geq \mathbf{0}$, it is optimal. Note, however, that the sufficient condition does not hold for the above problem.

## 14.9 CONCLUSION

In this chapter we have introduced the group theoretic approach to integer programming, an area of considerable current research. Although we have included much of the material currently available, our examination has not been exhaustive. The methods are evidently reasonably efficient, because commercial computer programs using them are in operation.

**Selected Supplemental References**

Section 14.4
[161]
Section 14.6
[112]
Section 14.7
[55]
Section 14.8
[55]
General
[34], [110], [115], [126], [127], [129], [167], [168], [179], [242], [257], [258], [259], [260], [289], [296]

## 14.10 PROBLEMS

**1.** Solve the example problems of the chapter using *only* the group of cut constraints. An example on which this technique does not work is the following, taken from Shapiro [257]:

$$\begin{aligned}
\text{Maximize } z = -\tfrac{117}{6}x_1 - x_2 - \tfrac{3}{2}x_3 - \tfrac{19}{6}x_4 - \tfrac{4}{3}x_5 - \tfrac{1}{6}x_6 & \\
\text{subject to:} \quad x_1 + 3x_2 + \tfrac{3}{2}x_3 + \tfrac{1}{3}x_4 - \tfrac{47}{6}x_5 - \tfrac{1}{6}x_6 + x_7 \qquad & = \tfrac{185}{2} \\
\tfrac{1}{2}x_1 + 3x_2 + \tfrac{1}{2}x_3 + \tfrac{1}{6}x_4 - \tfrac{11}{3}x_5 + \tfrac{1}{6}x_6 \qquad + x_8 \qquad & = 53 \\
\tfrac{1}{2}x_1 + x_2 + \tfrac{1}{2}x_3 - \tfrac{1}{6}x_4 - \tfrac{4}{3}x_5 - \tfrac{1}{6}x_6 \qquad + x_9 & = 33 \\
x_1, \ldots, x_9 & \geq 0 \\
\text{and integer} &
\end{aligned}$$

**2.** Prove that the cut constraints which may be generated from the noninteger optimum do indeed form a finite Abelian group.

**3.** Consider the following constraint set:

$$\begin{aligned}
2x_1 + 2x_2 + 3x_3 &= 7 \\
2x_1 + 4x_2 + x_3 &= 9
\end{aligned}$$

a. Show that, with $x_1$ and $x_2$ as the basic variables, the determinant of the basis is 4, but the order of the group is 2.

b. Show that if the inequalities in both constraints are replaced by inequalities ($\leq$), the order of the group is 4.

**4.** Indicate why no component of $\boldsymbol{x}_B$ appears in expression (14.6).

**5.** Write the group problem corresponding to Shapiro's example given in problem 1. Show that either the second or third group constraint may be dropped.

**6.** Show that the constraints of the group theoretic algorithm which may be constructed for the example in section 14.4 are as follows:

a. $\frac{1}{4}x_3 + \frac{1}{2}x_4 = \frac{3}{4} \pmod 1$

b. $\frac{1}{2}x_3 = \frac{1}{2} \pmod 1$

c. $\frac{3}{4}x_3 + \frac{1}{2}x_4 = \frac{1}{4} \pmod 1$

d. $0 = 0 \pmod 1$

Then, assuming that $x_3, x_4 \geq 0$, show that constraints a and c have identical solution sets, and that constraints b and d have solution sets of which the solution set of constraint a (or c) is a proper subset. (Constraint a [or c] generates the complete group and is said to be a generating element of the group, while constraint b [or d] generates only a subgroup.)

**7.** Compare the group of cut constraints for the example problem (14.5) with the group of constraints presented in the previous exercise.

**8.** Complete the calculations for Example 14.1.

**9.** Given the following group equation, indicate which element corresponds to 0, 1, 2, 3, and 4.

$$x_1 + 2x_2 + 3x_3 + 4x_4 = 3 \bmod 5$$

In how many different ways may a correspondence be made? The isomorphic correspondence of a group to itself is called an automorphism, and is useful for deriving faces of the corner polyhedron.

**10.** Solve the following group problems using Hu's method presented in section 14.4.

a. Minimize $z = 8x_1 + 2x_2 + 9x_3 + 6x_4$

subject to: $4x_1 + 8x_2 + 5x_3 + x_4 = 5 \pmod 9$

b. Minimize $z = 7x_1 + 6x_2 + 2x_3 + 8x_4 + 8x_5$

subject to: $5x_1 + 2x_2 + 6x_3 + 7x_4 + 8x_5 = 9 \pmod{17}$

c. Minimize $z = 5x_1 + 4x_2 + 8x_3 + x_4 + 9x_5$

subject to: $4x_1 + 9x_2 + 3x_3 + 7x_4 + 2x_5 = 8 \pmod{11}$

d. Minimize $z = 5x_1 + x_2 + 3x_3 + 7x_4 + 2x_5$

subject to: $3x_1 + 7x_2 + 2x_3 + 6x_4 + 5x_5 = 11 \pmod{12}$

e. Minimize $z = 9x_1 + 2x_2 + 7x_3 + x_4 + 8x_5$

$$\text{subject to:} \quad 5x_1 + 6x_2 + 10x_3 + 2x_4 + 5x_5 = 11 \pmod{18}$$
$$x_j \geq 0 \text{ and integer, } j = 1, \ldots, 5 \text{ in all cases}$$

**11.** Another sufficient condition (Hu) [158] for assuring that the solution to the group problem will be optimal for the integer problem is the following:

$$\boldsymbol{B}^{-1}\boldsymbol{b} - (D - 1)\boldsymbol{w} \geq \boldsymbol{0}$$

where $\boldsymbol{w}$ is an $m$ vector, each entry of which is the maximum element of the corresponding row of $\boldsymbol{B}^{-1}\boldsymbol{N}$.

a. Prove that this condition is sufficient.
b. Apply it to the example problems of the chapter.

**12.** Can you show that the three sufficient conditions assuring that the optimal solution to the group problem will be optimal for the integer program (the two given in section 14.5 as well as that in the preceding) are different?

**13.** Given the following problem:

$$\begin{aligned}
\text{Minimize } z = \; & 12x_1 + 14x_2 + 9x_3 + 15x_4 \\
\text{subject to:} \; & -\tfrac{31}{11}x_1 + \tfrac{12}{11}x_2 + \tfrac{16}{11}x_3 - \tfrac{3}{11}x_4 + x_5 = \tfrac{48}{11} \\
& \tfrac{3}{2}x_1 - \tfrac{1}{2}x_2 + \tfrac{5}{2}x_3 - x_4 + x_6 = \tfrac{15}{2} \\
& x_1, \ldots, x_6 \geq 0 \text{ and integer}
\end{aligned}$$

a. What is the order of the group?
b. Are any of the sufficient conditions for the group problem fulfilled?
c. Solve the problem using the group theoretic algorithm.

**14.** Given the following problem:

$$\begin{aligned}
\text{Minimize } z = \; & 8x_1 + 10x_2 + 11x_3 \\
\text{subject to:} \; & -\tfrac{3}{2}x_1 + x_2 + 5x_3 + x_4 = 6.5 \\
& \tfrac{7}{2}x_1 + 4x_2 + \tfrac{1}{2}x_3 + x_5 = 8.0 \\
& \tfrac{5}{2}x_1 + \tfrac{5}{2}x_2 + \tfrac{3}{2}x_3 + x_6 = 10.5 \\
& x_1, \ldots, x_6 \geq 0 \text{ and integer}
\end{aligned}$$

a. What is the order of the group?
b. Are any of the sufficient conditions for the group problem fulfilled?
c. Solve the problem using the group theoretic algorithm.

**15.** Given the following problem:

$$\begin{aligned}
\text{Maximize } z = \; & 2x_1 + 4x_2 \\
\text{subject to:} \; & 3x_1 + 2x_2 + x_3 = 9 \\
& x_1 + 3x_2 + x_4 = 8 \\
& x_1, \ldots, x_4 \geq 0 \text{ and integer}
\end{aligned}$$

a. What is the order of the group?
b. Are any of the sufficient conditions fulfilled?
c. Solve the problem using the group theoretic algorithm.
d. Graph the constraint set, and indicate the linear programming polyhedron, the integer programming polyhedron, and the corner polyhedron.

**16.** Of problems 1, 13, 14, and 15, which groups are cyclic and which groups are noncyclic?

**17.** It is frequently possible to express group constraints for noncyclic groups which imply the values for certain variables in the optimal solution to the group problem. Furthermore, such values can then be substituted into the other equations in which they appear, sometimes reducing noncyclic problems to cyclic problems. It is often possible to deduce the optimal group problem solution directly. Show that this is indeed the case for problem 14 and the example of section 14.8.

**18.** Dominance considerations may be used to eliminate variables from consideration in the group theoretic problem. For Shapiro's problem given in problem 1, show that all variables except $x_5$ and $x_6$ are dominated in the group problem. For problem 15 show that $x_4$ is dominated.

**19.** For problem 25 of Chapter 13, show that the same cut as Martin's accelerated Euclidean algorithm is achieved using one of the group of Gomory cut constraints generated from the indicated source constraint. The member of the group that should be used is the one which transforms the source constraint to an all-integer constraint in a pivot iteration.

## Appendix: Some Definitions from Group Theory

### Definition 1

*A group is a set of elements which has a single operation defined (which we shall call "addition"—with quotation marks) upon its elements, so that:*

1. *The "sum" of any two elements is an element of the group (closure).*
2. *An identity element exists (a null element).*
3. *Every member of the group has an inverse, also a member of the group, such that the "sum" of an element and its inverse is the identity element (inverse).*
4. *The operation "addition" satisfies the associative law, that is,* $(a + b) + c = a + (b + c)$ *(associativity).*

### Definition 2

*A finite group has a finite number of elements.*

### Definition 3

*The order of a finite group is the number of elements contained in the group.*

**Definition 4**

*A group is additive if its operation is defined as ordinary addition.*

**Definition 5**

*An Abelian group is one in which the operation "addition" is commutative.*

**Definition 6**

*A cyclic group is one in which successive "addition" of at least one element to itself generates the entire group. Some examples of Abelian groups are given below:*

1. *The residue class of integers modulo 5 with respect to ordinary addition form a cyclic group of order 5. Thus $2 = 7 \bmod 5$. The group elements are 0, 1, 2, 3, 4. Any element other than zero generates the group. Thus the group is cyclic.*
2. *The residue of class of vectors $\binom{0}{0}, \binom{0}{1}, \binom{1}{0}, \binom{1}{1}$ with respect to ordinary addition modulo $\binom{2}{2}$ form a group of order 4. This group is* not *a cyclic group since no single vector generates the whole group.*
3. *The numbers $i(=\sqrt{-1})$, $-1$, $-i$, $1$ form a cyclic group of order 4. "Addition" in this instance is defined as multiplication (thus the group is* not *an additive group). Either element $i$ or $-i$ generates the entire group.*

**Definition 7**

*Two groups are isomorphic if there is a one-to-one correspondence between corresponding elements. The residue class of integers modulo 4 under addition is isomorphic to the multiplicative group $i$, $-1$, $-i$, $1$. An isomorphism is:*

$$0 \Longleftrightarrow 1, \quad 1 \Longleftrightarrow i, \quad 2 \Longleftrightarrow -1, \quad 3 \Longleftrightarrow -i$$

# 15

# Branch and Bound Algorithms of Integer Programming

## 15.1 INTRODUCTION

In this chapter and the next we shall consider a class of algorithms for solving integer linear programming problems, both all-integer and mixed. This class of algorithms is perhaps one of the simplest in concept: branch and bound or implicit enumeration. This chapter will cover branch and bound algorithms, and the next will consider implicit enumeration algorithms. We shall now discuss some aspects common to both method types.

For an all-integer programming problem, suppose that each variable $x_j$ has an integer lower bound $h_j$ and an integer upper bound $u_j$. Complete enumeration of all possible solutions would consist of examining all possible combinations of $x_j$ $(= h_j, h_j + 1, \ldots, u_j - 1, u_j)$ with all possible combinations of every other variable. The number of such combinations can be tremendous for problems of even modest size, but as we shall see, only a (usually very small) fraction of the total number of possible solutions needs to be examined for the methods to be considered.

A framework for presenting the various methods is useful. First, we assume that the problem is to find a *maximum* value of an objective function. In each method a sequence of problems is considered, evaluated (and in some respect, solved), and for each problem some (integer) variables are restricted to have certain values. Such variables are appropriately called *specified variables* (also *fixed* or *stopped variables*). A set of specified variables which describes a problem will be referred to as a partial solution, or a partial problem, because, given a partial solution, a particular solution to the original problem will be implied (by rules particular to the method). Similarly, variables which are not so specified are called *free* or *unspecified variables.* We use the convention that the left-most element of a partial solution is the first specified variable of that solution, the second left-most element is the second specified variable, and so on. A second partial solution is said to be a continuation of a first partial solution if all the elements of the first partial solution are the left-most elements of the second partial solution. Thus, the partial solution ($x_3 = 4, x_2 = 6, x_6 = 1$) is a continuation of the partial solution ($x_3 = 4, x_2 = 6$). A second partial solution is said to be an immediate continuation of a partial solution if the set of specified variables of the second partial solution exceeds the set of specified variables of the first partial solution by precisely one element, and if the second partial solution is a continuation of the first. A completion of a partial solution is a continuation of that solution for which all unspecified (integer) variables take on integer values.

There is a *slight* difference between the two types of methods. Branch and bound methods fix or bound variables by rounding integer variables from fractional linear programming solution values in a sequential manner, whereas implicit enumeration techniques derive their values for variables primarily from the logic of the problem constraints. Within that framework, both methods depend to a great extent on heuristic rules.

The branch and bound methods introduced in this chapter may be categorized as mixed-integer, primal dual feasible starting solution, branch and bound methods. The procedure for these methods is to begin with a noninteger optimum. Then, the following procedure is employed:

1. Given a list of active partial solutions that have not yet been explored, select one to examine next. If appropriate, terminate the search and conclude that the optimal solution is the best one found so far, or that no optimal feasible solution exists. Otherwise, go to step 2.
2. Examine the partial solution selected, and draw some conclusions about its possible continuations that may exclude certain successor partial solutions from consideration. This includes adding the successor partial solutions, as appropriate, to the list. Then go to step 1.

The above framework is simplified, and there are numerous options which may be exercised in its execution. We shall now particularize it to the methods involved.

## 15.2 INTEGER PROGRAMMING USING BRANCH AND BOUND

A classic paper by A. H. Land and A. G. Doig [187] using a branch and bound principle appeared in 1960. The method is elegant in its simplicity; it requires the use of the simplex method, and can solve both all-integer and mixed-integer problems. In this method, the noninteger optimum is first achieved. Then (A) a variable whose solution value is fractional is considered. There are two possibilities:

1. Either it must not exceed the next lower integer, or

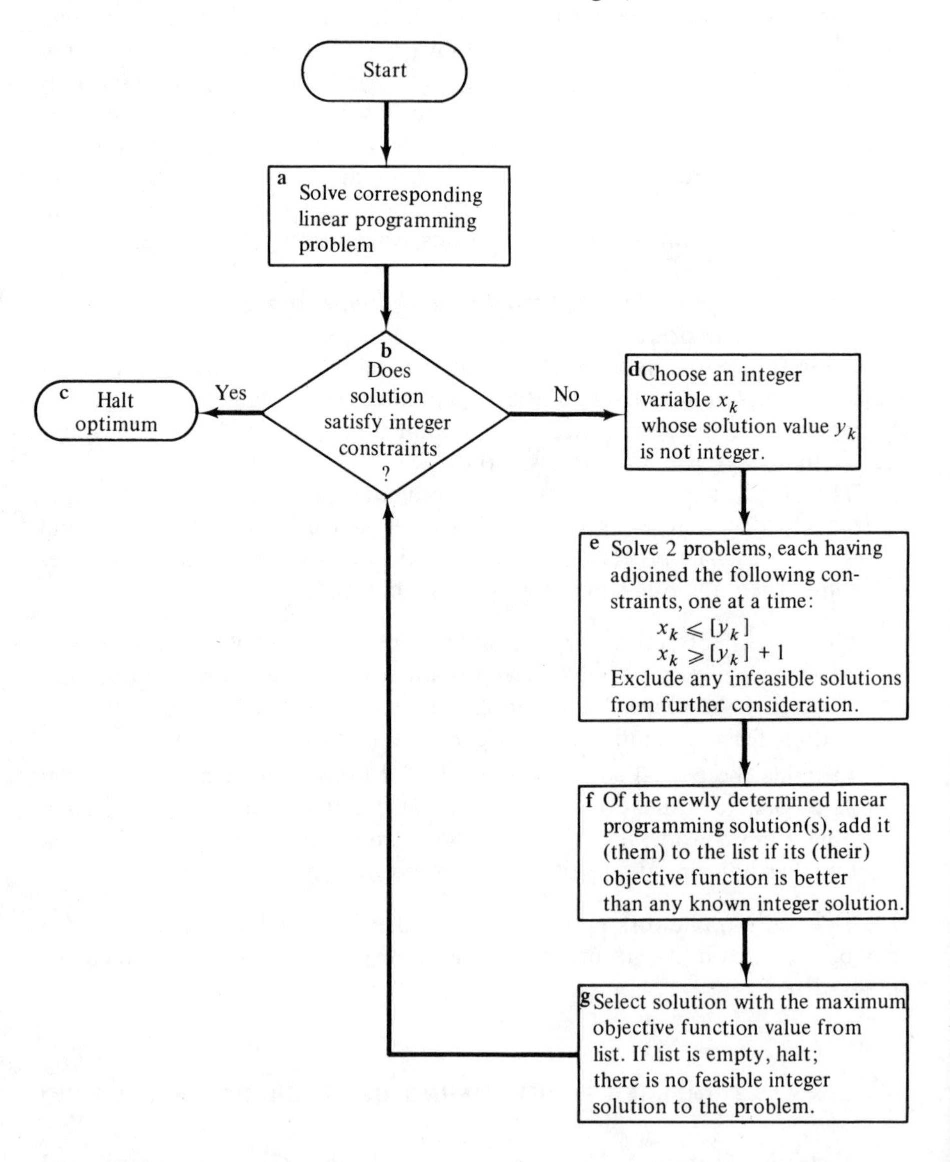


**Figure 15.1**
Flow Chart of the Branch and Bound Algorithm

2. It must equal or exceed the next higher integer.

Accordingly, each constraint is adjoined—one at a time—to the original problem, and the problem re-solved. The resulting objective function values and constraints are recorded in a table, provided that the objective function values are greater than any feasible integer solution found so far. Then the solution in the table with the maximum objective function value is selected and removed from the table. If the corresponding solution satisfies the integer constraints, it is optimal. Otherwise, go to point (A) above and proceed. If the table is emptied before a feasible solution is found, then no feasible integer solution to the problem exists. The original Land and Doig algorithm adjoined equality constraints; the use of inequality constraints was introduced by Dakin [60]. The flow chart for the Land and Doig procedure using inequalities is presented in Figure 15.1.

A few comments are in order. If a feasible integer solution is known a priori, it may be stored on the list, and may serve to reduce computations significantly by eliminating solutions from consideration. The procedure successively partitions the solution sets; the list of solutions can be thought of as part of a tree, with the noninteger solution as the origin. The tree can become large, although its growth is not rapid. Whenever two solutions are added (step **f** in the flow chart of Fig. 15.1), the one being examined is deleted. One or both may be infeasible, and therefore not added. Use of a bounded variable algorithm which automatically handles lower and upper bounds implicitly (see Chap. 8) simplifies computations significantly. Further, when relatively few variables are required to be integral, the number of partial solutions that must be considered is generally small. Two examples will now be given to illustrate the algorithm.

EXAMPLE 15.1: (Same problem as Example 13.2)

$$\begin{aligned} \text{Maximize } z &= 5x_1 + 2x_2 \\ \text{subject to:} \quad & 2x_1 + 2x_2 + x_3 = 9 \\ & 3x_1 + x_2 + x_4 = 11 \\ & x_1, x_2, x_3, x_4 \geq 0 \text{ and integer} \end{aligned}$$

The optimal noninteger solution is given in Tableau 1.

**Tableau 1**

| | | $x_1$ | $x_2$ | $x_3$ | $x_4$ |
|---|---|---|---|---|---|
| $z$ | 18.75 | 0 | 0 | .25 | 1.5 |
| $x_2$ | 1.25 | 0 | 1 | .75 | −.50 |
| $x_1$ | 3.25 | 1 | 0 | −.25 | .50 |

optimal noninteger solution to the problem

Since it is not integer, we proceed and arbitrarily choose $x_2$ as the variable to be made integer. By adjoining $x_2 \leq 1$ and using the method of upper bounds, with $\bar{x}_2$ as the complement of $x_2$, we obtain the solution given in Tableau 2. (To perform the iteration, replace $x_2$ by $1 - \bar{x}_2$ and perform a dual simplex iteration.)

**Tableau 2**

| | | $x_1$ | $\bar{x}_2$ | $x_3$ | $x_4$ |
|---|---|---|---|---|---|
| $z$ | 18.67 | 0 | 0.33 | 0 | 1.67 |
| $x_3$ | 0.33 | 0 | −1.33 | 1 | −0.67 |
| $x_1$ | 3.33 | 1 | −0.33 | 0 | 0.33 |

optimal solution to partial problem ($x_2 \leq 1$)
$\bar{x}_2$ is the complement of $x_2$ in the bound constraint $x_2 \leq 1$

By adjoining $x_2 \geq 2$ to the solution of Tableau 1 and performing the necessary iteration, we obtain the solution given in Tableau 3.

**Tableau 3**

| | | $x_1$ | $x_2'$ | $x_3$ | $x_4$ |
|---|---|---|---|---|---|
| $z$ | 16.5 | 0 | 3 | 2.5 | 0 |
| $x_4$ | 1.5 | 0 | −2 | −1.5 | 1 |
| $x_1$ | 2.5 | 1 | 1 | 0.5 | 0 |

optimal solution to partial problem ($x_2 \geq 2$)
$x_2'$ is the excess of $x_2$ above 2

Neither of these solutions is integer, but since both are feasible they are stored in the list. The solution of Tableau 2 is next selected from the list. By choosing $x_1$ as the branching variable and adjoining $x_1 \leq 3$, we obtain the solution of Tableau 4.

**Tableau 4**

| | | $\bar{x}_1$ | $\bar{x}_2$ | $x_3$ | $x_4$ |
|---|---|---|---|---|---|
| $z$ | 17 | 5 | 2 | 0 | 0 |
| $x_3$ | 1 | −2 | −2 | 1 | 0 |
| $x_4$ | 1 | −3 | −1 | 0 | 1 |

optimal solution to the partial ($x_2 \leq 1, x_1 \leq 3$)
$\bar{x}_j$ is the complement of $x_j$

Since the solution of Tableau 4 is integer, no solutions with objective functions less than 17 need be considered, and so the partial solution corresponding to Tableau 3 may be deleted from the list. By adjoining the constraint $x_1 \geq 4$ to Tableau 2, we find that there is no feasible solution to that problem. Hence, only the solution in Tableau 4 is added to the list. Next, the solution of Tableau 4 is selected from the list. Since it is integer, it is optimal. A tree for this problem's solution is given in Figure 15.2.

EXAMPLE 15.2:
Solve Example 15.1 with the following constraint added:

$$20x_1 - 10x_2 \geq 51$$

The solution closely follows that of Example 15.1. Designating $x_5$ as the slack of this constraint, Tableau 1 has the following row attached:

| | | $x_1$ | $x_2$ | $x_3$ | $x_4$ | $x_5$ |
|---|---|---|---|---|---|---|
| $x_5$ | 1.5 | 0 | 0 | −12.5 | 15 | 1 |

Similarly, Tableau 2 has the following row attached:

| | | $x_1$ | $x_2$ | $x_3$ | $x_4$ | $x_5$ |
|---|---|---|---|---|---|---|
| $x_5$ | 5.67 | 0 | −16.67 | 0 | 6.67 | 1 |

The solution corresponding to Tableau 3 is no longer feasible; hence, only the solution of Tableau 2 is added to the list. The solution of Tableau 2 appropriately augmented is then taken from the list. First the constraint $x_1 \leq 3$ is added, and Tableau 4′ results and is added to the list.

**Tableau 4′**

| | | $\bar{x}_1$ | $\bar{x}_2$ | $x_3$ | $x_4$ | $x_5$ |
|---|---|---|---|---|---|---|
| $z$ | 16.8 | 9 | 0 | 0 | 0 | .2 |
| $x_3$ | 1.20 | −6 | 0 | 1 | 0 | −.2 |
| $x_4$ | 1.10 | −5 | 0 | 0 | 1 | −.1 |
| $\bar{x}_2$ | .1 | −2 | 1 | 0 | 0 | −.1 |

optimal solution to partial problem ($x_2 \leq 1, x_1 \leq 3$)
$\bar{x}_j$ is the complement of $x_j$

Then the constraint $x_1 \geq 4$ is added to Tableau 2 appropriately augmented, and no feasible solution can be found. Next the solution in Tableau 4′ is taken from the list and considered. We choose $\bar{x}_2$ (or, equivalently, $x_2$) for

Example 15.1:

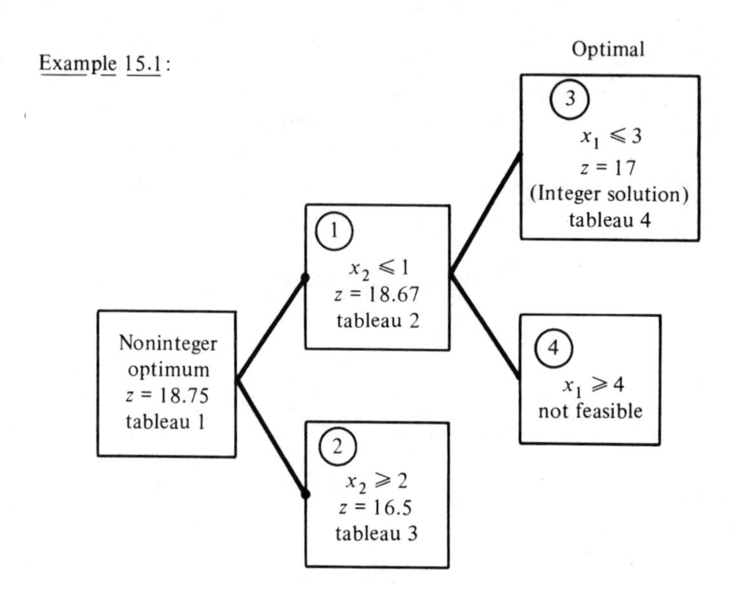


Example 15.2:

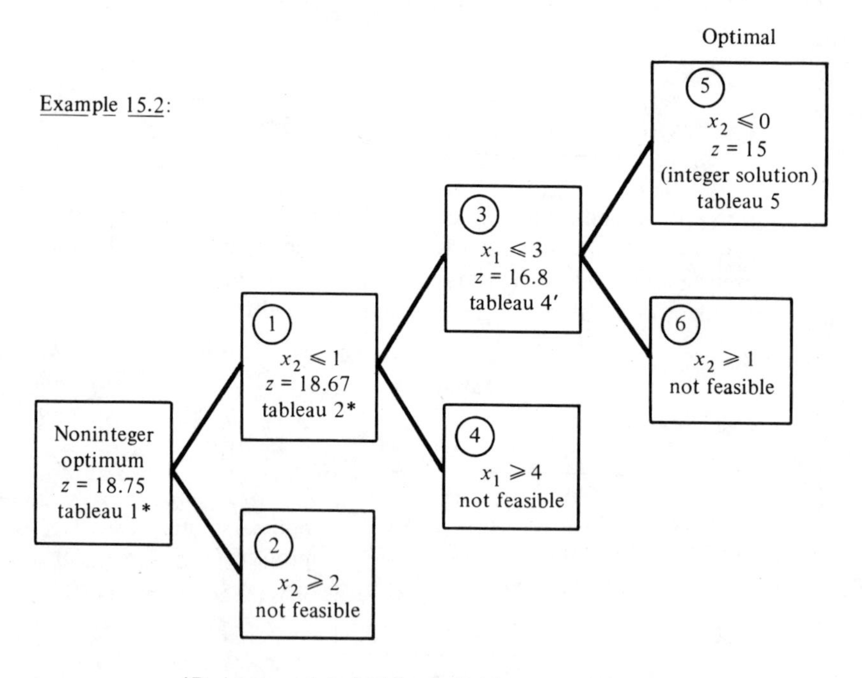


*Designates original tableau suitably augmented.

**Figure 15.2**

Tree Diagrams or Graphs of the Solutions for Examples 15.1 and 15.2.

branching. We first adjoin $\bar{x}_2 \geq 1$ (or, equivalently, $x_2 \leq 0$). The solution to this problem, given in Tableau 5, is then placed on the list.

**Tableau 5**

| | | $\bar{x}_1$ | $\bar{x}_2$ | $x_3$ | $x_4$ | $x_5$ |
|---|---|---|---|---|---|---|
| $z$ | 15 | 5 | 2 | 0 | 0 | 0 |
| $x_3$ | 3 | 2 | 2 | 1 | 0 | 0 |
| $x_4$ | 2 | 3 | 1 | 0 | 1 | 0 |
| $x_5$ | 9 | −20 | 10 | 0 | 0 | 1 |

optimal solution to partial problem ($x_2 \leq 0, x_1 \leq 3$)
$\bar{x}_j$ is the complement of $x_j$

Next the constraint $\bar{x}_2 \leq 0$ (or $x_2 \geq 1$) is adjoined to Tableau 4′, and it is found that no feasible solution exists. The next solution (Tableau 5) on the list is considered. Since it is integer, the problem is solved. A tree of the optimal solution for this problem is given in Figure 15.2.

The method works well when a good integer solution is made available early, either from the method or external to it, and when the auxiliary linear programming solutions may be developed at low cost. It is the author's belief that the computational efficiency of this method has appeared unfavorable in the past because of the cumbersome procedures for solving auxiliary linear programming problems that were in use when it was originally developed, although there is still some question concerning the potentially large number of partial solutions that must be solved and stored.

Many improvements to the Land and Doig branch and bound algorithm have been proposed. We shall consider certain of these below.

### Theoretical Considerations

Convergence of the Land and Doig method may be proven by showing that no unconsidered solution could possibly be better. At each stage two linear programming problems are solved, and the result is clearly optimal in terms of unspecified variables. By systematically examining only the maximum solutions and continuing until a maximum integer solution is found (ignoring any infeasible solutions), it can be shown that an optimal integer solution (if one exists) will be found. If no optimal integer solution exists, then that will be discovered as the method proceeds.

### Driebeek's Method for Mixed-Integer Programming

Norman Driebeek [81] developed a method for solving mixed-integer programming problems that is related to the method of Land and Doig. It appears practical for large problems with relatively few small integer variables. The partial solutions in this method are not really partial, since all integer variables are specified, but the noninteger variable values are found using linear programming. The procedure is to formulate the problem in such a way that every integer variable at every possible integer level is a zero-one integer variable. This is accomplished by substituting [311]:

$$x_i = \sum_{j=1}^{n} j\alpha_{ij}$$

where $\alpha_{ij} = 0$ or 1, and

$$\sum_{j=0}^{n} \alpha_{ij} = 1$$

The variable $x_i$ is an integer variable that is eliminated in the formulation by the above substitution. (The binary representation given in section 16.1 is more economical.) After the substitution has been made, $\alpha_{ij} = 1$ implies that $x_i = j$. Having set up the problem in terms of the $\alpha_{ij}$ variables, the optimal noninteger solution is determined via linear programming methods. Then a table of penalties is developed, as follows:

1. For each nonbasic variable $\alpha_{ij}$, the "true penalty" (really a lower bound to the penalty) for achieving $\alpha_{ij} = 1$ or $x_i = j$ is the optimal reduced cost of the variable $\alpha_{ij}$.
2. For each basic variable $\alpha_{ij}$, the "pseudo penalty" of increasing $\alpha_{ij}$ to 1 or setting $x_i$ to $j$ is the minimum change in the objective function by setting $\alpha_{ij} = 1$. The minimum change in the objective function is the amount by which the objective function changes in the first dual simplex iteration, that is

$$(a_{r0} - 1) \underset{k=1,\ldots,n}{\text{Max}} \left\{ \frac{c_k}{a_{rk}} \,\middle|\, a_{rk} < 0 \right\}$$

where $c_k$ is the reduced cost of $x_k$ and $\alpha_{ij}$ is basic in row $r$ at value $a_{r0}$.

Once the table is developed, an integer solution is selected for which the *larger* of:

1. The sum of the true penalties associated with that solution;
2. The largest of the pseudo penalties associated with that solution;

is minimum. The corresponding solution is solved for; if it is feasible, it is compared with the best available feasible solution and the better one retained. The procedure is continued until all promising solutions have been explored.

Driebeek also proposed using an arbitrary cutoff (90 per cent of noninteger optimum for a maximum, and analogously for a minimum) to abandon the search for a feasible integer solution. That is, if a solution's objective function value, as *overestimated* by the difference between the noninteger optimal objective function value and the penalty, is less than 90 per cent of the noninteger optimal value, the precise value is not computed and the solution is dropped from further consideration. If no solution is found with such a cutoff, then the cutoff is lowered, e.g., from 90 per cent to 80 per cent, and the solution repeated. Such repeated solutions (readjusting the cutoff), if necessary, are of course expensive computationally because previous calculations must be repeated.

EXAMPLE 15.3:

Solve Example 15.1 requiring that only $x_1$ and $x_2$ be integer using the Driebeek method.

$$\begin{aligned} \text{Maximize } z &= 5x_1 + 2x_2 \\ \text{subject to:} \quad & 2x_1 + 2x_2 + x_3 = 9 \\ & 3x_1 + x_2 + x_4 = 11 \end{aligned}$$

(all $x$'s are nonnegative; $x_1$, $x_2$ are integer)

From the first constraint it is seen that $x_2$ can be at most 4; from the second constraint it is seen that $x_1$ can be at most 3. The starting solution is given in Tableau 1. (Remember the variable $\alpha_{ij} = 1$ means that $x_i = j$.)

**Tableau 1**

| | | $\alpha_{10}$ | $\alpha_{11}$ | $\alpha_{12}$ | $\alpha_{13}$ | $\alpha_{20}$ | $\alpha_{21}$ | $\alpha_{22}$ | $\alpha_{23}$ | $\alpha_{24}$ | $x_3$ | $x_4$ |
|---|---|---|---|---|---|---|---|---|---|---|---|---|
| $z$ | 0 | 0 | −5 | −10 | −15 | 0 | −2 | −4 | −6 | −8 | 0 | 0 |
| $x_3$ | 9 | 0 | 2 | 4 | 6 | 0 | 2 | 4 | 6 | 8 | 1 | 0 |
| $x_4$ | 11 | 0 | 3 | 6 | 9 | 0 | 1 | 2 | 3 | 4 | 0 | 1 |
| $\alpha_{10}$ | 1 | 1 | 1 | 1 | 1 | 0 | 0 | 0 | 0 | 0 | 0 | 0 |
| $\alpha_{20}$ | 1 | 0 | 0 | 0 | 0 | 1 | 1 | 1 | 1 | 1 | 0 | 0 |

The optimal noninteger solution is given in Tableau 2.

**Tableau 2**

| | | $\alpha_{10}$ | $\alpha_{11}$ | $\alpha_{12}$ | $\alpha_{13}$ | $\alpha_{20}$ | $\alpha_{21}$ | $\alpha_{22}$ | $\alpha_{23}$ | $\alpha_{24}$ | $x_3$ | $x_4$ |
|---|---|---|---|---|---|---|---|---|---|---|---|---|
| $z$ | 18 | 9 | 6 | 3 | 0 | 0 | 0 | 0 | 0 | 0 | 1 | 0 |
| $\alpha_{24}$ | $\frac{3}{8}$ | $-\frac{3}{4}$ | $-\frac{1}{2}$ | $-\frac{1}{4}$ | 0 | 0 | $\frac{1}{4}$ | $\frac{1}{2}$ | $\frac{3}{4}$ | 1 | $\frac{1}{8}$ | 0 |
| $x_4$ | $\frac{1}{2}$ | −6 | −4 | −2 | 0 | 0 | 0 | 0 | 0 | 0 | $-\frac{1}{2}$ | 1 |
| $\alpha_{13}$ | 1 | 1 | 1 | 1 | 1 | 0 | 0 | 0 | 0 | 0 | 0 | 0 |
| $\alpha_{20}$ | $\frac{5}{8}$ | $\frac{3}{4}$ | $\frac{1}{2}$ | $\frac{1}{4}$ | 0 | 1 | $\frac{3}{4}$ | $\frac{1}{2}$ | $\frac{1}{4}$ | 0 | $-\frac{1}{8}$ | 0 |

The table of penalties is as follows (an asterisk indicates a pseudo penalty).

| *Level* | $x_1$ | $x_2$ |
|---|---|---|
| 0 | 9 | 3* |
| 1 | 6 | 0 |
| 2 | 3 | 0 |
| 3 | 0* | 0 |
| 4 | — | $\frac{15}{2}$* |

*Examples of computation of penalties:* The penalty for $x_1$ at level zero, a true penalty because $\alpha_{10}$ is nonbasic, is 9, the shadow price of $\alpha_{10}$. The penalty for $x_2$ at level 4, a pseudo penalty because $\alpha_{24}$ is basic, is $(\frac{3}{8} - 1)(3/-\frac{1}{4}) = \frac{5}{8}(12) = \frac{15}{2}$.

Using the penalties, we examine the following solutions:

$$x_1 = 3, \quad x_2 = 3 \text{ (a potential zero penalty)—not feasible}$$

$$x_1 = 3, \quad x_2 = 2 \text{ (a potential zero penalty)—not feasible}$$

$$x_1 = 3, \quad x_2 = 1 \text{ (a potential zero penalty)—feasible, } z = 17$$

(Driebeek uses the simplex method to find each such solution.) No unexplored solution has a potential penalty of less than 1 (the difference between the noninteger optimum's value of 18 and the integer solution's value of 17, i.e., the smallest is 3); hence, the solution $x_1 = 3$, $x_2 = 1$, $z = 17$ is optimal. Driebeek has had good computational experience with the method, although no computational results are known to have been published.

The method is shown to converge using the same kind of argument that Land and Doig use, and the use of penalties follows from linear programming considerations.

## 15.3 IMPLEMENTING BRANCH AND BOUND METHODS

A number of computer programs have been written implementing branch and bound methods. Meaningful comparisons, if made, have yet to be reported. (Geoffrion and Marsten [103] present a collection of comparative results which, they suggest, should not be taken too seriously.) This section indicates the important aspects of these programs and the algorithmic aspects of the approaches used. All of the approaches are essentially the inequality form of the Land and Doig method. The methods are made effective (and different) by the heuristics used in choosing which solution to solve and which variable to consider next.

Common to all of the approaches is the requirement for a good linear programming package. Of course, the linear programming system must solve linear programming problems quickly and efficiently. Utilization of upper and lower bound procedures permits economical representation of the bounds used in the branch and bound procedures. Because it is necessary to solve a number of similar linear programming problems, it is convenient to have restart procedures that permit finding the solution, starting from a known optimal solution to a similar problem. Both dual iteration capabilities and parametric capabilities permit even greater flexibility in finding optimal solutions to the problems after having changed the bounds. It would appear that a program employing the product form of the inverse method would be essential for solving integer programming problems, primarily because of its capability to recover bases of predecessor partial solutions at very low cost.

A convenience of a branch and bound algorithm is its ability to terminate computations at any time and utilize the best integer solution available at that time. An upper bound on the difference between the objective function of the best integer solution available at that time and the true optimal solution is available: it is the difference between the best integer solution found and the best noninteger solution not yet continued. This bound may be improved by using Driebeek's penalties [81]. For a given basic variable, we have $x_{Bi} = a_{i0}$. Suppose $a_{i0}$ is not integer. Then the minimum penalty to force $x_{Bi}$ to $[a_{i0}]$ is

$$(a_{i0} - [a_{i0}]) \operatorname*{Min}_{a_{ij}>0} \left\{\frac{a_{0j}}{a_{ij}}\right\}$$

and the minimum penalty to force $x_{Bi}$ to $\langle a_{i0} \rangle$ is

$$(1 - a_{i0} + [a_{i0}]) \operatorname*{Min}_{a_{ij}<0} \left\{\frac{a_{0j}}{(-a_{ij})}\right\}$$

The penalties are not additive among basic variables. But since every basic variable which should be integer (and is not) must become integer, we know that the objective function must decrease at least the following amount:

$$\text{(15.1)} \qquad \operatorname*{Max}_{i\in I}\left\{\operatorname*{Min}_{j\in N}\left\{\begin{array}{ll}(a_{0j})\dfrac{(a_{i0} - [a_{i0}])}{a_{ij}}, & \text{for } a_{ij} > 0 \\ (a_{0j})\dfrac{(1 - a_{i0} + [a_{i0}])}{(-a_{ij})}, & \text{for } a_{ij} < 0\end{array}\right\}\right\}$$

where $I$ is the set of basic variables which should be integer (but which are not) and $N$ is the set of nonbasic variables. We now illustrate, using an example.

EXAMPLE 15.4:

Suppose that an integer solution $x_1 = 3$, $x_2 = 0$, $z = 15$, was available initially for Example 15.1. Suppose that after the linear programming solution was at hand, it was decided to stop. Obviously, no integer solution could have an objective function greater than 18.75. Incorporating Driebeek's penalties, we have

$$\text{Max}\,\{(\text{Min}\,\{\tfrac{1}{12}, \tfrac{9}{4}\}, \text{Min}\,\{\tfrac{3}{4}, \tfrac{3}{4}\})\} = \tfrac{3}{4}$$

which means that the best integer solution cannot have an objective function value greater than $18\frac{3}{4} - \frac{3}{4} = 18$. Hence, the solution $x_1 = 3$, $x_2 = 0$ is at most 3 units (18 minus 15) worse than the optimal integer solution.

Tomlin [282] showed that if the nonbasic variable that would enter the basis (i.e., the one having the minimum ratio) were an integer variable, the corresponding reduction in the objective function may possibly be increased to the smallest of the following values:

1. The same calculation rounding up the entering integer variable to one, if it is less than one.
2. The calculation (as in 1) for some other nonbasic integer variable.
3. The original Driebeek penalty for some nonbasic continuous variable.

Although Tomlin's penalties are different from Driebeek's for Example 15.4, the bound of 18 does not change. The same is true for the expanded penalties given by expression (15.2).

Tomlin [282] further showed that the penalty may possibly be increased by considering the Gomory mixed-integer cut constraint (see Chap. 13) constructed for the constraint under consideration. It is not necessary to construct the constraint; we merely compute the penalty, as follows:

$$P_i = \underset{j}{\text{Min}} \begin{cases} \dfrac{(f_{i0})a_{0j}}{a_{ij}} & \text{for } a_{ij} > 0 \quad j \in N - J \\ \dfrac{(1 - f_{i0})a_{0j}}{(-a_{ij})} & \text{for } a_{ij} < 0 \quad j \in N - J \\ \dfrac{(f_{i0})a_{0j}}{f_{ij}} & \text{for } f_{ij} \leq f_{i0} \quad j \in J \\ \dfrac{(1 - f_{i0})a_{0j}}{(1 - f_{ij})} & \text{for } f_{ij} > f_{i0} \quad j \in J \end{cases} \tag{15.2}$$

where $J$ is the set of integer nonbasic variables*, $N - J$ is the set of continuous nonbasic variables, and $f_{ij}$ is the fractional part of $a_{ij}$, i.e., $a_{ij} - [a_{ij}]$. The result gives the minimum penalty for making the basic variable in constraint $i$ integer. This requires about as much computational effort as the original computation for a penalty. Following the earlier development, although the penalties are not additive, every integer variable must be integer at an integer optimum. Hence, at least the maximum of the penalties developed in expression (15.2) must be incurred in achieving an integer completion of a solution.

* $J$ excludes any variables with $a_{ij}$ integer.

There are two slightly different schemes used in the branch and bound procedure, and we shall discuss both:

1. Methods that store solved partial problems on the solution tree.
2. Methods that store unsolved partial problems on the solution tree.

### Methods that Store Solved Partial Problems on the Solution Tree

Methods that store solved partial problems on the solution tree include the inequality form of the Land and Doig algorithm as well as the method used by Bénichou *et al.* [26] of IBM France. The former method was explained earlier and a flow chart of that procedure given in Fig. 15.1. The procedure for the IBM approach differs from that in Fig. 15.1 in essentially 3 blocks: ***d***, ***f***, and ***g***.

Bénichou *et al.* partition the tree of solutions into 3 sets:

1. The set of solutions which will *never* be selected for continuation (in block ***g***).
2. The set of solutions which may or may not be selected for continuation (in block ***g***), but whose likelihood of being selected is relatively low—an inactive set.
3. The set of solutions which may or may not be selected for continuation (in block ***g***), but whose likelihood of being selected is relatively high—an active set.

The sets are defined as follows: Let $z'$ be the objective function value of the best noninteger solution *not yet continued*; let $z^{\infty}$ be the objective function value of the worst integer solution that would ever be considered; let $z_k$ be the objective function value of a partial solution $k$; let $E(z_k)$ be the estimated objective function value of an integer completion to a partial solution $k$ (we shall shortly define what is meant by estimated). By reference to the previous discussion, we see that:

Set 1 includes all $z_k < z^{\infty}$.

Set 2 includes all partial solutions not in sets 1 and 3.

Set 3 includes all $z_k \geq \text{Max}\,(z' - \delta_1, z^{\infty} + \delta_2)$ such that $E(z_k) \geq z''$ where $\delta_1$, $\delta_2$, $z''$, and $z^{\infty}$ are parameters that must be specified. In addition, $z''$ is a lower bound on the estimated objective function value of a partial solution that may be considered.

The sets are dynamic in that solutions may move from one set to another prior to their being continued. Only solutions in set 3 may be continued. Solutions in set 1 are forgotten and never considered again.

A number of rules for selecting a solution from set 3 are given by Bénichou *et al.* Before we introduce those rules, we define pseudocosts and estimations as used by Bénichou *et al.* Whenever an upper bound is *lowered* (and the problem re-solved), we have a deterioration in the objective function. The *per unit* or pseudocost of *lowering* the upper bound on $x_j$ is $PCL_j$. Similarly, when a lower bound is *upped*, we have a *per unit* or pseudocost of *upping* the bound on $x_j$ of $PCU_j$.

The pseudocosts are the average per unit deterioration in the objective function, whereas Driebeek's penalties are a conservative estimate of the total deterioration of the objective function. For clarity, we illustrate on Example 15.1. Consider the pseudocosts on $x_2$ which may be computed from Tableaus 1, 2, and 3. Since we lower the upper bound on $x_2$ going from Tableau 1 to 2, we have

$$PCL_j = \frac{\text{Deterioration in objective function}}{\text{Amount by which } x_j \text{ must decrease}}$$

or

$$PCL_2 = \frac{18.75 - 18.67}{.25} = .33$$

Similarly, by going from Tableau 1 to Tableau 3, we up the (lower) bound on $x_2$. We thus have

$$PCU_j = \frac{\text{Deterioration in objective function}}{\text{Amount by which } x_j \text{ must increase}}$$

or

$$PCU_2 = \frac{18.75 - 16.5}{.75} = 3$$

Obviously, pseudocosts are not known until after an iteration is taken. Bénichou *et al.* assume that the pseudocosts are constant (although they recognize that is not strictly true). They also assume that the pseudocosts are additive, to the extent that the effect of a pseudocost of one variable can be added to the effect of a pseudocost of another variable (which is also not strictly true) in achieving integrality. Using this concept, they define the estimated objective function of the completion of a partial solution as

$$E(z_k) = z_k - \sum_j \min\{PCL_j(f_j^k), \quad PCU_j(1 - f_j^k)\} \tag{15.3}$$

where $f_j^k$ is the fractional part of the integer variable $j$ in partial solution $k$. The summation is taken over all basic variables that are required to be integers. The motivation behind the estimation is to determine the least expensive continuation, in terms of pseudocosts, without considering questions of feasibility. Bénichou *et al.* do not discuss what to do if pseudocosts are not available, although it seems reasonable to use Driebeek penalties on a per unit basis. They also do not discuss the averaging of subsequent measurements of pseudocosts, but that too seems reasonable, as does computation of a sample measure of dispersion, e.g., variance, and use of such a measure in a confidence interval sense.

To select a solution for continuation, Bénichou *et al.* use different sets of rules for deciding which node to continue. One of the candidates just generated in the most recent branching process should be selected for continuation. If there are two, the following criteria are used:

1. Choose the solution with the best estimation (15.3).
2. Choose the solution with the best objective function value.
3. Choose the solution for which the pseudocost at the previous branch was smaller.

They do not indicate which of these (or in what order) to use. If only one solution was generated at the last branch, then that solution will be continued. If the last branch did not produce a solution, they invoke the following rules:

1. Choose the solution with the best estimation.
2. Choose the last solution generated.
3. Choose the last solution generated until an integer solution is found, then switch to rule 1.

Here, too, they give no idea of relative performance of the different rules.

In choosing the variable upon which to branch, they consider both fixed and dynamic priorities: the fixed include user-specified orders as well as decreasing order of the absolute objective function values. Thus, using a fixed priority, the first variable in the list which is noninteger is branched upon. The dynamic priority proposed is the variable which has the *maximum* value of

$$\text{Min}\,\{PCL_j(f_j^k), \quad PCU_j(1 - f_j^k)\}$$

the reasoning paralleling the discussion of deterioration of the objective function due to branching. They further qualify the dynamic branching variable to be nonquasi-integer if possible (otherwise, quasi-integer). A solution is quasi-integer if all its solution values are within some tolerance $\epsilon$ of integer values.

Some tests of the various rules are described, although none of them includes a dynamic priority, and set 3 for test purposes includes all solutions better than any known integer solution (set 2 is empty). The results quoted are favorable, and the method seems capable of solving realistic large integer programming problems.

### Methods that Store Unsolved Partial Problems on the Solution Tree

Methods that store unsolved partial problems on the solution tree differ from the methods just described in that, when a problem is solved, it is then decided which variable is to be branched upon. Both branches are stored in the tree for possible future solution. In addition, when the solution is stored, a tentative evaluation of the stored solution is made. Generally this type of method requires solution of fewer linear programming problems, whereas the earlier type requires maintenance of a smaller list. We shall describe some of the salient features of different methods of this category, but first we shall indicate the procedure.

1. (Selection of a solution). Choose a partial linear programming problem (initially the linear programming problem without any added constraints) from the list of partial problems and solve it. If the list is empty, stop; the optimum is the best integer solution which has been found. (If no integer solution has been found, then there is no feasible integer solution to the problem.) There are three possible outcomes (see steps 2, 3, and 4).
2. The solution is infeasible. Discard it, and go to step 1.
3. The solution satisfies all the integer requirements. If the solution has a greater objective function value than the previously best known integer solution, replace the previous integer solution by the new solution. Discard any solutions from the list having lower tentative objective function evaluations. Go to step 1.
4. The solution does not satisfy all of the integer requirements. Choose an integer variable which is not integer in this solution and branch upon it. Perform a tentative evaluation on each of these solutions. Store the two new solutions in the list of waiting solutions, then go to step 1.

Researchers who have contributed to this approach include Beale and Small [21], Gutterman [142], Roy, Benayoun, and Tergny [241], and Tomlin [282], among others. The methods proposed by these researchers have been implemented in computer programs. In the evaluation stage the methods use the Driebeek penalties to determine a bound on the integer solution of a branch. Tomlin [282] uses the improved penalty calculation described above to obtain improved bounds, at essentially the same cost of computation as the Driebeek penalties.

In choosing the branching variable, the methods choose a variable that makes the worst penalty as bad as possible. In other words, the variable selected for branching is the one whose larger penalty is greatest, so that, hopefully, the associated solution will have an objective function value *so bad* that it will be eliminated from consideration, thus helping to keep down the size of the list of solutions. One branch for this variable has the penalty just computed; the other has a penalty which may be calculated in the usual manner, i.e., using the Driebeek penalty. Roy *et al.* [241], however, observed that, as in the previous subsection, the minimum penalty of *any* continuation of a partial solution is the maximum (over the variables) of the minimum of the two Driebeek penalties for each variable, and gives a tighter result. (Of course, Tomlin's improved bound could possibly be used to increase this penalty.)

The next partial problem chosen for solution is normally the partial problem with the potentially largest objective function value. (Remember, we are assuming a maximizing problem.) However, in the interest of computational expediency, if the partial problem just stored on the tree having the smaller penalty has a potential objective function value "reasonably" close to the partial problem having the largest objective function value, then the former may be selected.

Roy *et al.* provide two subsidiary considerations for this choice:

1. The number of branchings which have occurred along the path from the original noninteger optimum. The greater this number is, the better are the chances for obtaining an integer solution.
2. The number of integer variables which fortuitously have integer solutions, *even though* they have not been branched upon. Here, too, the larger this number is, the better are the chances for obtaining an integer solution.

Gutterman [142] proposed, in a manner analogous to the discussion in the next chapter, both explicit and implicit treatment of the constraint that any feasible solution must have an objective function larger than that of the best known integer solution. Similarly, he also proposed using the objective function value of the best known integer solution, and dual variables to conditionally tighten the bounds that are used.

We now illustrate some of the above concepts via examples.

EXAMPLE 15.5:

Solve the problem of Example 15.1 using a branch and bound method that stores *solved* partial problems on the solution list. (This parallels the original branch and bound method; the changes are caused by the rules used.) Our solution illustrates some of the rules proposed in the current section. We commence with Tableau 1 from Example 15.1 (repeated here for convenience), omitting the identity matrix.

**Tableau 1**

| | | $x_3$ | $x_4$ |
|---|---|---|---|
| $z$ | 18.75 | 0.25 | 1.5 |
| $x_2$ | 1.25 | 0.75 | −0.5 |
| $x_1$ | 3.25 | −0.25 | 0.5 |

Using Tableau 1, the Driebeek penalties are as follows:

$x_2 \geq 2$, 2.25; $x_2 \leq 1$, .0833
$x_1 \geq 4$, .75; $x_1 \leq 3$, .75

(the numbers following the inequalities are minimum penalties for bounds imposed)

Tomlin's improvements to the penalties are as follows:

$x_2 \geq 2$, 2.25; $x_2 \leq 1$, .25 (minimum penalties to integer
$x_1 \geq 4$, .75; $x_1 \leq 3$, 1.5 solution for bounds imposed)

Tomlin's penalties (15.2), based on Gomory's mixed-integer cut but determined from the signs of the coefficients $a_{ij}$, are as follows, for comparison purposes:

$x_2 \geq 2$, 2.25; $x_2 \leq 1$, .75
$x_1 \geq 4$, .75; $x_1 \leq 3$, 2.25

Hence, two of the penalties are increased. We shall not consider these latter bounds further. We shall set $z'' = z^{\infty} = -\infty$ initially, and subsequently set them to the objective function value of the best integer solution available. We do not have pseudocosts available and hence use Tomlin's improved Driebeek penalties. We choose the variable with the greatest minimum penalty and branch on $x_1$. We first adjoin $x_1 \geq 4$ to Tableau 1. The corresponding solution is infeasible. We then adjoin $x_1 \leq 3$ to Tableau 1. The optimal solution to this subproblem is given in Tableau 2.

**Tableau 2**

| | | $x_3$ | $\bar{x}_1$ |
|---|---|---|---|
| $z$ | 18 | 1 | 3 |
| $x_2$ | 1.5 | 0.5 | −1 |
| $x_4$ | 0.5 | −0.5 | 2 |

$\bar{x}_1$ is the complement of $x_1$ in the bound constraint $x_1 \leq 3$

The solution given in Tableau 2 is therefore added to the list. Because it is the only solution on the list it is selected immediately. The Driebeek penalties for it are as follows:

$x_2 \geq 2$, 1.5; $x_2 \leq 1$, 1 (minimum penalties for
$x_4 \geq 1$, 1; $x_4 \leq 0$, .75 bounds imposed)

Tomlin's improvements to the penalties are as follows:

$x_2 \geq 2$, 3; $x_2 \leq 1$, 1 (minimum penalties to integer
$x_4 \geq 1$, 1; $x_4 \leq 0$, 3 solution for bounds imposed)

In this instance, there is a tie. We arbitrarily choose $x_2$ and first adjoin $x_2 \leq 1$ to Tableau 2. The solution is given in Tableau 3.

**Tableau 3**

| | | $\bar{x}_2$ | $\bar{x}_1$ |
|---|---|---|---|
| $z$ | 17 | 2 | 5 |
| $x_3$ | 1 | −2 | −2 |
| $x_4$ | 1 | −1 | 1 |

$\bar{x}_1$ is the complement of $x_1$ in $x_1 \leq 3$
$\bar{x}_2$ is the complement of $x_2$ in $x_2 \leq 1$

This solution is integer, hence we set $z^\infty = 17$ and add the solution of Tableau 3 to the list. We should next adjoin $x_2 \geq 2$ to Tableau 2. However, the maximum objective function for an integer continuation of that branch, using Tomlin's penalties, is $18 - 3 = 15$, which is below $z^\infty = 17$. Next the solution of Tableau 3 is taken from the list. Since it is integer, it is optimal.

EXAMPLE 15.6:

Solve the problem of Example 15.1 using a branch and bound method that stores *unsolved* problems on the solution tree. We begin with Tableau 1 from the preceding example. Using the Tomlin penalties, we store in the list the problem of Tableau 1, appending the constraint $x_2 \geq 2$. The maximum objective function value for an integer solution along that branch is $18.75 - 2.25 = 16.50$. We therefore consider the other branch, $x_2 \leq 1$. Using the observation of Roy *et al.* on page 430, an upper limit to the objective function value of an integer continuation is $18.75 - .75 = 18$. We adjoin $x_2 \leq 1$ to Tableau 1 and solve the resulting problem, which leads to Tableau 4.

**Tableau 4**

| | | $\bar{x}_2$ | $x_4$ |
|---|---|---|---|
| $z$ | 18.67 | 0.33 | 1.67 |
| $x_3$ | 0.33 | 1.33 | −0.67 |
| $x_1$ | 3.33 | −0.33 | 0.33 |

$\bar{x}_2$ is the complement of $x_2$ in the bound constraint $x_2 \leq 1$

We compute the Driebeek penalties for branching from Tableau 4.

$x_3 \geq 1$, 1.67; $x_3 \leq 0$, 0.083
$x_1 \geq 4$, 0.67; $x_1 \leq 3$, 1.67
(minimum penalties for bounds imposed)

Using Tomlin's improvement, we have the following penalties:

$x_3 \geq 1$, 1.67; $x_3 \leq 0$, 0.33
$x_1 \geq 4$, 0.67; $x_1 \leq 3$, 1.67
(minimum penalties to integer solutions for bounds imposed)

We therefore choose to branch on $x_1$. First we store the solution of Tableau 4 with the added constraint $x_1 \leq 3$ on the list and solve the problem of Tableau 4, having added $x_1 \geq 4$. No feasible solution exists. The problem just stored on the list is then retrieved and solved, with the resulting solution given in Tableau 3 of Example 15.5. That solution is integer, with an objective function value of 17. The only unsolved problem remaining on the list has a maximum objective function solution value of 16.50; hence, the optimal solution is that given in Tableau 3.

We have illustrated some specific rules for the two types of branch and bound algorithms. The methods described have been used to develop some very successful computer programs. However, relatively little comparison among algorithms has as yet been undertaken. It seems reasonable that the methods proposed, coupled with a heuristic scheme for finding a good starting integer solution, and revising implied bounds, hold promise for further development of new computer programs to solve integer programming problems—both all- and mixed-integer—efficiently.

### Selected Supplemental References

Section 15.2
[60], [81], [187], [231], [278], [285], [311]
Section 15.3
[21], [26], [72], [142], [241], [281], [282]
General
[2], [22], [71], [85], [134], [141], [153], [163], [170], [177], [192], [199]

## 15.4 PROBLEMS

Use the various branch and bound methods introduced in sections 15.2 and 15.3 to solve problems 1 through 10.

**1.** Example 13.6.

**2.** Example 16.5.

**3.**

$$
\begin{aligned}
\text{Maximize } z = 4x_1 + 5x_2 + 9x_3 + 5x_4 & \\
\text{subject to:} \quad x_1 + 3x_2 + 9x_3 + 6x_4 &\leq 16 \\
6x_1 + 6x_2 + 7x_4 &\leq 19 \\
7x_1 + 8x_2 + 18x_3 + 3x_4 &\leq 44 \\
x_1, x_2, x_3, x_4 &\geq 0 \\
x_1, x_2 \text{ integer} &
\end{aligned}
$$

**4.**

$$\begin{aligned} \text{Maximize } z = {} & 5x_1 + 8x_2 + 2x_3 + 7x_4 \\ \text{subject to: } & 2x_1 + 8x_2 + 5x_3 + 5x_4 \leq 15 \\ & 5x_1 + 3x_2 \qquad\quad + 9x_4 \leq 21 \\ & 5x_1 + 2x_2 + 3x_3 + 7x_4 \leq 18 \\ & x_1, x_2, x_3, x_4 \geq 0 \\ & x_1, x_2 \text{ integer} \end{aligned}$$

**5.**

$$\begin{aligned} \text{Maximize } z = {} & 7x_1 + 9x_2 + x_3 + 6x_4 \\ \text{subject to: } & 8x_1 + 2x_2 + 4x_3 + 2x_4 \leq 16 \\ & 4x_1 + 8x_2 + 2x_3 \qquad\quad \leq 20 \\ & 7x_1 \qquad\quad + 6x_3 + 2x_4 \leq 11 \\ & x_1, x_2, x_3, x_4 \geq 0 \\ & x_1, x_2 \text{ integer} \end{aligned}$$

**6.** Solve problem 4 with $x_1$ and $x_3$ integer.

**7.** Solve problem 5 with $x_2$ and $x_4$ integer.

**8.** Solve problem 3 with all variables constrained to be integer.

**9.** Solve problem 4 with all variables constrained to be integer.

**10.** Solve problem 5 with all variables constrained to be integer.

**11.** Solve the following problems using the various branch and bound methods introduced in sections 15.2 and 15.3:

a. problem 2 of Chapter 13.
b. problem 7 of Chapter 13.
c. problem 8 of Chapter 13.
d. problem 9 of Chapter 13.
e. problem 10 of Chapter 13.
f. problem 11 of Chapter 13.
g. problem 12 of Chapter 13.
h. problem 25 of Chapter 13.
i. problem 1 of Chapter 14.
j. problem 13 of Chapter 14.
k. problem 14 of Chapter 14.
l. problem 15 of Chapter 14.

**12.** Explain why the noninteger optimum to Example 15.3 is different from that of Example 15.1, even though the problems are identical.

**13.** Prove that the Driebeek true penalty costs are additive.

**14.** Develop a scheme for computing the improved Tomlin penalties as computed in Example 15.5. (Hint: Use the computation of expression (15.2), taking into account the particular bound being computed. Expression (15.2) gives a minimum penalty $P_i$ for integerizing the variable basic in row $i$—it does not distinguish between the minimum penalty of rounding the basic variable up and the minimum penalty of rounding the basic variable down.)

**15.** G. L. Thompson [278], in the stopped simplex method, extended the Land and Doig algorithm by making the search lexicographic, and by using specially developed bounds to limit search. The procedure is as follows:

a. Rewrite the problem so that the objective function is to minimize one variable, for convenience the variable with the greatest index. This can be done in one of two ways, and we consider only one (for the other, see Thompson [278]):
Append the constraint

$$\sum_{j=1}^{n} c_j x_j - x_{n+1} \leq 0$$

which makes the problem degenerate. Precautions must be taken to avoid cycling.

b. Solve for the optimal linear programming solution to the prepared problem. If it is integer, halt. If there is no feasible solution, the problem has no feasible solution. Otherwise, round the objective function variable up to the nearest integer. Go to step c.
c. Stop (in our terms, specify or fix) the objective function variable at that value and minimize the first indexed variable. If there is no feasible solution, halt; the problem has no integer solution. Otherwise, go to step e.
d. Is there a feasible solution? If yes, go to step e. If no, go to step f.
e. If the solution is integer, halt; an optimum has been attained. If it is not integer, round down the variable to the nearest integer, stop the variable, and minimize the first unstopped variable. Go to step d.
f. Determine, using linear programming, the minimum integer value of the objective function variable (by rounding up) for which such a stopped solution has a feasible solution. Maintain a list of such values. Go to step g.
g. Has this variable (other than the objective function variable) been rounded up to its present value? If so, unstop it, focus on next previous stopped variable, and go to step g. Otherwise, round the variable up to the next integer and stop it. Minimize the first unstopped variable and go to step d. If the variable is the objective function variable, increase it to the minimum value on the list, delete that value on the list, and go to step c.

Solve Example 15.1 using the stopped simplex method.

# 16

# Implicit Enumeration, Surrogate Constraints, and a Partitioning Method for Mixed Integer Programming Problems

## 16.1 INTRODUCTION

In this chapter we present implicit enumeration methods, as well as surrogate constraints and partitioning methods. The enumerative framework is the second part of that described in section 15.1, hence the reader is encouraged to reread that section before proceeding.

The implicit enumeration approaches of this chapter are primarily, though not exclusively, in terms of zero-one integers. It should be pointed out that zero-one integer and all-integer problems can be classified as special cases of each other. This is obviously true when representing a zero-one integer variable as a general integer variable: only an upper bound constraint (explicit or implicit) need be utilized. To represent a general integer variable in the form of zero-one variables is almost as obvious; an integer variable $x_j$ can be replaced by a sum of zero-one variables:

$$y_1 + y_2 + \cdots + y_q$$

or, more economically, as a sum of zero-one variables using powers of 2 as coefficients:

$$y_1 + 2y_2 + 4y_3 + \cdots + 2^{q-1}y_q$$

Thus, all-integer methods and zero-one integer methods may be used interchangeably.

Before we introduce any of the methods, it is useful to identify the kinds of considerations that will serve to eliminate potential solutions from consideration. All of the methods choose test values of certain variables and then explore the consequences of the choices on other variables using upper and lower bounds on variables, essentially utilizing Theorem 12.3 and some results of linear programming.

## 16.2 IMPLICIT ENUMERATION FOR PROBLEMS INVOLVING ZERO-ONE VARIABLES—BALAS' ADDITIVE ALGORITHM AND A GENERALIZATION

In this section we shall consider implicit enumeration methods for solving zero-one integer programming problems. The methods are in the same spirit as those considered in Chapter 15, but are primarily applicable to problems involving variables which can only be zero or one. The methods covered in this section and the next may be categorized as zero-one integer, dual feasible starting solution, implicit enumeration algorithms.

We begin by describing the problem and reviewing the method of Balas [4]. Then we present an algorithm that generalizes that of Balas.

The zero-one integer linear programming problem may be formulated as follows:

$$\begin{aligned} &\text{Minimize } z = \sum_{j=1}^{n} c_j x_j \\ (16.1) \qquad &\text{subject to:} \quad \sum_{j=1}^{n} a_{ij} x_j \le b_i \quad i = 1, \dots, m \\ &x_j = 0, 1 \end{aligned}$$

Notice that in this section and the next two, we are assuming a minimization problem. It is possible to rearrange any such problem so that $c_j \geq 0$ by making the substitution $x_j = 1 - \bar{x}_j$ for $c_j < 0$, where $\bar{x}_j$ is the complement of $x_j$ in the constraint $x_j \leq 1$. By doing this, we assure ourselves that a basic dual feasible solution, which is used as a starting point, is available. The entries $c_j$, $a_{ij}$ are not required to be integer.

We now repeat the notation:

$$x^+ = \begin{Bmatrix} x \text{ if } x > 0 \\ 0 \text{ if } x \leq 0 \end{Bmatrix} \qquad x^- = \begin{Bmatrix} 0 \text{ if } x > 0 \\ x \text{ if } x \leq 0 \end{Bmatrix}$$

A solution to a problem is the assigning of zeros and ones to the variables. Consider the following partial solution:

$$(j_1{+}{+}, j_2{-}{-}, j_3{-}, j_4{-}, j_5{+}, j_6{-}, j_7{+}{+}, \dots, j_p{-})$$

with entries interpreted as follows: $j_k{+}{+}$ ($j_k{-}{-}$) means that $x_{j_k}$ has been selected to be set equal to one (zero) in accordance with choice rules that are to be presented; $j_k+$ ($j_k-$) means that $x_{j_k}$ equal to one (zero) is implied because of a partial solution of which it is a continuation. It may also mean that $x_{j_k}$ has been set equal to one (zero) after all possible continuations of $(j_1, \ldots, j_k{-}{-})$ $((j_1, \ldots, j_k{+}{+}))$ have been (implicitly) enumerated. Thus, the partial solution $(3{+}{+}, 5-, 6+, 2{-}{-})$ means that $x_3$ was first set equal to one by choice, then $x_5$ was seen to be zero (either by deduction or by having considered all consequences of $x_5$ being one). Similarly, $x_6$ must be one; then $x_2$ was chosen to be zero. The unsigned element $j_k$ in a partial solution is meant to be any one of the characters $j_k{-}{-}, j_k{+}{+}, j_k-, j_k+$. (We do not make use of the choice rule $j_k{-}{-}$ here, although the rule is certainly permissible.) A variable is said to be assigned in a partial solution if it is one of the characters in the partial solution; otherwise it is said to be free. (The notation and terminology has been evolved from that used by Balas [4], Geoffrion [102], Glover [116], and Graves and Whinston [131].)

Finally, every variable $x_j$ shall be assumed to have a lower bound $h_j$ and an upper bound $u_j$ such that

$$h_j \leq x_j \leq u_j$$

It can be seen that this is perfectly general by noting that $h_j = 0$, $u_j = 1$ designates the zero-one problem when no other information is known, and $h_j = 0$, $u_j = M$ ($M$ arbitrarily large) designates the lower and upper bounds in a more general problem.

#### The Additive Algorithm of Balas [4]

After making the earlier indicated transformation on the problem (viz., so that all $c_j \geq 0$), Balas sets up all the constraints as inequalities of the form

$$\sum_{j=1}^{n} a_{ij}x_j \leq b_i$$

He achieves this form by multiplying inequalities with the reverse inequality by $-1$, and using two inequalities to represent an equality. Then, if all $b_i \geq 0$, the optimal solution is $x_j = 0$, $j = 1, \ldots, n$. We shall present the algorithm assuming that a partial solution $(j_1, \ldots, j_p)$ is available, noting that the initial partial solution has no variables assigned. Then the following procedure is used:

1. a) If the current partial solution satisfies the problem constraints, complete the partial solution by setting all free variables to zero. Set $z^* =$ the objective function value of the solution and save the solution. No other completion of the partial solution can be better than the present one; hence, no other continuations need be examined. Go to step 5.
   b) If the partial solution does not satisfy the problem constraints, go to

step 2. Denote the objective function value of the current partial solution as $z$.

2. For *free variables* $x_j$ such that $z + c_j < z^*$ ($z^*$ is initialized at $\infty$ if no feasible solution is known initially), and such that some coefficient $a_{ij} < 0$ for at least one constraint

$$b_i - \sum_{k=1}^{p} a_{ij_k} x_{j_k} < 0$$

(we shall denote the set of such variables as the set $N$), check the relationships

$$\sum_{j \in N} a_{ij}^- \leq b_i - \sum_{k=1}^{p} a_{ij_k} x_{j_k} (<0) \tag{16.2}$$

An intuitive interpretation of (16.2) is that by restricting attention to variables which can help satisfy violated constraints, we may check to see whether any constraints can never be satisfied. If any such relationship (16.2) is violated, there is no feasible continuation; hence, go to step 5. Otherwise, go to step 3.

3. a) If *all* relationships (16.2) are satisfied as *strict* inequalities, determine which free variable (if set to one) of the set $N$ would reduce the total infeasibility (the sum of the absolute values of the amount by which all the constraints are violated) most, i.e., choose $j_{p+1} \in N$ and set $x_{j_{p+1}} = 1$ such that

$$\sum_{i=1}^{m} \left( b_i - \sum_{k=1}^{p+1} a_{ij_k} x_{j_k} \right)^-$$

is maximized. The new partial solution is $(j_1, \ldots, j_p, j_{p+1}++)$. Go to step 1.

b) If, for any subset of constraints, the relationship (16.2) holds as equalities, denote by the set $F$ all free variables $x_j$ such that $a_{ij} < 0$ for at least one constraint of the subset. Check the relationship

$$\sum_{j \in F} c_j < z^* - z \tag{16.3}$$

Then go to step 4.

4. a) If (16.3) is satisfied, then $x_j = 1$, $j \in F$ is the only possible optimal feasible continuation of the present partial solution. It may or may not be feasible, however. The next partial solution to test is therefore $(j_1, \ldots, j_p, j_{p+1}+, \ldots, j_{p+q}+)$, where $j_{p+1}, \ldots, j_{p+q} \in F$. Go to step 1.

b) If (16.3) is not satisfied, then there is no possible optimal feasible continuation of the present partial solution. Go to step 5.

5. a) Consider the present partial solution. Find the right-most element $j_k++$ and delete all elements to the right of it. Replace it by $j_k-$. That is the new partial solution. Go to step 1.
   b) If there is no element $j_k++$ in the present partial solution, the (implicit) enumeration is complete and the optimal solution (which has been saved in step 1a) has objective function value $z^*$. If $z^*$ is $\infty$, there is no feasible solution.

REMARKS:

The above algorithm is essentially the one presented by Balas [4]. Note that it yields only one optimum (the algorithm could readily be altered to yield all optima, as Balas points out). Note also that we have made no attempt to implement alterations to the algorithm, such as those indicated by Glover and Zionts [122], among others. In addition, a notation different from that of Balas has been used here in an attempt to simplify the presentation.

#### A Generalized Additive Algorithm (Zionts [310])

Remarkably, by using the results of Theorem 12.3 to generate upper and lower bounds on the variables, together with the Balas structure of implicit enumeration, and by taking advantage of the simplification that can be obtained, a simpler and more powerful algorithm may be achieved. The resulting algorithm includes special tests developed by other authors, including Fleishmann [92] and Geoffrion [102]. Conceptually, it is convenient to pose the problem in Balas' framework with a few alterations:

1. Equality constraints are represented as such.
2. A constraint of the form

$$\sum_{j=1}^{n} c_j x_j \leq z^* - \epsilon \quad \text{where} \quad 0 \leq \epsilon \leq \min_j \{c_j\}$$

   is added where $z^*$ is the minimum objective function for a feasible solution found thus far. The effect of this constraint is to imbed all checks involving the objective function within the general constraint checking procedure. If $\epsilon = 0$ is used, then all alternate optima will be found. Initially, $z^* = M$ (or $\infty$) (where $M$ is a sufficiently large number, e.g., $\sum c_j$).
3. Replace steps 2, 3, and 4 of Balas' method by the following:

   2′. Use the results of Theorem 12.3 to generate upper and lower bounds, as appropriate, for each zero-one variable in every constraint.

   a) If a lower bound greater than zero (but less than or equal to one) is found for some variable $x_k$, then $x_k$ is implied to be one in all continuations of the present partial solution. Thus, $k+$ augments the current partial solution. Go to step 1.

b) If a lower bound greater than one is found for some variable $x_k$, then there is no feasible continuation. Go to step 5.

c) If an upper bound less than one but not less than zero is found for some variable $x_k$, then $x_k$ is implied to be zero in all continuations of the present partial solution. Then $k-$ augments the current partial solution. Go to step 1.

d) If an upper bound less than zero is found for some variable $x_k$, then there is no feasible continuation. Go to step 5.

e) If all upper bounds are at least one and all lower bounds are at most zero, then no tighter bounds are available. Go to step 3′.

3′. Determine which free variable would reduce the total infeasibility (the sum of the absolute values of the amount by which all constraints are violated) most, i.e., choose $j_{p+1}$ and set $x_{j_{p+1}} = 1$ such that

$$\sum_{i \notin E} \left| \left( b_i - \sum_{k=1}^{p+1} a_{ij_k} x_{j_k} \right)^- \right| + \sum_{i \in E} \left| \left( \sum_{k=1}^{p+1} a_{ij_k} x_{j_k} - b_i \right) \right|$$

is minimized where $E$ is the set of equalities. The new partial solution is $(j_1, \ldots, j_p, j_{p+1} ++)$. Go to step 1. (This is equivalent to Balas' choice step, except that equalities are represented as such.)

4′. Step 4 of Balas' method is deleted.

One refinement worth mentioning is the use of alternate criteria in step 3′. Two promising possibilities are:

1. Choosing the free variable with the minimum $c_j$.
2. a) Where the total infeasibility can be reduced (see step 3′), choose the variable with the minimum cost per unit of infeasibility reduction.
   b) Where the total infeasibility cannot be reduced, use the rule given in 1 above.

The first possibility is computationally inexpensive, whereas the second is computationally expensive. The second might be used in the early stages of computation until an integer solution is found. Then one of the other strategies may be used.

Actually, the calculations in step 2′ can be simplified tremendously by elimination of tests which cannot occur, thus avoiding repetitious calculations.* (Only two variables in each constraint are checked; if the bound on either is decisive, then other variables are checked.) A flow chart for such a method is given in the appendix to this chapter. When referring to the flow chart, remem-

* Professor Earl McCoy, University of Alabama, has suggested some of the ideas for the simplification.

ber that the method it portrays is equivalent to that given above. An example problem will be solved by both Balas' and the generalized methods.

EXAMPLE 16.1: (Problem 1 of Balas [4])

$$
\begin{aligned}
\text{Minimize } z = 5x_1 + 7x_2 + 10x_3 + 3x_4 + x_5 \\
\text{subject to:} \quad -x_1 + 3x_2 - 5x_3 - x_4 + 4x_5 &\leq -2 \\
2x_1 - 6x_2 + 3x_3 + 2x_4 - 2x_5 &\leq 0 \\
x_2 - 2x_3 + x_4 + x_5 &\leq -1 \\
x_1, \ldots, x_5 &= 0, 1
\end{aligned}
$$

Using Balas' method, we have the following partial solutions:

1. ($\emptyset$). In step 3a add 3++.
2. (3++). In step 3a add 2++.
3. (3++, 2++) $z^* = 17$; feasible. In step 5a, backtrack.
4. (3++, 2−). In step 2, relation (16.2) is violated by constraint 2; therefore, backtrack (step 5a).
5. (3−). In step 2, relation (16.2) is violated by constraint 3; therefore, backtrack (step 5b). $x_3 = 1$, $x_2 = 1$ is the optimal solution.

Using the generalized method, we have the following partial solutions:

1. ($\emptyset$). From constraint 3, the lower bound on $x_3$ is seen to be 1.
2. (3+). From constraint 2, the lower bound on $x_2$ is seen to be 1.
3. (3+, 2+) $z = 17$; feasible. No backtracking is possible. The problem is solved.

#### Theoretical Considerations

To show that implicit enumeration generates an optimal solution if one exists, note that if none of the tests were ever decisive we would have complete enumeration of all solutions.

To illustrate, for Example 16.1, we might have

1. 3++, 2++, 1++, 5++, 4++
2. 3++, 2++, 1++, 5++, 4−
3. 3++, 2++, 1++, 5−, 4++
4. 3++, 2++, 1++, 5−, 4−
5. 3++, 2++, 1−, 5++, 4++

.
.
.

32. 3−, 2−, 1−, 5−, 4−

Instead, using Balas' method we first eliminated numbers 1 through 7, by showing that 8 was feasible and of lower cost than any solution 1 through 7. Then we showed that numbers 9 through 16 were all infeasible, and finally we showed that numbers 17 through 32 were all infeasible. This was accomplished by examining only four solutions. The generalized algorithm involves essentially the same reasoning.

### Generalizing the Zero-One Additive Approach to All-Integer Problems

The generalization of the zero-one additive algorithm to solve all-integer problems is straightforward and has been proposed by Krolak [185] and others. (Such approaches are, of course, all-integer, rather than zero-one integer.) All of the tests on bounds carry over (from Theorem 12.3), but are not additive in this case. What is required is a scheme for representing partial solutions, analogous to that used in our exposition of Balas' additive algorithm, so that only one active solution need be remembered at a time. Such a scheme might include something like the following:

1. When a variable is to be set to some value, it is set to the largest (or alternatively smallest) feasible value.
2. When backtracking and considering another value for the next previous choice variable, decrement (increment) that variable for the next partial solution. If the variable happens to be equal to its minimum (maximum) feasible value, treat it as a variable whose value is implied and backtrack to the next previous choice variable.
3. Recompute the bounds after each choice step.

Computational results using all-integer implicit enumeration algorithms have been meager to date, and the results do not appear very promising.

## 16.3 SURROGATE CONSTRAINTS

To accelerate the solution for zero-one problems, Glover [116] proposed the use of surrogate constraints, a positive linear combination of constraints which yields a constraint that is "strong" in some to be defined sense. (Use of surrogate constraints in a linear programming context to identify redundant constraints had earlier been considered by Thompson, Tonge, and Zionts [279].) Surrogate constraints in zero-one integer programming have been further studied and used to advantage by Balas [5] and Geoffrion [101]. Their methods are closely related, and we shall present a variation that seems to combine good features of each. First, we define the term "strength of a surrogate constraint" as used by Balas [5].

**Definition 16.1**

*A surrogate constraint is a constraint* $\boldsymbol{y}'\boldsymbol{A}\boldsymbol{x} \leq \boldsymbol{y}'\boldsymbol{b}$, *where* $\boldsymbol{A}\boldsymbol{x} \leq \boldsymbol{b}$ *is the constraint set and* $\boldsymbol{y} \geq \boldsymbol{0}$ *is a vector of appropriate order.*

**Definition 16.2**

*Given two surrogate constraints,* $\boldsymbol{a}_0\boldsymbol{x} \leq \boldsymbol{b}_0$, *and* $\boldsymbol{a}_1\boldsymbol{x} \leq \boldsymbol{b}_1$, *the stronger surrogate constraint has the* larger *objective function value when minimizing an objective function, subject* only *to the surrogate constraint and nonnegativity constraints.*

The following theorem is a variation on those presented by Balas [5], Geoffrion [101], and Glover [116].

**THEOREM 16.1**

FOR A GIVEN LINEAR PROGRAMMING PROBLEM (THE CONTINUOUS ANALOG OF THE ZERO-ONE PROBLEM AS GIVEN IN EXPRESSION (16.1)) THE OPTIMAL DUAL SOLUTION YIELDS MULTIPLIERS FOR CONSTRUCTING A STRONGEST SURROGATE CONSTRAINT.

PROOF:

The problem is

$$\text{(16.4)} \qquad \begin{aligned} &\text{Minimize } z = \boldsymbol{c}'\boldsymbol{x} \\ &\text{subject to:} \quad \boldsymbol{A}\boldsymbol{x} \leq \boldsymbol{b} \end{aligned}$$

where $\boldsymbol{x}$ is a vector of zeroes and ones and the continuous analog in maximizing form is

$$\text{(16.5)} \qquad \begin{aligned} &\text{Maximize } z = -\boldsymbol{c}'\boldsymbol{x} \\ &\text{subject to:} \quad \boldsymbol{A}\boldsymbol{x} \leq \boldsymbol{b} \\ &\qquad\qquad\qquad \boldsymbol{x} \leq \boldsymbol{1} \\ &\qquad\qquad\qquad \boldsymbol{x} \geq \boldsymbol{0} \end{aligned}$$

The dual problem is

$$\text{(16.6)} \qquad \begin{aligned} &\text{Minimize } z = \boldsymbol{b}'\boldsymbol{y} + \boldsymbol{1}'\boldsymbol{u} \\ &\text{subject to:} \quad \boldsymbol{A}'\boldsymbol{y} + \boldsymbol{u} \geq -\boldsymbol{c} \end{aligned}$$

Let an optimal solution to the dual be $\boldsymbol{y}^*$, $\boldsymbol{u}^*$, and its corresponding objective function value be

$$z^* = \boldsymbol{b}'\boldsymbol{y}^* + \boldsymbol{1}'\boldsymbol{u}^*$$

The corresponding surrogate constraint is

$$\boldsymbol{y}^{*\prime}\boldsymbol{A}\boldsymbol{x} + \boldsymbol{u}^{*\prime}\boldsymbol{x} \leq \boldsymbol{y}^{*\prime}\boldsymbol{b} + \boldsymbol{u}^{*\prime}\boldsymbol{1} = z^*$$

Now we must prove that the optimal solution to problem (16.7) below is minimum for $\boldsymbol{y} = \boldsymbol{y}^*$ and $\boldsymbol{u} = \boldsymbol{u}^*$.

$$\text{(16.7)} \qquad \begin{array}{ll} \text{Maximize} & -\mathbf{c}'\boldsymbol{x} \\ \text{subject to:} & (\boldsymbol{y}'\boldsymbol{A} + \boldsymbol{u}')\boldsymbol{x} \leq (\boldsymbol{y}'\boldsymbol{b} + \boldsymbol{u}'\mathbf{1}),\ \boldsymbol{x} \geq \mathbf{0} \end{array}$$

Let $\boldsymbol{x}^*$ be the optimal solution to (16.5); then by duality we have that $-\mathbf{c}'\boldsymbol{x}^* = (\boldsymbol{y}^{*\prime}\boldsymbol{A} + \boldsymbol{u}^{*\prime})\boldsymbol{x}^* = \boldsymbol{y}^{*\prime}\boldsymbol{b} + \boldsymbol{u}^{*\prime}\mathbf{1} = \mathbf{z}^*$. We may also write

$$\text{(16.8)} \qquad -\mathbf{c}'\boldsymbol{x} \leq (\boldsymbol{y}^{*\prime}\boldsymbol{A} + \boldsymbol{u}^{*\prime})\boldsymbol{x} \leq \boldsymbol{y}^{*\prime}\boldsymbol{b} + \boldsymbol{u}^{*\prime}\mathbf{1}$$

The optimal solution to (16.7) with $\boldsymbol{y}^*$ and $\boldsymbol{u}^*$ is $\boldsymbol{x}^*$. For any other surrogate constraint generated by any nonnegative multipliers $\boldsymbol{y}$ and $\boldsymbol{u}$, clearly any feasible solution to (16.5) including $\boldsymbol{x}^*$ is feasible. In addition some solutions which are not feasible solutions to (16.5) may be feasible for (16.7); hence the optimal solution for (16.7) in such a case can possibly be greater than $-\mathbf{c}'\boldsymbol{x}^*$, but never less, thereby proving the theorem.

Surrogate constraints are employed in a number of ways. The following is based on Balas [5], Geoffrion [101], and Glover [116], but most closely follows Geoffrion:

1. The objective function is adjoined as a constraint, requiring that any feasible solution have an objective function value at least as good as the best found so far.
2. The corresponding linear programming problem is solved and a surrogate constraint added. If the solution is integer, halt; it is optimal. Then the generalized procedure is used, but just prior to each choice step a new surrogate constraint is generated by solving a linear programming problem holding any assigned variables fixed. If the primal solution is integral, then that is an optimal completion to the partial solution. It may then be recorded and backtracking utilized. If there is no feasible solution, then there is no feasible continuation. Some specified number of most recent surrogate constraints are retained and treated as part of the constraint set.
3. When backtracking, delete any surrogate constraints conditional upon partial solutions being deleted.

Geoffrion [101] has reported excellent computational results solving a linear programming problem after each partial solution (which he defines as being just prior to a trial choice—choosing a $j{+}{+}$), but points out that, on a specific problem, checking less frequently decreases the solution time even further. An example is given to help clarify how surrogate constraints are constructed for a problem.

EXAMPLE 16.2:

Develop a surrogate constraint for the problem of Example 16.1 which is repeated here for convenience:

$$\begin{aligned} \text{Minimize} \quad & z = 5x_1 + 7x_2 + 10x_3 + 3x_4 + x_5 \\ \text{subject to:} \quad & -x_1 + 3x_2 - 5x_3 - x_4 + 4x_5 \leq -2 \\ & 2x_1 - 6x_2 + 3x_3 + 2x_4 - 2x_5 \leq 0 \\ & x_2 - 2x_3 + x_4 + x_5 \leq -1 \\ & x_1, \ldots, x_5 = 0, 1 \end{aligned}$$

Solving the linear programming problem corresponding to Example 16.1 with $x_j = 0$ or 1 replaced by $0 \leq x_j \leq 1$, we have the optimal noninteger primal solution:

$$x_1 = 0,\ x_2 = \frac{1}{3},\ x_3 = \frac{2}{3},\ x_4 = 0,\ x_5 = 0, \text{ and } z = 9$$

The solution to the dual problem is as follows:

$$y_1 = 0,\ y_2 = 2\frac{2}{3}, \text{ and } y_3 = 9$$

Thus the surrogate constraint is found by multiplying 0 times the first inequality, $2\frac{2}{3}$ times the second inequality and 9 times the third inequality. This yields the following surrogate constraint:

$$5\frac{1}{3}x_1 - 7x_2 - 10x_3 + 14\frac{1}{3}x_4 + 3\frac{2}{3}x_5 \leq -9$$

from which we can deduce that $x_3$ is 1 and $x_4$ is zero using the generalized algorithm. Although the surrogate constraint implies values for two variables, using the surrogate constraint with the generalized algorithm requires the same number of partial solutions as not using the generalized algorithm. Of course, if the generalized method with the surrogate constraints were being used, a surrogate constraint would not be required for this problem at all.

Regarding computational efficiency, Geoffrion [101] reports that, in all but one of thirty test problems, the time required to solve the problem with a surrogate constraint added every time is less than when none are added, and the one case is inconclusive, since neither method found the optimal in the time available. Further, Geoffrion used only the basic tests of Balas and not the improved tests presented here. It would appear that the use of surrogate constraints is justified by the computational advantage gained. Certain questions, such as the frequency with which surrogate constraints are added, require further investigation.

For completeness, we indicate briefly the approaches of Balas [5], Geoffrion [101], and Glover [116], all of which are very much like the implicit enumeration scheme with surrogate constraints given above.

#### The Balas Filter Method [5]

The Balas Filter method first constructs the surrogate constraint defined in Definition 16.1. Then, within the surrogate constraint,

$$\sum_{j=1}^{n} a_j x_j \leq b \qquad (b < 0)$$

the variables are arranged in such a manner that all $a_j$ negative have lower indices than any $a_j$ positive. The following conditions must hold:

1. For $a_k < 0$, $a_i < 0$, and $i > k$ we must have $c_i/a_i \leq c_k/a_k$. This implies that of any two variables that help to satisfy the surrogate constraint, the more "efficient" variable is first.
2. For $a_k > 0$, $a_i > 0$, $i > k$ we must have $a_k < a_i$. This implies that of any two variables that do not help to satisfy the surrogate constraint, the one that increases infeasibility least is first.

Then, using only the rearranged surrogate constraint, the method generates a tree of partial solutions in a specified order. The best partial solution (i.e., the one with the minimum objective function value) is selected from the tree. There are two possibilities:

1. If it is not feasible with respect to the surrogate constraint, it is used to generate continuations in the tree by branching on a free variable. (If there is no feasible continuation with respect to the surrogate constraint, the branch is terminated.) Then the best partial solution remaining in the tree is examined, and the process repeated.
2. If the solution is feasible with respect to the surrogate constraint, it is checked for feasibility with respect to all the constraints. If it satisfies those constraints, it is the optimal integer solution; otherwise, that solution (after first checking the complete set of constraints for variables implied to be zero or one, and for the possibility of no feasible continuation) is extended by branching on a free variable and the new solutions are stored in the tree. Then the best partial solution remaining in the tree is examined, and the process repeated.

Balas' surrogate constraint problem for the problem of Example 16.2 is arranged in the following manner:

$$\begin{aligned} &\text{Minimize} \quad 7x_2 + 10x_3 + \quad x_5 + \quad 5x_1 + \quad 3x_4 \\ &\text{subject to:} \ -7x_2 - 10x_3 + 3\frac{2}{3}x_5 + 5\frac{1}{3}x_1 + 14\frac{1}{3}x_4 \leq -9 \end{aligned}$$

(The branching proceeds from left to right.) The method uses tests which are an improvement over those of the additive algorithm though not as comprehensive as those of the generalized algorithm.

### Geoffrion's Improved Implicit Enumeration Approach [101]

As mentioned earlier, our exposition is most closely related to that of Geoffrion. There are two relatively minor differences in Geoffrion's approach which we have not indicated. First, Geoffrion's surrogate constraint is defined as

$$\mathbf{y}'\mathbf{A}\mathbf{x} + \mathbf{c}'\mathbf{x} \leq \mathbf{y}'\mathbf{b} + \bar{z}$$

where $\bar{z}$ is an upper bound to the objective function value of the integer optimum ($\mathbf{y}$ is the optimal solution to the dual, yielding the strongest constraint by his definition [slightly different from ours]). Second, he uses a relatively simple set of infeasibility tests in implementing his scheme, noting that any improved rules could, of course, be used.

### Glover's Multiphase Dual Method [116]

Glover's multiphase dual algorithm uses only one surrogate constraint, which is derived in a heuristic manner by choosing the constraint multipliers arbitrarily. The constraint is modified if it is found that any of the constraints which comprise it cannot possibly be binding in a continuation of the current partial solution. The multiplier of such a constraint in forming the surrogate constraint is set to zero until appropriate backtracking occurs, at which time the appropriate multiplier in the surrogate constraint is restored to its earlier value. In addition, the continuous optimal continuation of a partial solution is solved for as the problem solution proceeds—subject only to the surrogate constraint and the zero-one bound constraints. The other tests of the multiphase dual method may be seen as special cases of the generalized tests given here.

### Other Computational Devices

Numerous other devices have been used in zero-one integer programming schemes. These include choosing a "better" origin than the dual feasible solution used by most schemes. Of course, tests such as those used by Balas and others on the objective function would not be valid in general. The refined use of bounds into which the additive algorithm was generalized earlier would still be valid, although the flow chart in the appendix would have to be altered as indicated. Other devices, such as indexing the variables for each constraint in the order of their desirability (i.e., the minimum value of $c_j/a_{ij}$ $[a_{ij} \neq 0]$ is most desirable), are also used. To justify an increase in computational and/or storage requirements, it is necessary to show a corresponding improved efficiency in the solution process.

## 16.4 AGGREGATING CONSTRAINTS AS A METHOD FOR SOLVING ZERO-ONE PROBLEMS*

The ultimate in surrogate constraints, if achievable, is a single constraint which has the same feasible integer solution set as the original set of constraints. Such a surrogate or aggregate constraint can indeed be developed, and in this section we shall consider methods for constructing such aggregate constraints. Once such a constraint has been developed, we merely need to solve an integer programming problem having a single constraint—a knapsack problem.

To form the aggregate constraint, we generate a linear combination of the original problem constraints. The procedure to be developed combines two constraints into one, and then combines the resulting constraint with a third constraint, and so on, until a single constraint is achieved.

The example used to illustrate aggregating methods will be Example 16.1 (Balas [4]). Since it is necessary to work with equalities, slack variables are added and the constraints written as follows:

$$\begin{aligned}
&\text{Minimize } z = 5x_1 + 7x_2 + 10x_3 + 3x_4 + x_5 \\
&\text{subject to } f_1\colon -x_1 + 3x_2 - 5x_3 - x_4 + 4x_5 + x_6 = -2 \\
&\qquad f_2\colon 2x_1 - 6x_2 + 3x_3 + 2x_4 - 2x_5 + x_7 = 0 \\
&\qquad f_3\colon x_2 - 2x_3 + x_4 + x_5 + x_8 = -1 \\
&\qquad x_j = 0 \text{ or } 1 \qquad j = 1, \ldots, 5 \\
&\qquad x_j \geq 0 \text{ and integer } j = 6, \ldots, 8
\end{aligned}$$

We have enumerated in Table 16.1 the 32 $(= 2^5)$ possible trial solutions to the problem. Representing the solution vector as $x_1, \ldots, x_5$, only solutions (01100) and (11100) satisfy the constraints.

Combining the constraints to achieve a single constraint is no trivial matter; most combinations do not preserve the original solution set. For example, if we simply add the three constraints in the example we have the following:

$$x_1 - 2x_2 - 4x_3 + 2x_4 + 3x_5 + x_6 + x_7 + x_8 = -3$$

* This section was written in collaboration with Kenneth E. Kendall, School of Management, State University of New York at Buffalo.

**Table 16.1**

All Possible Solutions to Example Problem
(A circle indicates that the particular constraint is satisfied)

| | | | | | VALUE OF LEFT-HAND SIDES (EXCLUDING $x_6, x_7, x_8$) | | | | | | | | VALUE OF LEFT-HAND SIDES (EXCLUDING $x_6, x_7, x_8$) | | |
|---|---|---|---|---|---|---|---|---|---|---|---|---|---|---|---|
| $x_1$ | $x_2$ | $x_3$ | $x_4$ | $x_5$ | $f_1$ | $f_2$ | $f_3$ | $x_1$ | $x_2$ | $x_3$ | $x_4$ | $x_5$ | $f_1$ | $f_2$ | $f_3$ |
| 0 | 0 | 0 | 0 | 0 | 0 | 0 | 0 | 1 | 0 | 0 | 0 | 0 | −1 | 2 | 0 |
| 0 | 0 | 0 | 0 | 1 | 4 | −2 | 1 | 1 | 0 | 0 | 0 | 1 | 3 | 0 | 1 |
| 0 | 0 | 0 | 1 | 0 | −1 | 2 | 1 | 1 | 0 | 0 | 1 | 0 | −2 | 4 | 1 |
| 0 | 0 | 0 | 1 | 1 | 3 | 0 | 2 | 1 | 0 | 0 | 1 | 1 | 2 | 2 | 2 |
| 0 | 0 | 1 | 0 | 0 | −5 | 3 | −2 | 1 | 0 | 1 | 0 | 0 | −6 | 5 | −2 |
| 0 | 0 | 1 | 0 | 1 | −1 | 1 | −1 | 1 | 0 | 1 | 0 | 1 | −2 | 3 | −1 |
| 0 | 0 | 1 | 1 | 0 | −6 | 5 | −1 | 1 | 0 | 1 | 1 | 0 | −7 | 7 | −1 |
| 0 | 0 | 1 | 1 | 1 | −2 | 3 | 0 | 1 | 0 | 1 | 1 | 1 | −3 | 5 | 0 |
| 0 | 1 | 0 | 0 | 0 | 3 | −6 | 1 | 1 | 1 | 0 | 0 | 0 | 2 | −4 | 1 |
| 0 | 1 | 0 | 0 | 1 | 7 | −8 | 2 | 1 | 1 | 0 | 0 | 1 | 6 | −6 | 2 |
| 0 | 1 | 0 | 1 | 0 | 2 | −4 | 2 | 1 | 1 | 0 | 1 | 0 | 1 | −2 | 2 |
| 0 | 1 | 0 | 1 | 1 | 6 | −6 | 3 | 1 | 1 | 0 | 1 | 1 | 5 | −4 | 3 |
| 0 | 1 | 1 | 0 | 0 | −2 | −3 | −1 | 1 | 1 | 1 | 0 | 0 | −3 | −1 | −1 |
| 0 | 1 | 1 | 0 | 1 | 2 | −5 | 0 | 1 | 1 | 1 | 0 | 1 | 1 | −3 | 0 |
| 0 | 1 | 1 | 1 | 0 | −3 | −1 | 0 | 1 | 1 | 1 | 1 | 0 | −4 | 1 | 0 |
| 0 | 1 | 1 | 1 | 1 | 1 | −3 | 1 | 1 | 1 | 1 | 1 | 1 | 0 | −1 | 1 |

The two feasible solutions to the problem are: $x_1 = 0, x_2 = 1, x_3 = 1, x_4 = 0, x_5 = 0$ with slack variable values $x_6 = 0, x_7 = 3, x_8 = 0$, and $x_1 = 1, x_2 = 1, x_3 = 1, x_4 = 0, x_5 = 0$ with slack variable values $x_6 = 1, x_7 = 1, x_8 = 0$.

The resulting solution set contains the original feasible solution set, as well as some additional solutions which do not satisfy the original constraints, namely solutions (00100), (01101), (01110), (10100), and (11110).

The Elmaghraby and Wig [87] method requires that *all* coefficients appearing in either constraint (being combined) be *strictly positive.* This can be achieved in general by *first* making all coefficients *nonnegative* by replacing any $x_j$ having a negative coefficient by $1 - \bar{x}_j$, where $\bar{x}_j$ is the complement of $x_j$. Then, to make the coefficients strictly positive, add one constraint to the second to form a new constraint; add the first constraint to twice the second to form a second new constraint. The two new constraints have all coefficients strictly positive, and are equivalent to the original constraints. (This construction of constraints is a method developed by Mathews [208].) Designating the new constraints as

$$\sum_{j=1}^{n} a_{ij}x_j = b_i, \quad i = 1, 2$$

we now wish to find multipliers for the constraints for aggregation purposes. Let 1 be the multiplier for the first constraint. Let $\lambda$ be the multiplier for the second constraint where $\lambda$ is an integer satisfying expression (16.9)

$$\lambda > b_2 \operatorname*{Max}_j \left\{\frac{a_{1j}}{a_{2j}}\right\} \tag{16.9}$$

The composite constraint is then

$$\sum_{j=1}^{n} (a_{1j} + \lambda a_{2j})x_j = b_1 + \lambda b_2$$

The process is repeated using the composite constraint and the third constraint to construct a new composite constraint, and so on, until all constraints have been aggregated.

EXAMPLE 16.3:

Aggregate the constraints of the problem of Example 16.1 using the Elmaghraby-Wig method for aggregating constraints.

Since the process is recursive, we only have to work with two equations at a time. The first step is to eliminate the negative coefficients of our first two equations by substituting $1 - \bar{x}_j$ for all $x_j$ that have negative coefficients in equations of $f_1$ and $f_2$:

$$\begin{aligned} -(1-\bar{x}_1) + 3x_2 - 5(1-\bar{x}_3) - (1-\bar{x}_4) + 4x_5 + x_6 &= -2 \\ 2x_1 - 6(1-\bar{x}_2) + 3x_3 + 2x_4 - 2(1-\bar{x}_5) + x_7 &= 0 \end{aligned}$$

or, equivalently:

$$\begin{aligned} f'_1:&\quad \bar{x}_1 + 3x_2 + 5\bar{x}_3 + \bar{x}_4 + 4x_5 + x_6 = 5 \\ f'_2:&\quad 2x_1 + 6\bar{x}_2 + 3x_3 + 2x_4 + 2\bar{x}_5 + x_7 = 8 \end{aligned}$$

Eliminating the zero coefficients as described above, we have the following:

$f''_1 \equiv f'_1 + f'_2$:

$$2x_1 + \bar{x}_1 + 3x_2 + 6\bar{x}_2 + 3x_3 + 5\bar{x}_3 + 2x_4 + \bar{x}_4 + 4x_5 + 2\bar{x}_5 + x_6 + x_7 = 13$$

$f''_2 \equiv f'_1 + 2f'_2$:

$$4x_1 + \bar{x}_1 + 3x_2 + 12\bar{x}_2 + 6x_3 + 5\bar{x}_3 + 4x_4 + \bar{x}_4 + 4\bar{x}_5 + 4x_5 + x_6 + 2x_7 = 21$$

The multiplier is calculated, using expression (16.9), as

$$\lambda > b_2 \operatorname*{Max}_{j} \left\{\frac{a_{1j}}{a_{2j}}\right\} = 21\left(\frac{1}{1}\right)$$

We therefore choose $\lambda = 22$.

We now generate the composite constraint by multiplying $f''_1$ by one, $f''_2$ by $\lambda$:

$$\begin{aligned} f''_1:&\quad 2x_1 + \bar{x}_1 + 3x_2 + 6\bar{x}_2 + 3x_3 + 5\bar{x}_3 + 2x_4 + \bar{x}_4 + \\ &\quad 4x_5 + 2\bar{x}_5 + x_6 + x_7 = 13 \\ \lambda f''_2:&\quad 88x_1 + 22\bar{x}_1 + 66x_2 + 264\bar{x}_2 + 132x_3 + 110\bar{x}_3 + 88x_4 + 22\bar{x}_4 + \\ &\quad 88x_5 + 88\bar{x}_5 + 22x_6 + 44x_7 = 462 \end{aligned}$$

and add them, giving us:

$$f_0' \equiv f_1'' + \lambda f_2'': \; 90x_1 + 23\bar{x}_1 + 69x_2 + 270\bar{x}_2 + 135x_3 + 115\bar{x}_3 + 90x_4 + 23\bar{x}_4 + 92x_5 + 90\bar{x}_5 + 23x_6 + 45x_7 = 475$$

We repeat these steps using the above equation and our third original equation, which is rewritten to eliminate negative coefficients:

$$f_3': \; x_2 + 2\bar{x}_3 + x_4 + x_5 + x_8 = 1$$

After forming the new equation, $f_0' + f_3'$ and $f_0' + 2f_3'$, we choose a multiplier of 478, resulting in the following composite constraint:

$$43110x_1 + 11017\bar{x}_1 + 34008x_2 + 129330\bar{x}_2 + 64665x_3 + 56999\bar{x}_3 + 44067x_4 + 11017\bar{x}_4 + 45025x_5 + 43110\bar{x}_5 + 11017x_6 + 21555x_7 + 957x_8 = 228482$$

We can cut the number of variables in half by the substitution of $(1 - x_j)$ for all $\bar{x}_j$, and we obtain our final composite constraint:

$$36883x_1 - 95322x_2 + 7666x_3 + 33053x_4 + 1915x_5 + 11017x_6 + 21555x_7 + 957x_8 = -22991$$

There are only two feasible solutions to this equation. They are exactly the same as those obtained when we enumerated the original three constraints.

### An Improved Method for Generating a Composite Constraint

A number of improvements have been proposed. Glover and Woolsey [121] discovered a superior method for combining equations into a composite constraint having smaller coefficients than that generated by the Elmaghraby and Wig method. Kendall and Zionts [174] have improved upon this method by further reducing the size of the coefficients and the number of variables which, in turn, reduces the computer storage requirements. We describe one of the methods proposed by Kendall and Zionts, which has the advantage of there being *no restrictions* on the signs of the entries in the constraints. This implies that no complements of variables are required, thereby economizing on computer storage during the aggregation (and possibly solution) process.

The framework of this method is similar to that of the Elmaghraby-Wig method. Two constraints are combined, the result is combined with a third constraint, and so on. However, neither of the two multipliers is permitted to be one.

The following theorem for combining two constraints is the basis of the method.

**THEOREM 16.2**

GIVEN TWO CONSTRAINTS OF THE FOLLOWING FORM:

$$\sum_{j=1}^{n} a_{1j}x_j = b_1 \tag{16.10}$$

$$\sum_{j=1}^{n} a_{2j}x_j = b_2 \tag{16.11}$$

$$x_j \geq 0 \text{ and integer}$$

$$x_1, \ldots, x_p \leq 1 \qquad (p < n)$$

THEN EQUATIONS (16.10) AND (16.11) ARE EQUIVALENT TO THE EQUATION (16.12)

$$\lambda_1 \sum_{j=1}^{n} a_{1j}x_j + \lambda_2 \sum_{j=1}^{n} a_{2j}x_j = \lambda_1 b_1 + \lambda_2 b_2 \tag{16.12}$$

WHERE THE MULTIPLIERS $\lambda_i > 0$ SATISFY THE FOLLOWING CONDITIONS:

1. $\lambda_1 > b_2 - \sum_{j=1}^{n} a_{2j}^- - \underset{a_{2j}\neq 0}{\text{Min}}\{|a_{2j}|\}$

   $\lambda_2 > b_1 - \sum_{j=1}^{n} a_{1j}^- - \underset{a_{1j}\neq 0}{\text{Min}}\{|a_{1j}|\}$

2. $\dfrac{b_2 - \sum_{j=1}^{n} a_{2j}^-}{\lambda_1}$

   AND

   $\dfrac{b_1 - \sum_{j=1}^{n} a_{1j}^-}{\lambda_2}$

   ARE NOT INTEGERS.

3. $\lambda_1$ AND $\lambda_2$ ARE RELATIVELY PRIME. (THE GREATEST COMMON DIVISOR OF $\lambda_1$ AND $\lambda_2$ IS ONE.)

PROOF:

Equations (16.10) and (16.11) clearly imply (16.12). To prove that equation (16.12) implies equations (16.10) and (16.11), we note that

$$\sum_{j=1}^{n} a_{1j}x_j = b_1 - \frac{\left(\sum_{j=1}^{n} a_{2j}x_j - b_2\right)\lambda_2}{\lambda_1}$$

$\sum_{j=1}^{n} a_{1j}x_j$ and $b_1$ are both integer, hence $\left(\sum_{j=1}^{n} a_{2j}x_j - b_2\right)\lambda_2\Big/\lambda_1$ is integer. From the condition that $\lambda_1$ and $\lambda_2$ must be relatively prime, we see that $\left(\sum_{j=1}^{n} a_{2j}x_j - b_2\right)\Big/\lambda_1$ must be integer. Using expression (16.12) we have

$$\frac{\sum_{j=1}^{n} a_{2j}x_j - b_2}{\lambda_1} = \frac{b_1 - \sum_{j=1}^{n} a_{1j}x_j}{\lambda_2} = q$$

where $q$ is an integer.

We now prove $q$ is zero by contradiction. First assume $q > 0$. This implies that $\left(b_1 - \sum_{j=1}^{n} a_{1j}x_j\right)\Big/\lambda_2$ is a positive integer. We may now write

$$\frac{b_1 - \sum_{j=1}^{n} a_{1j}^- + \sum_{j=1}^{n} a_{1j}^- - \sum_{j=1}^{n} a_{1j}x_j}{\lambda_2} = q \geq 1 \tag{16.13}$$

Since $\left(b_1 - \sum_{j=1}^{n} a_{1j}^-\right)\Big/\lambda_2$ is not integer (because of condition 2) and hence not zero, neither is

$$\frac{\sum_{j=1}^{n} a_{1j}^- - \sum_{j=1}^{n} a_{1j}x_j}{\lambda_2}$$

By noting that

$$\sum_{j=1}^{n} a_{1j}^- - \sum_{j=1}^{n} a_{1j}x_j = \sum_{j=1}^{n} a_{1j}^-(1 - x_j) - \sum_{j=1}^{n} a_{1j}^+x_j$$

we observe that

$$\sum_{j=1}^{n} a_{1j}^- - \sum_{j=1}^{n} a_{1j}x_j \leq -(\operatorname*{Min}_{a_{1j}\neq 0}\{|a_{1j}|\}) \tag{16.14}$$

Substituting expression (16.14) into (16.13) gives

$$b_1 - \sum_{j=1}^{n} a_{1j}^- - \operatorname*{Min}_{a_{1j}\neq 0}\{|a_{1j}|\} \geq \lambda_2$$

a contradiction to condition 1. A very similar argument shows that $q$ cannot be negative. Hence, $q$ must be zero, in which case both constraints (16.10) and (16.11) are satisfied. This completes the proof.

In actual practice, the smallest value of $\lambda$'s that satisfy the above criteria are chosen in order to help keep the magnitude of coefficients in the composite constraint small. The method is outlined in detail in the flow chart in Figure 16.1.

EXAMPLE 16.4:

Solve the problem of Example 16.3 by the Kendall-Zionts method for aggregating constraints.

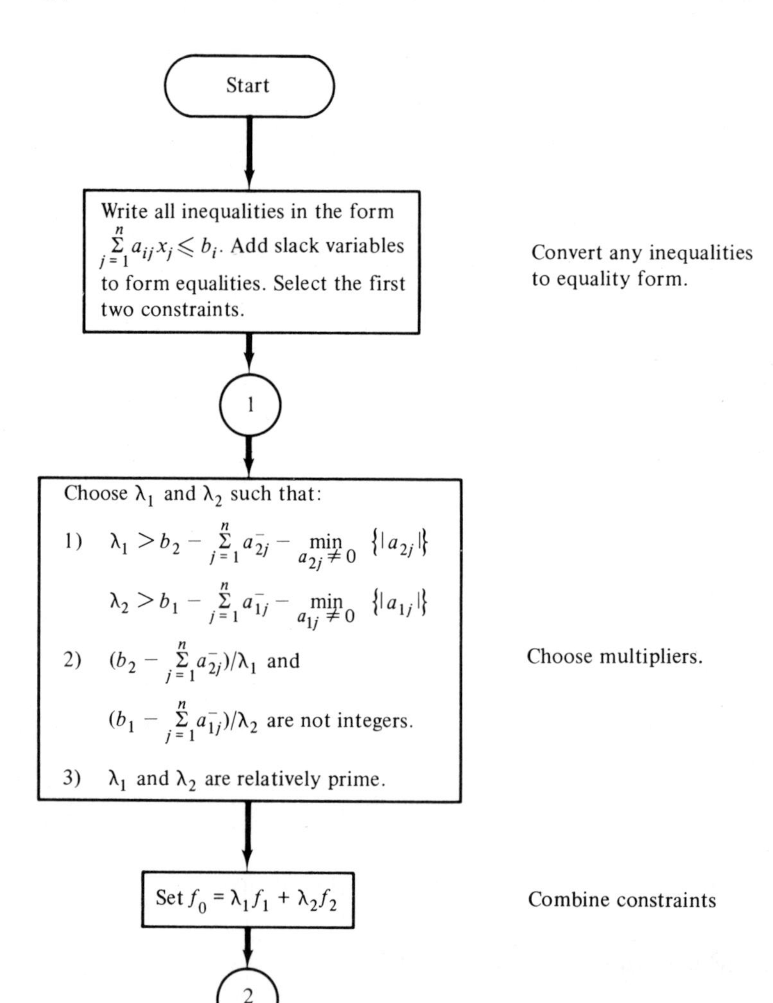


**Figure 16.1**

Flow Chart for the Kendall-Zionts Method for Aggregating Constraints

The Kendall-Zionts method allows us to begin with the original equations $f_1$ and $f_2$:

$$\begin{aligned} f_1&: -x_1 + 3x_2 - 5x_3 - x_4 + 4x_5 + x_6 &= -2 \\ f_2&: 2x_1 - 6x_2 + 3x_3 + 2x_4 - 2x_5 + x_7 &= 0 \end{aligned}$$

In the above equations Min $\{|a_{1j}|\} = 1$ and Min $\{|a_{2j}|\} = 1$. Also, we have $b_1 - \sum_{j=1}^{n} a_{1j}^{-} = -2 + 7 = 5$ and $b_2 - \sum_{j=1}^{n} a_{2j}^{-} = 0 + 8 = 8$.

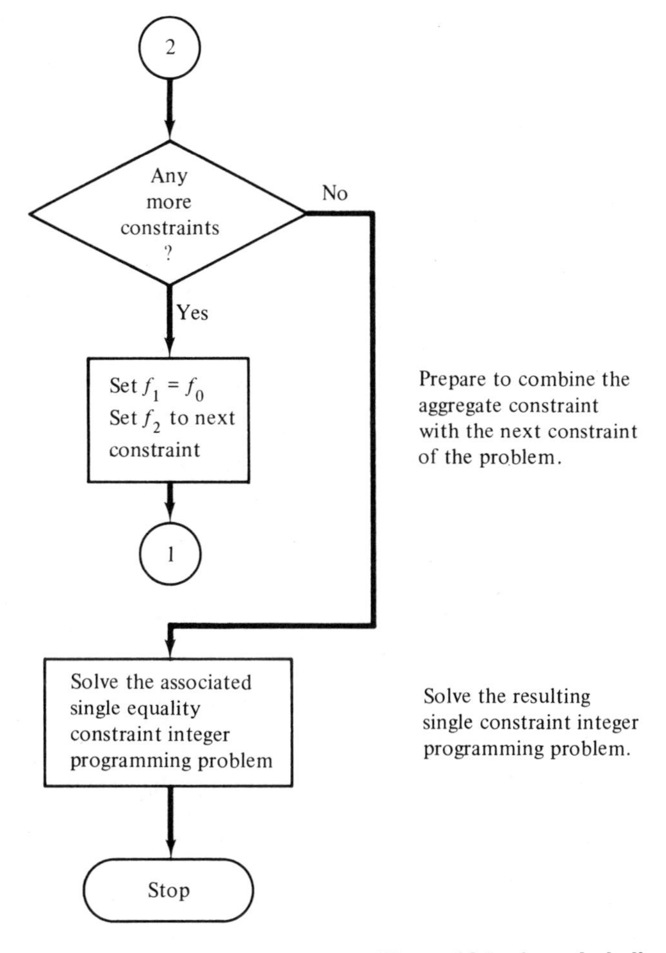


**Figure 16.1 (concluded)**

The multipliers must therefore satisfy:

1. $\lambda_1 > 8 - 1$ and $\lambda_2 > 5 - 1$, that is, $\lambda_1 \geq 8$ and $\lambda_2 \geq 5$.
2. $8/\lambda_1$ and $5/\lambda_2$ are not integers.
3. $\lambda_1$ and $\lambda_2$ are relatively prime.

The natural choice is $\lambda_1 = 9$ and $\lambda_2 = 7$, because those are the smallest numbers that satisfy the above criteria.
We therefore have the following:

$$\lambda_1 f_1: \; -9x_1 + 27x_2 - 45x_3 - 9x_4 + 36x_5 + 9x_6 = -18$$
$$\lambda_2 f_2: \; 14x_1 - 42x_2 + 21x_3 + 14x_4 - 14x_5 + 7x_7 = 0$$

Adding $\lambda_1 f_1$ and $\lambda_2 f_2$ we arrive at the following composite equation:

$$f_0\colon\ 5x_1 - 15x_2 - 24x_3 + 5x_4 + 22x_5 + 9x_6 + 7x_7 = -18$$

which we will call $f_0$ in subsequent steps.
We next combine the aggregate constraint with the third constraint of the problem:

$$f_3\colon\ x_2 - 2x_3 + x_4 + x_5 + x_8 = -1$$

We choose the multipliers so that the following criteria are satisfied:

1. $\lambda_0 > 1 - 1$ and $\lambda_3 > 21 - 5$, that is, $\lambda_0 \geq 1$ and $\lambda_3 \geq 17$.
2. $1/\lambda_0$ and $21/\lambda_3$ are not integers.
3. $\lambda_0$ and $\lambda_3$ are relatively prime.

Choosing the smallest feasible multipliers, $\lambda_0 = 2$ and $\lambda_3 = 17$, we obtain

$$\begin{aligned} \lambda_0 f_0\colon\ & 10x_1 - 30x_2 - 48x_3 + 10x_4 + 44x_5 + 18x_6 + 14x_7 && = -36 \\ \lambda_3 f_3\colon\ & \qquad\quad\ 17x_2 - 34x_3 + 17x_4 + 17x_5 \qquad\qquad\qquad + 17x_8 && = -17 \end{aligned}$$

Finally, we add the above equations, which gives us our final aggregated constraint:

$$10x_1 - 13x_2 - 82x_3 + 27x_4 + 61x_5 + 18x_6 + 14x_7 + 17x_8 = -53$$

This yields exactly the same solution set as our original constraints, but has much smaller coefficients and right-hand side than those obtained using the Elmaghraby and Wig method. The constraint is also better than the constraint obtained using the similar Glover and Woolsey method [121].

Kendall and Zionts [174] also provide a computationally attractive scheme for generating smaller multipliers, based on a framework (which is not as computationally attractive) for generating even smaller multipliers. An aggregate constraint for Example 16.4 using the computationally attractive scheme is found to be

$$10x_1 - 17x_2 - 74x_3 + 23x_4 + 57x_5 + 18x_6 + 14x_7 + 13x_8 = -49$$

Although the improvements may seem small for this example, they are generally more substantial when the coefficients in the constraints being combined are large. This occurs either as a result of the original problem having large coefficients, or as a result of combining constraints.

### Other Developments in Aggregating Constraints

Other methods for aggregating constraints have been proposed by Glover and Woolsey [121] and Bradley [40, 41]. Some allow for considerable flexibility in choosing multipliers, but systematic comparison of the methods has yet to be made. It is clear that the order in which constraints are combined makes a difference. (For instance, in Example 16.4, if we first combine constraints 1 and 3 and then combine constraint 2 with the result, we generate an aggregate constraint having larger coefficients than that obtained in the example.) Thus, assuming smaller coefficients in the aggregate constraint are desirable, further study of the order of the constraints in the aggregation process is appropriate.

The methods described in this section are for zero-one variables. However, they may be extended to bounded integer variables, so the method is applicable to general all-integer problems.

### Advantages and Disadvantages of Aggregating Schemes

All of the aggregation schemes have the same disadvantage—the coefficients become too large to be stored in one computer word as an integer. Since it is necessary to maintain accuracy in an integer programming problem, a multiple precision package of subroutines (subroutines that store the integer in more than one computer word and execute the basic operations of add, subtract, and multiply in that manner) must be utilized. This set of subroutines would be the major part of a computer program using an aggregation approach.

One advantage of these methods is the simplicity of solving the single constraint problem that is generated once the constraints are aggregated, even though solving the resulting problem is more difficult than solving a knapsack problem having an inequality constraint. A second major advantage derives from the recursive nature of the algorithms. It is necessary to store only two constraints in computer memory. Further, the objective function isn't considered until all of the constraints are combined into one.

The possibility of writing an efficient computer program using a constraint aggregation algorithm and a multiprecision package of subroutines is not far-fetched. Even though multiprecision requires more than one computer word per coefficient, the storage requirements for the coefficients of the aggregated constraint will be *much* less than the storage requirements for all of the constraints of the original problem. This implies that such a program would be particularly valuable to a user restricted by the amount of available core storage. This method would allow a person using a timesharing (and consequently "core-sharing") system or a limited memory minicomputer to solve reasonably large problems which would be impossible to solve using presently available integer programming codes. It appears that constraint aggregation methods warrant serious study as a feasible alternative for solving integer programming problems.

## 16.5 BENDERS' PARTITIONING PROCEDURE FOR SOLVING MIXED-INTEGER PROGRAMMING PROBLEMS

J. F. Benders [25] proposed solving a mixed-integer programming problem utilizing a partitioning procedure which is very much in the spirit of the type of algorithms being considered in Chapters 15 and 16. It partitions the problem into two parts—an integer part and a continuous linear programming part—and uses an all-integer method on the integer part. More generally, the scheme is such that some of the terms in the constraints may be nonlinear. We shall treat only the case of integer variables.

Let the problem be

$$\begin{aligned} &\text{Maximize } z_1 = \boldsymbol{c}_1'\boldsymbol{x} + \boldsymbol{c}_2'\boldsymbol{w} \\ &\text{subject to:} \quad \boldsymbol{A}_1\boldsymbol{x} + \boldsymbol{A}_2\boldsymbol{w} \leq \boldsymbol{b} \\ &\qquad\qquad \boldsymbol{x}, \boldsymbol{w} \geq \boldsymbol{0} \\ &\qquad\qquad \boldsymbol{w} \text{ integer} \end{aligned} \tag{16.15}$$

It is convenient to assume that the above problem is bounded, which can be ensured via use of a regularization constraint (see Chap. 4). Benders observed that the problem could be partitioned as follows:

(16.16)

$$\underset{\substack{\boldsymbol{w} \geq \boldsymbol{0} \\ \text{and integer}}}{\text{Maximize}}\ z_1 = \boldsymbol{c}_2'\boldsymbol{w} + \left(\text{Maximum } z_2 = \boldsymbol{c}_1'\boldsymbol{x} \text{ subject to } \boldsymbol{A}_1\boldsymbol{x} \leq \boldsymbol{b} - \boldsymbol{A}_2\boldsymbol{w},\ \boldsymbol{x} \geq \boldsymbol{0}\right)$$

A naive way of solving (16.16) is to consider all possible integer values of $\boldsymbol{w}$ and then (with values of $\boldsymbol{w}$ specified) solve the linear programming problem corresponding to the right-hand expression of (16.16) for each. This is, in effect, a form of explicit enumeration of integer solutions. Benders proposed a procedure in a sense dual to the Dantzig-Wolfe decomposition algorithm presented in Chapter 8; in his procedure constraints instead of variables are added as the computation proceeds. To motivate development of the method, observe that the right-hand expression of (16.16) can be expressed in the dual form; namely, problems (16.17) and (16.18) are dual.

$$\begin{aligned} &\text{Maximize } z_2 = \boldsymbol{c}_1'\boldsymbol{x} \\ &\text{subject to:} \quad \boldsymbol{A}_1\boldsymbol{x} \leq \boldsymbol{b} - \boldsymbol{A}_2\boldsymbol{w} \\ &\qquad\qquad \boldsymbol{x} \geq \boldsymbol{0} \end{aligned} \tag{16.17}$$

$$\begin{aligned} &\text{Minimize } z_2 = (\boldsymbol{b} - \boldsymbol{A}_2\boldsymbol{w})'\boldsymbol{y} \\ &\text{subject to:} \quad \boldsymbol{A}_1'\boldsymbol{y} \geq \boldsymbol{c}_1 \\ &\qquad\qquad \boldsymbol{y} \geq \boldsymbol{0} \end{aligned} \tag{16.18}$$

(In both expressions (16.17) and (16.18) $\boldsymbol{w}$ is specified.) Consider now an optimal linear programming solution to (16.17) and (16.18). Denoting the solution as

$x^*$, $y^*$, we must have, by the duality theorem (Theorem 5.1), that

$$c_1'x^* = (b - A_2w)'y^*$$

Further, from (16.16) and the equivalence of (16.17) and (16.18), the following must be true:

$$\text{Maximum } c_1'x + c_2'w = \underset{\substack{w \geq 0 \\ \text{and integer}}}{\text{Maximum}} \{c_2'w + \text{Minimum } \{(b - A_2w)'y\}\} \tag{16.19}$$

where $y$ is a solution to the convex set (16.18).

A less naive way of solving the mixed integer problem (16.15) is to enumerate the $p$ extreme point solutions $y_k$, $k = 1, \ldots, p$ of (16.18) (which do not depend on the values of $w$) and then solve the all integer problem:

$$\begin{aligned} &\text{Maximize } c_2'w + z_2 \\ &\text{subject to: } z_2 \leq (b - A_2w)'y_k \quad k = 1, \ldots, p \\ &\qquad\qquad w \geq 0 \text{ and integer} \end{aligned} \tag{16.20}$$

The solution $w$ will be the optimal values of the integer variables; the solution for the noninteger variables $x$ may be found by solving (16.17) using the solution to (16.20). In general $p$ will be sufficiently large as to make this solution method unattractive.

Bender's partitioning method consists of generating solutions $y_k$ one at a time, the idea being that relatively few solutions $y_k$ will be needed. A solution $y_k$ is generated by solving (16.18) using a specific solution value $w$. When a solution $y_k$ is generated, the constraint $z_2 \leq (b - A_2w)'y_k$ is used to augment constraint set $Q$ of (16.22) (see below). The optimal solution $w$ of (16.22) is used to solve (16.18) for a new $y_k$, and so on, until an optimal solution to (16.15) is achieved.

In the event that the solution to (16.18) is infeasible, there is either an infinite optimal solution or no feasible solution to (16.15). In the event that the solution to (16.18) is unbounded, there is no feasible solution to (16.15) for the particular solution $w$ used in solving (16.18). In such a case, $w$ for any value of $x$ violates one or more constraints of (16.15), and hence such constraint(s) must be reflected in (16.20). Accordingly, designate the components of the incoming unbounded vector in (16.18) as $v$, (corresponding to $y$), and then add the single constraint

$$(b - A_2w)'v \geq 0 \tag{16.21}$$

to constraint set $Q$ of (16.22). Then continue to solve alternately (16.22) and

(16.20) until an optimal solution is achieved. (The constraint (16.21) may be thought of as a surrogate constraint of sorts).

We now present the algorithm. Let $Q$ be the set of constraints on $\boldsymbol{w}$, initially consisting of only a regularization constraint.

1. Solve the following problem using an integer programming solution method which assures that all variables in the vector $\boldsymbol{w}$ are integer.

$$\begin{aligned} &\text{Maximize } z_1 \leq \boldsymbol{c}_2'\boldsymbol{w} + z_2 \\ &\text{subject to: } \boldsymbol{w}, z_2 \text{ constrained by the constraints of the set } Q. \\ &z_2 \leq M, \quad M \text{ sufficiently large} \end{aligned} \tag{16.22}$$

   If there is no feasible solution, then the original problem has no feasible solution. Otherwise, designate the solution as $\boldsymbol{w}^*$ and proceed to step 2.
2. Using the solution value $\boldsymbol{w}^*$ found in step 1, solve problem (16.18). If the solution is unbounded, adjoin the constraint $(\boldsymbol{b} - A_2\boldsymbol{w})'\boldsymbol{v} \geq 0$ (where $\boldsymbol{v}$ is the vector defined earlier) to the set $Q$ and go to step 1. (An example of the calculation of $\boldsymbol{v}$ is shown in Example 16.5.) Otherwise, designate the solution as $\boldsymbol{y}^*$ and go to step 3.
3. If the value of $z_1$ found in the most recent step 1 exactly equals $\boldsymbol{c}_2'\boldsymbol{w}^* + (\boldsymbol{b} - A_2\boldsymbol{w}^*)'\boldsymbol{y}^*$, stop; the solution $\boldsymbol{x}^*$, $\boldsymbol{w}^*$ ($\boldsymbol{x}^*$ corresponding to $\boldsymbol{y}^*$ can be obtained from the simplex tableau) is optimal. Otherwise, adjoin to the set $Q$ the constraint

$$z_2 \leq (\boldsymbol{b} - A_2\boldsymbol{w})'\boldsymbol{y}^*$$

   and go to step 1.

This completes the statement of the algorithm.

Upper and lower bounds on the objective function value of an optimum are provided. The value of $z_1$ found in step 1 serves as an upper bound, since additional constraints in $Q$ can only decrease successive values of $z_1$. The value obtained for the objective function in step 2 plus $\boldsymbol{c}_2'\boldsymbol{w}$ gives a lower bound on the objective function. An example is now presented.

EXAMPLE 16.5:

$$\begin{aligned} \text{Maximize } z_1 &= 4x_1 + 3x_2 + 5w_1 \\ \text{subject to: } \quad & 2x_1 + 3x_2 + w_1 \leq 12 \\ & 2x_1 + x_2 + 3w_1 \leq 12 \\ & x_1, x_2, w_1 \geq 0 \\ & w_1 \text{ integer} \end{aligned}$$

The following constraint is an arbitrary regularization constraint (for integer variables only):

$$w_1 \leq 20$$

Then, following step 1, solve the following problem using an integer programming algorithm.

$$(16.23) \quad \begin{aligned} \text{Maximize} \quad & z_1 \leq 5w_1 + z_2 \\ \text{subject to:} \quad & w_1 \leq 20 \\ & z_2 \leq M, \ M \text{ large} \\ & w_1 \geq 0 \text{ and integer} \end{aligned}$$

The solution is $w_1 = 20$, $z_2 = M$, and $z_1 = 100 + M$.
Step 2. Solve the problem: Minimize $z_2 = (12 - w_1)y_1 + (12 - 3w_1)y_2$ for $w_1 = 20$, or

$$(16.24) \quad \begin{aligned} & \text{Minimize } z_2 = -8y_1 - 48y_2 \\ & \text{subject to:} \quad 2y_1 + 2y_2 \geq 4 \quad \text{(the slack variable is } y_3\text{)} \\ & \qquad\qquad\quad 3y_1 + \ \ y_2 \geq 3 \quad \text{(the slack variable is } y_4\text{)} \end{aligned}$$

The solution is unbounded* for $y_2$; the vector $\boldsymbol{v}$ has components corresponding as 1 for $y_2$, and 0 for $y_1$ (2 for $y_3$, and 1 for $y_4$, which do not appear in $\boldsymbol{v}$) because as $y_2$ increases 1 unit, $y_1$ does not change (but $y_3$ increases 2 units, and $y_4$ increases 1 unit). (Attempting to introduce $y_2$ into the basis is legitimate in the criss-cross method, although, since the iteration is unbounded, no iteration can be taken—see Chap. 6). Designating $\boldsymbol{v}$ as $\begin{pmatrix} 0 \\ 1 \end{pmatrix}$ we have from (16.21) $(12 - w_1, 12 - 3w_1)\begin{pmatrix} 0 \\ 1 \end{pmatrix} \geq 0$, or

$$3w_1 \leq 12$$

Adjoining this constraint to the set $Q$, we have:

$$\text{Step 1.} \quad \begin{aligned} \text{Maximize} \quad & z_1 \leq 5w_1 + z_2 \\ \text{subject to:} \quad & w_1 \leq 20 \\ & 3w_1 \leq 12 \\ & z_2 \leq M \\ & w_1 \geq 0 \text{ and integer} \end{aligned}$$

The solution is $w_1 = 4$, $z_2 = M$, and $z_1 = 20 + M$.

* It is also unbounded for $y_1$.

Step 2. Making use of the solution $w_1 = 4$, the problem is to minimize $z_2 = 8y_1$, subject to the constraints of (16.24). A finite optimal solution is $y_1 = 0$, $y_2 = 3$, with an objective function of zero ($y_1 = 0$, $y_2$ any non-negative real number is also optimal).
Step 3. The bounds of the optimal solution are not equal; viz., $20 + M \neq 20$. Thus, the optimal objective function is between 20 and $20 + M$. Hence, adjoin to the set $Q$ the constraint for the solution $y_1 = 0$, $y_2 = 3$.

$$z_2 \leq \begin{pmatrix} 12 - w_1 \\ 12 - 3w_1 \end{pmatrix}' \begin{pmatrix} 0 \\ 3 \end{pmatrix} \quad \text{or} \quad z_2 \leq 36 - 9w_1$$

Step 1.

(16.25)

$$\begin{aligned} &\text{Maximize } z_1 \leq 5w_1 + z_2 \\ &\text{subject to:} \quad w_1 \leq 20 \\ &\qquad 3w_1 \leq 12 \\ &\qquad z_2 \leq M \\ &\qquad z_2 \leq 36 - 9w_1 \\ &\qquad w_1 \geq 0 \text{ and integer} \end{aligned}$$

The optimal solution is $w_1 = 0$, $z_2 = 36$, and $z_1 = 36$.
Step 2. Making use of the solution $w_1 = 0$, the problem is to minimize $z_2 = 12y_1 + 12y_2$, subject to the constraints of (16.24). An optimal solution is $y_1 = 2$, $y_2 = 0$, with objective function value $z_2 = 24$.
Step 3. The bound calculations yield $36 \neq 0 + 24$. Hence, this solution is not optimal. The bounds are now 24 and 36, and the new constraint is

$$z_2 \leq \begin{pmatrix} 12 - w_1 \\ 12 - 3w_1 \end{pmatrix}' \begin{pmatrix} 2 \\ 0 \end{pmatrix} \quad \text{or} \quad z_2 \leq 24 - 2w_1$$

Step 1. Adjoining the new constraint to (16.25), the solution is $w_1 = 2$, $z_2 = 18$, and $z_1 = 28$.
Step 2. Making use of this solution, the objective function is minimize $z_2 = 10y_1 + 6y_2$, subject to the constraints of (16.24). The solution is $y_1 = \frac{1}{2}$, $y_2 = \frac{3}{2}$, and the objective function value is 14.
Step 3. The bound calculations yield $28 \neq 10 + 14$. Hence, this solution is not optimal. The bounds are now 24 and 28. The new constraint using $y_1 = \frac{1}{2}$, $y_2 = \frac{3}{2}$ is

$$z_2 \leq 24 - 5w_1$$

Step 1. Adjoining the new constraint plus the one added in the last step 1, to (16.25), there are four alternate solutions.

a) $w_1 = 0$, $z_2 = 24$, and $z_1 = 24$
b) $w_1 = 1$, $z_2 = 19$, and $z_1 = 24$

c) $w_1 = 2$, $z_2 = 14$, and $z_1 = 24$
d) $w_1 = 3$, $z_2 = 9$, and $z_1 = 24$

Step 2. For solution a, we have $y_1 = 2$, and the objective function $z_2 = 24$. For solution b, we have $y_1 = \frac{1}{2}$, $y_2 = \frac{3}{2}$, and the objective function $z_2 = 19$. For solution c, we have $y_1 = \frac{1}{2}$, $y_2 = \frac{3}{2}$, and the objective function $z_2 = 14$. For solution d, we have $y_1 = 0$, $y_2 = 3$, and the objective function $z_2 = 9$.
Step 3. For each of the four solutions, the bounds are both 24; hence the solutions are optimal. They are: (a) $w_1 = 0$, $x_1 = 6$, $x_2 = 0$; (b) $w_1 = 1$, $x_1 = 4$, $x_2 = 1$; (c) $w_1 = 2$, $x_1 = 2$, $x_2 = 2$; and (d) $w_1 = 3$, $x_1 = 0$, $x_2 = 3$. Coincidentally, all optimal values of the noninteger variables turn out to be integers.

Benders' partitioning method for solving mixed-integer programming problems appears to be a viable means of using an all-integer method to solve a mixed-integer problem. Its performance is, of course, highly dependent upon the all-integer algorithm utilized. Benders reported encouraging computational experience, and, more recently, Geoffrion and Marsten [103] report favorable computational experience.

**Selected Supplemental References**

Section 16.2
[4], [77], [92], [102], [122], [131], [165], [172], [184], [191], [196], [224], [225], [229], [250], [253], [266], [295], [310].
Section 16.3
[5], [101], [104], [116], [120].
Section 16.4
[40], [41], [87], [121], [174], [208].
Section 16.5
[25].
General
[44], [45], [94], [113], [119], [137], [228], [251], [270].

## 16.6 PROBLEMS

Solve problems 1 through 5 using the various implicit enumeration schemes introduced in sections 16.2 and 16.3.

**1.**

$$\text{Minimize } z = 3x_1 + 7x_2 + 9x_3 + 2x_4 + 6x_5$$

$$\begin{aligned} \text{subject to:} \quad & 3x_1 + x_2 + 4x_3 + 5x_4 + 8x_5 \geq 10 \\ & 5x_1 + 2x_2 + 9x_3 + x_4 + 5x_5 \geq 12 \\ & 3x_1 + x_2 + 3x_3 + 2x_4 + 2x_5 \geq 2 \\ & x_1, x_2, x_3, x_4, x_5 = 0, 1 \end{aligned}$$

**2.** Minimize $z = 4x_1 + 6x_2 + 2x_3 + 6x_4 + 7x_5 + 7x_6$

subject to:
$$
\begin{aligned}
6x_1 - 4x_2 \qquad - 6x_4 \qquad - 2x_6 &\geq 0\\
3x_1 + x_2 + 7x_3 + 7x_4 + 8x_5 + 8x_6 &\geq 17\\
-9x_1 + 4x_2 + 5x_3 + 2x_4 + 8x_5 + 8x_6 &\geq 15\\
x_1, x_2, x_3, x_4, x_5 &= 0, 1
\end{aligned}
$$

**3.** Minimize $z = 8x_1 + 4x_2 + 8x_3 + 9x_4 + 9x_5 + 9x_6 + 8x_7$

subject to:
$$
\begin{aligned}
7x_1 + 3x_2 - 5x_3 - 5x_4 - 2x_5 \qquad + 5x_7 &\geq 5\\
-5x_1 + x_2 + x_3 - 9x_4 - 4x_5 - 2x_6 + 9x_7 &\geq 1\\
-7x_1 + 6x_2 + 7x_3 + 3x_4 - 9x_5 - 4x_6 - 2x_7 &\geq 2\\
9x_1 + x_2 - 3x_3 - 3x_4 - x_5 - 8x_6 - 8x_7 &\geq 5\\
x_1, \ldots, x_7 &= 0, 1
\end{aligned}
$$

(This problem has no feasible solution.)

**4.** Solve problem 3 omitting the last constraint.

**5.** Solve problem 3 omitting the second constraint.

**6.** Complete the calculations in Example 16.3.

**7.** Solve the following problems using the methods for aggregating constraints presented in section 16.4:

a. problem 1.
b. problem 2.
c. problem 3.
d. problem 4.
e. problem 5.
f. problem 2 of Chapter 13.
g. problem 7 of Chapter 13.
h. problem 8 of Chapter 13.
i. problem 9 of Chapter 13.
j. problem 10 of Chapter 13.
k. problem 25 of Chapter 13.
l. problem 1 of Chapter 14.
m. problem 13 of Chapter 14.
n. problem 14 of Chapter 14.
o. problem 15 of Chapter 14.

For some of the problems it may be necessary to use the substitution of section 16.1.

**8.** Solve the following problems using Benders' partitioning scheme:

a. problem 3 of Chapter 15.
b. problem 4 of Chapter 15.
c. problem 5 of Chapter 15.
d. problem 6 of Chapter 15.
e. problem 7 of Chapter 15.

**9.** Consider the following problem:

$$\text{Minimize } z = \sum_{j=1}^{20} x_j$$

$$\text{subject to: } \sum_{j=1}^{20} x_j \geq 6.5$$

$$x_j = 0, 1 \qquad j = 1, \ldots, 20$$

The problem is trivial. Explain the difference in solving the problem by implicit enumeration with and without surrogate constraints.

**10.** Using an implicit enumeration algorithm, find both feasible solutions to the problem of Example 16.1.

**11.** In the discussion of Glover's multiphase dual method, it was stated that the optimal solution of the following linear programming problem is to be found:

$$\begin{aligned} \text{Minimize } z &= \sum_{j=1}^{n} c_j x_j \\ \text{subject to: } & \sum_{j=1}^{n} a_j x_j \le b \\ & 0 \le x_j \le 1 \qquad j = 1, \ldots, n \\ & c_j \ge 0 \qquad j = 1, \ldots, n \\ & b < 0 \end{aligned}$$

Show how such a solution can be found in a very simple manner. (Hint: Use the ordering in Balas' filter method. Then work only with variables $x_j$ for which $a_j < 0$.)

**12.** Solve the problems of Examples 16.3 and 16.4 using the other permutations of the constraints.

**13.** Given a problem in the form (16.1), assuming $c_j \ge 0$, suppose that the optimal noninteger solution is first found. H. Salkin [250] has found that using as a starting solution the optimal noninteger solution, rounding all fractional variables of that solution to one, is an effective means of accelerating the implicit enumeration process. Apply Salkin's scheme to the following problems using the generalized additive algorithm (taking care to alter the logic as indicated in the appendix) and compare the results with the original solutions:

a. Example 16.1.
b. Example 16.2.
c. Problem 1.
d. Problem 2.
e. Problem 3.
f. Problem 4.
g. Problem 5.

**14.** Kendall and Zionts [174] also propose the following method:

a. Enumerate a few values of $g_i(x) = b_i - \sum_{j=1}^{n} a_{ij} x_j$ for $0 \le x_j \le u_j$

beginning with the solution $x_j = u_j$ if $a_{ij} < 0$ and $x_j = 0$, otherwise, so that $g_i(x)$ strictly decreases and so that no values in the sequence are omitted. The first few values are easy to enumerate because we increment or decrement the variable whose $|a_{ij}|$ is minimum, and so on using a lexicographical ordering scheme.

b. Choose $\lambda_1 > \text{Min}\,\{g_2(x)\}$ (over those enumerated), $\lambda_2 > \text{Min}\,\{g_1(x)\}$ (over those enumerated)

$\lambda_1 \ne \text{any } g_2(x)$, $\lambda_2 \ne \text{any } g_1(x)$

$$\lambda_1 \text{ and } \lambda_2 \text{ are relatively prime, and } \lambda_1 > \frac{b_2 - \sum_{j=1}^{n} a_{2j}^{-} u_j}{2}$$

and

$$\lambda_2 > \frac{b_1 - \sum_{j=1}^{n} a_{1j}^{-} u_j}{2}$$

Use this method to find the aggregate constraint indicated on page 458 for the problem of Example 16.4.

## APPENDIX: A FLOW CHART FOR THE GENERALIZED BALAS ADDITIVE ALGORITHM USING BOUNDS ON VARIABLES*

### Dictionary of Terms

*ISW*—0 if no implied variable has been detected, 1 if one or more implied variables have been detected.

$\sum a_{ij}^{+}$—Sum of positive coefficients of *free variables* in row $i$.

$\sum a_{ij}^{-}$—Sum of negative coefficients of *free variables* in row $i$.

$a_{ip}$—The most *positive* element for a free variable in row $i$; otherwise zero.

$a_{is}$—The most *negative* element for a free variable in row $i$; otherwise zero.

$b_i' = b_i - \sum_{k=1}^{p} a_{ij_k}$

* This flow chart is taken by permission from the *Naval Research Logistics Quarterly*, 19, 165–181, 1972.

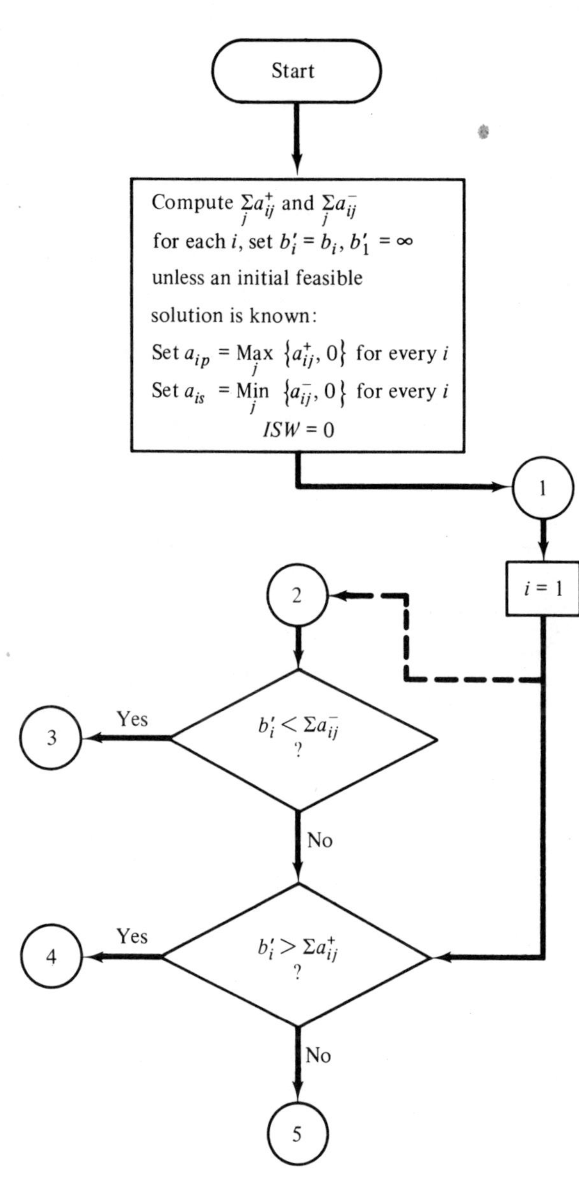


It is assumed that all $c_j \geqslant 0$. (To relax this assumption, it is only necessary to alter one branch: use the dotted line after ① rather than the solid line.)

Initialize and compute partial sums. The starting partial solution is the empty set. The first constraint is the objective function constraint.

Is there no feasible continuation?

For equalities, is there no feasible continuation? For inequalities, is the constraint conditionally nonbinding?

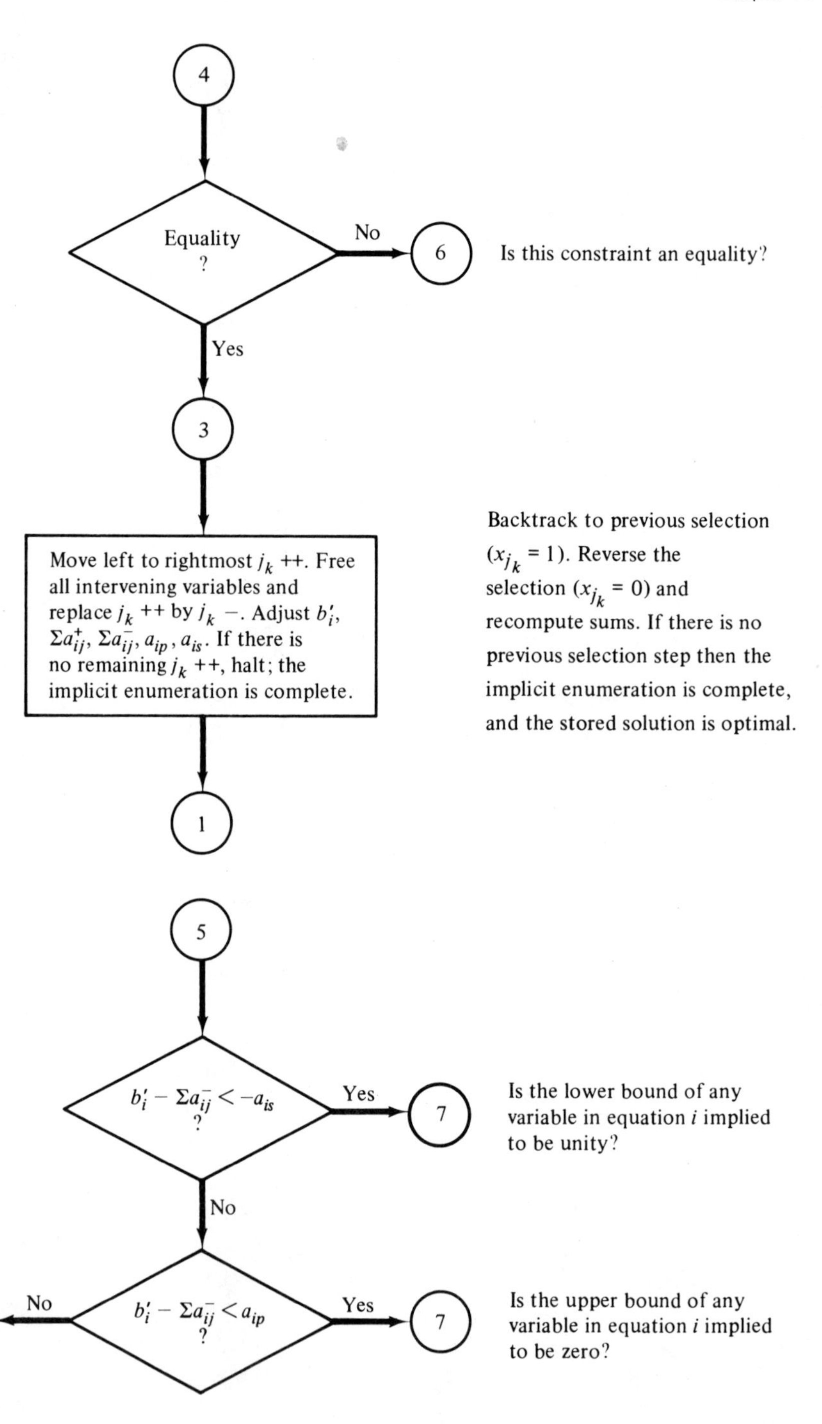


Is this constraint an equality?

Backtrack to previous selection ($x_{j_k} = 1$). Reverse the selection ($x_{j_k} = 0$) and recompute sums. If there is no previous selection step then the implicit enumeration is complete, and the stored solution is optimal.

Is the lower bound of any variable in equation $i$ implied to be unity?

Is the upper bound of any variable in equation $i$ implied to be zero?

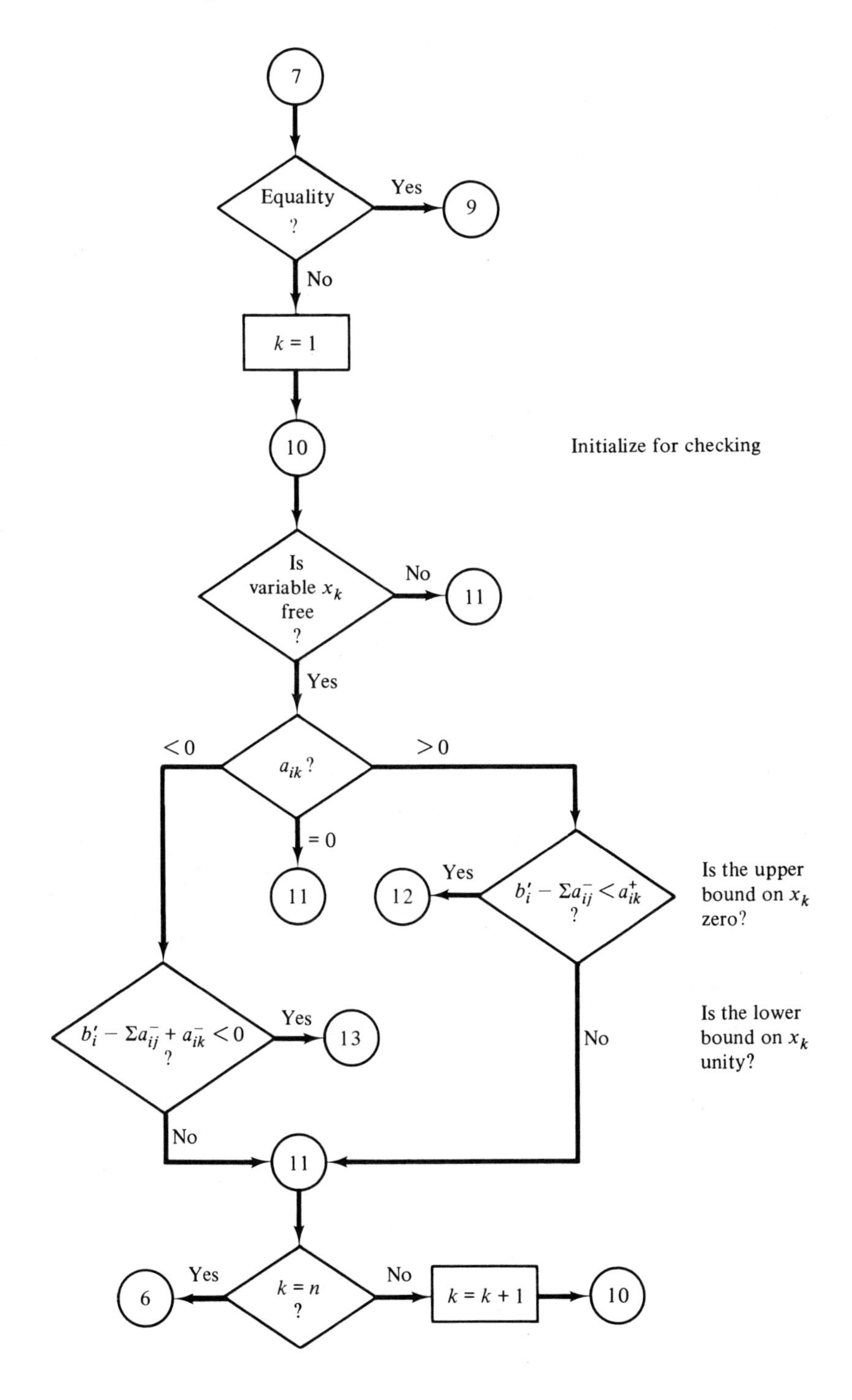
7
Equality
?
Yes
9
No
$k = 1$
10
Initialize for checking
Is
variable $x_k$
free
?
No
11
Yes
$<0$
$a_{ik}$?
$>0$
$= 0$
11
Yes
12
$b_i' - \Sigma a_{ij}^- < a_{ik}^+$
?
Is the upper
bound on $x_k$
zero?
$b_i' - \Sigma a_{ij}^- + a_{ik}^- < 0$
?
Yes
13
No
Is the lower
bound on $x_k$
unity?
No
11
Yes
6
$k = n$
?
No
$k = k + 1$
10

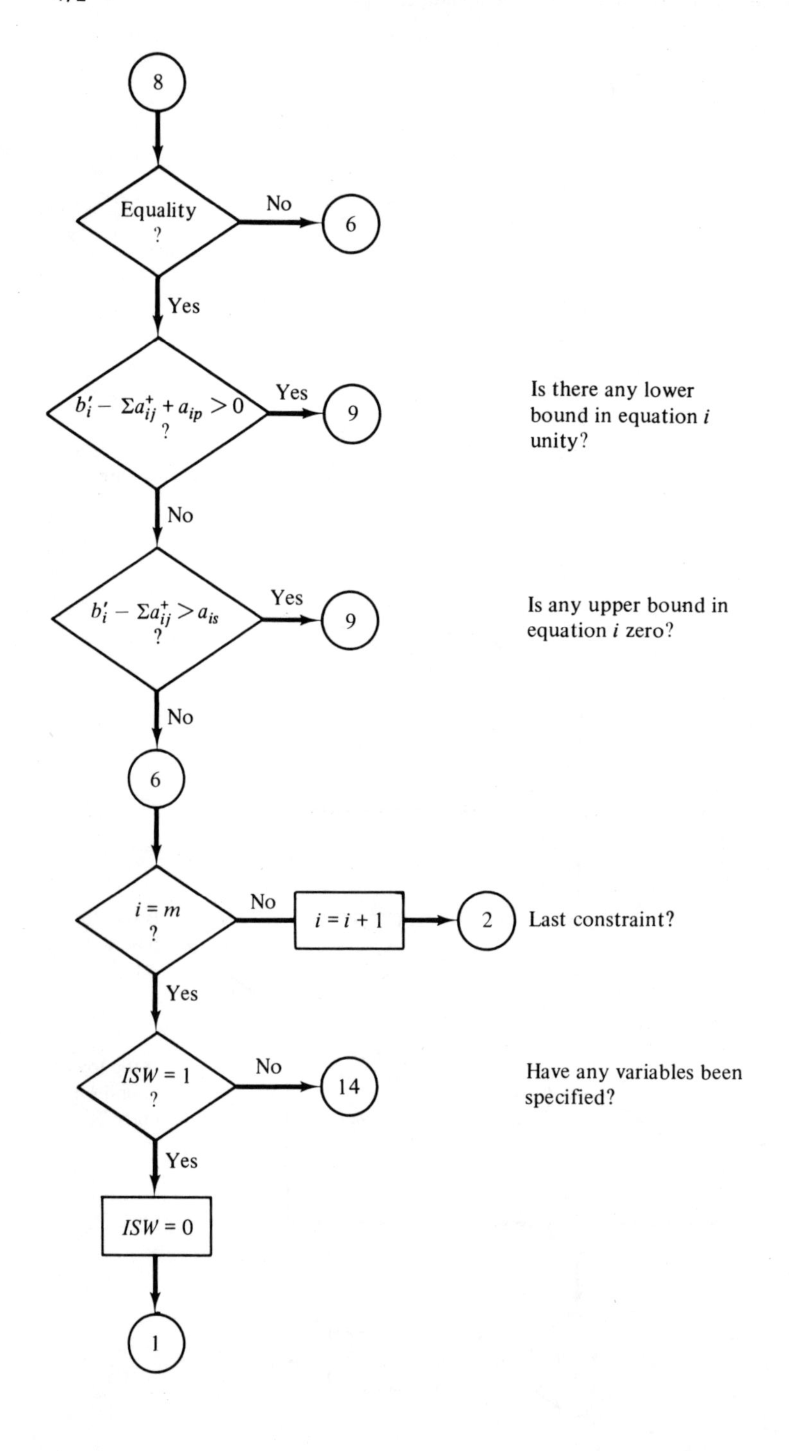
8
Equality ?
No
6
Yes
$b_i' - \Sigma a_{ij}^+ + a_{ip} > 0$ ?
Yes
9
Is there any lower bound in equation $i$ unity?
No
$b_i' - \Sigma a_{ij}^+ > a_{is}$ ?
Yes
9
Is any upper bound in equation $i$ zero?
No
6
$i = m$ ?
No
$i = i + 1$
2
Last constraint?
Yes
$ISW = 1$ ?
No
14
Have any variables been specified?
Yes
$ISW = 0$
1

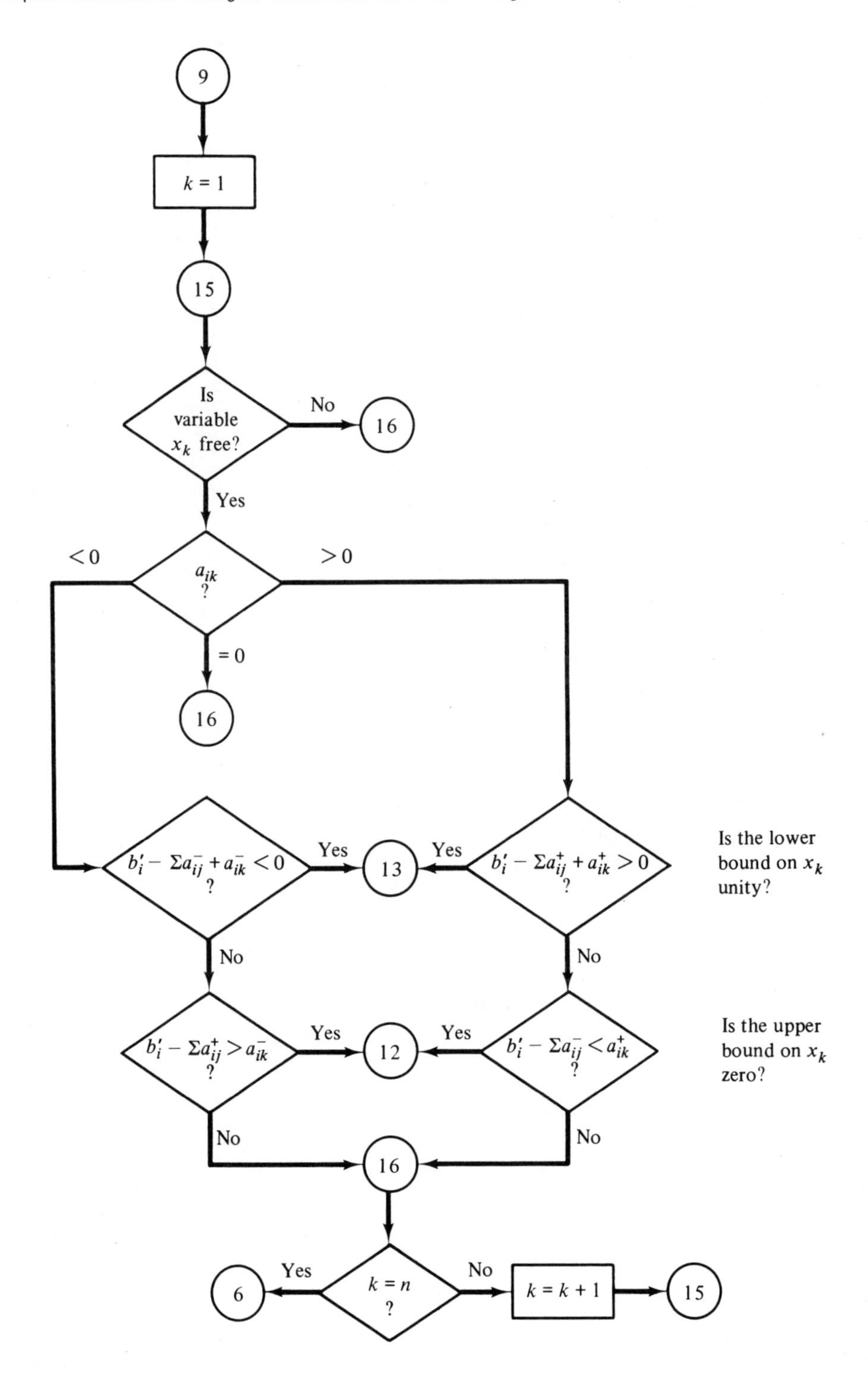
9
$k = 1$
15
Is variable $x_k$ free?
No
16
Yes
$a_{ik}$ ?
$<0$
$>0$
$=0$
16
$b_i' - \Sigma a_{ij}^- + a_{ik}^- < 0$ ?
Yes
13
Yes
$b_i' - \Sigma a_{ij}^+ + a_{ik}^+ > 0$ ?
Is the lower bound on $x_k$ unity?
No
No
$b_i' - \Sigma a_{ij}^+ > a_{ik}^-$ ?
Yes
12
Yes
$b_i' - \Sigma a_{ij}^- < a_{ik}^+$ ?
Is the upper bound on $x_k$ zero?
No
No
16
Yes
6
$k = n$ ?
No
$k = k + 1$
15

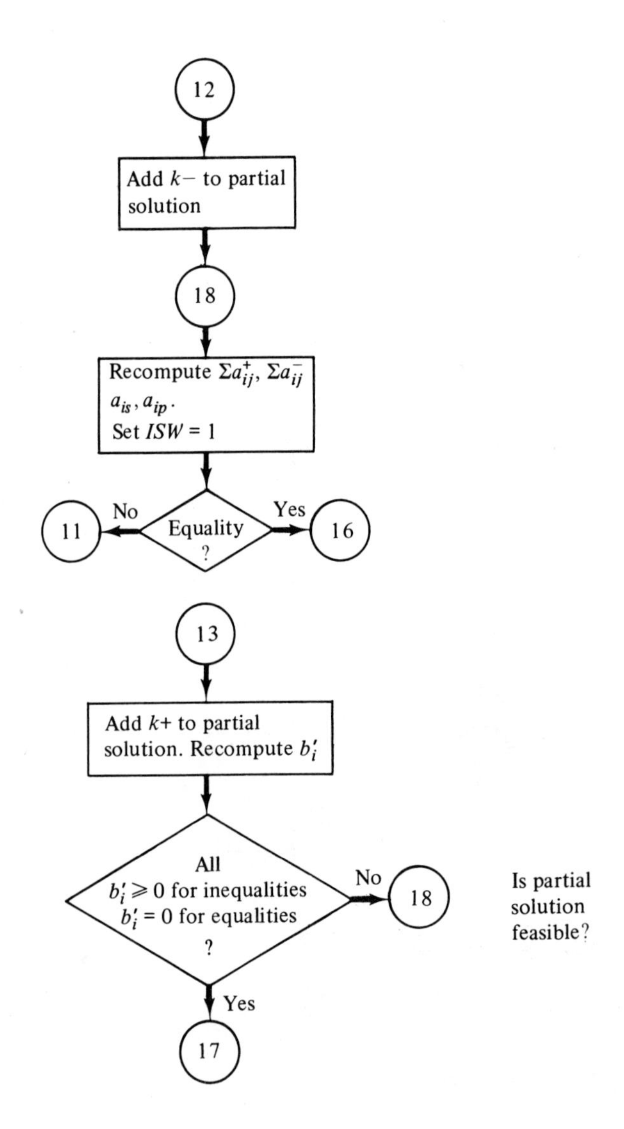


Is partial solution feasible?

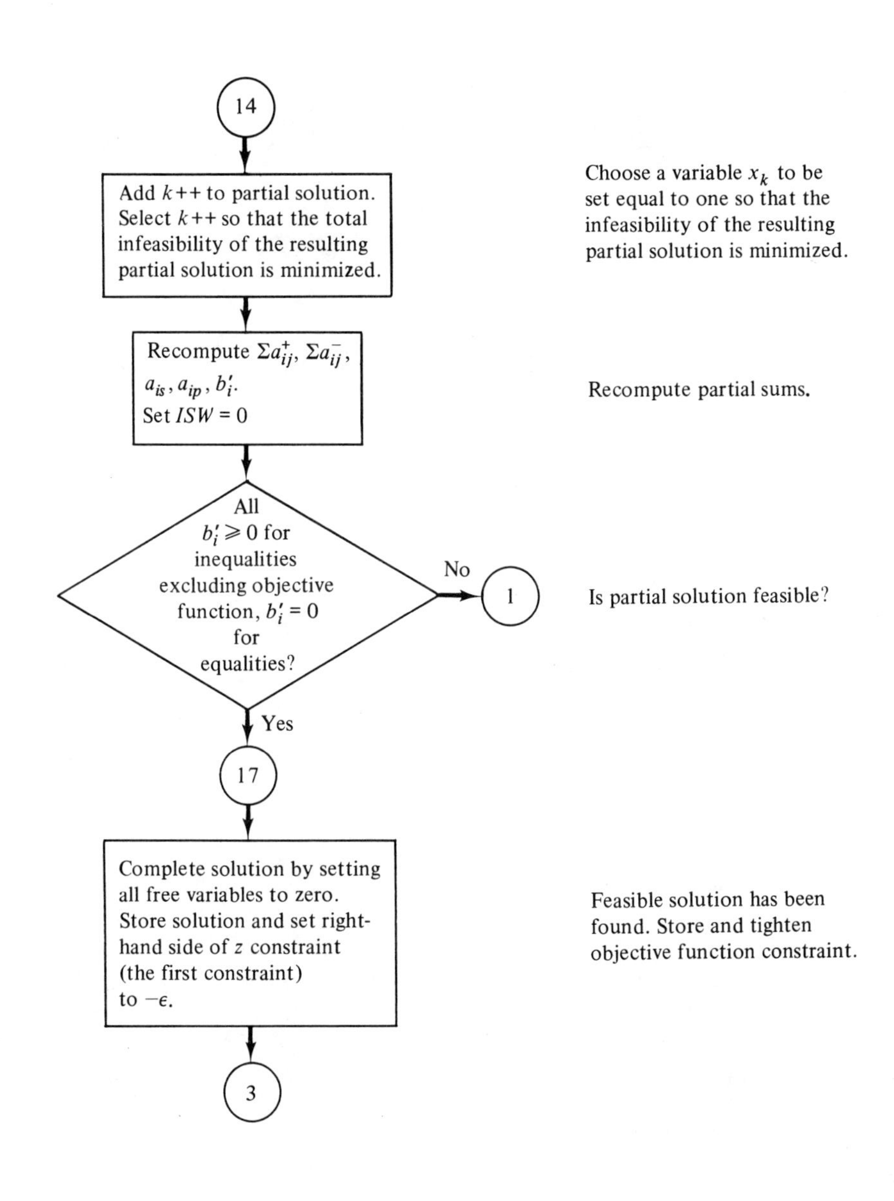
14
Add $k$++ to partial solution. Select $k$++ so that the total infeasibility of the resulting partial solution is minimized.
Choose a variable $x_k$ to be set equal to one so that the infeasibility of the resulting partial solution is minimized.
Recompute $\Sigma a_{ij}^+$, $\Sigma a_{ij}^-$, $a_{is}$, $a_{ip}$, $b_i'$. Set $ISW = 0$
Recompute partial sums.
All $b_i' \geq 0$ for inequalities excluding objective function, $b_i' = 0$ for equalities?
No
1
Is partial solution feasible?
Yes
17
Complete solution by setting all free variables to zero. Store solution and set right-hand side of $z$ constraint (the first constraint) to $-\epsilon$.
Feasible solution has been found. Store and tighten objective function constraint.
3

# 17

# Some Practical Aspects of Solving Integer Programming Problems

***R. E. D. Woolsey****

In this chapter we consider some of the practical aspects of solving integer programming problems, including the importance of a feasible integer solution. We also discuss certain formulation devices and tricks that appear useful in solving integer programming problems.

## 17.1 INTRODUCTION

In most fields there is a gap between the theory and the practice. The gap in integer programming is especially wide. Algorithms have been proven to converge, but for certain algorithms used on specific problems, the convergence is so slow that, for practical purposes, the algorithms effectively do not converge. For example, there exists a class of three constraint, three variable problems which can be made to require an arbitrarily large number of Gomory cut constraints using Gomory's all-integer method. (See, for example, Thompson [278], p. 180.) There also exist classes of zero-one implicit enumeration problems which are solved very rapidly using implicit enumeration methods; yet, when minor modifications are made in the formulation, the solution times become excessive.

Part of the problem in integer programming stems from the inherent differ-

* Colorado School of Mines, Golden, Colorado.

ence between integer programming and linear programming. Linear programming problems are concerned with maximizing or minimizing a linear form over a convex set, whereas integer programming problems are concerned with maximizing or minimizing a linear form over what is most definitely a nonconvex set (except in trivial cases). At least at the present level of technology, integer programming problems are many orders of magnitude more difficult than linear programming problems. As researchers and technicians improve the computational feasibility of integer programming solution methods, the balance will change. Second, until recently, many operations research texts introduced integer programming methods along with linear programming, queuing theory, simulation, and other topics, leaving the innocent reader to conclude that integer programming methods were as reliable and dependable as the others. Thus, many overzealous operations researchers found themselves in the unfortunate position of spending considerable computer effort on the solution of integer programming problems, not realizing the low probability that the solution process would converge to an optimum within a reasonable amount of time. Fortunately, the textbook writers are becoming more candid with their readers. To quote Wagner [292] (p. 447), ". . . so far these (integer programming) models have been of only limited practical importance."

Many of the early algorithms employ dual methods. (Such algorithms are categorized as all-integer, primal dual feasible or dual feasible, cut methods.) Unfortunately, however, a dual algorithm does not produce a feasible solution until the problem is completely solved. Thus, regardless of how much time has been spent using a dual algorithm, unless an optimal solution has been obtained, almost nothing has been achieved. Assuming the problem is a maximizing problem, a dual feasible solution, though not integer, does give an *upper* bound on the objective function value of an optimal solution. Thus, after spending considerable effort and computer time trying to solve a problem, operations researchers were left with only a bound on the objective function to show for their efforts.

Primal integer solution methods, by contrast, have the advantage of generating a feasible solution relatively early in the process. However, once stationary iterations are encountered, the solution progress is generally very slow.

Implicit enumeration approaches tend to work well for zero-one problems of reasonable size. It was proposed that arbitrary all-integer problems could be formulated and solved as zero-one problems using the transformation given in Chapter 16, namely

$$y = x_1 + 2x_2 + \cdots + 2^{q-1}x_q$$

Implicit enumeration schemes work well provided that the total number of variables is not larger than about 130. Therefore, the above transformation may be employed in an all-integer problem with a relatively small number of integer variables which are not zero or one. Nonetheless, it is possible to have problems which cause implicit enumerative algorithms to enumerate *totally* (see, for example, problem 9 of Chap. 16), although surrogate constraints may help considerably.

An advantage of both primal algorithms and implicit enumeration (and branch and bound) algorithms is that a feasible solution tends to be discovered early, and is improved upon by subsequent solutions. This consideration is very important in practice.

The above discussion indicates some of the problems associated with the different solution methods. Problem-solvers, of course, must come up with answers, regardless of deficiencies in the solution methods—*with or without the use of integer programming.* (see [299].) Therefore, with this in mind, we shall now consider three ways in which integer programming methods can be of greater value to the problem-solver:

1. Formulation techniques.
2. Convergence acceleration techniques.
3. Heuristic techniques for getting a "good" starting solution.

These are arranged in apparent order of importance based on the author's experience in actual use.

## 17.2 FORMULATION TECHNIQUES

It has been our experience that the formulation of an integer programming problem may greatly affect problem solution time. We shall now consider some formulation techniques.

### Dual Algorithms

Glenn Martin observed in 1962 that the performance of the Gomory fractional algorithm was spectacularly improved when redundant constraints were added to fixed-cost problems. These redundant constraints were added in an ad hoc manner and the problem solution time decreased substantially.

The performance of Gomory's fractional algorithm is also dependent on the constraint ordering, because of the choice of the lexicographically minimum column (suitably defined) in selecting the variable to enter the basis. However, no well-defined rules for ordering seem to be consistently efficient, even though permuting constraints may help accelerate the solution in some circumstances.

### Zero-One Implicit Enumeration Methods

The order of rows is also important for implicit enumeration methods, but for a somewhat different reason than for dual methods. Most computer programs are written to scan constraints from top to bottom, and to scan variables from left to right. Thus, we should order the constraints and variables so that those most likely to be decisive are scanned first.

The first classification of constraints likely to be decisive, based on experience, is equality constraints. Particularly decisive are constraints of the following form:

$$\sum_{j=1}^{n} x_j = k \text{ (where } k \ll n)$$

($\ll$ means much less than.) They should be listed first, followed by the remaining equality constraints.

The next most decisive classification of constraints is inequalities which enforce logical conditions. We have already given examples of such problems in Chapter 12, but we include another example here. Suppose, in a capital budgeting problem, we have two projects. Corresponding to the projects (1 and 2) are variables $x_1$ and $x_2$, respectively. Each variable is one if the respective project is adopted and zero otherwise. Supposing that project 1 has a present value of \$100, project 2 has a present value of \$150, and both projects together have a present value of \$500, we have the following as part of the objective function (following Watters [294] and others):

$$\text{Maximize } z = 100x_1 + 150x_2 + 250x_{12}$$

where $x_{12}$ is a zero-one variable which is one only if both $x_1$ and $x_2$ are one. The following constraints enforce these conditions:

$$\begin{aligned} -x_1 - x_2 + 2x_{12} &\le 0 \quad (\text{or } x_{12} \le .5x_1 + .5x_2) \\ x_1 + x_2 - x_{12} &\le 1 \quad (\text{or } x_{12} \ge x_1 + x_2 - 1) \end{aligned}$$

The remaining inequality constraints should be ordered in decreasing order of infeasibility (the amount by which a constraint is violated when all zero-one variables are given integer values so that the slack of the constraint has the most negative value possible.) Thus, given the following constraints:

| | | INFEASIBILITY |
|---|---|---|
| 1. | $x_1 + 2x_2 + 6x_3 + 5x_4 \le 13$ | 1 |
| 2. | $x_1 + 4x_2 + 5x_3 + x_4 \le 11$ | 0 |
| 3. | $5x_1 + 6x_2 + x_3 + 4x_4 \le 12$ | 4 |
| 4. | $2x_1 + x_2 + 2x_3 + 2x_4 \le 2$ | 5 |

we would arrange them in the following order (after the equality and logical constraints): 4, 3, 1, 2. That this argument is heuristic at best may be seen by considering the following constraint:

$$100{,}000x_1 + \sum_{j=2}^{100} x_j \le 100$$

Although this constraint very likely ranks first among a set of integer inequality constraints using this priority, it implies only that $x_1$ is zero. Thus, a strategy that may be used in addition to a sensible ordering of constraints is computer programmed logic that ignores constraints that cannot be decisive. (Such logic is built into the generalized implicit enumeration flow chart of the Appendix to Chap. 16.)

### Branch and Bound Methods

The ordering of variables in branch and bound methods can be particularly important, because many of the methods branch on the "next" free variable that is required to be integer. Therefore, zero-one variables should appear, from left to right, in order of increasing costs for a minimizing problem or in order of decreasing profits for a maximizing problem. The positioning of variables is undoubtedly important for algorithms that branch on the "next" integer, but it should not matter at all for algorithms which employ penalties for the selection of the branching variable (as described in Chap. 15). That it is important for other branch and bound algorithms has been empirically confirmed on BBMIP, a branch-and-bound computer code based on the method of Land and Doig [187], as well as GEIPP [301].

The ordering of constraints should be the same as recommended for implicit enumeration methods, for the same reason just given for the ordering of variables. As might be expected, methods which employ penalties will generally be unaffected by the ordering.

## 17.3 ALGORITHMIC CHANGES

### The Mesa Procedure—Restricting the Objective Function Value

The Mesa procedure derives its name from the New Mexican topographical feature having the same name. Many cutting-plane integer programming solutions have graphs similar to that given in Figure 17.1. Much of the progress towards an optimal solution, as measured by the objective function value progress, occurs in the early iterations. Then the solution reaches a plateau where the objective function increases very slowly or remains constant for many iterations before changing.

Use of Mesa constraints has accelerated many integer programming solutions. Assume the Gomory fractional method is being used to solve a maximizing problem. Once the noninteger optimal solution has been found, the cut constraints *reduce* the objective function, eventually producing an optimum. A Mesa constraint is an inequality constraint on the objective function; in

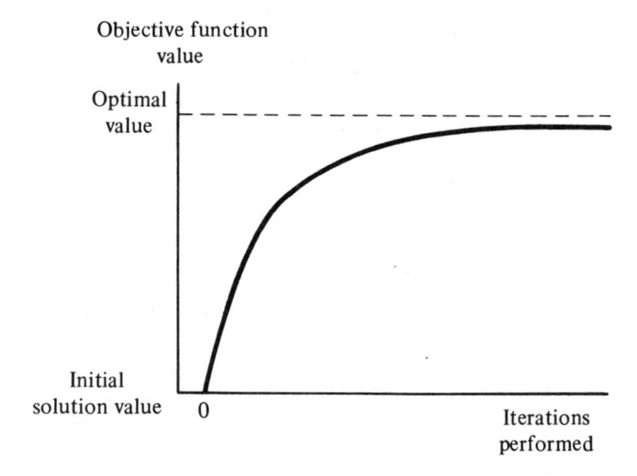


**Figure 17.1**
A Typical Graph of the Objective Function of an Integer Programming Problem Solved by a Cutting-Plane Method

this case

$$\sum_j c_j x_j \leq \bar{z}$$

where $\bar{z}$ is some value of the objective function previously considered. Such constraints are added and the problem restarted at a number of strategic points to accelerate the solution.

An important question is when and for what values of $\bar{z}$ should a Mesa constraint be added. If, upon reaching a plateau having an objective function value $z_1$, we adjoin a Mesa constraint

$$\sum_j c_j x_j \leq z_1$$

and then restart the solution, we shall, of course, within a few iterations (one, if we pivot on the Mesa constraint first), reach a solution having an objective function value of $z_1$. However, the solution will be massively dual degenerate because, except for the reduced cost of the slack variable of the Mesa constraint, all of the reduced costs will be zero. Thus, the solution is returned to its old plateau. In order to avoid this return, it is desirable to choose an objective function value slightly larger than the value of the solution on which we stopped. We may still reach a massively dual degenerate solution, but part of the region satisfying this constraint will be infeasible. Intuitively, at least, there will be a better chance to progress toward an optimal integer solution. The second consideration is when such constraints should be added. Experience indicates (perhaps for the same reasons) that it is best to add the Mesa constraints before the solution bogs down on a plateau. As with other devices presented earlier, experience with Mesa constraints has not been uniformly successful, although on average it has been quite good. Finally, Mesa constraints do not seem to help much on the solution to problems which do not exhibit a plateau effect.

An illustrative example of the success of Mesa is its application to problem 5 of Haldi's fixed-charge problems [146], as included in problem 1 at the end of the chapter. By reference to Table 17.1, we see that, using Gomory's fractional algorithm, the problem requires 3300 iterations to achieve an objective function value of 80. (The optimal objective function value is 76.) It appears that the number of iterations and time required per unit improvement in the objective function grow exponentially. Also included in Table 17.1 is the effect of applying a Mesa constraint after each of the five iteration numbers given. The effect of the Mesa constraint is significant for this problem. Although generally effective, Mesa constraints have been shown to have different levels of effectiveness when used with different methods.

**Table 17.1**

Typical Plateau Behavior and Mesa Sensitivity*

| GOMORY'S FRACTIONAL ALGORITHM | | | EFFECT OF ADDING MESA CONSTRAINT AFTER $k$ ITERATIONS | | | |
|---|---|---|---|---|---|---|
| *Number of iterations* $k$ | *Objective function value* $z(k)$ | *Time thru* $k$ *iter. (secs.)* | *Iter. from Mesa* | *Time from Mesa (secs.)* | *Total iter. with Mesa* | *Total time with Mesa (secs.)* |
| 78 | 84 | 0.493 | 481 | 4.625 | 559 | 5.118 |
| 89 | 83 | 0.521 | 343 | 3.490 | 432 | 4.011 |
| 100 | 82 | 0.580 | 327 | 3.414 | 427 | 3.994 |
| 700 | 81 | 6.272 | 298 | 3.101 | 998 | 9.373 |
| 3300 | 80 | 30.955 | 260 | 2.291 | 3560 | 33.174 |

* Times and iterations are to an optimum. (The problem—Haldi's problem 5—was run on a CDC-3600 computer using Gomory's fractional algorithm.)

### The Bang-for-the-Buck Procedure—Minimizing the Cost per unit of Infeasibility Reduction

The bang-for-the-buck procedure was given in Chap. 16 (with a less exciting name) as a suggested way of accelerating zero-one implicit enumeration methods. We repeat it here for emphasis, and indicate how successful it has been in one application. Recall that in the Balas additive algorithm, when a trial variable is selected to be set to one, the variable that minimizes total infeasibility (defined as the sum of the absolute amounts by which each constraint is violated) is selected. The bang-for-the-buck procedure chooses instead, where total infeasibility may be reduced (an example in which total infeasibility may *not* be reduced by setting some variable to one is given as problem 2 at the end of the chapter), the variable whose cost per unit of infeasibility reduction is minimum.

When used together with the strategy for ordering rows given in section 17.2, this method has led to spectacular differences in running time. For example, a 10 by 10 lock box problem of the Denver and Rio Grande Railroad, consisting of 20 constraints and 110 variables, required 4 hours 38 minutes on a PDP-10 computer when solved by the Balas additive algorithm. (The framework of the lock box problem is given in the problems, and its formulation is asked there.) After reordering the rows and altering the program to use the above algorithm for selecting variables, the problem solution was found in 3.7 seconds on the same computer. A 20 by 20 lock box problem (40 rows, 420 variables) was solved in 16.1 seconds, and a 50 by 50 lock box problem (100 rows, 2550 variables) required 23.98 seconds, all using the same altered program.

The author has had fantastic luck with this procedure on this problem. Zionts has applied the same method to a television spot commercial integer programming problem with a general, but not spectacular, improvement. Because of this method's increase in computational requirements in choosing the variable to set to one, it may be helpful, as suggested in Chapter 16, to use the improved choice rule together with the original Balas rule. For example, the improved choice rule may be utilized in the early stages of solution as well as when the Balas rule does not lead to many implied values for variables.

## 17.4 HEURISTIC METHODS FOR "GOOD" STARTING SOLUTIONS

In this section we consider methods for getting "good" starting solutions, which may then be input to some other algorithm for further solution, or used as possible near-optimal solutions for further evaluation.

### The Method of Lawler and Bell [191]

First we consider the method of Lawler and Bell which, strictly speaking, is a method for finding an optimal solution. It is an implicit enumeration method for problems in which all variables are zero or one. The problems may be assumed to have the same form as (16.1). Consider the explicit enumeration scheme of using a binary number, each bit corresponding to one of the $n$ zero-one variables. Then, beginning with zero, each of the $2^n$ possible solutions is examined. The strength of Lawler and Bell's procedure is that it gives three conditions under which sequences of solutions in the enumeration scheme may be skipped. These conditions are easily tested, and usually many of the possible solutions are eliminated.

This method has not usually been considered a useful integer programming technique, primarily because it does not appear to be as effective as the Balas additive algorithm and related algorithms.

Although the method may have drawbacks in integer programming, we may still consider it as a method for finding a good solution. Considerable computational exprience has indicated that the Lawler-Bell method gives a "good" feasible solution to zero-one problems quickly and at little expense. We must rearrange the problem first to help assure that the solution is good and is found relatively quickly. We do this by first reordering the constraints as indicated in section 17.2. Then we choose a most binding inequality constraint (on a per cent infeasible basis) and reorder the variables in that constraint in the same manner as in the Balas filter method (see section 16.3). (We may instead wish to use a surrogate constraint.) We then apply the Lawler-Bell algorithm to the problem until we get the first feasible solution, or until we exceed some specified time or iteration limit. At that time we take the best solution that has been found, which we then use as a starting solution for a second phase.

### A Multidirectional Approach

This approach is like a popular cold remedy—a combination of active ingredients, not unlike a doctor's prescription. The basic idea is to mix approaches in some sensible way. For example, primal codes typically get feasible solutions quickly, whereas they usually slow down tremendously after some number of iterations. Once the solution process has slowed, we may take advantage of the current feasible integer solution and its associated objective function, and solve for the optimal fractional solution. Then we proceed with a branch and bound method or a dual method such as Gomory's fractional method, making use of the bound given by the feasible integer solution. It is of course possible to bounce back and forth between the primal and the dual (or branch and bound) methods, but it does not appear that the primal process will be helped by a bound from the branch and bound or dual method.

Another type of blend of approaches is to use a dual method together with a branch and bound method. The branch and bound method might be used first to find a feasible integer solution. Then the objective function bound constraint may be imposed upon the problem, and the dual approach used. However, problems of massive dual degeneracy that may occur with such an approach should not be taken lightly.

Another approach is to mix pivot choice algorithms within a given cut method. Research in this area conducted by the author, Dave Summers, and C. Krabek of Control Data Corporation indicates that *any* combination of pivot choice rules is generally superior to *any* one choice rule. (This does not appear to be true for the class of set-covering problems, for which the choice of the most negative right-hand side as the cut-generating constraint consistently seems to do very well.)

**Other Methods for Generating Starting Solutions**

First there is the possibility of rounding a noninteger optimum. The major problem is that it may not be possible to preserve feasibility (let alone optimality) when rounding a solution.

The second method, developed by Müller-Merbach [214], is called the method of "cautious approach." The procedure is to start at a feasible integer solution, such as the origin. Then we have the following steps:

1. Compute the maximum feasible improvement in the objective function that can be attained by *increasing* one or more variables. Do this for all combinations of variables by increasing each variable in the combination the same integral number of units. (If there are $n$ variables, there are $2^n$ combinations. For large problems a simplification is obviously necessary.)
2. Choose the maximum improvement found in step 1. If the maximum is zero, stop; a solution has been attained. Otherwise, increase each variable associated with the maximum improvement by *one*. Then, using the new solution point as a starting point, go to step 1.

EXAMPLE 17.1:

Find a solution to the problem of Example 13.2 by the method of cautious approach.

$$\begin{aligned} \text{Maximize } z = \; & 5x_1 + 2x_2 \\ \text{subject to:} \quad & 2x_1 + 2x_2 \leq 9 \\ & 3x_1 + x_2 \leq 11 \\ & x_1, x_2 \geq 0 \text{ and integer} \end{aligned}$$

Begin with $x_1 = 0$, $x_2 = 0$. In the $x_1$ direction, we may increase $x_1$ by 3 units, with an increase in $z$ of 15. In the $x_2$ direction, we may increase $x_2$ by 4 units, with an increase in $z$ of 8. In the $x_1, x_2$ direction, we may increase $x_1$ and $x_2$ by 2 units, with an increase in $z$ of 14.
We summarize this as:

Step 1. $x_1$: $\Delta x_1 = 3$, $\Delta z = 15$; $x_2$: $\Delta x_2 = 4$, $\Delta z = 8$;
$x_1, x_2$: $\Delta x_1 = \Delta x_2 = 2$, $\Delta z = 14$

We choose $x_1$, the variable which gives the maximum improvement in the objective function value, and move one unit in the $x_1$ direction.
We summarize this as:

Step 2. $\Delta x_1 = 1$; new solution $x_1 = 1$, $z = 5$.

We then have the following steps:

Step 1. $x_1: \Delta x_1 = 2, \Delta z = 10; \quad x_2: \Delta x_2 = 3, \Delta z = 6;$

$x_1, x_2: \Delta x_1 = \Delta x_2 = 1, \quad \Delta z = 7$

Choose the maximum $\Delta z$: $x_1$.

Step 2. $\Delta x_1 = 1$; new solution $x_1 = 2, z = 10$.

Step 1. $x_1: \Delta x_1 = 1, \Delta z = 5; \quad x_2: \Delta x_2 = 1, \Delta z = 2;$

$x_1, x_2: \Delta x_1 = \Delta x_2 = 1, \Delta z = 7$

Choose the maximum $\Delta z$: $x_1, x_2$.

Step 2. $\Delta x_1 = 1, \Delta x_2 = 2$; new solution $x_1 = 3, x_2 = 1, z = 17$.

Step 1. $x_1: \Delta x_1 = 0, \Delta z = 0; \quad x_2: \Delta x_2 = 0, \Delta z = 0;$

$x_1, x_2: \Delta x_1 = \Delta x_2 = 0, \Delta z = 0$

Step 2. Stop. The solution is $x_1 = 3, x_2 = 1$ (which happens to be optimal).

As described by Müller-Merbach [214], Kreuzberger [181, 182, 183] proposed a method of pattern moves. The method begins by using logic similar to steps 1 and 2 of the method of cautious approach, except that the largest step is taken in step 2. When no further steps can be taken, every possible pair of variables is considered. (This step is likely to be prohibitively expensive for large problems). The most profitable of the feasible trades (i.e., increasing one variable and decreasing another) is considered, and an adjustment is made (if profitable). Then the process is repeated until no additional feasible trades are possible.

EXAMPLE 17.2:

Use the method of Kreuzberger to find a solution to the problem of Example 17.1.

The first iteration generates the solution $x_1 = 3$, $x_2 = 0$, $z = 15$. The second iteration generates the solution $x_1 = 3$, $x_2 = 1$, $z = 17$, which is optimal. In order to illustrate the second part of the solution, assume that the first iteration process yields the solution $x_1 = 1$, $x_2 = 3$, $z = 11$. The only pair of variables is $(x_1, x_2)$. The increase in $x_1$ must be at least $\frac{2}{5}$ of the decrease in $x_2$ (in order to increase the objective function). Since we can replace up to two units of $x_2$ by the same number of units of $x_1$, we increase $x_1$ and decrease $x_2$ by 2. The solution is then $x_1 = 3$, $x_2 = 1$, $z = 17$. Now the test shows that no additional trades are profitable.

Echols and Cooper [84] and Drebes and Cooper [78] present highly sophisticated heuristics for developing good solutions which use methods similar to

Müller-Merbach and Kreuzberger, but add a routine for systematically re-searching for a better solution after all pairwise trades that offer improvement have been made. Both papers present encouraging computational experience, but the authors admit that their methods need further research, although they seem to perform well for most of the problems solved. On a set of five 25 by 50 integer programming problems, however, the time required for calculations was exorbitant.

## 17.5 SOME FINAL WORDS OF ADVICE

This chapter has been, at best, a collection of tricks to afford the serious integer programmer some reasonable chance of solving integer programs. One should not overlook the alternative of buying time on one of the proprietary codes, such as Scientific Control System's UMPIRE on the Univac 1108, Control Data Corporation's OPHELIE on the 6600, or IBM's Extended Mathematical Programming System on the 360–370. In a number of cases brought to this author, commercial solutions using such methods were obtained to problems with up to 600 columns and 400 rows for machine time cost of less than $25.

Recently, Pierre Blondeau [35] of the Houston CDC Data Center reported that the OPHELIE mixed-integer code solved: (1) an investment planning model with 2164 rows, 2087 columns, 117 zero-one variables, and 14,588 nonzero elements in 1500 cpu seconds; and (2) a chemical investment problem with 397 rows, 102 continuous variables, 240 zero-one variables, and 2616 nonzero elements in 278 cpu seconds. Both of the foregoing were solved on a CDC 6600.

As a final word of warning, I should like to append two guidelines from [300] to help the neophyte:

1. Just because the words OPTIMAL SOLUTION appear on the printout does *not* mean that the solution *is* optimal. (It is of interest to note that, on the OPHELIE branch and bound mixed integer algorithm for the CDC 6600, the printout says OPTIMAL PRESUMPTION upon reaching the continuous solution.)
2. When a given integer algorithm finds an "optimal solution," and this solution is also obtained by at least two other codes, the probability that this is indeed the optimum solution is at least .5.

These guidelines, the result of some years of bitter experience, are passed on with some hope that they may be of value. Along with these two rules, the author strongly suggests that many problems are formulated as integer programming problems because they *can* be, not because they *should* be. For examples of such, the author can only point to the celebrated paper by Woolsey [299].

**Selected Supplemental References**

Section 17.3
[271], [304]
Section 17.4
[78], [84], [181], [182], [183], [191], [214]
General
[44], [146], [147], [284], [299], [300], [301], [302], [303]

## 17.6 PROBLEMS

**1.** The following fixed-charge (or fixed-cost) problem of Haldi [146] is one of a number that are extremely difficult to solve using Gomory's fractional algorithm.

$$
\begin{aligned}
\text{Minimize } z = \quad & x_1 + x_2 + x_3 \\
\text{subject to:} \quad & 20\delta_1 + 30\delta_2 + x_1 + 2x_2 + 2x_3 \leq 180 \\
& 30\delta_1 + 20\delta_2 + 2x_1 + x_2 + 2x_3 \leq 150 \\
& -60\delta_1 + x_1 \leq 0 \\
& -75\delta_2 + x_2 \leq 0 \\
& \delta_1 \leq 1 \\
& \delta_2 \leq 1 \\
& x_1, x_2, x_3, \delta_1, \delta_2 \geq 0 \text{ and integer}
\end{aligned}
$$

a. Give an intuitive explanation for why the problem is so difficult to solve this way.
b. Can you deduce some "redundant" constraints on the objective function? (Hint: Suppose $x_1$, $x_2$, $\delta_1$, and $\delta_2$ were zero.)
c. Use a branch and bound method to solve the problem, treating $x_1$, $x_2$, and $x_3$ as continuous variables instead of integers.
d. Use Benders' partitioning method to solve the problem, treating $x_1$, $x_2$, and $x_3$ as continuous variables instead of integers.
e. Use Gomory's mixed-integer method to solve the problem, treating $x_1$, $x_2$, and $x_3$ as continuous variables instead of integers.
f. Why is it reasonable to relax the integrality requirements of $x_1$, $x_2$, and $x_3$?

**2.** Show that the following problem is an example in which setting no variable to unity reduces infeasibility, even though there is an optimal integer solution, and hence the bang-for-the-buck procedure of section 17.3 cannot be used.

$$\begin{aligned}
\text{Minimize } z = \quad & 2x_1 + 3x_2 \\
\text{subject to:} \quad & -2x_1 + 8x_2 \geq 1 \\
& 8x_1 - 2x_2 \geq 1 \\
& x_1, x_2 \text{ are 0 or 1}
\end{aligned}$$

**3.** An $m$ by $m$ lock box problem is the problem faced by a company which has $m$ locations at which it receives checks in payment for goods and services. It has the choice of maintaining an account at a bank for clearing purposes (a lock box) in one or more of the locations. Each account or lock box has a fixed cost associated with it—the cost of the banking services, plus time delays, which are translated into costs, in forwarding the checks from one location to another location. Formulate the associated problem of minimizing total costs as an integer programming problem. (Hint: Assume each location $i$ has one [super] check which must go to exactly one lock box $j$ at some cost $c_{ij}$. Thus, $x_{ij}$ [a zero-one variable] is the number of [super] checks that go from location $i$ to lock box $j$. Each lock box $j$ has a fixed cost $f_j$ if it is used at all; its capacity may be assumed to be as large as necessary to handle as many checks as required. The formulation has $2m$ constraints and $m(m + 1)$ variables.)

**4.** Solve the following problem using the Gomory all-integer method, with the first negative right-hand side element generating the cut constraint. Then permute the constraints and solve the problem again in the same way. The difference in solution paths is a simple illustration of the phenomena observed by Trauth and Woolsey [304].

$$\begin{aligned}
\text{Minimize } z = \; & x_1 + x_2 + 2x_3 \\
\text{subject to:} \quad & x_1 \qquad\;\; + 2x_3 \geq 4 \\
& \qquad\; x_2 + \;\; x_3 \geq 15 \\
& x_1, x_2, x_3 \geq 0 \text{ and integer}
\end{aligned}$$

**5.** Discuss the difficulties associated with using approximation methods for generating good solutions to zero-one problems.

**6.** Find a good solution to each of the following problems using the approximation methods described in section 17.4:

a. example 13.4.
b. example 13.5.
c. example 13.9.
d. problem 7 of Chapter 13.
e. problem 8 of Chapter 13.
f. problem 9 of Chapter 13.
g. problem 10 of Chapter 13.
h. problem 25 of Chapter 13.
i. problem 1 of Chapter 14.
j. problem 13 of Chapter 14.
k. problem 14 of Chapter 14.
l. problem 15 of Chapter 14.
m. problem 1 of Chapter 15.
n. problem 2 of Chapter 15.
o. problem 3 of Chapter 15.

**7.** The Mickey Mouse method was developed by the author as a heuristic for getting a starting feasible solution for integer programming problems. Based on a paper of Riesco and Thomas [235], it may be outlined as follows. Given that we have a problem of the form

$$\text{Minimize } z = \sum_{j=1}^{n} c_j x_j$$

$$\text{subject to:} \quad \sum_{j=1}^{n} a_{ij} x_j \geq b_i, \quad i = 1, 2, 3, \ldots, m$$

Step 1. Assign some arbitrary upper bound, $u_j$, to each $x_j$, which gives a feasible solution and a starting value of the objective function $z = \sum_{j=1}^{n} c_j u_j$. Let $\hat{x} = (u_1, u_2, \ldots, u_n)$.

Step 2. For each row, determine the value $(v_i)$ of the row from

$$v_i = \sum_{j=1}^{n} a_{ij} x_j - b_i$$

where $x_j$ = present value of $j^{\text{th}}$ variable.

Step 3. For any $v_i = 0$, set all variables in that row with $a_{ij} > 0$ to their present value and underline them, as they may not be further reduced. If all variables are underlined, stop.

Step 4. For each variable not underlined, find $x_j^*$ from

$$x_j^* = \operatorname*{Max}_{j} \left\{ \operatorname*{Max}_{i} \left\{ x_j - \left[ \frac{v_j}{a_{ij}} \right], 0 \right\} \right\}$$

In the above, $x_j^*$ is the minimum value to which $x_j$ may be reduced and stay feasible.

Step 5. From the $x_j^*$ values found in step 4, select the variable to be reduced by

$$j_0 = j \text{ such that } c_j(x_j - x_j^*) \text{ is maximum}$$

In short, select the variable to be reduced to its $x_j^*$ as the variable which reduces the objective function the most. If the max $c_j(x_j - x_j^*) = 0$, stop; the variable is at its lower bound.

Step 6. Replace $x_{j_0}$ with $x_{j_0}^*$, calculate the new $\hat{z}$, and go to step 2.

a. Find a starting solution to the following problem from Trauth and Woolsey using this method, with $\hat{x} = (20\ 20\ 20\ 20)$ as a starting solution.

$$
\begin{aligned}
\text{Minimize } z = {} & 13x_1 + 15x_2 + 14x_3 + 11x_4 \\
\text{subject to:} \quad & 4x_1 + 5x_2 + 3x_3 + 6x_4 \geq 96 \\
& 20x_1 + 21x_2 + 17x_3 + 12x_4 \geq 200 \\
& 11x_1 + 12x_2 + 12x_3 + 7x_4 \geq 101 \\
& x_1, x_2, x_3, x_4 \geq 0 \text{ and integer}
\end{aligned}
$$

b. Discuss the difficulty of finding solution $\hat{x}$ in step 1 in general.
c. Use the Mickey Mouse method to find a starting solution to the problems given in problem 6.

# References

[1] Abadie, J., ed., *Integer and Nonlinear Programming*. New York: American Elsevier Publishing Company, 1970.

[2] Agin, N., "Optimum Seeking with Branch and Bound," *Management Science*, 13, B176–185, 1966.

[3] Alter, R., and B. Lientz, "A Note on a Problem of Smirnov—A Graph Theoretic Interpretation," *Naval Research Logistics Quarterly*, 17, 407–408, 1970.

[4] Balas, E., "An Additive Algorithm for Solving Linear Programs with Zero-One Variables," *Operations Research*, 13, 517–545, 1965.

[5] ———, "Discrete Programming by the Filter Method," *Operations Research*, 15, 915–957, 1967.

[6] ———, "The Dual Method for the Generalized Transportation Problem," *Management Science*, 12, 555–568, 1966.

[7] ———, "Duality in Discrete Programming," *Technical Report No. 67-5*, Operations Research House, Stanford University, California, July 1967, rev. August 1967.

[8] ———, "Duality in Discrete Programming: II. The Quadratic Case," *Management Science*, 16, 14–32, 1969.

[9] ———, "An Infeasibility-Pricing Decomposition Method for Linear Programs," *Operations Research*, 14, 847–873, 1966.

[10] ———,"Integer Programming and Convex Analysis," paper presented to the Thirty-ninth National Meeting of the Operations Research Society of America, Dallas, Texas, May 5, 1971.

[11] ———, "Intersection Cuts—A New Type of Cutting Planes for Integer Programming," *Operations Research*, 19, 19–39, 1971.

[12] ———, "Solution of Large-Scale Transportation Problems Through Aggregation," *Operations Research*, 13, 82–93, 1965.

[13] ———, V. J. Bowman, F. Glover, and D. Sommer, "An Intersection Cut from the Dual of the Unit Hypercube," *Operations Research*, 19, 40–44, 1971.

[14] Balas, E., and P. L. Ivanescu, "On the Generalized Transportation Problem," *Management Science*, 11, 188–202, 1965.

[15] Balinski, M. L., "Integer Programming: Methods, Uses, Computation," *Management Science*, 12, 253–313, 1965.

[16] ———, and R. E. Gomory, "A Mutual Primal-Dual Simplex Method," in *Recent Advances in Mathematical Programming*, ed. R. L. Graves and P. Wolfe. New York: McGraw-Hill Book Company, 1963.

[17] Barnett, S., "A Simple Class of Parametric Linear Programming Problems," *Operations Research*, 16, 1161–1165, 1968.

[18] ———, and G. J. Kynch, "Exact Solution of a Simple Cutting Problem," *Operations Research*, 15, 1051–1056, 1967.

[19] Baugh, C. R., T. Ibaraki, and S. Muroga, "Results in Using Gomory's All-Integer Integer Algorithm to Design Optimum Logic Networks," *Operations Research*, 19, 1090–1096, 1971.

[20] Beale, E. M. L., "Cycling in the Dual Simplex Algorithm," *Naval Research Logistics Quarterly*, 2, 269–276, 1955.

[21] ———, and R. E. Small, "Mixed Integer Programming by a Branch and Bound Technique," in *Proceedings of the IFIP Congress 1965*, 2, ed. W. A. Kalenich. Washington, D. C.: Spartan Press, 1965.

[22] Bellmore, M., G. Bennington, and S. Lubore, "A Network Isolation Algorithm," *Naval Research Logistics Quarterly*, 17, 461–470, 1970.

[23] Bellmore, M., W. D. Eklof, and G. L. Nemhauser, "A Decomposable Transshipment Algorithm for a Multiperiod Transportation Problem," *Naval Research Logistics Quarterly*, 16, 517–524, 1969.

[24] Bellmore, M., and G. L. Nemhauser, "The Traveling Salesman Problem: A Survey," *Operations Research*, 16, 538–558, 1968.

[25] Benders, J. F., "Partitioning Procedures for Solving Mixed Variables Programming Problems," *Numerische Mathematik*, 4, 238–252, 1962.

[26] Bénichou, M., J. M. Gauthier, P. Girodet, G. Hentèges, G. Ribière, and O. Vincent, "Experiments in Mixed Integer Linear Programming," presented at the Seventh International Mathematical Programming Symposium, The Hague, The Netherlands, September 1970.

[27] Ben-Israel, A., and A. Charnes, "An Explicit Solution of a Special Class of Linear Programming Problems," *Operations Research*, 16, 1166–1175, 1968.

[28] ———, "On Some Problems of Diophantine Programming," *Cahiers du Centre D'Etudes de Recherche Opérationelle*, 4, 21–280, 1962.

[29] Ben-Israel, A., and P. D. Robers, "A Decomposition Method for Interval Linear Programming," *Management Science*, 16, 374–387, 1970.

[30] Ben-Israel, A., A. Charnes, A. P. Hunter, and P. D. Robers, "On the Explicit Solution of a Special Class of Linear Economic Models," *Operations Research*, 18, 462–470, 1970.

[31] Bennett, J. M., "An Approach to Some Structural Linear Programming Problems," *Operations Research*, 14, 636–645, 1966.

[32] ———, and D. R. Green, "An Approach to Some Structured Linear Programming Problems," *Operations Research*, 17, 749–750, 1969.

[33] Bennington, G., and S. Lubore, "Resource Allocation for Transportation," *Naval Research Logistics Quarterly*, 17, 471–484, 1970.

[34] Birkhoff, G., and S. MacLane, *A Survey of Modern Algebra* (rev. ed.). New York: The Macmillan Company, 1953.

[35] Blondeau, P., private communication, August 1970.

[36] Boot, J. C. G., *Mathematical Reasoning in Economics and Management Science*. Englewood Cliffs, N.J.: Prentice-Hall, Inc., 1967.

[37] ———, "Price Determination Based on Quality: An Application of Minimax," *Statistica Neerlandica*, 19, 41–53, Spring 1965.

[38] Bowman, V. J., Jr., and G. L. Nemhauser, "A Finiteness Proof for Modified Dantzig Cuts in Integer Programming," *Naval Research Logistics Quarterly*, 17, 309–314, 1970.

[39] Bradley, G. H., "Equivalent Integer Programs and Canonical Problems," *Management Science*, 17, 354–366, 1971.

[40] ———, "Heuristic Solution Methods and Transformed Integer Linear Programming Problems," Administrative Sciences, Yale, Report No. 43, March 1971, The Proceedings of the Fifth Annual Princeton Conference on Information Science and Systems.

[41] ———, "Transformation of Integer Programs to Knapsack Problems," Administrative Sciences, Yale, Report No. 37, November 1970.

[42] Breuer, M. A., "The Formulation of Some Allocation and Connection Problems as Integer Programs," *Naval Research Logistics Quarterly*, 13, 83–96, 1966.

[43] Briskin, L. E., "A Note on Trauth and Woolsey's Integer Linear Programming Article," *Management Science*, 16, 651, 1970.

[44] Cabot, A. V., "An Enumeration Algorithm for Knapsack Problems," *Operations Research*, 18, 306–311, 1970.

[45] ———, and A. P. Hurter, Jr., "An Approach to Zero-One Integer Programming," *Operations Research*, 16, 1206–1211, 1968.

[46] Charnes, A., "Optimality and Degeneracy in Linear Programming," *Econometrica*, 20, 160–170, 1952.

[47] ———, and W. W. Cooper, *Management Models and Industrial Applications of Linear Programming*, Vols. I and II. New York: John Wiley & Sons, 1961.

[48] ———, "Structural Sensitivity Analysis in Linear Programming and an Exact Product Form Inverse," *Naval Research Logistics Quarterly*, 15, 517–522, 1968.

[49] ———, J. K. DeVoe, D. B. Learner, and W. Reinecke, "A Goal Programming Model for Media Planning," *Management Science*, 14, B423–430, 1968.

[50] Charnes, A., W. W. Cooper, D. B. Learner, and E. F. Snow, "Note on an Application of a Goal Programming Model for Media Planning," *Management Science*, 14, 431–436, 1968.

[51] Charnes, A., F. Glover, and D. Klingman, "The Lower Bounded Distribution Model," *Naval Research Logistics Quarterly*, 18, 277–282, 1971.

[52] ———, "A Note on a Distribution Problem," *Operations Research*, 18, 1213–1215, 1970.

[53] Charnes, A., and D. Klingman, "The Distribution Problem with Upper and Lower Bounds on Node Requirements," *Management Science*, 16, 638–641, 1970.

[54] Charnes, A., and W. M. Raike, "One-Pass Algorithms for Some Generalized Network Problems," *Operations Research*, 14, 914–924, 1966.

[55] Chen, D. S., "A Group Theoretic Algorithm for Solving Integer Linear Programming Problems," unpublished Doctoral Dissertation, Department of Industrial Engineering, State University of New York at Buffalo, September 1970.

[56] ———, and S. Zionts, "An Exposition of the Group Theoretic Approach to Integer Linear Programming," *Opsearch*, 9, 75–102, 1972.

[57] Contini, B., and S. Zionts, "Restricted Bargaining for Organizations with Multiple Objectives," *Econometrica*, 36, 397–414, 1968.

[58] Cremeans, J. E., R. A. Smith, and G. R. Tyndall, "Optimal Multicommodity Network Flows with Resource Allocation," *Naval Research Logistics Quarterly*, 17, 269–280, 1970

[59] Daellenbach, H. G., and E. J. Bell, *User's Guide to Linear Programming*. Englewood Cliffs, N.J.: Prentice-Hall, Inc., 1970.

[60] Dakin, R. J., "A Tree-Search Algorithm for Mixed Integer Programming Problems," *Computer Journal*, 8, 250–255, 1965.

[61] Dalton, R. E., and R. W. Llewellyn, "An Extension of the Gomory Mixed-Integer Algorithm to Mixed-Discrete Variables," *Management Science*, 12, 569, 1966.

[62] Dantzig, G. B., *Linear Programming and Extensions*. Princeton, N.J.: Princeton University Press, 1963.

[63] ———, "Note on Solving Linear Programs in Integers," *Naval Research Logistics Quarterly*, 6, 75–76, 1959.

[64] ———, "On the Significance of Solving Linear Programming Problems with Some Integer Variables," *Econometrica*, 28, 30–44, 1960.

[65] ———, "Upper Bounds, Secondary Constraints, and Block Triangularity," *Econometrica*, 23, 174–183, 1955.

[66] ———, L. R. Ford, Jr., and D. R. Fulkerson, "A Primal-Dual Algorithm for Linear Programs," in *Linear Inequalities and Related Systems, Annals of Mathematics Studies, No. 38*, ed. H. W. Kuhn and A. W. Tucker. Princeton, N.J.: Princeton University Press, 1956.

[67] Dantzig, G. B., A. Orden, and P. Wolfe, "Generalized Simplex Method for Minimizing a Linear Form Under Linear Inequality Constraints," Rand Corporation, Santa Monica, Cal., 1954.

[68] Dantzig, G. B., and R. M. VanSlyke, "Generalized Upper Bounding Techniques," *Journal of Computer and System Sciences*, 1, 213–226, 1967.

[69] Dantzig, G. B., and P. Wolfe, "Decomposition Principle for Linear Programs," *Operations Research*, 8, 101–111, 1960.

[70] Davis, E. W., and G. E. Heidorn, "An Algorithm for Optimal Project Scheduling under Multiple Resource Constraints," *Management Science*, 17, B803–816, 1971.

[71] Davis, P. S., and T. L. Ray, "A Branch-Bound Algorithm for the Capacitated Facilities Location Problem," *Naval Research Logistics Quarterly*, 16, 331–344, 1969.

[72] Davis, R. E., D. A. Kendrick, and M. Weitzman, "A Branch-and-Bound Algorithm For Zero-One Mixed Integer Programming Problems," *Operations Research*, 19, 1036–1044, 1971.

[73] Desler, J. F., and S. L. Hakimi, "A Graph-Theoretic Approach to a Class of Integer Programming Problem," *Operations Research*, 17, 1017–1033, 1969.

[74] D'Esopo, D. A., and B. Lefkowitz, "Note on an Integer Linear Programming Model for Determining a Minimum Embarkation Fleet," *Naval Research Logistics Quarterly*, 11, 79–82, 1964.

[75] Dickson, J. C., and F. P. Frederick, "A Decision Rule for Improved Efficiency in Solving Linear Programming Problems with the Simplex Algorithm," *Communications of the ACM*, 3, 509–512, 1960.

[76] Doulliez, P. J., and M. R. Rao, "Capacity of a Network with Increasing Demands and Arcs Subject to Failure," *Operations Research*, 19, 905–915, 1971.

[77] Dragan, I., "Un Algorithme Lexicographique pour la Résolution des Programmes Linéaires en Variables Binaires," *Management Science*, 16, 246–252, 1969.

[78] Drebes, C., and L. Cooper, "Investigations in Integer Linear Programming by Direct Search Methods," Report No. COO-1493-10, Department of Applied Mathematics and Computer Science, School of Engineering and Applied Science, Washington University, St. Louis, Missouri, undated.

[79] Dresher, M., *Games of Strategy—Theory and Applications*. Englewood Cliffs, N.J.: Prentice-Hall, Inc., 1961.

[80] Dreyfus, S. E., "An Appraisal of Some Shortest-Path Algorithms," *Operations Research*, 17, 395–412, 1969.

[81] Driebeek, N. J., "An Algorithm for the Solution of Mixed Integer Programming Problems," *Management Science*, 12, 576–587, 1966.

[82] ———, *Applied Linear Programming*. Reading, Mass.: Addison-Wesley Publishing Company, 1969.

[83] Dwyer, P. S., "The Direct Solution of the Transportation Problem with Reduced Matrices," *Management Science*, 13, 77–96, 1966.

[84] Echols, R. E., and L. Cooper, "The Solution of Integer Linear Programming Problems by Direct Search," Report No. AM66-1, Department of Applied Mathematics and Computer Science, Washington University, 1966.

[85] Efroymson, M. A., and T. L. Ray, "A Branch-Bound Algorithm for Plant Location," *Operations Research*, 14, 361–368, 1966.

[86] Elmaghraby, S. E., "The Theory of Networks and Management Science. Part I," *Management Science*, 17, 1–34, 1970.

[87] ———, and M. K. Wig, "On the Treatment of Stock Cutting Problems as Diophantine Programs," North Carolina State University and Corning Glass Research Center, Report No. 61, May 1970.

[88] Enrick, N. L., "Statistical Control Applications in Linear Programming," *Management Science*, 11, B177–186, 1965.

[89] Fabian, T., "Blast Furnace Burdening and Production Planning—A Linear Programming Example," *Management Science*, 14, B1–27, 1967.

[90] Farbey, B. A., A. H. Land, and J. D. Murchland, "The Cascade Algorithm for Finding All Shortest Distances in a Directed Graph," *Management Science*, 14, 19–28, 1967.

[91] Fitzpatrick, G. R., J. Bracken, M. J. O'Brien, L. G. Wentling, and J. C. Whiton, "Programming the Procurement of Airlift and Sealift Forces: A Linear Programming Model for Analysis of Least-Cost Mix of Strategic Development System," *Naval Research Logistics Quarterly*, 14, 241–256, 1967.

[92] Fleishman, B., "Computational Experience with the Algorithm of Balas," *Operations Research*, 15, 153–155, 1967.

[93] Florian, M., and P. Robert, "A Direct Search Method to Locate Negative Cycles in a Graph," *Management Science*, 17, 307–310, 1971.

[94] Florian, M., P. Trepant, and G. McMahon, "An Implicit Enumeration Algorithm for the Machine Sequencing Problem," *Management Science*, 17, B782–792, 1971.

[95] Ford, L. R., Jr., and D. R. Fulkerson, *Flows in Networks*. Princeton, N.J.: Princeton University Press, 1962.

[96] Fulkerson, D. R., and P. Wolfe, "An Algorithm for Scaling Matrices," *SIAM Review*, 4, 142–147, 1962.

[97] Garfinkel, R. S., and G. L. Nemhauser, "The Set-Partitioning Problem: Set Covering with Equality Constraints," *Operations Research*, 17, 848–856, 1969.

[98] Gass, S. I., *Linear Programming* (3rd ed.). New York: McGraw-Hill Book Company, 1969.

[99] Gassner, B. J., "Cycling in the Transportation Problem," *Naval Research Logistics Quarterly*, 11, 43–58, 1964.

[100] Geoffrion, A. M., "Elements of Large-Scale Mathematical Programming (Parts I and II)," *Management Science*, 16, 652–691, 1970.

[101] ———, "An Improved Implicit Enumeration Approach for Integer Programming," *Operations Research*, 17, 437–454, 1969.

[102] ———, "Integer Programming by Implicit Enumeration and Balas' Method," Rand Corporation RM4783–RP, Santa Monica, Cal., February 1966.

[103] ———, and R. E. Marsten, "Integer Programming: A Framework and State-of-the-Art Survey," *Management Science*, 18, 465–491, 1972.

[104] Geoffrion, A. M., and A. B. Nelson, "User's Instructions for 0–1 Integer Programming Code RIP30C," Rand Corporation Memorandum RM 5627-PR, Santa Monica, Cal., May 1968.

[105] Ghare, P. M., D. C. Montgomery, and W. C. Turner, "Optimal Interdiction Policy for a Flow Network," *Naval Research Logistics Quarterly*, 18, 37–46, 1971.

[106] Gilmore, P. C., and R. E. Gomory, "A Linear Programming Approach to the Cutting Stock Problem, Part I," *Operations Research*, 9, 849–859, 1961.

[107] ———, "A Linear Programming Approach to the Cutting Stock Problem, Part II," *Operations Research*, 11, 863–888, 1963.

[108] ———, "Multistage Cutting Stock Problems of Two or More Dimensions," *Operations Research*, 13, 94–120, 1965.

[109] ———, "The Theory and Computation of Knapsack Functions," *Operations Research*, 14, 1045–1074, 1966.

[110] Glover, F., "An Algorithm for Solving the Linear Integer Programming Problem over a Finite Additive Group, with Extensions to Solving General Linear and Certain Nonlinear Integer Programs," Working Paper 27, Operations Research Center, University of California, Berkeley, June 1966.

[111] ———, "A Bound Escalation Method for the Solution of Integer Linear Programs," *Centre D'Etudes de Recherche Opérationnelle*, 6, 131–168, 1964.

[112] ———, "Faces of the Gomory Polyhedron," in *Integer and Nonlinear Programming*, ed. J. Abadie. New York: American Elsevier Publishing Company, 1970.

[113] ———, "Flows in Arborescences," *Management Science*, 17, 568–586, 1971.

[114] ———, "Generalized Cuts in Diophantine Programming," *Management Science*, 13, 254–268, 1966.

[115] ———, "Integer Programming over a Finite Additive Group," Graduate School of Business, University of Texas, April 1968.

[116] ———, "A Multiphase-Dual Algorithm for the Zero-One Integer Programming Problem," *Operations Research*, 13, 879–919, 1965.

[117] ———, "A New Foundation for a Simplified Primal Integer Programming Algorithm," *Operations Research*, 16, 727–740, 1968.

[118] ———, "A Note on Extreme-Point Solutions and a Paper by Lemke, Salkin, and Spielberg," *Operations Research*, 19, 1023–1024, 1971.

[119] ———, "A Note on Linear Programming and Integer Feasibility," *Operations Research*, 16, 1212–1216, 1968.

[120] ———, "Surrogate Constraints," *Operations Research*, 16, 741–749, 1968.

[121] ———, and R. E. Woolsey, "Aggregating Diophantine Equations," Management Science Report Series, No. 70–4, University of Colorado, October 1970.

[122] Glover, F., and S. Zionts, "A Note on the Additive Algorithm of Balas," *Operations Research*, 13, 546–549, 1965.

[123] Gomory, R. E., "An Algorithm for Integer Solutions to Linear Programs," in *Recent Advances in Mathematical Programming*, ed. R. L. Graves and P. Wolfe. New York: McGraw-Hill Book Company, 1963.

[124] ———, "An Algorithm for the Mixed Integer Problem," Rand Corporation, P-1885, Santa Monica, Cal., June 1960.

[125] ———, "An All-Integer Integer Programming Algorithm," in *Industrial Scheduling*, ed. J. F. Muth and G. L. Thompson. Englewood Cliffs, N.J.: Prentice-Hall, Inc., 1963.

[126] ———, "On the Relation Between Integer and Non-Integer Solution to Linear Programs," Proceedings of National Academy of Science, 53, 260–265, 1965.

[127] ———, "Some Polyhedra Related to Combinatorial Problems," *Linear Algebra and Its Applications*, 2, 451–558, 1969.

[128] ———, and A. F. Hoffman, "On the Convergence of an Integer Programming Process," *Naval Research Logistics Quarterly*, 10, 121–123, 1963.

[129] Gorry, G. A., and J. F. Shapiro, "An Adaptive Group Theoretic Algorithm for Integer Programming Problems," *Management Science*, 17, 285–306, 1971.

[130] Graves, G. W., "A Complete Constructive Algorithm for the General Mixed Linear Programming Problem," *Naval Research Logistics Quarterly*, 12, 1–34, 1965.

[131] ——, and A. Whinston, "A New Approach to Discrete Mathematical Programming," *Management Science*, 15, 177–190, 1968.

[132] Graves, R. L., "A Principal Pivoting Simplex Algorithm for Linear and Quadratic Programming," *Operations Research*, 15, 482–494, 1967.

[133] ———, and P. Wolfe, eds., *Recent Advances in Mathematical Programming*. New York: McGraw-Hill Book Company, 1963.

[134] Greenberg, H., "A Branch-Bound Solution to the General Scheduling Problem," *Operations Research*, 16, 353–361, 1968.

[135] ———, *Integer Programming*. New York: Academic Press, 1971.

[136] ———, "A Note on a Modified Primal-Dual Algorithm to Speed Convergence in Solving Linear Programs," *Naval Research Logistics Quarterly*, 16, 271–273, 1969.

[137] ———, and R. L. Hegerich, "A Branch Search Algorithm for the Knapsack Problem," *Management Science*, 16, 327–332, 1970.

[138] Grigoriadis, M. D., "A Dual Generalized Upper Bounding Technique," *Management Science*, 17, 269–284, 1971.

[139] ———, and W. F. Walker, "A Treatment of Transportation Problems by Primal Partition Programming," *Management Science*, 14, 565–599, 1968.

[140] Grinold, R. C., "Symmetric Duality for Continuous Linear Programs," *SIAM*, 18, 84–97, 1970.

[141] Gross, D., and R. M. Soland, "A Branch and Bound Algorithm for Allocation Problems in which Constraint Coefficients Depend Upon Decision Variables," *Naval Research Logistics Quarterly*, 16, 157–174, 1969.

[142] Gutterman, M. M., "Efficient Implementation of a Branch and Bound Algorithm," paper presented at the Seventh Mathematical Programming Symposium, The Hague, The Netherlands, September 1970.

[143] Hadley, G., *Linear Algebra*. Reading, Mass.: Addison-Wesley Publishing Company, 1961.

[144] ———, *Linear Programming*. Reading. Mass.: Addison-Wesley Publishing Company, 1962.

[145] Haessler, R. W., "A Heuristic Programming Solution to a Nonlinear Cutting Stock Problem," *Management Science*, 17, B793–802, 1971.

[146] Haldi, J., "25 Integer Programming Test Problems," Working Paper No. 43, Graduate School of Business, Stanford University, April 1964 (rev.).

[147] ———, and L. M. Isaacson, "A Computer Code for Integer Solutions to Linear Programs," *Operations Research*, 13, 946–960, 1965.

[148] Hammer, P. L., and S. Rudeanu, *Boolean Methods in Operations Research*. Berlin: Springer-Verlag, 1968.

[149] Harris, M. Y., "A Mutual Primal-Dual Linear Programming Algorithm," *Naval Research Logistics Quarterly*, 17, 199–206, 1970.

[150] Hartman, J. K., and L. S. Lasdon, "A Generalized Upper Bounding Method for Doubly Coupled Linear Programs," *Naval Research Logistics Quarterly*, 17, 411–430, 1970.

[151] Heesterman, A. R. G., and J. Sandee, "Special Simplex Algorithm for Linked Problems," *Management Science*, 11, 420–428, 1965.

[152] Held, M., and R. M. Karp, "The Traveling-Salesman Problem and Minimum Spanning Trees," *Operations Research*, 18, 1138–1162, 1970.

[153] Hillier, F. S., "A Branch-and-Scan Algorithm for Pure Integer Linear Programming with General Variables," *Operations Research*, 17, 638–679, 1969.

[154] ———, and G. J. Lieberman, *Introduction to Operations Research*. San Francisco, Cal.: Holden-Day, Inc., 1968.

[155] Hoffman, A. J., "Cycling in the Simplex Algorithm," *National Bureau of Standards Report 2974*, 7, December 16, 1953.

[156] Holladay, J., "Some Transportation Problems and Techniques for Solving Them," *Naval Research Logistics Quarterly*, 11, 15–42, 1964.

[157] Hooi-Tong, L., "On a Class of Directed Graphs—With an Application to Traffic-Flow Problems," *Operations Research*, 18, 87–94, 1970.

[158] Hu, T. C., *Integer Programming and Network Flows*. Reading, Mass.: Addison-Wesley Publishing Company, 1969.

[159] ———, "A Decomposition Algorithm for Shortest Paths in a Network," *Operations Research*, 16, 91–102, 1968.

[160] ———, "Minimum-Cost Flows in Convex-Cost Networks," *Naval Research Logistics Quarterly*, 13, 1–10, 1966.

[161] ———, "On the Asymptotic Integer Algorithm," MRC Technical Summary Report No. 946, Mathematics Research Center, University of Wisconsin, October 1968.

[162] ———, "Revised Matrix Algorithms For Shortest Paths," *SIAM*, 15, 207–218, 1967.

[163] Ignall, E., and L. Schrage, "Application of the Branch and Bound Technique to Some Flow-Shop Scheduling Problems," *Operations Research*, 13, 400–412, 1965.

[164] Isaacs, R., *Differential Games*. New York: John Wiley & Sons, 1965.

[165] Ivanescu, P. L., "Some Network Flow Problems Solved with Pseudo-Boolean Programming," *Operations Research*, 13, 388–399, 1965.

[166] Jensen, P. A., "Optimum Network Partitioning," *Operations Research*, 19, 916–932, 1971.

[167] Jeroslow, R. G., "Comments on Integer Hulls of Two Linear Constraints," *Operations Research*, 19, 1061–1069, 1971.

[168] ———, and K. O. Kortanek, "On an Algorithm of Gomory," *SIAM*, 21, 55–60, 1971.

[169] Johnson, E. L., "Networks and Basic Solutions," *Operations Research*, 14, 619–623, 1966.

[170] Jones, A. P., and R. M. Soland, "A Branch-And-Bound Algorithm for Multi-level Fixed-Charge Problems, *Management Science*, 16, 67–76, 1969.

[171] Kabe, D. G., "Note on a Coverage Problem," *Operations Research*, 13, 1027–1028, 1965.

[172] Kalymon, B. A., "Note Regarding A New Approach to Discrete Mathematical Programming," *Management Science*, 17, 777, 1971.

[173] Kaul, R. N., "An Extension of Generalized Upper Bounded Techniques for Linear Programming," ORC 65–27, Operations Research Center, University of California, Berkeley, 1965.

[174] Kendall, K., and S. Zionts, "Solving Integer Programming Problems by Aggregating Constraints," paper presented to the Joint Meeting of the Operations Research Society of America, The Institute of Management Sciences, American Institute of Industrial Engineers, Atlantic City, New Jersey, November 8–10, 1972.

[175] Klein, M., "A Primal Method for Minimal Cost Flows with Applications to the Assignment and Transportation Problems," *Management Science*, 14, 205–220, 1967.

[176] Kohn, R. E., "Application of Linear Programming to a Controversy on Air Pollution Control," *Management Science*, 17, B609–621, 1971.

[177] Kolesar, P. J., "A Branch and Bound Algorithm for the Knapsack Problem," *Management Science*, 13, 723–735, 1967.

[178] ———, "Linear Programming and the Reliability of Multicomponent Systems," *Naval Research Logistics Quarterly*, 14, 317–328, 1967.

[179] Kortanek, K. O., and R. Jeroslow, "An Exposition on the Constructive Decomposition of the Group of Gomory Cuts and Gomory's Round-Off Algorithm," Technical Report #39, Department of Operations Research, Cornell University, Ithaca, New York, January 1968.

[180] Kortanek, K. O., D. Sodaro, and A. L. Soyster, "Multi-Product Production Scheduling Via Extreme Point Properties of Linear Programming," *Naval Research Logistics Quarterly*, 15, 287–300, 1968.

[181] Kreuzberger, H., "Ein Näherungsverfahren zur Bestimmung ganzzahliger Lösungen bei linearen Optimierungsproblemen," *Ablauf-und Planungsforschung*, 9, 137–152, 1968.

[182] ———, "Numerische Erfahrungen mit einem heuristischen Verfahren zur Lösung ganzzahliger linearer Optimierungsprobleme," *Elektronische Datenverarbeitung*, 12, 289–306, 1970.

[183] ———, "Verfahren zur Lösung ganzzahliger linearer Optimierungsaufgaben," Dissertation, Darmstadt, 1969.

[184] Krolak, P. D., "Computational Results of an Integer Programming Algorithm," *Operations Research*, 17, 743–748, 1969.

[185] Laderman, J., L. Gleiberman, and J. F. Egan, "Vessel Allocations by Linear Programming," *Naval Research Logistics Quarterly*, 13, 315–320, 1966.

[186] Lagemann, J. J., "A Method for Solving the Transportation Problem," *Naval Research Logistics Quarterly*, 14, 89–100, 1967.

[187] Land, A. H., and A. G. Doig, "An Automatic Method of Solving Discrete Programming Problems," *Econometrica*, 28, 497–520, 1960.

[188] Lasdon, L. S., *Optimization Theory for Large Systems*. New York: The Macmillan Company, 1970.

[189] ———, and R. C. Terjung, "An Efficient Algorithm for Multi-Item Scheduling," *Operations Research*, 19, 946–969, 1971.

[190] Lawler, E. L., "Covering Problems: Duality Relations and a New Method of Solution," *SIAM*, 14, 1115–1132, 1966.

[191] ———, and M. D. Bell, "A Method for Solving Discrete Optimization Problems," *Operations Research*, 14, 1098–1112, 1966.

[192] Lawler, E. L., and D. E. Wood, "Branch-and-Bound Methods: A Survey," *Operations Research*, 14, 699–719, 1966.

[193] Lemke, C. E., "The Dual Method of Solving the Linear Programming Problem," *Naval Research Logistics Quarterly*, 1, 36-47, 1954.

[194] ———, and A. Charnes, "Extremal Problems in Linear Inequalities," *USDAF Research Project, Technical Report No. 36*, Carnegie Institute of Technology, Department of Mathematics, Pittsburgh, Pa., 1953.

[195] Lemke, C. E., H. M. Salkin, and K. Spielberg, "Set Covering by Single-Branch Enumeration with Linear-Programming Subproblems," *Operations Research*, 19, 998–1022, 1971.

[196] Lemke, C. E., and K. Spielberg, "Direct Search Algorithms for Zero-One and Mixed-Integer Programming," *Operations Research*, 15, 892–915, 1967.

[197] Lewis, R. W., E. F. Rosholdt, and W. L. Wilkinson, "A Multi-Mode Transportation Network Model," *Naval Research Logistics Quarterly*, 12, 261–274, 1965.

[198] Little, J. D. C., "The Synchronization of Traffic Signals by Mixed-Integer Linear Programming," *Operations Research*, 14, 568–594, 1966.

[199] ———, K. G. Murty, D. W. Sweeney, and C. Karel, "An Algorithm for the Traveling Salesman Problem," *Operations Research*, 11, 972–989, 1963.

[200] Llewellyn, R. W., *Linear Programming*. New York: Holt, Rinehart and Winston, 1964.

[201] Loucks, D. P., C. S. ReVelle, and W. R. Lynn, "Linear Programming Models for Water Pollution Control," *Management Science*, 14, B166–181, 1967.

[202] Lourie, J. R., "Topology and Computation of the Generalized Transportation Problem," *Management Science*, 11, 177–187, 1965.

[203] Luce, R. D., and H. Raiffa, *Games and Decisions*. New York: John Wiley & Sons, 1964.

[204] Machol, R. E., "An Application of the Assignment Problem," *Operations Research*, 18, 745, 1970.

[205] Mangasarian, O. L., "Linear and Nonlinear Separation of Patterns by Linear Programming," *Operations Research*, 13, 444–452, 1965.

[206] Marshall, K. T., and J. W. Suurballe, "A Note on Cycling in the Simplex Method," *Naval Research Logistics Quarterly*, 16, 121–137, 1968.

[207] Martin, G. T., "An Accelerated Euclidean Algorithm for Integer Linear Programming," in *Recent Advances in Mathematical Programming*, ed. R. L. Graves and P. Wolfe. New York: McGraw-Hill Book Company, 1963.

[208] Mathews, G. B., "On the Partition of Numbers," *Proceedings of the London Mathematical Society*, 28, 486–490, 1897.

[209] McMillan, Claude, Jr., *Mathematical Programming*. New York: John Wiley & Sons, 1970.

[210] Meyer, C. F., "Production-Distribution Network with Transit Billing," *Management Science*, 14, B204–218, 1967.

[211] Meyer, M., "Applying Linear Programming to the Design of Ultimate Pit Limits," *Management Science*, 16, B121–135, 1969.

[212] Mills, G., "A Decomposition Algorithm for the Shortest-Route Problem," *Operations Research*, 14, 279–291, 1966.

[213] Mizukami, K., "Optimum Redundancy for Maximum System Reliability by the Method of Convex and Integer Programming," *Operations Research*, 16, 392–406, 1968.

[214] Müller-Merbach, H., "Approximation Methods for Integer Programming," paper presented at the Seventh Mathematical Programming Symposium, The Hague, The Netherlands, September 1970.

[215] ———, "An Improved Starting Algorithm for the Ford- Fulkerson Approach to the Transportation Problem," *Management Science*, 13, 97–104, 1966.

[216] Murchland, J. D., "A New Method for Finding All Elementary Paths in a Complete Directed Graph," Report LSE-TNT-22, Transport Network Theory Unit, London School of Economics, October 1965.

[217] Nemhauser, G. L., and W. B. Widhelm, "A Modified Linear Program for Columnar Methods in Mathematical Programming," *Operations Research*, 19, 1051–1060, 1971.

[218] Netter, J. P., "An Algorithm to Find Elementary Negative-Cost Circuits with a Given Number of Arcs—The Traveling Salesman Problem," *Operations Research*, 19, 234–236, 1971.

[219] Norman, R. Z., and M. O. Rabin, "An Algorithm for the Minimum Cover of a Graph," *Proceedings of the American Mathematical Society*, 10, 315–319, 1959.

[220] Owen, G., *Game Theory*. Philadelphia, Penna.: W. B. Saunders Company, 1968.

[221] Paranjape, S. R., "The Simplex Method: Two Basic Variables Replacement," *Management Science*, 12, 135–141, 1965.

[222] Parikh, S. C., and W. S. Jewell, "Decomposition of Project Networks," *Management Science*, 11, 444–459, 1965.

[223] Penn, A. I., "A Generalized LaGrange-Multiplier Method for Constrained Matrix Games," *Operations Research*, 19, 933–945, 1971.

[224] Peterson, C. C., "Computational Experience with Variants of the Balas Algorithm Applied to the Selection of R & D Projects," *Management Science*, 13, 736–750, 1967.

[225] Pierce, J. F., "Application of Combinatorial Programming to a Class of All-Zero-One Integer Programming Problems," *Management Science*, 15, 191–209, 1968.

[226] Plane, D. R., and C. McMillan, Jr., *Discrete Optimization*. Englewood Cliffs, N.J.: Prentice-Hall, Inc., 1971.

[227] Pratt, J. W., H. Raiffa, and R. Schlaifer, *Introduction to Decision Theory*. New York: McGraw-Hill Book Company, 1965.

[228] Pritsker, A. A. B., L. J. Walters, and P. M. Wolfe, "Multiproject Scheduling with Limited Resources: A Zero-One Programming Approach," *Management Science*, 16, 93–108, 1969.

[229] Rao, A., "Balas' and Glover's Algorithm: A Comparison," paper presented at the Thirty-second National Meeting of the Operations Research Society of America, Chicago, Illinois, November 1–3, 1967.

[230] Rao, M. R., and S. Zionts, "Allocation of Transportation Units to Alternative Trips—A Column Generation Scheme with Out-of-Kilter Subproblems," *Operations Research*, 16, 52–63, 1968.

[231] Rebelein, P. R., "An Extension of the Algorithm of Driebeek for Solving Mixed Integer Programming Problems," *Operations Research*, 16, 193–197, 1968.

[232] Reinfeld, N. V., and W. R. Vogel, *Mathematical Programming*. Englewood Cliffs, N. J.: Prentice-Hall, Inc., 1958.

[233] Reiter, S., and D. B. Rice, "Discrete Optimizing Solution Procedures for Linear and Nonlinear Integer Programming Problems," *Management Science*, 12, 829–850, 1966.

[234] Rhys, J. M. W., "A Selection Problem of Shared Fixed Costs and Network Flows," *Management Science*, 17, 200–207, 1971.

[235] Riesco, A., and M. E. Thomas, "A Heuristic Solution Procedure for Linear Programming Problems with Special Structure," *AIIE Transactions*, 1, 157–163, 1969.

[236] Rothfarb, B., and I. T. Frisch, "Common Terminal Multicommodity Flow with a Concave Objective Function," *SIAM*, 18, 489–502, 1970.

[237] ———, "On the 3-Commodity Flow Problem," *SIAM*, 17, 46–58, 1969.

[238] Rothfarb, W., N. P. Shein, and I. T. Frisch, "Common Terminal Multicommodity Flow," *Operations Research*, 16, 202–204, 1968.

[239] Rothschild, B., and A. Whinston, "Feasibility of Two Commodity Network Flows," *Operations Research*, 14, 1121–1129, 1966.

[240] ———, "On Two Commodity Network Flows," *Operations Research*, 14, 377–387, 1966.

[241] Roy, B., R. Benayoun, and J. Tergny, "From S.E.P. Procedure to the Mixed Ophelie Program," in *Integer and Nonlinear Programming*, ed. J. Abadie. Amsterdam: North-Holland, 1970.

[242] Rubin, D. S., "On the Unlimited Number of Faces in Integer Hulls of Linear Programs with a Single Constraint," *Operations Research*, 18, 940–946, 1970.

[243] Russell, E. J., "Extension of Dantzig's Algorithm to Finding an Initial Near-Optimal Basis for the Transportation Problem," *Operations Research*, 17, 187–191, 1969.

[244] Rutenberg, D. P., "Generalized Networks, Generalized Upper Bounding and Decomposition of the Convex Simplex Method," *Management Science*, 16, 388–401, 1970.

[245] Saigal, R., "Compact Basis Inverse Representation for Dynamic Linear Programs," paper presented at the 36th National Meeting of the Operations Research Society of America, Miami Beach, Florida, November 1969.

[246] ———, "A Constrained Shortest Route Problem," *Operations Research*, 16, 205–208, 1968.

[247] Sakarovitch, M., and R. Saigal, "An Extension of Generalized Upper Bounding Techniques for Structured Linear Programs," *SIAM*, 15, 906–914, 1967.

[248] Saksena, J. P., and A. J. Cole, "The Bounding Hyperplane Method of Linear Programming," *Operations Research*, 19, 1–18, 1971.

[249] Saksena, J. P., and S. Kumar, "The Routing Problem with 'K' Specified Nodes," *Operations Research*, 14, 909–913, 1966.

[250] Salkin, H. M., "On the Merit of the Generalized Origin and Restarts in Implicit Enumeration," *Operations Research*, 18, 549–554, 1970.

[251] Schrage, L., "Solving Resource-Constrained Network Problems by Implicit Enumeration—Nonpreemptive Case," *Operations Research*, 18, 263–278, 1970.

[252] Schroeder, R. G., "Linear Programming Solutions to Ratio Games," *Operations Research*, 18, 300–305, 1970.

[253] Senju, S., and Y. Toyoda, "An Approach to Linear Programming with 0–1 Variables," *Management Science*, 15, B196–207, 1968.

[254] Seppanen, J. J., and J. M. Moore, "Facilities Planning with Graph Theory," *Management Science*, 17, B242–253, 1971.

[255] Shakun, M. F., ed., "Game Theory and Gaming," *Management Science*, 18, Part II, P1-P126, 1972.

[256] Shanno, D. F., and R. L. Weil, "'Linear' Programming with Absolute-Value Functionals," *Operations Research*, 19, 120–123, 1971.

[257] Shapiro, J. F., "Dynamic Programming Algorithm for the Integer Programming Problem—I: The Integer Programming Problem Viewed as a Knapsack Type Problem," *Operations Research*, 16, 103–121, 1968.

[258] ———, "Generalized LaGrange Multipliers in Integer Programming," *Operations Research*, 19, 68–76, 1971.

[259] ———, "Group Theoretic Algorithms for the Integer Programming Problem II: Extension to a General Algorithm," *Operations Research*, 16, 928–947, 1968.

[260] ———, "Turnpike Theorems for Integer Programming Problems," *Operations Research*, 18, 432–440, 1970.

[261] ———, and H. M. Wagner, "A Finite Renewal Algorithm for the Knapsack and Turnpike Models," *Operations Research*, 15, 319–341, 1967.

[262] Sharp, W. F., "A Linear Programming Algorithm for Mutual Fund Portfolio Selection," *Management Science*, 13, 499–510, 1967.

[263] Shubik, M., ed., *Game Theory and Related Approaches to Social Behavior*. New York: John Wiley & Sons, 1964.

[264] Simonnard, M., *Linear Programming*, trans. by W. S. Jewell. Englewood Cliffs, N.J.: Prentice-Hall, Inc., 1966.

[265] Smith, S. B., "Planning Transistor Production by Linear Programming," *Operations Research*, 13, 132–139, 1965.

[266] Spielberg, K., "Plant Location with Generalized Search Origin," *Management Science*, 16, 165–178, 1969.

[267] Spivey, W. A., and R. M. Thrall, *Linear Optimization*. New York: Holt, Rinehart and Winston, 1970.

[268] Steckhan, H., "A Theorem on Symmetric Traveling Salesman Problems," *Operations Research*, 18, 1163–1167, 1970.

[269] Steinberg, D. I., "The Fixed Charge Problem," *Naval Research Logistics Quarterly*, 17, 217–236, 1970.

[270] Steinman, H., and R. Schwinn, "Computational Experience with a Zero/One Programming Problem," *Operations Research*, 17, 917–919, 1969.

[271] Summer, D., "ILP-2, Integer Linear Program," Control Data Corporation, Data Centers Division, Minneapolis, Minn.

[272] Szwarc, W., "The Transportation Paradox," *Naval Research Logistics Quarterly*, 18, 185–202, 1971.

[273] ———, "The Truck Assignment Problem," *Naval Research Logistics Quarterly*, 14, 529–558, 1967.

[274] Taha, H. A., and G. L. Curry, "Classical Derivation of the Necessary and Sufficient Conditions for Optimal Linear Programs," *Operations Research*, 19, 1045–1050, 1971.

[275] Talacko, J. V., and R. T. Rockafellar, "A Compact Simplex Algorithm and a Symmetric Algorithm for General Linear Programs," unpublished paper, Marquette University, Milwaukee, Wis., August 1960.

[276] Thomas, J., "Linear Programming Models for Production-Advertising Decisions," *Management Science*, 17, B474-484, 1971.

[277] Thompson, G. E., *Linear Programming*. New York: The Macmillan Company, 1971.

[278] Thompson, G. L., "The Stopped Simplex Method: I. Basic Theory for Mixed Integer Programming; Integer Programming," *Revue Française de Recherche Opérationnelle*, 159–182, 1964.

[279] ———, F. M. Tonge, and S. Zionts, "Techniques for Removing Non-binding Constraints and Extraneous Variables from Linear Programming Problems," *Management Science*, 12, 588–608, 1966.

[280] Tillman, F. A., and J. M. Liittschwager, "Integer Programming Formulation to Constrained Reliability Problems," *Management Science*, 13, 887–899, 1967.

[281] Tomlin, J. A., "Branch and Bound Methods for Integer and Nonconvex Programming," in *Integer and Nonlinear Programming*, ed. J. Abadie. Amsterdam: North-Holland, 1970.

[282] ———, "An Improved Branch-and-Bound Method for Integer Programming," *Operations Research*, 19, 1070–1074, 1971.

[283] ———, "Minimum-Cost Multicommodity Network Flows," *Operations Research*, 14, 45–51, 1966.

[284] Trauth, C. A., Jr., and R. E. Woolsey, "Integer Linear Programming: A Study in Computational Efficiency," *Management Science*, 15, 481–493, 1969.

[285] Tuan, N. P., "A Flexible Tree-Search Method for Integer Programming Problems," *Operations Research*, 19, 115–119, 1971.

[286] Tucker, A. W., "Combinatorial Theory Underlying Linear Programs," in *Recent Advances in Mathematical Programming*, ed. R. L. Graves and P. Wolfe. New York: McGraw-Hill Book Company, 1963.

[287] Tyndall, W. F., "A Duality Theorem for a Class of Continuous Linear Programming Problems," *SIAM*, 13, 644–666, 1965.

[288] ———, "An Extended Duality Theorem for Continuous Linear Programming Problems," *SIAM*, 15, 1294–1298, 1968.

[289] Van Der Waerden, B. L., *Modern Algebra*, 2nd ed. New York: Frederick Unger Publishing Company, 1950.

[290] von Neumann, J., and O. Morgenstern, *Theory of Games and Economic Behavior*, 2nd ed. Princeton, N.J.: Princeton University Press, 1947.

[291] Waespy, C. M., "Linear-Programming Solutions for Orbital-Transfer Trajectories," *Operations Research*, 18, 635–653, 1970.

[292] Wagner, H. M., *Principles of Operations Research*. Englewood Cliffs, N.J.: Prentice-Hall, Inc., 1969.

[293] Walker, W. E., "A Method for Obtaining the Optimal Dual Solution to a Linear Program Using the Dantzig-Wolfe Decomposition," *Operations Research*, 17, 368–370, 1969.

[294] Watters, L. J., "Reduction of Integer Polynomial Programming Problems to Zero-One Linear Programming Problems," *Operations Research*, 15, 1171–1174, 1967.

[295] Weingartner, H. M., and D. N. Ness, "Methods for the Solution of the Multi-Dimensional 0/1 Knapsack Problem," *Operations Research*, 15, 83–103, 1967.

[296] White, W. W., "Integer Linear Programming: Relations Between Discrete and Continuous Solutions," ORC 66–27, University of California, Berkeley, 1966.

[297] Whiton, J. C., "Some Constraints on Shipping in Linear Programming Models," *Naval Research Logistics Quarterly*, 14, 257–260, 1967.

[298] Wolfe, P., and L. Cutler, "Experiments in Linear Programming," in *Recent Advances in Mathematical Programming*, ed. R. L. Graves and P. Wolfe. New York: McGraw-Hill Book Company, 1963.

[299] Woolsey, R. E. D., "A Candle to St. Jude or Four Real World Applications of Integer Programming," *Interfaces*, 2, February 1971.

[300] ———, "Comment On Briskin's Note," *Management Science*, 17, 500–501, 1971.

[301] ———, ed., "GEIPP," General Electric Information Systems Division, Phoenix, Arizona.

[302] ———, P. Ryan, and B. Holcomb, "ARRIBA," Control Data, Data Centers Division, Minneapolis, Minn.

[303] Woolsey, R. E. D., and C. A. Trauth, Jr., "IPSC, A Machine Independent Integer Linear Program," Sandia Corporation Research Report SC-RR-66-433, July 1966.

[304] ———, "MESA, A Heuristic Integer Linear Programming Technique," Sandia Corporation Research Report SC-RR-68-299, July 1968.

[305] Yen, J. Y., "Finding the *K* Shortest Loopless Paths in a Network," *Management Science*, 17, 712–716, 1971.

[306] ———, "On Hu's Decomposition Algorithm for Shortest Paths in a Network," *Operations Research*, 19, 983–985, 1971.

[307] Young, R. D., "A Primal (All-Integer) Integer Programming Algorithm," *Journal of Research, National Bureau of Standards B, Mathematics and Mathematical Physics*, 69B (3), 213–250, July-September 1965.

[308] ———, "A Simplified Primal (All-Integer) Integer Programming Algorithm," *Operations Research*, 16, 750–782, 1968.

[309] Zionts, S., "The Criss-Cross Method for Solving Linear Programming Problems," *Management Science*, 15, 426–445, 1969.

[310] ———, "Generalized Implicit Enumeration Using Bounds on Variables for Solving Linear Programs with Zero-One Variables, *Naval Research Logistics Quarterly*, 19, 165–181, 1972.

[311] ———, "On An Algorithm for the Solution of Mixed Integer Programming Problems," *Management Science*, 15, 113–116, 1968.

[312] ———,"Scaling Linear-Programming Matrices," (unpublished) Project No. (90-17-102) (1), Applied Research Laboratory, U.S. Steel Corporation, Monroeville, Penna., October 28, 1963.

[313] ———, "Size-Reduction Techniques of Linear Programming and Their Application," unpublished doctoral dissertation, Carnegie Institute of Technology, Pittsburgh, Pa., 1965.

[314] ———, "Some Empirical Tests of the Criss-Cross Method," *Management Science*, 19, 406–410, 1972.

[315] ———, "Toward a Unifying Theory of Integer Linear Programming," *Operations Research*, 17, 359–367, 1969.

[316] Zschau, E. V. W., "A Primal Decomposition Algorithm for Linear Programming," Working Paper No. 91, Graduate School of Business, Stanford University, Stanford, Cal., January 1967.

# Index